Advances in Experimental Medicine and Biology

Volume 1147

More information about this series at http://www.springer.com/series/5584

Alexander Birbrair
Editor

Pericyte Biology in Disease

Springer

Editor
Alexander Birbrair
Department of Radiology
Columbia University Medical Center
New York, NY, USA

Department of Pathology
Federal University of Minas Gerais
Belo Horizonte, MG, Brazil

ISSN 0065-2598 ISSN 2214-8019 (electronic)
Advances in Experimental Medicine and Biology
ISBN 978-3-030-16907-7 ISBN 978-3-030-16908-4 (eBook)
https://doi.org/10.1007/978-3-030-16908-4

This Springer imprint is published by the registered company Springer Nature Switzerland AG
The registered company address is: Gewerbestrasse 11, 6330 Cham, Switzerland

Preface

This book's initial title was "Pericyte Biology: Development, Homeostasis and Disease." However, due to the current great interest in this topic, we were able to assemble more chapters than would fit in one book, covering pericyte biology under distinct circumstances. Therefore, the book was subdivided into three volumes entitled: *Pericyte Biology-Novel Concepts*; *Pericyte Biology in Different Organs*; and *Pericyte Biology in Disease*.

This book *Pericyte Biology in Disease* presents contributions by expert researchers and clinicians in the multidisciplinary areas of medical and biological research. The chapters provide timely detailed overviews of recent advances in the field. This book describes the major contributions of pericytes to the biology of different organs in physiological and pathological conditions. Further insights into the biology of pericytes will have important implications for our understanding of organ development, homeostasis, and disease. The authors focus on the modern methodologies and the leading-edge concepts in the field of cell biology. In recent years, remarkable progress has been made in the identification and characterization of pericytes in several tissues using state-of-the-art techniques. These advantages facilitated the identification of pericyte subpopulations and definition of the molecular basis of pericytes role within different organs. Thus, the present book is an attempt to describe the most recent developments in the area of pericyte behavior, which is one of the emergent hot topics in the field of molecular and cellular biology today. Here, we present a selected collection of detailed chapters on what we know so far about the pericytes in various tissues and under distinct pathophysiological conditions. Fifteen chapters written by experts in the field summarize the present knowledge about the roles of pericytes in disease.

Ander Izeta and colleagues from Tecnun-University of Navarra discuss the role of pericytes in cutaneous wound healing. Anirudh Sattiraju and Akiva Mintz from Columbia University Irving Medical Center describe the multifaceted role of pericytes in glioblastoma and their potential use for therapeutic interventions. Jiha Kim from North Dakota State University compiles our understanding of pericytes in breast cancer. Aaron W. James and colleagues from Johns Hopkins University update us on pericytes in sarcomas and other mesenchymal tumors. Pritinder Kaur and colleagues from Curtin University summarize current knowledge on pericytes in metastasis. Mayana Zatz and colleagues from the University of São Paulo address the importance of pericytes in amyotrophic lateral sclerosis. Alla B. Salmina and colleagues from Krasnoyarsk State Medical University focus on pericytes in Alzheimer's disease. Francisco J. Rivera and colleagues from Universidad Austral de Chile introduce our current knowledge about pericytes in multiple sclerosis. Turgay Dalkara and colleagues from Hacettepe University describe pericytes role in ischemic stroke. Franck P.G. Lebrin and colleagues from Leiden University Medical Center discuss pericytes in hereditary hemorrhagic telangiectasia. Annika Keller and colleagues from Zurich University Hospital update us on pericytes in primary familial brain calcification. Katherine L. Hayes from the University of Massachusetts Medical School summarizes our current understanding on pericytes in type 2 diabetes. Volha Summerhill and Alexander Orekhov from Skolkovo Innovative Center compile our knowledge on pericytes in atherosclerosis. Bushra Shammout and Jill R. Johnson from Aston University address the role of pericytes in chronic lung disease. Finally, Sara Benedetti and colleagues from University College London give an overview of pericytes in muscular dystrophies.

It is hoped that the articles published in this book will become a source of reference and inspiration for future research ideas. I would like to express my deep gratitude to my wife Veranika Ushakova and Mr. Murugesan Tamilsevan from Springer, who helped at every step of the execution of this project.

This book is dedicated to the memory of my grandfather Pavel Sobolevsky, PhD, a renowned mathematician, who passed away during the creation of this piece.

My grandfather Pavel Sobolevsky z"l, PhD (March 26, 1930–August 16, 2018)

New York, NY, USA
Belo Horizonte, MG, Brazil

Alexander Birbrair

Contents

Contributors

Luis Alarcon-Martinez Department of Neuroscience and Centre de Recherche du Centre Hospitalier de l'Université de Montréal, Université de Montréal, Montréal, QC, Canada

Sara Benedetti Great Ormond Street Institute of Child Health, University College London, London, UK

NIHR Great Ormond Street Hospital Biomedical Research Centre, London, UK

Alexander Birbrair Department of Radiology, Columbia University Medical Center, New York, NY, USA

Department of Pathology, Federal University of Minas Gerais, Belo Horizonte, MG, Brazil

Natale Cavaçana Human Genome and Stem Cell Research Center, Biosciences Institute, University of São Paulo, São Paulo, Brazil

Leslie Chang Division of Surgical Pathology, Department of Pathology, Johns Hopkins University, Baltimore, MD, USA

Giuliana Castello Coatti Department of Genetics and Genome Sciences, Case Western Reserve University, Cleveland, OH, USA

Turgay Dalkara Department of Neurology, Faculty of Medicine, Hacettepe University, Ankara, Turkey

Institute of Neurological Sciences and Psychiatry, Hacettepe University, Ankara, Turkey

Department of Radiology, Massachusetts General Hospital, Harvard University, Boston, MA, USA

Taichi Ezaki Department of Anatomy and Developmental Biology, School of Medicine, Tokyo Women's Medical University, Tokyo, Japan

Francesca Daniela Franzoso Department of Neurosurgery, Clinical Neuroscience Center, Zürich University Hospital, Zürich University, Zürich, Switzerland

Georgios Galaris Department of Internal Medicine (Nephrology), Einthoven Laboratory for Experimental Vascular Medicine, Leiden University Medical Center, Leiden, The Netherlands

Araika Gutiérrez-Rivera Tissue Engineering Group, Bioengineering Area, Instituto Biodonostia, Hospital Universitario Donostia, San Sebastián, Spain

Katherine L. Hayes Department of Surgery and Diabetes Center of Excellence, University of Massachusetts Medical School, Worcester, MA, USA

Bryan Hinrichsen Laboratory of Stem Cells and Neuroregeneration, Institute of Anatomy, Histology and Pathology, Faculty of Medicine, Universidad Austral de Chile, Valdivia, Chile

Center for Interdisciplinary Studies on the Nervous System (CISNe), Universidad Austral de Chile, Valdivia, Chile

Haizea Iribar Tissue Engineering Group, Bioengineering Area, Instituto Biodonostia, Hospital Universitario Donostia, San Sebastián, Spain

Ander Izeta Tissue Engineering Group, Bioengineering Area, Instituto Biodonostia, Hospital Universitario Donostia, San Sebastián, Spain

Department of Biomedical Engineering and Sciences, School of Engineering, Tecnun-University of Navarra, San Sebastián, Spain

Aaron W. James Division of Surgical Pathology, Department of Pathology, Johns Hopkins University, Baltimore, MD, USA

Jill R. Johnson Biosciences Department, School of Life and Health Sciences, Aston University, Birmingham, UK

Pritinder Kaur Curtin Health Innovation Research Institute, Curtin University, Perth, WA, Australia

Annika Keller Department of Neurosurgery, Clinical Neuroscience Center, Zürich University Hospital, Zürich University, Zürich, Switzerland

Jiha Kim Department of Biological Sciences, North Dakota State University, Fargo, ND, USA

Yulia K. Komleva Department of Biochemistry, Medical, Pharmaceutical & Toxicological Chemistry, Krasnoyarsk State Medical University named after Prof. V.F. Voino-Yasenetsky, Krasnoyarsk, Russia

Research Institute of Molecular Medicine & Pathobiochemistry, Krasnoyarsk State Medical University named after Prof. V.F. Voino-Yasenetsky, Krasnoyarsk, Russia

Franck P. G. Lebrin Department of Internal Medicine (Nephrology), Einthoven Laboratory for Experimental Vascular Medicine, Leiden University Medical Center, Leiden, The Netherlands

Physics for Medicine, ESPCI, INSERM U1273, CNRS, Paris, France

MEMOLIFE Laboratory of Excellence and PSL Research University, Paris, France

Olga L. Lopatina Department of Biochemistry, Medical, Pharmaceutical & Toxicological Chemistry, Krasnoyarsk State Medical University named after Prof. V.F. Voino-Yasenetsky, Krasnoyarsk, Russia

Research Institute of Molecular Medicine & Pathobiochemistry, Krasnoyarsk State Medical University named after Prof. V.F. Voino-Yasenetsky, Krasnoyarsk, Russia

Carolyn A. Meyers Division of Surgical Pathology, Department of Pathology, Johns Hopkins University, Baltimore, MD, USA

Akiva Mintz Department of Radiology, Columbia University Irving Medical Center, New York, NY, USA

Shunichi Morikawa Department of Anatomy and Developmental Biology, School of Medicine, Tokyo Women's Medical University, Tokyo, Japan

Louise Anne Moyle Institute of Biomaterials and Biomedical Engineering, Donnelly Centre for Cellular and Biomolecular Research, University of Toronto, Toronto, ON, Canada

Alexander Orekhov Laboratory of Angiopathology, Institute of General Pathology and Pathophysiology, Moscow, Russia

Institute for Atherosclerosis Research Moscow, Skolkovo Innovative Center, Moscow, Russia

Zalitha Pieterse Curtin Health Innovation Research Institute, Curtin University, Perth, WA, Australia

Francisco J. Rivera Laboratory of Stem Cells and Neuroregeneration, Institute of Anatomy, Histology and Pathology, Faculty of Medicine, Universidad Austral de Chile, Valdivia, Chile

Center for Interdisciplinary Studies on the Nervous System (CISNe), Universidad Austral de Chile, Valdivia, Chile

Institute of Molecular Regenerative Medicine, Paracelsus Medical University Salzburg, Salzburg, Austria

Spinal Cord Injury and Tissue Regeneration Center Salzburg (SCI-TReCS), Paracelsus Medical University Salzburg, Salzburg, Austria

Alla B. Salmina Department of Biochemistry, Medical, Pharmaceutical & Toxicological Chemistry, Krasnoyarsk State Medical University named after Prof. V.F. Voino-Yasenetsky, Krasnoyarsk, Russia

Research Institute of Molecular Medicine & Pathobiochemistry, Krasnoyarsk State Medical University named after Prof. V.F. Voino-Yasenetsky, Krasnoyarsk, Russia

Anirudh Sattiraju Department of Radiology, Columbia University Irving Medical Center, New York, NY, USA

Michelle A. Scott Division of Plastic Surgery, Department of Surgery, Children's Hospital of Pennsylvania, Philadelphia, PA, USA

Bushra Shammout Biosciences Department, School of Life and Health Sciences, Aston University, Birmingham, UK

Maria Elena Silva Laboratory of Stem Cells and Neuroregeneration, Institute of Anatomy, Histology and Pathology, Faculty of Medicine, Universidad Austral de Chile, Valdivia, Chile

Center for Interdisciplinary Studies on the Nervous System (CISNe), Universidad Austral de Chile, Valdivia, Chile

Institute of Pharmacy, Faculty of Sciences, Universidad Austral de Chile, Valdivia, Chile

Devbarna Sinha Peter MacCallum Cancer Centre, Melbourne, VIC, Australia

Volha Summerhill Laboratory of Angiopathology, Institute of General Pathology and Pathophysiology, Moscow, Russia

Institute for Atherosclerosis Research Moscow, Skolkovo Innovative Center, Moscow, Russia

Francesco Saverio Tedesco Department of Cell and Developmental Biology, University College London, London, UK

Great Ormond Street Institute of Child Health, University College London, London, UK

Jérémy H. Thalgott Department of Internal Medicine (Nephrology), Einthoven Laboratory for Experimental Vascular Medicine, Leiden University Medical Center, Leiden, The Netherlands

Muge Yemisci Department of Neurology, Faculty of Medicine, Hacettepe University, Ankara, Turkey

Institute of Neurological Sciences and Psychiatry, Hacettepe University, Ankara, Turkey

Yvette Zarb Department of Neurosurgery, Clinical Neuroscience Center, Zürich University Hospital, Zürich University, Zürich, Switzerland

Mayana Zatz Human Genome and Stem Cell Research Center, Biosciences Institute, University of São Paulo, São Paulo, Brazil

Chapter 1
Pericytes in Cutaneous Wound Healing

Shunichi Morikawa, Haizea Iribar, Araika Gutiérrez-Rivera, Taichi Ezaki, and Ander Izeta

Abstract Most of the studies on cutaneous wound healing are focused on epidermal closure. This is obviously important, as the epidermis constitutes the main barrier that separates the inner organism from the environment. However, dermal remodeling is key to achieve long-lasting healing of the area that was originally wounded. In this chapter, we summarize what is known on the stromal components that strongly influence the outcome of healing and postulate that dedifferentiation of stably differentiated cells plays a major role in the initial response to wounding, as well as in long-term wound remodeling. Specifically, we explore the available evidence implicating skin pericytes, endothelial cells, Schwann cells, and macrophages as major players in a complex symphony of cellular plasticity and signaling events whose balance will promote healing (by tissue regeneration or repair) or fibrosis.

Keywords Pericytes · Schwann cell precursors · Dermis · Dedifferentiation Remodeling · Regeneration · Scar · Revascularization · Reinnervation Macrophages · Wound healing · Injury response · Reprogramming · Neural crest Boundary cap

This chapter is dedicated to the memory of the late Dr. Shunichi Morikawa, an original thinker and pioneering scientist who largely increased our current understanding of the role of pericytes in cutaneous wound healing.

S. Morikawa · T. Ezaki
Department of Anatomy and Developmental Biology, School of Medicine, Tokyo Women's Medical University, Tokyo, Japan

H. Iribar · A. Gutiérrez-Rivera
Tissue Engineering Group, Bioengineering Area, Instituto Biodonostia, Hospital Universitario Donostia, San Sebastián, Spain

A. Izeta (✉)
Tissue Engineering Group, Bioengineering Area, Instituto Biodonostia, Hospital Universitario Donostia, San Sebastián, Spain

Department of Biomedical Engineering and Sciences, School of Engineering, Tecnun-University of Navarra, San Sebastián, Spain
e-mail: ander.izeta@biodonostia.org

A. Birbrair (ed.), *Pericyte Biology in Disease*, Advances in Experimental Medicine and Biology 1147, https://doi.org/10.1007/978-3-030-16908-4_1

Shunichi Morikawa and Haizea Iribar contributed equally to this work.

Introduction

Pericytes constitute a heterogeneous group of cells, somewhat loosely defined by their perivascular location, as the mural cells of blood microvessels (Armulik et al. 2011). For this reason, the literature has to be carefully revised and different terms of search must be used to grasp the vast knowledge accumulated on their putative roles in tissue repair and fibrosis. To increase the confounding factors, adult cells may dedifferentiate and transdifferentiate in response to wounding as well as in response to tissue disaggregation and cell isolation, and the boundaries between so-called terminally differentiated cell populations blur. The consensus in the field is that pericytes are highly plastic cells (Birbrair et al. 2017). As a result, if we could sample a wound and look at the continuum of cells active at the wound bed, virtually at any time we would encounter a number of cells that may represent intermediate states among cell type A and cell type B, apart from myriad cell types that infiltrate the wound, replicate, or die. This complex picture must be carefully delineated. In this chapter, we aim to dissect the role of pericyte fate and plasticity in wound closure and remodeling. To this end, we discuss the different aspects of vascular formation, peripheral innervation, and role of macrophages in cutaneous wound healing and thus we explore the available evidence implicating pericytes, endothelial cells, Schwann cells, and macrophages as the major players in promoting wound healing or fibrosis.

Vascular Formation in Wound Healing

The Circulatory System in the Skin

The blood supply to the skin stems from arteries in the subcutis layer. Branches from these arteries run upwards to form two plexuses of anastomosing vessels, one sitting deep in the dermis (the cutaneous plexus) and the other more superficial (the subpapillary plexus) (Braverman 2000; Young et al. 2014). The venous and lymphatic drainages run parallel to the arterial supply (Fig. 1.1).

The deep cutaneous plexus or *rete cutaneous* (Sorrell and Caplan 2004) sits at the junction between the dermis and hypodermis. It supplies blood to the dermal fat layer as well as the reticular dermis and epidermal appendages (hair follicles, sebaceous and sweat glands). The superficial subpapillary plexus or *rete subpapillare* lies just beneath the dermal papillae, and supplies the capillaries in the dermal papillae.

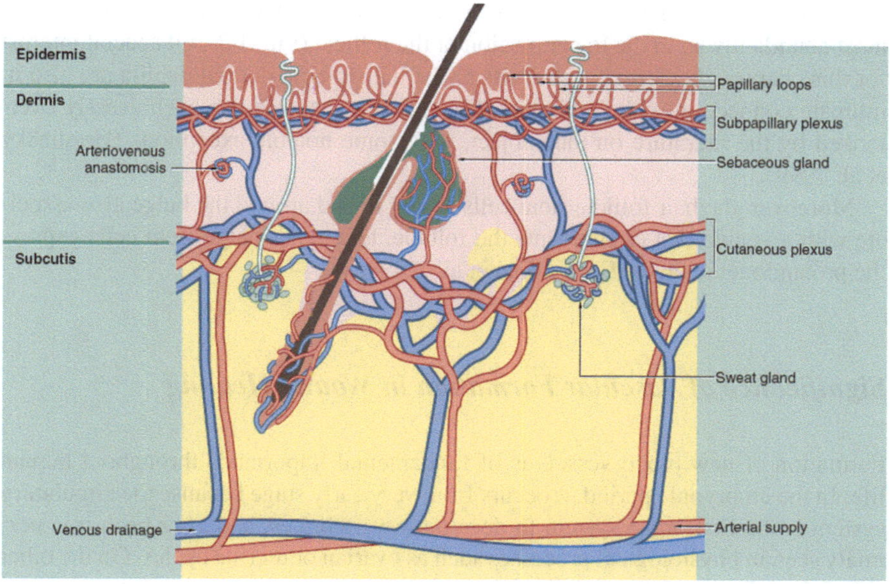

Fig. 1.1 The skin circulation system. The arteries supplying the skin are located deep in the subcutis, from which they give rise to branches passing upwards to form two plexuses of anastomosing vessels. The deeper plexus lies at the junction of the subcutis and dermis and is known as the cutaneous plexus; the more superficial plexus lies at the junction between papillary and reticular dermis and is known as the subpapillary or superficial plexus. The venous drainage of the skin is arranged into plexuses broadly corresponding to the arterial supply. The skin has a rich lymphatic drainage which forms plexuses corresponding to those of the blood vascular system. Reprinted from Young et al. (2014) with permission

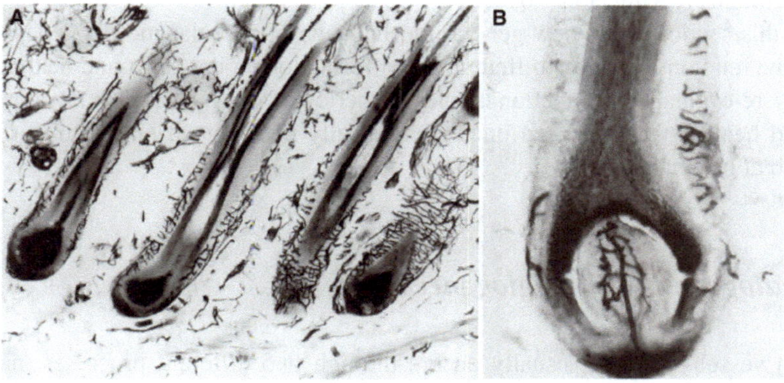

Fig. 1.2 Vascularization of hair follicles. Parallel, longitudinally oriented vessels extend from the base of the bulb to the pilary canal. (**a**) Numerous capillary networks around eyebrow hair follicles demonstrated with alkaline phosphatase. (**b**) A tuft of blood vessels inside the dermal papilla of an eyebrow follicle demonstrated with the alkaline phosphatase technique. Reprinted from Montagna and Parakkal (1974) with permission

The hair follicle is a deeply vascularized organ (Montagna and Parakkal 1974), most abundantly so in the lower portion of the follicle (Fig. 1.2). Of special interest for this chapter, the dermal stem cells of the hair follicle dermal papilla are also in intimate connection with capillaries (Fig. 1.2b), a fact that has been grossly overlooked by the literature on the subject, with some notable exceptions (Hordinsky et al. 1999).

Moreover, the hair follicle stem cells sitting in and around the bulge also associate with a venule that circumvents the follicle, and upper bulge stem cells express the proangiogenic factor *Egfl6* (Xiao et al. 2013).

Significance of Vascular Formation in Wound Healing

Formation of new blood vessels is of fundamental importance throughout human life. In the embryonic period, it occurs from very early stage because the circulatory system is the first organ system to develop in vertebrates, and after birth it is normally seen in physiological processes such as ovarian or uterine cycles. On the other side, it is well known that pathological vascular formation is closely related to tumor progression and metastasis, or at the onset of diabetic retinopathy.

In wound healing, the newly formed blood vessels efficiently and rapidly supply all over the wounded site various factors needed for repair, such as oxygen, nutrients, cytokines, inflammatory cells, and matrix molecules (Eming et al. 2007a; Johnson and Wilgus 2014; Polverini 2011). Elucidating the detailed mechanism of vascular formation during wound healing is meaningful not only for vascular biology, but also for the therapeutic perspective in developing strategies to cope with impaired wound healing caused by poor vascular formation seen in diabetes or peripheral vascular diseases (Eming et al. 2007a; Johnson and Wilgus 2014; Polverini 2011; King et al. 2014).

In this section we review general mechanisms of vascular morphogenesis, discuss the relationship of the different vessel formation modes to wound healing, and finally re-evaluate the functional roles of pericytes during vascular formation in wound healing that has been once energetically investigated but is almost disregarded at present time.

Vasculogenesis and Angiogenesis

Blood vessel formation usually occurs through two different processes, namely, vasculogenesis and angiogenesis. Vasculogenesis, a term coined by Risau and colleagues in 1988, refers to the primal, de novo formation of blood vessels (Drake 2003; Risau et al. 1988), whereas angiogenesis (a name first proposed by Hertig in 1935) means the growth of secondary blood vessels from preexisting vasculature (Fig. 1.3; Conway et al. 2001).

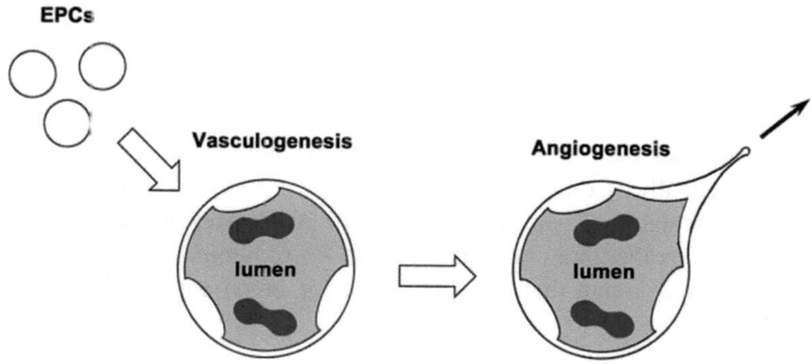

Fig. 1.3 Vasculogenesis and angiogenesis. In embryonic period, blood vessels are formed de novo by vasculogenesis. In the process of vasculogenesis, endothelial precursor cells (EPCs) of angioblasts and hemangioblasts that emerge from mesenchyme differentiate into endothelial cells (ECs) to create primary blood vessels with luminal surface. Then, angiogenesis occurs in primary vessels and ECs further start to grow to expand the vessel network

During embryonic development, endothelial progenitor cells (EPCs) differentiate to endothelial cells (ECs) and assemble to form a primitive blood vessel network. This process is called vasculogenesis. Then, angiogenesis occurs and ECs of the primary vessel network proliferate and remodel to produce secondary blood vessels growing from the primary network. Through these serial events, the vessel network becomes denser and expanded. Thus, in development, vasculogenesis and angiogenesis work together in a coordinated manner.

On the other hand, because the basic body vasculature is already completed in adulthood, it was once considered that EPCs are absent from adult individuals. Upon injury, blood vessels were thought to remodel by using preexisting ECs of the vasculature (Drake 2003; Balaji et al. 2013). Thus, vasculogenesis by EPCs was thought to be confined to developmental stages and angiogenesis by ECs to be occurring after birth. However, we now know that EPCs are preserved after birth, and that they retain the potential to generate new blood vessels (Asahara et al. 1997; Shi et al. 1994). We now define the neovascularization performed by EPCs in the adulthood as postnatal vasculogenesis (PV: Drake 2003; Ribatti et al. 2001).

PV is involved in various scenarios of vascular formation, such as in female reproductive organs, ischemic tissues, tumors, and wound healing situations (Asahara et al. 1999; Bauer et al. 2005; Velazquez 2007).

Types of Postnatal Vascular Formation and Their Implication in Wound Healing

Angiogenesis is classified into two major subtypes according to the mechanism of producing new blood vessels, namely, sprouting angiogenesis (SA) and intussusceptive angiogenesis (IA). In this section, we review the two types of angiogenesis as well as postnatal vasculogenesis, and discuss their relevance to wound healing.

In addition, two other mechanisms of vascular formation have been reported. They are called “vessel co-option” and “vascular mimicry.” However, they are observed almost exclusively in tumor angiogenesis, and we do not dig deeper into their detailed mechanism here. Briefly, in vessel co-option, tumor cells gain blood supply by hijacking the normal blood vessels in the close vicinity, and the co-opted vessels are gradually modified to become abnormally impaired vessels, characteristic to tumor vasculature (Ziyad and Iruela-Arispe 2011). Meanwhile, vascular mimicry literally refers the lining of tumor cells on vessel luminal surface. Tumor cells “mimic” ECs and create their own channels for blood supply (Dunleavey and Dudley 2012).

Sprouting Angiogenesis

Sprouting angiogenesis (SA) was identified fairly long time ago and has been eagerly studied by many researchers, for some time being regarded as almost the only mechanism to produce new blood vessels (De Spiegelaere et al. 2012).

Mechanistically, the process of SA is usually explained by enzymatic degradation of the vascular basement membrane (BM) followed by detachment of perivascular cells from the blood vessel wall and, finally, “sprouting” of activated ECs (Ausprunk and Folkman 1977; Herbert and Stainier 2011; Ribatti and Crivellato 2012; Rundhaug 2005).

More recently, a novel model of vascular EC sprout conformation has been introduced and is accumulating increasing attention (Fig. 1.4). The model illustrates that vascular EC sprouts are composed of specialized ECs called “tip cell” and “stalk cell” (Blancas et al. 2013; De Smet et al. 2009; Gerhardt et al. 2003): Tip cells are located at the forefront of endothelial sprouting and characterized by numerous fine threadlike cytoplasmic processes (filopodia), while stalk cells are located behind tip cells. Tip cells can perceive the gradient of various angiogenic factors including vascular endothelial growth factor (VEGF) by using filopodia, and are considered to guide the vessel growth. In contrast to tip cells that hardly proliferate, stalk cells are identified as frequently dividing cells and are considered to extend the length of growing EC sprouts. Molecular mechanisms of the tip and stalk cell model are well studied and the various signaling pathways that regulate it have gradually become clear. For example, the differentiation of ECs into tip or stalk cells is firmly controlled via local Notch/Dll4 signaling (De Smet et al. 2009).

Relevance of Sprouting Angiogenesis to Skin Wound Healing

SA has been found in various scenarios of tissue repair, including cutaneous wound healing (Amselgruber et al. 1999; Cliff 1963; Morikawa and Ezaki 2011). Especially in skin wound healing, SA is reported to work as the main mechanism of vascular formation surpassing other types (Cliff 1963; Morikawa and Ezaki 2011; Kilicaslan et al. 2013; Paku et al. 2011). More recently, Chong et al. identified what they called

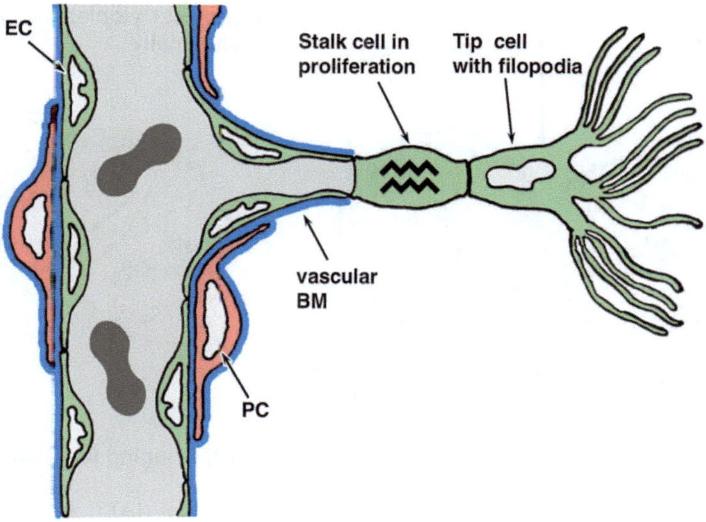

Fig. 1.4 Sprouting angiogenesis: Tip-stalk cell model. Sprouting angiogenesis (SA) is a manner of angiogenesis in which ECs are literally "sprout up" or "bud" from preexisting vessel wall. EC located at the forefront of vascular sprout is called "tip cell" that can sense the angiogenic signals by their numerous long filopodia similar to those seen in axonal growth cones, and navigate vessel growth. Located just behind the tip cell is "stalk cell", another important EC population that constitutes vascular sprouts. Different from tip cells that hardly proliferate, stalk cells show highly proliferative potential and thus have a function to physically extend the growing sprouts. Tip and stalk cells are devoid of vascular basement membrane (BM), and pericyte (PC) covering

"tortuous microvessels" at the wound bed, with aberrant cell shapes, increased permeability, and altered flow dynamics (Chong et al. 2017). This novel type of transient wound vessel was reported to sprout more often than standard capillaries.

Importantly, we propose that the mechanism underlying the SA that occurs in skin wound healing diverges from the current tip-stalk cell model. This possibility is further elaborated below.

Intussusceptive Angiogenesis

Intussusceptive angiogenesis (IA) is another type of angiogenesis that has been recognized much more recently than SA (Djonov et al. 2003). Similar to SA, and under the basic control of the VEGF signal, IA occurs in a wide variety of normal and pathological vascular formation scenarios, including prenatal development, ovarian cycle, tumors, etc. Some studies suggest that IA predominantly works during development (Djonov et al. 2003; Kurz et al. 2003).

Mechanistically, IA occurs by longitudinal splitting of the vascular lumen, resulting in the increase of blood vessel number (Fig. 1.5). The process of IA starts from the insertion of the connective tissue column, the so-called transluminal connective

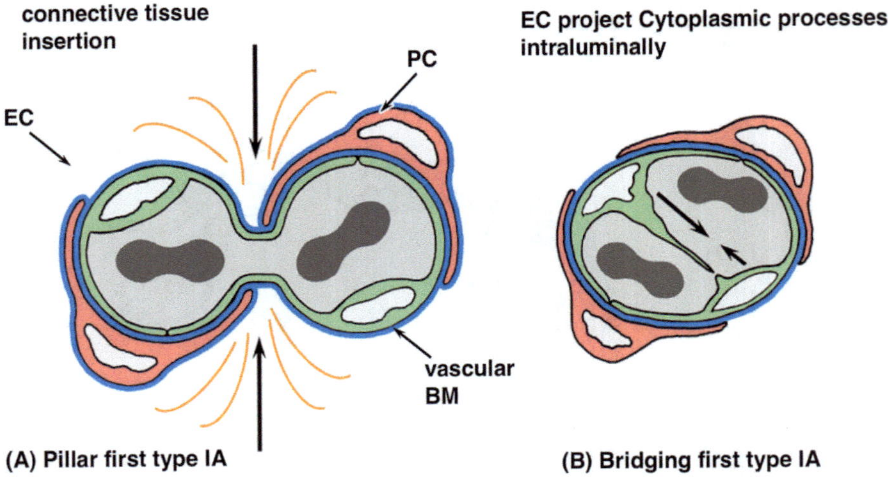

Fig. 1.5 Intussusceptive angiogenesis. Intussusceptive angiogenesis (IA) is another manner of angiogenesis in which a vessel is longitudinally split into two vessels. IA is further classified into two types, namely, "pillar first IA" (**a**) and "bridging first IA" (**b**). In pillar first type, connective tissue pillar is formed at first, and then the pillar is inserted to vessels to split them (arrows). On the other hand, in bridging first type, ECs first extend cytoplasmic processes intraluminally to split the lumen first (EC bridging; arrows), and then connective tissue pillar is formed to finally separate the vessels

tissue pillar, into a vessel, which then completely splits the vessel into two, to form two daughter vessels ("pillar first IA": Fig. 1.5a, Djonov et al. 2003). However, a variant process of IA has been found in which intraluminal protrusion of EC occurs in the first place, then, once the "bridging" of intraluminal EC protrusion is formed, connective tissue pillar actually divides the vessel (Fig. 1.5b; "bridging first IA," Paku et al. 2011; Egginton et al. 2001).

In contrast to SA that usually involves EC proliferation, IA does not necessarily need EC proliferation but rather increased EC size and flattened shape (Egginton et al. 2001; Styp-Rekowska et al. 2011), and consequently it enables a lower metabolic cost than SA (Djonov et al. 2003) and a prompt increase in blood supply.

Relevance of Intussusceptive Angiogenesis to Skin Wound Healing

Although IA has been identified in various scenarios of tissue repair (Kilicaslan et al. 2013; Frontczak-Baniewicz and Walski 2002; Patan et al. 2001), in skin wound healing new blood vessels are likely formed by SA. In contrast, IA is rarely found during skin wound healing, although its relative abundance may increase at the later stages of healing (Kilicaslan et al. 2013; S.M., personal observation). Kilicaslan et al. (2013) reported that in day-5 skin wounds, SA was observed five times as abundantly as IA, while in day 7 SA was observed only threefold increased as compared to IA. Further, between the two types of IA, bridging first IA is liable to occur in skin wound healing (Kilicaslan et al. 2013; S.M., personal observation).

SA is invasive, and from the earlier stages of tissue repair ECs can break into immature avascular area where fibrin is still rich and collagen is poor. In contrast, IA, particularly pillar first type IA, needs extensive connective tissue synthesis to form tissue pillar before vascular formation (Paku et al. 2011). Although in several aspects it may be considered advantageous to using SA, IA might not be suitable for initial vascular formation at the earlier stages of wound healing, where migration and accumulation of collagen-producing fibroblasts have just started.

Intriguingly, Kilicaslan and colleagues reported that IA could be induced in skin wound healing by EGF treatment. Occurrence of IA increased from about 20% to 50% on day 5 after treatment, and on day 7 the percentage even slightly exceeded the occurrence of SA. Although the detailed mechanism of how EGF enhances the occurrence of IA should be further clarified, the information provided by the work may be of great importance from a therapeutic point of view (Kilicaslan et al. 2013).

Postnatal Vasculogenesis

EPCs are generally classified into two populations, namely, bone marrow-derived EPCs and tissue-resident EPCs.

A population of bone marrow-derived circulating mononuclear cells serve as EPCs in adulthood, and work in PV (Nishimura and Asahara 2005). Currently, EPCs isolated from blood are classified into two groups according to the time needed for culturing them. So-called early EPCs make colonies within several days of culture and "late EPCs" take several weeks to make colonies. Early EPCs have a limited capacity of proliferation, whereas late EPCs show vigorous proliferative nature and have an alternate name as "outgrowth cell" (Marcola and Rodrigues 2015; Tagawa et al. 2015). Bone marrow-derived EPCs have been defined by positivity for CD34, CD133, and VEGFR2 membrane markers, as well as *Ulex europaeus* lectin binding and LDL uptake. However it is also suggested that outgrowth cells are not derived from CD133+ cells (Timmermans 2007), opening up the possibility of these being two independent cell subpopulations.

Between the two EPC subpopulations, outgrowth cells are thought to actually become ECs (Marcola and Rodrigues 2015; Adams and Alitalo 2007; Rehman et al. 2003) (Fig. 1.6). Outgrowth cells incorporate into EC lining of blood vessels to circumferentially enlarge vessels (Fig. 1.6a), or may alternatively be recruited at the forefront of the vascular sprouting to carry on further vessel growth (Fig. 1.6b).

In contrast, it is suggested that early EPCs, the other EPC subpopulation, do not differentiate into ECs and are retained in the perivascular space, where they are thought to produce angiogenic factors and support vessel growth and thus they are called by the alternative name "angiogenic cells" (Fig. 1.6c). These perivascular angiogenic cells are reported to have monocytic phenotype. Roles of macrophages in vascular formation are discussed in detail in other sections of the chapter.

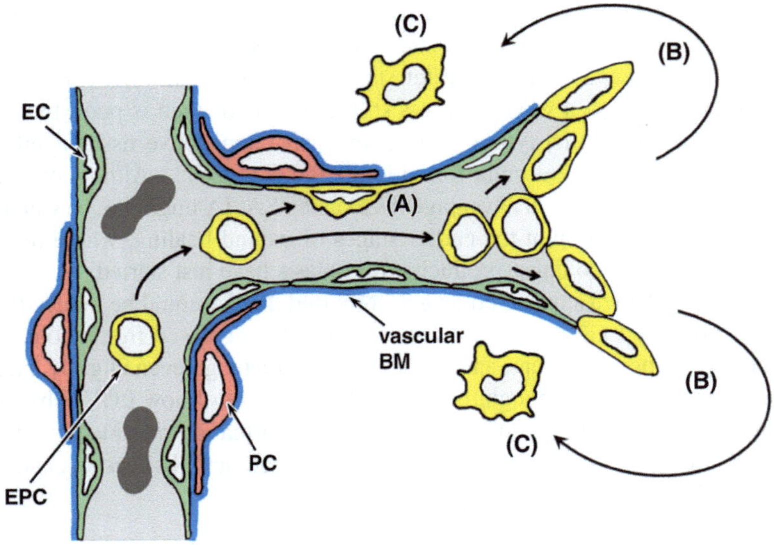

Fig. 1.6 Postnatal vasculogenesis. Endothelial progenitor cells (EPCs) are involved in vascular formation even after birth. They are thought to exist in bone marrow-forming niches. They are also considered to reside in local tissues. Among EPCs that come from bone marrow to the site of vascular formation via blood stream, so-called outgrowth cells can be incorporated into EC lining to increase the surface area of vessels (**a**), or move to the forefront of vascular sprouts to further extend the vessels (**b**). EPCs located at the forefront of the sprouts are supposedly devoid of vascular BM. Meanwhile, another population of EPCs called "angiogenic cells" come out from the vessel wall and stay at perivascular space (**c**), and produce angiogenic factors to support vessel growth

Apart from the aforementioned bone marrow-derived circulating EPCs, several studies suggest the existence of EPCs that reside in local tissues. For example, pericytes (PCs) qualify among the candidates for tissue-resident EPCs. Some PC populations isolated from vasa vasorum of murine femoral artery may have a potential to differentiate into ECs (Kabara et al. 2014). The authors observed that the isolated PCs (multipotent PCs; mPCs) formed tubular structures in culture in response to VEGF, and after the subsequent incubation with VEGF followed by TGF beta the cellular structures differentiated into blood vessel-like tubular structures lined of CD31+ ECs in the luminal surface and covered by αSMA+ PCs.

Finally, other authors have suggested that in PV during skin wound healing, newly formed ECs arise from Nestin-expressing dermal stem cells of the hair follicle bulge area (Aki et al. 2010; Amoh et al. 2005). Of note, many details are still open to question regarding these candidates for tissue-resident EPCs, and therefore we encourage careful verification of the literature on this subject.

Relevance of Postnatal Vasculogenesis to Skin Wound Healing

It has been well documented that PV plays an important role during skin wound healing (Asahara et al. 1999). In our opinion, SA is likely the main mechanism of vascular formation in skin wound healing, and it is very possible that a substantial part of ECs located at the forefront of vascular sprouts in the process of SA is derived not from preexisting ECs but from EPCs.

In wound healing, local hypoxia after the trauma is thought to start the PV process, and after that various factors such as VEGF and basic FGF are suggested to be involved in mobilizing EPCs from the bone marrow to the wound sites via circulation (King et al. 2014). Among the known signals to stimulate migration of EPCs from their niches into blood circulation, upregulation of NO levels mediated by endothelial nitric oxide synthase (eNOS) of marrow vessels is critical (Gallagher et al. 2007; Lee et al. 1999). Of note, eNOS function is impaired in diabetic conditions such as hyperglycemia or insulin resistance. Focusing on eNOS function is thus critical to fully elucidate the mechanism of poor PV in diabetes, which results in delayed wound healing.

Is the Tip-Stalk Cell Model Applicable to All Vascular Formation?

The studies on the tip-stalk cell model represent high-quality data and the model is widely accepted. For these reasons, we see an increasing momentum to apply this basic concept to clinically related studies such as tumor angiogenesis (Dufraine et al. 2008).

However, there is a significant fact that we must not overlook: the observation of tip cells that locate at the leading front of EC sprouting and guide the vascular growth by using filopodia is confined to a few situations, namely, (i) to prenatal developmental processes, (ii) when focusing on adult mammal neovascularization to developing central nervous system (CNS; including developing retinas), and (iii) to some in vitro studies (Hetheridge et al. 2012). To our knowledge, filopodial tip cell was only observed after birth in studies on vascular formation during experimental tumor progression, in which CNS tumor cells (C6 glioma) were implanted.

These facts led us to come up with the following questions: Is the current tip-stalk cell model applicable to all types of neovascularization? Alternatively, is the model unique to prenatal development?

Vascular Formation in Prenatal Development

In development, the location and timing of vascular formation are tightly controlled. Blood vessels are formed de novo together with surrounding tissue in a settled manner that is spatially and temporally reproducible. In contrast, neovascularization in postnatal life (i.e., in wound healing or tumor growth) is promoted by unpredictable, nongenetically programmed local factors (such as hypoxia, proinflammatory substances) and occurs within fully differentiated tissues (Kilarski and Gerwins 2009).

Further, it is known that nerves and blood vessels grow in a coordinated manner during development (Carmeliet and Tessier-Lavigne 2005). Vessels can attract axons to track alongside by specific signaling molecules such as artemin and neurotrophin 3 (Gerhardt et al. 2003). On the other direction, nerves (and Schwann cells) can affect vessel growth by producing VEGF (Mukouyama et al. 2002). Interestingly, the endothelial tip cell that presents numerous filopodia closely resembles the axonal growth cone that also has numerous filopodia. They are similar not only in their morphology but also functionally, and common signaling molecules are expressed on their filopodia. Thus in development, blood vessels and nerves grow in keeping a special relationship with each other, but after birth, especially in regeneration, they might not necessarily behave in the same manner.

Coincidentally, when vascular sprouts in postnatal vascularization accompanied by retinopathy induced by experimental hypoxia were closely observed, typical filopodia were hardly detectable on the tip cells of the sprouts (Chan-Ling et al. 2004).

In summary, we should acknowledge the possibility that vascular formation is regulated by different mechanisms between development and postnatal life.

Peculiarity of the CNS Vasculature

In addition, we should also note that the vascular system of CNS is quite unique in comparison to the majority of the other tissues in the body. In CNS, neurons and glia are packed together and form a very compact structure containing fewer connective tissues than other organs. Especially, at the capillary level, ECs lose connective tissue with the sole exception of basement membrane, and form the blood–brain barrier (BBB) to strictly control the passage of substances from the bloodstream to neurons, together with astrocytes and pericytes (Abbott et al. 2010). In the retina, ECs also form a similar structure as the BBB, called (inner) blood–retinal barrier (BRB), together with Muller cells and pericytes (Campbell and Humphries 2011).

Outside the CNS, and specifically in the skin, capillaries do not form a structure like BBB or BRB, and run within much wider connective tissue space, which contains a variety of cells (i.e., fibroblasts, macrophages, Schwann cells) that are different from those that constitute the BBB/BRB. As reviewed in other sections, these cells take part in significant molecular pathways for tissue repair and regeneration.

In CNS development, it is suggested that a network made by astrocytes functions as a template for blood vessel network formation. Blood vessels grow and develop into networks along with the astrocyte template, while astrocytes produce molecules such as VEGF and ECM, and support vessel growth (Gerhardt et al. 2003; Watanabe and Raff 1988; West et al. 2005).

In the tissue outside CNS where astrocytes are lacking, we need to determine which cell populations serve a similar role as astrocytes undertake in the CNS vasculature.

Characteristics of the Vascular EC Sprouts in Postnatal Vascular Formation

Through the re-examination of past studies in the last 50 years, we eventually concluded that the vascular formation by tip-stalk cell mechanism is hardly seen in postnatal vascular formation. Meanwhile, we analyzed the differential features of vascular EC sprouts in postnatal vascular formation.

In the following sections, we compare the differences between the tip-stalk cell model and postnatal vascular sprouts. We then propose a re-evaluation of the current ideas about how postnatal vascular formation occurs.

Absence of Filopodia on the Leading EC

In comparison to tip cells that are characterized by numerous and long filopodia, ECs at the forefront of the sprouts in postnatal vessel formation usually lack them. In some cases, tiny cytoplasmic protrusions are detected in forefront ECs at the ultrastructural level (Haas et al. 2000; Rhodin and Fujita 1989), but typical filopodia of tip cells that are clearly identifiable even at the light microscopic level cannot be seen at the forefront ECs of sprouts in postnatal vascular formation. For tip cells that are specialized to guide the sprout growth, numerous long filopodia might be important to receive angiogenic signals effectively. Meanwhile, in postnatal EC sprouts, pericytes are usually found attaching to leading ECs. These PC-like cells are considered to guide the sprout growth instead of leading ECs here, suggesting that leading ECs do not have to develop specialized filopodial structure.

Morphology of Vascular EC Sprouts Found in Postnatal Vascular Formation

In addition to lacking filopodia on the leading EC, postnatal vascular sprouts have other unique morphological features (Fig. 1.7). In addition to the differences in length that reflect the consecutive growing stages, EC sprouts are morphologically classified into two groups according to the shape of leading ECs, as "pointed end" (Fig. 1.7a1, b1) and "blunt" type (Fig. 1.7a2, a3, b2–b4). These two types can be

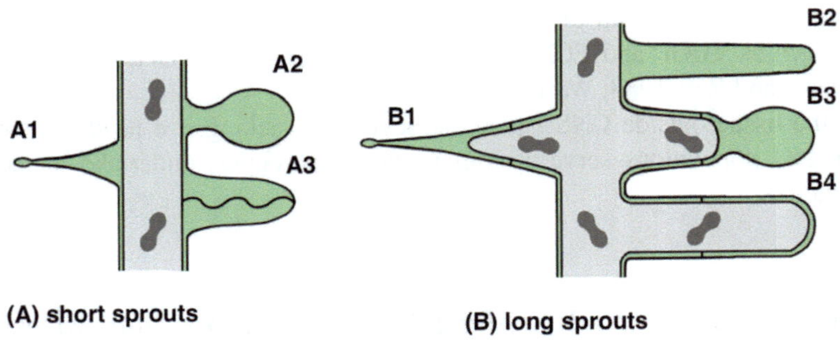

Fig. 1.7 Morphology of vascular EC sprouts found in postnatal vascular formation. Vascular sprouts of various lengths are found in postnatal vascular formation. Short sprouts (**a**) are thought to reflect the initial phase of sprouting, whereas long sprouts (**b**) reflect the further growing phase. Both short and long sprouts are basically classified into two types according to the shape of leading ECs, as "pointed end" (**a1**, **b1**) and "blunt" type (**a2**, **a3**, **b2–b4**). Different morphology of leading ECs is suggested to reflect the transitional physiological states of EC. In both types, short sprouts are composed of either the cytoplasmic projection of a cell (**a1**, **a3**) or a protrusion of a whole-cell body (**a2**). Some short sprouts are composed of two ECs making inter-endothelial junctions each other but have no lumen yet (**a3**). Both pointed-end and blunt long sprouts are basically composed of multiple ECs and have a lumen behind the leading ECs (**b1**, **b3**, **b4**), though some long blunt sprouts especially found in skin wound healing are composed of unusually elongated single ECs that bridge one vessel with another (**b2**). Considering their composition such as singular EC and lacking lumen, the types of long sprouts should be still in the initial phase of sprouting. Some long blunt sprouts in which leading EC is very flattened are alternatively referred to as "saccular sprout," which may also reflect the different physiological states of ECs

found both in the short sprouts at the initial phase (Fig. 1.7a1–a3) and in longer sprouts of further growing-phase stage (Fig. 1.7b1–b4). The different morphology of leading ECs is thought to reflect the transitional physiological states of EC movement, namely, pointed-end sprouts are in relation to rapid migration of leading EC, whereas blunt sprouts are seen in slow migration (Rhodin and Fujita 1989).

In both pointed-end and blunt types, short sprouts as those illustrated in Fig. 1.7a1–a3 are often found as a cytoplasmic projection of a cell (Amselgruber et al. 1999; Morikawa and Ezaki 2011; Rhodin and Fujita 1989) or a protrusion of a whole-cell body (Morikawa and Ezaki 2011). Some short sprouts are composed of two ECs that make inter-endothelial junctions between each other, but a vascular lumen is not formed (Fig. 1.7a3; Cavallo et al. 1973), or show only slit-like immature lumen (Spanel-Borowski et al. 1987). Meanwhile, long sprouts of both the pointed-end (Fig. 1.7b1) and blunt types (Fig. 1.7b2–b4) are usually composed of multiple ECs and have a fully formed lumen behind the leading ECs (Cliff 1963; Morikawa and Ezaki 2011; Rhodin and Fujita 1989; Schoefl 1963). Among blunt sprouts, the ones in which the leading EC is very flattened are sometimes referred to as "saccular sprout" (Fig. 1.7b4) (Cliff 1963; Rhodin and Fujita 1989; Baluk et al. 2003). The significance of this morphology is uncertain but it may indicate a different physiological state of ECs.

Notably, in skin wound healing, sprouts made by a single EC in which the EC cell body becomes large and unusually elongated are found approaching another vessel, presumably for bridging (Fig. 1.7b2) (Morikawa and Ezaki 2011). This type of long sprouts are possibly in an initial phase of growth as they are composed of a single EC and have no luminal surface. In tumor angiogenesis, vascular sprouts may progress without EC proliferation at the initial stage (Ausprunk and Folkman 1977). Omitting EC proliferation is desirable because it enables prompt sprout progression at a lower metabolic cost. In wound healing, elongation of the leading EC might also be employed for that purpose.

Proliferation Potential of Leading ECs

In the tip-stalk cell model, filopodial tip cells sense the angiogenic stimuli and guide sprout growth, whereas stalk cells located just behind the tip cell have another significant function to proliferate and increase the number of EC to physically extend the sprouts (Gerhardt et al. 2003). Tip cells are specialized for vessel guidance and not involved in sprout extension, and are known to hardly proliferate. However, in postnatal vascular sprouts, the proliferation potential of leading ECs has been observed in studies by using electron microscopy (Rhodin and Fujita 1989), autoradiography combined with electron microscopy (Cavallo et al. 1973), and BrdU incorporation (Morikawa and Ezaki 2011). Moreover, mitotic figures of the leading EC have been directly captured at the ultrastructural level (Morikawa and Ezaki 2011; Cavallo et al. 1973). Therefore, the functional role assigned to leading ECs might be different between developmental and postnatal vascular sprouts, or between sprouts found in CNS and other peripheral tissues.

Association of Pericytes with Leading ECs

The most striking difference between postnatal vascular sprouts and the tip-stalk cell model is the guiding structure. As aforementioned in previous sections, in the tip-stalk cell model, tip cells themselves guide the sprouts by sensing the angiogenic stimuli by using filopodia. On the other hand, in postnatal vascular sprouts, pericytic non-EC stromal cells are almost always found in association with leading ECs. Here they might have significant roles to promote sprout growth by providing guidance of sprouts and inducting EC proliferation. One of the key factors in promoting this seems to be MFG-E8 (Motegi and Ishikawa 2017; Uchiyama et al. 2014).

In the next section, we review and re-evaluate the concept of PC-driven vascular formation by quickly looking back at the history of studies dedicated to this subject.

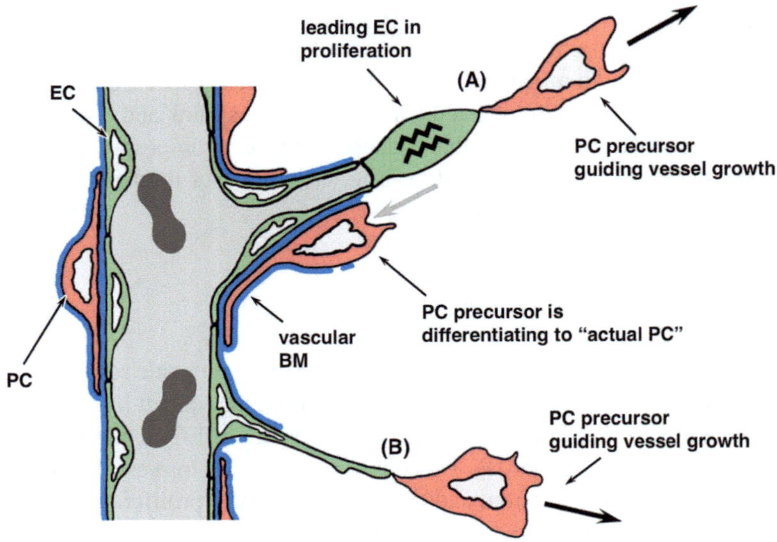

Fig. 1.8 Pericyte-driven angiogenesis. A number of studies suggest that PCs guide SA during vascular formation of postnatal life. PC precursors without vascular BM coating are found closely associating with ECs located at the forefront of vascular sprouts (**a**). The leading ECs are often proliferating. Because these PC precursors produce VEGF, they can induce EC proliferation and promote their migration (black arrow). Some PC precursors were found to be associating with the cytoplasmic process of ECs projected toward stromal space, suggesting the scene of initial phase of vascular sprouting (**b**). Note that the leading EC in (**a**) and the cytoplasmic projection of EC (**b**) are devoid of vascular BM. Sometimes PCs partially covered by vascular BM are seen. They are thought to be in the later stage of angiogenesis and are in the middle of differentiating from precursors to actual "pericytes"

Re-evaluation of Pericyte-Driven Postnatal Vascular Formation

The concept of PC-driven vascular formation has been proposed many times in the past in different settings (Amselgruber et al. 1999; Morikawa and Ezaki 2011; Rhodin and Fujita 1989; Baluk et al. 2003; Morikawa et al. 2002; Nehls et al. 1992; Nehls and Drenckhahn 1993; Virgintino et al. 2007). Because it was originally postulated in the process of SA and it is more likely that SA is the main mechanism of vascular formation in skin wound healing, here we first review the possible functional roles of PCs in SA (see also the summary in Fig. 1.8).

Pericyte-Driven Sprouting Angiogenesis

Rhodin and Fujita (1989) carefully analyzed the process of postnatal SA in the rat mesentery, by combining the intravital and transmission microscopy techniques. In their study, they observed that stromal fibroblasts were associated with the leading ECs of vascular sprouts, and guided the growth of vessels. They also observed that the fibroblasts (that usually lack vascular BM) were gradually covered by the vascular BM of newly formed vessels and eventually became PCs.

Successively, Nehls et al. (1992) found in the rat mesentery that the fibroblasts that guided the EC sprouting were positive for the PC marker desmin. Because microvascular PCs usually integrate within the vascular BM of the blood vessels in the resting state (intramural position; Ashton and de Oliveira 1966), stromal fibroblasts positive for desmin are thus considered "extramural PCs" differentiated from stromal PC precursors. PC precursors associated with ECs located at the forefront of vascular sprouts have later been detected with other PC markers, such as PDGFRβ, NG2 proteoglycan, endosialin, and αSMA (Morikawa and Ezaki 2011; Virgintino et al. 2007; Ozerdem and Stallcup 2003).

Pericyte-Driven Sprouting Angiogenesis in Skin Wound Healing

In the context of skin wound healing, PC precursors expressing VEGF associate with the dividing ECs located at the forefront of vascular sprouts (Morikawa and Ezaki 2011). These PC precursors thus induce EC proliferation and guide sprout growth during vascularization. It has been well documented that VEGF can induce EC proliferation and migration. Intriguingly, in skin wounds, unusually elongated ECs are seen proceeding from a vessel toward another vessel, presumably for vascular bridging (illustrated in Fig. 1.7b2), and they are surrounded by PCs at their leading tips (Morikawa and Ezaki 2011). VEGF expressing PCs might therefore facilitate EC elongation. VEGF production by PCs has also been observed in other vascular formation settings (Darland et al. 2003; Redmer et al. 2001; Reynolds and Redmer 1998). At the later stages of SA, PCs are thought to deactivate and function as stabilizers of the newly formed vessel wall, by stopping EC proliferation and migration and by vascular BM production as well. It is suggested that VEGF expression is lost in PCs that are shifted to this "static state" (Morikawa and Ezaki 2011) and, in turn, the potential of producing vascular BM components such as laminin or type IV collagen by PCs is turned on (Jeon et al. 1996).

In the currently accepted theoretical model of SA, PCs first disappear from the vessel wall and EC sprouting then starts. Later, the PCs return to the newly formed vessel wall to stabilize it. Therefore, the only functional role attributed to PCs is that of vessel maturation at the later stages of vessel formation. However, as it is well reflected in their transitional marker expression, their functional roles could be different according to the stages of vessel formation. During the process of differentiation from precursors to actual PCs, they may shift their functional roles in accordance with the stages of vascular formation. Our proposal is that pericytes promote EC proliferation and migration at the early stage of vascular formation, but in turn they promote the stabilization of newly formed vessel wall at the later stage. Therefore we advocate for a re-evaluation of the PC-driven sprouting angiogenesis, especially in the skin wound healing context. The basic concepts of PC-driven SA hypothesis that we here propose are summarized in Fig. 1.8.

The idea that apart from dividing from preexisting PC populations (Diaz-Flores et al. 2009), a fraction of PCs arising from stem cells during vessel formation is not original. This has been suggested by a number of past studies (Rhodin and Fujita 1989; Nehls et al. 1992; Nehls and Drenckhahn 1993; Crocker et al. 1970; Nakayasu 1988; Sims 1986), including one in the context of new vessel formation in skin wound healing (Sasaki et al. 2008). Some authors suggested that they are originated from circulating fibrocytes (Xueyong et al. 2008). The possible origins of PC precursors are discussed in detail in other sections of this chapter.

Pericyte-Driven Vascular Formation Other than Sprouting Angiogenesis

PCs may also play significant roles in the promotion of other types of vascular formation than SA. In IA (both of pillar first and bridging first types), PCs and/or their precursors are supposed to play a role in connective tissue pillar synthesis by ECM production (Egginton et al. 2001; Burri and Djonov 2002; Makanya et al. 2009). Bagley and colleagues found that PCs are closely associated with EPCs or capillary networks derived from EPCs, and suggested the possibility that guidance of vascular growth in postnatal vascularization is mediated by PCs (Bagley et al. 2005). Further, in various types (lung, breast, prostate) of tumor angiogenesis, the PCs form tubes without EC lining at first, and thereafter ECs form tubes running along with the PCs (Ozerdem and Stallcup 2003).

Putting together all available data, we suggest that PCs likely promote vascular formation in postnatal life in various scenarios. It has been postulated that the PCs may stabilize and induce blood vessel maturation through TGFβ signaling at the later stages of vascular formation (Crocker et al. 1970; Darland and D'Amore 2001; Hirschi and D'Amore 1996; Hirsh and Weitz 1999; Sato and Rifkin 1989). We postulate that PC function in vascular formation also includes a shift in their functional roles according to physiological conditions.

Pericyte to Macrophage Transition

Since the role of macrophages in wound healing will be discussed later in this chapter, we think it is worthwhile to remind readers that, at least in the CNS, perivascular cells have been known for a long time to be able to generate resident macrophages under certain conditions (Baron and Gallego 1972; Rezaie and Male 2002). A more recent study found Sox2+ PDGFRβ+ pericytes giving rise to Iba1+ microglia in response to ischemic stroke (Sakuma et al. 2016). Importantly, these were activated cells that could not be isolated from nonischemic areas, indicative of dedifferentiation.

Peripheral Innervation in Wound Healing

Cutaneous Innervation

The skin is highly innervated by a complex network of nerve fibers, composed by both sensory and autonomic (mostly sympathetic) neurons (Laverdet et al. 2015).

Autonomic nerve fibers are unmyelinated and constitute the minority of skin nerves. They are only found in the dermal layer, where they regulate blood and lymphatic vessel circulation and associate to skin appendages such as the arrector pili muscle (APM) and the sweat glands (Wang and Gibbons 2013).

Sensory nerves are more abundant and present in both the epidermis and dermis (Arthur and Shelley 1959). They form a heterogeneous plexus together with blood and lymphatic vessels in the deep and superficial compartments of the skin. Specifically, larger nerve trunks >25 μm below the dermoepidermal junction inter-

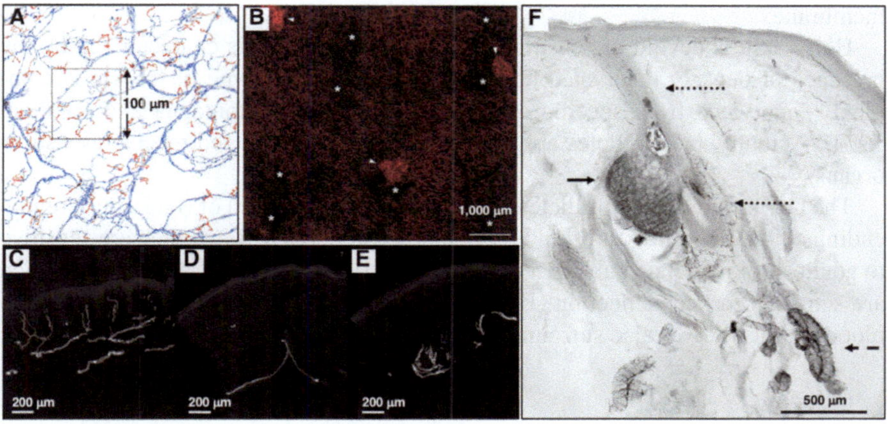

Fig. 1.9 Peripheral innervation of the skin. Cutaneous innervation follows a complex pattern. (**a**) Human forearm skin epidermis seen from above. The scheme depicts larger nerve trunks in the lower dermis that form smaller trunks in the upper dermis. One of the latter (enclosed in a square whose sides measure 100 μm in length) shows the typical penicillate nerve endings, both at the subepidermal (blue) and epidermal (red) levels. Adapted with permission from Arthur and Shelley (1959). However, the actual nerve density is much higher than depicted here. (**b**) Superficial dermal nerve plexus on a dermal sheet of 38 mm^2 stained with anti-PGP9.5, which labels all types of nerve fibers, showing the dense innervation pattern. Three touch domes (arrowheads) are located close to hair follicle openings (asterisks). Adapted with permission from Reinisch and Tschachler (2005). (**c–e**) Distribution of myelinated fibers (anti-MBP staining) in glabrous (**c**) and hairy (**d, e**) skin. In glabrous skin papillary dermis, myelinated fibers were homogeneously abundant and left dermal bundles to reach their targets. In hairy skin, myelinated fibers were irregularly distributed with higher density in the proximity of hair follicles: compare panel E versus panel D. Adapted with permission from Provitera et al. (2007). (**f**) Innervation of a hair follicle is shown (dotted arrows), with an attached sebaceous gland (solid arrow). A network of sensory nerve fibers surrounds the base of the hair follicle and extends parallel to the hair shaft up to the epidermal surface. A densely innervated sweat gland may be seen in the deeper dermal tissue (dashed arrow). Adapted with permission from Wang and Gibbons (2013)

twine with the superficial capillary plexus. In contrast, no obvious topographic relation can be seen between the terminal nerve endings and blood vessels at the uppermost 25 μm below the junction (Tschachler et al. 2004; Fig. 1.9).

The sensory fibers can be classified by their signal transduction velocity into C (unmyelinated, slow), Aδ? (lightly myelinated, intermediate velocity), and Aβ (highly myelinated, fast) type fibers (Djouhri 2016). Usually these nerve fibers participate in the formation of cutaneous sensitive/sensorial receptors responsible for detecting mechanical, thermal, and chemical stimuli. Some are naked nerve endings and others are composed of nerve fibers integrated within a connective capsule conforming a sensitive corpuscle (Laverdet et al. 2015).

Importantly, cutaneous nerves and receptors present qualitative and quantitative differences between glabrous and hairy skin (Zimmerman et al. 2014). In the glabrous skin four low-threshold mechanoreceptors with specific function and morphology are associated to Aδ fibers: Merkel complexes and Ruffini, Meissner, and Pacini corpuscles. Intra-epidermal free nerve endings (type C and Aδ) are also present in non-hairy skin, losing their myelin sheath when they enter the basement membrane.

Hairy skin is devoid of sensitive corpuscles but instead is characterized by the presence of densely innervated hair follicles, where each type of follicle is complexly innervated by a specific combination of nerves (Montagna and Parakkal 1974). Actually, hair follicles are considered highly specialized mechanosensitive organs.

The nerves around hair follicles in the mouse are composed of (i) circumferential endings of Aβ-field low-threshold mechanoreceptors (LTMRs), which are sensitive to gentle stroking (Bai et al. 2015), and (ii) lanceolate endings of Aδ-LTMRs which are sensitive to hair deflection (Rutlin et al. 2014; Fig. 1.10). Of note, similar to blood vessels, nerves also surround the bulge and dermal papilla stem cell niches.

Neuromodulators Affected by Denervation

The communication between peripheral nerves (and associated neural cells) and skin cell populations involves a variety of molecules (neuropeptides, neurohormones, and neurotrophins) that act as neurotransmitters, hormones, or paracrine factors. Those neuromodulators bind to their specific receptors, expressed in both neural and nonneural cells, and take part in a multitude of skin functions, with important implications in wound repair (Roosterman et al. 2006).

Evidences supporting the implication of neuromodulators in wound healing arise from observations made on diabetes. Numerous studies noticed that, in diabetic animals, reduced innervation correlated with lower expression of substance P (SP), calcitonin gene-related peptide (CGRP), neuropeptide-Y (NPY), vasoactive intestinal peptide (VIP), and nerve growth factor (NGF) (Anand 1996; Gibran et al. 2002; Kuncova et al. 2005; Pradhan et al. 2011).

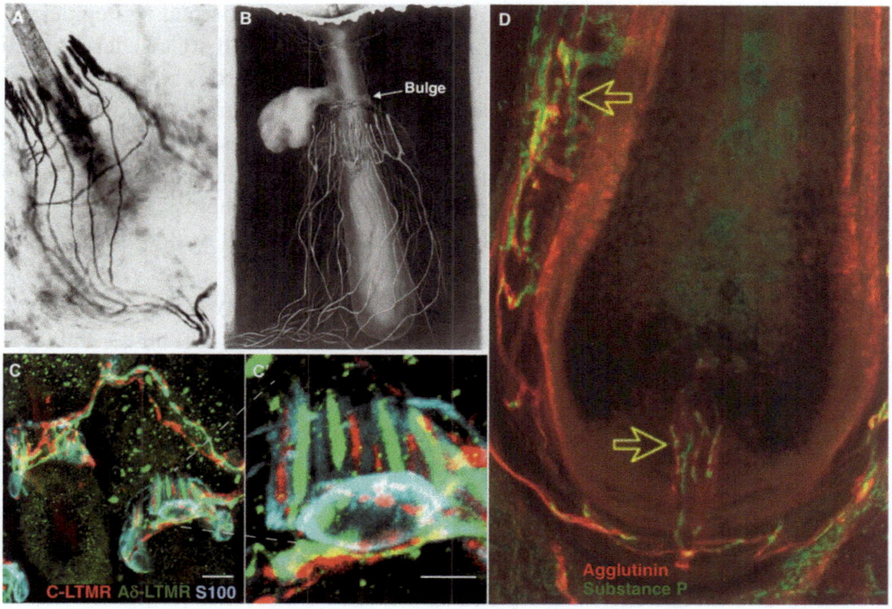

Fig. 1.10 Peripheral innervation of the hair follicle. (**a** and **b**) At the level of the bulge of large follicles a collar of "follicle end organs" surrounds the hair follicle (**a**). Thick frozen section prepared with Winkelmann's techniques. In small follicles, the lanceolate nerve endings surround the bulb (**b**). Reproduced with permission from Montagna and Parakkal (1974). (**c–c′**) A similar arrangement is found in laboratory mice. On this image of back skin from Th^{CreER};$Rosa26^{tdTomato}$;$TrkB^{tauEGFP}$ mice, C-LTMRs were labeled with tdTomato fluorescence (red), Aδ-LTMRs were labeled with anti-GFP (green), and terminal Schwann cells were stained with anti-S100 (cyan). (**c′**) shows higher magnification of the terminal Schwann cell in the middle of the lanceolate complex shown in (**c**). Scale bars, 10 µm in (**c**), 5 µm in (**c′**). Reproduced with permission from Li and Ginty (2014). (**d**) Substance P+ nerves can be seen in the dermal papilla as well as in the perifollicular vasculature and nerve plexi (open arrows). Section of human scalp from a healthy adult male stained with *Ulex europaeus* agglutinin (for blood vessels) and substance P (for nerves). Reproduced with permission from Hordinsky et al. (1999)

SP is a neuropeptide involved in the immunomodulatory responses in early wound healing stages; it induces neutrophil activation and infiltration (Richards et al. 1999) and leukocyte chemotaxis; and modulates the secretion of proinflammatory cytokines in distinct dermal cell populations, such as interleukin-1 (IL-1), IL-2, IL-6, TNF-α, and TGF-α (Delgado et al. 2005; Wei et al. 2012). Denervation via laser-induced neuropathy, capsaicin treatment, or nerve resection decreased the SP levels detected in the skin (Chiang et al. 2005; Gamse et al. 1986; Senapati et al. 1986). Rook et al. have shown that the delay in wound closure induced by morphine was due to the inhibition of SP secretion associated with a decreased expression of its specific receptor NK-1R (Rook et al. 2009). Indeed, it has been described in several wound models that exogenous SP administration or the induction of its production at the injured site improves healing (Delgado et al. 2005; Ishikawa et al. 2014; Spenny et al. 2002).

CGRP is ubiquitously expressed by unmyelinated subepidermal and epidermal sensory fibers and also by inflammatory and other dermal cells (Caviedes-Bucheli et al. 2008; Hagner et al. 2002). This neuropeptide has numerous biological activities that include a role in immune response, dermal cell proliferation, and angiogenesis induction (Brain and Grant 2004; Dallos et al. 2006; Hosoi et al. 1993; Mapp et al. 2012). The absence of CGRP or the administration of antagonists resulted in delayed wound closure with a reduced angiogenesis rate, whereas the exogenous administration of CGRP resulted in an acceleration of wound healing (Rook et al. 2009; Engin 1998; Toda et al. 2008).

An important growth factor with a crucial role in skin physiology is the nerve growth factor (NGF). NGF is immediately released in response to injury and presents important functions in tissue repair (Kawamoto and Matsuda 2004). In wound healing, NGF induces the differentiation of fibroblasts into contractile myofibroblasts and regulates their migration and activity (Micera et al. 2001; Palazzo et al. 2012). In wound models, the topical application of NGF improves healing; increases the survival and migration of irradiated skin fibroblasts and blood cells (Shi et al. 2003); and accelerates healing by inducing collagen production on fibroblasts (Nithya et al. 2003). NGF is also implicated in the control of innervation density, stimulating neuronal regeneration and secretion of neuropeptides (Lewin and Mendell 1993; Park et al. 2010). Thus, NGF is a neurotrophic factor that promotes cell survival, stimulates neurite outgrowth, and modulates cell differentiation, all of which are important for optimal wound healing.

Experimental Denervation a nd Impaired Wound Healing

Experimentally induced denervation assays support the role of peripheral innervation in wound healing. Capsaicin is an agonist of vanilloid receptor-1, expressed specifically in Aδ and C fibers and which causes denervation. In rats, chemical denervation by capsaicin treatment generated 43% loss of CGRP+ nerves and a consequent increase in wound areas, prolonged scab retention, and delayed reepithelialization. Hence the partial withdrawal of sensory innervation impairs cutaneous wound healing (Smith and Liu 2002). Previous work indicated that sensory nerves are involved in the neurogenic inflammatory reactions, and that capsaicin-mediated denervation may thus prevent the initial inflammatory response after wounding (Jancso et al. 1967).

Chemical denervation of rats with 6-hydroxydopamine (6-OHDA) produces an impaired wound response after full-thickness lesions. By day 14, only 48% of denervated rats were able to repair the wound as compared to 84% of controls (Kim et al. 1998). By using the same approach, Souza and colleagues described that animals denervated with 6-OHDA presented an accelerated wound contraction, with an increased number of αSMA+ myofibroblasts, reduction in mast cell migration, and delayed reepithelialization (Souza et al. 2005). When 6-OHDA was administrated after the acute inflammatory phase, animals displayed delayed contraction indicating that the denervation-induced absence of catecholamines also had a role at the final phase of healing.

When denervation is generated by spinal cord hemidissection, non-innervated skin areas present an impaired wound healing response. In this model, a delay in wound contraction and reepithelialization is evident mostly in the last stages, 14 days post-wounding (Fukai et al. 2005). In similar previous studies it was suggested that absence of neuropeptide secretion by nerve endings may be responsible for the retarded wound contraction in denervated areas (Engin et al. 1996). In accordance, local application of substance P improves healing in denervated rat skin areas (Ishikawa et al. 2014). Likewise, capsaicin induced the depletion of substance P and was also responsible for axonal NGF retrograde transport inhibition (Miller et al. 1982). Thus, the exogenous administration of NGF accelerates wound contraction and epithelialization (Li et al. 1980; Muangman et al. 2004).

Nerve-Associated Schwann Cells as Pro-regenerative Actors

Although most researchers attribute the main role of innervation in wound repair to the secretion of neuromediators (Ashrafi et al. 2016), in this chapter we propose the hypothesis that nerve-associated cells, specifically the terminal Schwann cells that ensheath the nerve endings, may have a prominent role as pro-regenerative actors.

This concept originates in metazoan phylogeny. Mammalian regeneration capacity is limited, as we have lost multitissue regenerative potential except for the distal digit tip (Han et al. 2008) and punched ear lobes in certain genetic backgrounds (Cheverud et al. 2012; Seifert et al. 2012a). However, some vertebrates (such as axolotl, salamander, and zebrafish) and invertebrates (hydra) possess an extraordinary regeneration capacity. In all cases the nerves seem to be required for a successful regeneration to occur (Brockes and Kumar 2008; Carlson and Conboy 2007; Kumar and Brockes 2012; Simoes et al. 2014; Stocum 2011).

During the well-studied limb regeneration in urodele amphibians, epithelial cells are organized and migrate toward the amputated region in order to close the wound and form the wound epidermis (WE). The maturation, by cell migration, of this specialized WE leads to apical epithelial cap formation (AEC). Under the AEC, cells within the skin mesenchyme are liberated from the compact ECM, which is degraded by proteases, and undergo dedifferentiation, followed by migration and reentering the cell cycle to generate an undifferentiated cell mass known as blastema. The blastema gives rise to a new extremity. The interactions between the AEC and blastema ensure the correct growth and patterning of the novel structure (Kumar and Brockes 2012; Simoes et al. 2014; Stocum 2011). When limbs are denervated apparently neither WE formation nor initial blastema morphogenesis is inhibited, but instead blastemal progenitor cell proliferation is notably affected (Brockes and Kumar 2008; Kumar and Brockes 2012; Stocum 2011).

Nerve requirement seems to be dose and time dependent, since the number of nerves positively correlates with regeneration rate and the moment in which denervation occurs exerts distinct phenotypes (Simoes et al. 2014; Seifert et al. 2012b; Singer 1952). Thus, the resection of brachial nerves in zebrafish fins avoided blastema formation, while a reduced amount of nerves was enough to generate a residual blastema that gave rise to a smaller but morphologically defined fin (Simoes et al. 2014).

Derived from observations made during denervation assays, it was proposed that nerves produce neurotrophic factors that control the regenerative process, the so-called neurotrophic hypothesis (Singer 1951, 1952, 1964). The fact that neural tissue promotes proliferation of blastema cells was demonstrated by the application of neural extracts to cultured blastemas and the observation of an increased mitotic rate and the restoration of protein synthesis. Later the identity of the active molecule was associated to a protein (Choo et al. 1978), possibly transferrin (Albert and Boilly 1988).

Supporting the idea of a nerve-derived pro-regenerative molecule, Kumar and Brockes described the existence of a diffusible signal, nAG protein, which acts as intermediary between regenerating axons and WE to induce blastema proliferation. The nAG is first secreted by axon ensheathing Schwann cells (SCs) followed by expression in both SCs and gland cells of the WE. This shift in nAG protein expression explains the influence of nerves in the proliferation of the early blastema and not in the later phases of the regeneration, in which WE effect is predominant (Brockes and Kumar 2008; Kumar et al. 2010). Thus, it was proposed that besides the mitogens produced by axons, such as transferrin, NGF, FGF2, or SP, that could be affecting blastema development (Anand et al. 1987; Mescher et al. 1997; Mescher and Kiffmeyer 1992), SCs associated to nerves were responsible for secreting factors that supported blastemal cell proliferation and accumulation. For instance, GGF-2 and other growth factors (FGF, IL-1, IL-2, and IL-6) secreted by axons and inflammatory cells are mitogenic for SCs in transfected mammalian peripheral nerves (Davies 2000), so that they could partly explain the nerve dependence of the SCs and the loss of their activity in denervated limb regeneration.

These phylogenetic studies have thus enlightened our knowledge of the tissue regeneration process, with potential applications not only in the case of amputations but also in tissue repair in general. As for other evidence implicating nerves and the associated Schwann cells in tissue regeneration, the recent work by Freda Miller and co-workers has described the role of Schwan cell precursors (SCPs) in digit tip regeneration, again underlining the parallels shared with the urodele limb regeneration (Johnston et al. 2016). In physiological conditions, Sox2+/S100β+ SCP cells are associated with the innervating axons along the digit tip mesenchyme. When digit tip is amputated, a PDGFRβ+ blastema-like structure is formed adjacent to the K14+ wound epithelium (WE). SCPs dissociate from the innervating axons and migrate into the wounded area where they secrete paracrine factors. SCPs remain in a dedifferentiated state within the regenerating blastema until axons reinnervate the regenerated digit tip, the moment in which they reassociate with axons. Denervation by sciatic nerve resection (that inhibited the axonal regrowth) prevented migration of SCP to the repairing tissue, probably due to the lack of mitogenic factors from axons that are required for SC functionality. Moreover, the specific ablation of

Sox2+ SCPs and the induction of genetic SOX2 ablation also impaired digit tip regeneration, demonstrating that SCPs are directly implicated in blastema-like mesenchymal cell dedifferentiation and highlighting the role of SOX2 in the process (Johnston et al. 2016).

Actually, the same research group previously described the implication of Sox2+ SCs in skin wound healing. They showed that Sox2+/S100β+ nerve terminal (NT) cells located around HFs in mouse dorsal skin contributed to the wound healing (Johnston et al. 2013). Interestingly, the major contribution to the healing wound was carried out by tissue-resident SCs that acquired SOX2 expression after injury, as described in peripheral nerve injury SC. In addition, when SOX2 was ablated a delay in wound closure was observed (Johnston et al. 2013). Both articles supported the role of peripheral SCs as promoters of mammalian skin repair and implicated SOX2 as a modulator of the SC dedifferentiation.

Based on those results and the observations made on amphibian limb regeneration, it makes sense that nerve-derived Sox2+ SCPs may proliferate and migrate to the damaged region induced by injury signals, where they secrete growth factors that regulate mesenchymal (blastema-like structure) cell proliferation. This idea is supported by the evidence obtained from peripheral nerve injury regeneration, in which also similar dedifferentiation and mitogenic mechanism have been described for SCs (Cattin et al. 2015; Jessen and Mirsky 2008; Napoli et al. 2012; Parrinello et al. 2010).

More recently the group of Lukas Sommer very neatly demonstrated that, upon wounding, dedifferentiated SCs reentered into cell cycle and populated the wound bed, where they secreted factors previously associated with wound healing and promoted myofibroblast differentiation by paracrine modulation of TGFβ signaling (Parfejevs et al. 2018a). Parfejevs et al. further proposed that postnatal multipotent neural crest stem cells (NCSCs) may be induced by injury or stress in several organs, by reprogramming differentiated cells such as SCs. These noxa would additionally activate a repair program in these adult NC-derived cells, which would then promote tissue repair or regeneration by paracrine signaling (Parfejevs et al. 2018b; Silva et al. 2018a).

Another interesting observation of mammalian tissue regeneration was made on Murphy Roths Large (MRL) mice, which have the capacity to heal the injured ear without scar formation and are characterized for developing a blastema-like structure that restores completely the cartilage, skin, hair follicles, and adipose tissue of the ear tip (Clark et al. 1998). MRL mice also show an enhanced capacity for peripheral nerve regeneration (Buckley et al. 2011). Additional studies demonstrated that after denervation, blastema formation, and chondrogenesis are prevented, wound areas increase, wound distal margins become necrotic, and as a result the ear holes lose the ability to re-epithelialize (Buckley et al. 2012). Denervation also had a notable negative effect on the ear wound healing mechanisms of the C57BL/6 strain, suggesting that innervation may be important for regeneration and also for normal wound repair (Buckley et al. 2011, 2012).

Schwann to Pericyte Transition

We have recently shown that Schwann cells and pericytes of human skin present strikingly similar gene expression profiles (Etxaniz et al. 2014). Some of the lineage-tracing lines that have been used in the mouse to follow up dermal precursor fate may also trace SCs and pericytes (Iribar et al. 2017). These data suggested a previously unrecognized relationship between these two cell types. Alternatively, the two cell subtypes might be considered as a single, highly dynamic cell precursor with environmental differences playing a role in distinctive cellular states (Etxaniz et al. 2014).

Of interest, SCs in the skin derive in development from a subpopulation of neural crest cells known as boundary cap (BC) cells (Gresset et al. 2015). BC cell subpopulations in development can be traced by the expression of *Prss56* and *Krox20* as lineage tracers. Although both Prss56-traced and Krox20-traced BC cell progenies migrated along nerves to the skin, only the first population remained as SCs, in contact with nerves. Strikingly, the Krox20-traced BC cell subpopulation delaminated from nerves in between E12.5 and E.13.5 and integrated into the forming vascular plexus as mural cells (Radomska and Topilko 2017). Thus, the Schwann to pericyte transition that we propose as a firm possibility upon skin injury is already demonstrated in development by Topilko et al., who named it as "glial to vascular switch" (Radomska and Topilko 2017). In fact, cephalic neural crest has long been known to give rise to pericytes (Etchevers et al. 2001; Trost et al. 2016), but this was certainly unexpected in the trunk.

Additionally, in the mouse incisor tooth organogenesis, Schwann cells give rise to MSCs that will in turn generate pulpar cells and odontoblasts (Kaukua et al. 2014). Whether this particular subset of MSCs sits in a perivascular location is currently unknown, but this seems highly likely given the fact that MSCs seem to correspond with pericytes and adventitial cells in most tissues (Corselli et al. 2010; Crisan et al. 2012).

In glioblastoma, CD133+ tumor stem cells of glial origin are able to generate pericytes. Stem cells migrate toward endothelial cells induced by SDF-1α/CXCR4 signaling cues and transdifferentiation into pericytes is enforced by TGFβ expression (Cheng et al. 2013). Notch1 signaling activation similarly reduced expression of Sox2 in glioblastoma stem cells and induced the expression of pericyte markers and angiogenesis-promoting factors (Guichet et al. 2015), indicating that different signaling may induce similar end results. The picture of the pathways involved is surely much more complex (Brooks et al. 2013). Importantly, this cancer stem cell to pericyte transition has not been observed in any other tumor type (Krishna Priya et al. 2016), indicating that it may be specific for the glial stem cells.

Macrophages in Wound Healing

Macrophages in Tissue Repair

Macrophages are needed for competent healing of several tissues (Lucas et al. 2010; Mirza et al. 2009; Summan et al. 2006; van Amerongen et al. 2007). Depending on the healing phase, they promote debridement of the injury area, cell proliferation, angiogenesis, ECM deposition, and remodeling, because they are able to acquire diverse functional phenotypes which are mainly determined by the microenvironment (Mahdavian Delavary et al. 2011; Novak and Koh 2013a).

Macrophages are critical for the regulation of all stages of tissue regeneration (van Amerongen et al. 2007; Arnold et al. 2007; Bergmann et al. 2006; Godwin et al. 2013; Li et al. 2012; Schlundt et al. 2018; Shi and Pamer 2011), and when their function is deregulated they contribute to impaired healing and fibrosis (Eming et al. 2007b; Loots et al. 1998; Mirza and Koh 2011; Mirza et al. 2013; Sindrilaru et al. 2011; Villalta et al. 2011; Wynn et al. 2011).

In the murine blood, two functional subsets of monocytes have been described: (i) an inflammatory subset, defined as CCR2+ CXC3CR1^{low} Ly6C+ (also known as GR1), and (ii) a noninflammatory subset or resident monocytes, defined as CCR2− CXC3CR1^{high} Ly6C− (Geissmann et al. 2003; Willenborg et al. 2012). In humans, most monocytes are CD14^{hi}CD16− and are referred to as "classical" monocytes or are CD14+CD16+ and are referred to as "nonclassical" monocytes (Passlick et al. 1989; Strauss-Ayali et al. 2007). Approximately 90% of human monocytes express the classical markers, whereas in mice the two populations of monocytes are approximately equally represented in the blood (Passlick et al. 1989; Mosser and Edwards 2008). Importantly, a recent study suggested that circulating nonclassical monocytes are recruited to cutaneous injuries, where they colonize perivascular niches and generate alternatively activated wound-healing macrophages (Olingy et al. 2017).

Neutrophils and macrophages are the major fraction of inflammatory cells recruited to the wound area. Neutrophils are the first inflammatory cells to be recruited (few hours postinjury), but their presence is timely restricted to the early stage of the wound healing. Pericytes seem to have a role in neutrophil extravasation, by modulating the secreted ECM in response to pro-fibrotic stimuli (Sava et al. 2015). In fact, a circulating proangiogenic subset of neutrophils (characterized by the expression of these markers: CD49d+ VEGFR1^{high} CXCR4^{high}) is recruited to injured tissue both in mice and humans, in response to VEGF-A signaling, by parallel activation of VEGFR1 on neutrophils and VEGFR2 on endothelial cells (Massena et al. 2015). Macrophages instead persist through all wound-healing response. Their overall number increases during the inflammation phase, peaks at tissue repair stage, and gradually decreases through the tissue-remodeling stage (Martin and Leibovich 2005; Fig. 1.11).

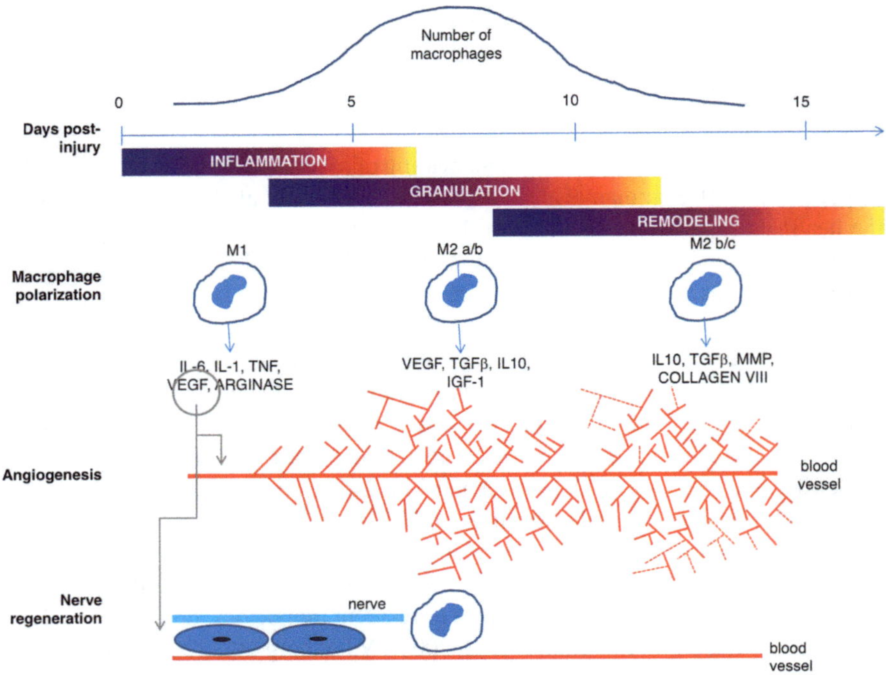

Fig. 1.11 Phases of tissue repair: inflammation, proliferation, and remodeling. Macrophage number increases in number early in the inflammatory phase, peaks in the granulation phase, and decreases during the remodeling. Macrophages in the inflammatory phase exhibit a proinflammatory phenotype and express VEGF that promotes angiogenesis. During the granulation phase, macrophages express VEGF and TGFβ, among others, promoting angiogenesis and extracellular matrix deposition. In this phase proinflammatory and anti-inflammatory signals coexist. Finally, in the remodeling phase tissue remodeling is mediated by expression of TGFβ, MMPs, and collagen VIII. Macrophages participate in angiogenesis, extracellular remodeling, and nerve regeneration indirectly, via cytokine secretion, and directly in the nerve regeneration bridge region. Adapted from Dunleavey and Dudley (2012), Cattin et al. (2015), and Koh et al. (2013)

Lineage-tracing studies have showed that most tissue-resident macrophage populations are derived from early embryonic hematopoietic progenitors that emerge in the yolk sac prior to hematopoietic stem cells (Gomez Perdiguero et al. 2015; Schulz et al. 2012) and persist into adulthood independently of blood monocyte input in the steady state (Schulz et al. 2012; Ginhoux et al. 2010; Yona et al. 2013). Although all tissues are populated at birth with fetal macrophages, replacement of these cells by hematopoietic stem cell (HSC)-derived progenitors gradually occurs with time. Adult tissue macrophages that are maintained by circulating precursors include the intestine (Bain et al. 2014), dermis (McGovern et al. 2014; Tamoutounour et al. 2013), heart (Epelman et al. 2014; Molawi et al. 2014), and pancreas macrophages (Calderon et al. 2015). In contraposition, other tissue-resident macrophage populations such as microglia, epidermal (Langerhans), liver (Kupffer), and alveolar macrophages exhibit negligible need for replacement in adulthood.

Independently of their developmental origin, it has been reported that the tissue itself plays a key role in controlling the persistence, recruitment, and differentiation of the monocytes (Scott et al. 2016). Moreover, macrophage populations exhibit distinct transcriptional signatures (Gautier et al. 2012; Lavine et al. 2014) and epigenetic marks that are specific to the tissue of residence. Other studies have remarked the key role of tissue-specific factors in the imprinting of the macrophage transcriptional program (Scott et al. 2016; Gibbings et al. 2015), but very little is known about the precise mechanisms governing these processes.

Macrophage Polarization

Macrophages undergo activation to fulfill specific functional roles during inflammation and its resolution. A complex mixture of cytokines, metabolites, plasma proteins, growth factors, and microbial ligands present in the inflammatory milieu and in the injured tissue microenvironment confers the phenotype to macrophages (Mosser and Edwards 2008; Gordon 2003; Jenkins et al. 2011; Jenkins et al. 2013; Mantovani et al. 2004).

The development into mature and fully activated macrophages is artificially classified into stages, but most likely will represent a continuum of cellular states. In the first phase (differentiation), recruited monocytes mature into macrophages, stimulated by growth factors such as GM-CSF or M-CSF. During recruitment, monocytes are exposed to varying concentrations of mediators, inducing a second phase of priming by cytokines: IFN-γ (M1 macrophages, see below) or IL-4 and IL-13 (M2 macrophages, see below). During the third phase of activation, macrophages reach a mature functional phenotype in response to microbial and opsonic stimuli such as antibody complexes. If the macrophage survives inflammation, it undergoes the final phase commonly referred to as deactivation or resolution (Gordon and Martinez 2010; Stout et al. 2005). In this last phase, the macrophage proinflammatory potential is deactivated, and goes through functional changes that permits it to clear debris and express general repair functions by expression of IL-10, TGF-β, and a multitude of anti-inflammatory mediators such as nucleotides, lipoxins, and glucocorticoids (Gordon and Martinez 2010).

M1 and M2 have been used to refer to the two extremes of a spectrum of possible phenotypes of macrophage activation (Mosser and Edwards 2008; Gordon and Martinez 2010; Mantovani et al. 2002, 2004). M1 macrophages correspond to the classically activated macrophages (inflammatory macrophages), emulating the Th1 nomenclature. M2 (alternatively activated; type II; Mø2; M2) has been proposed as a generic name for the various forms of macrophage activation other than the classic M1, based on the sharing of selected functional properties (e.g., low IL-12) and their general involvement in type II responses, immunoregulation, and tissue remodeling. However, in vivo, this is much more complex and macrophages display features of both phenotypes (Mantovani et al. 2002; Duffield 2003; Raes et al. 2002) and because of that polarization should be understood as a continuum of diverse functional states (Martinez and Gordon 2014; Martinez et al. 2008; Stout 2010).

Interferon-γ (IFN-γ), alone or in combination with microbial products [e.g., lipopolysaccharide (LPS)] or cytokines [e.g., tumor necrosis factor (TNF)], activates macrophages (Hamilton 2002; O'Shea and Murray 2008). Classical macrophage activation (M1) is characterized by high capacity to present antigen; high interleukin-12 (IL-12) and IL-23 production (Verreck et al. 2004) and consequent activation of a polarized type I response; and high production of toxic intermediates [nitric oxide (NO), reactive oxygen intermediates (ROI)].

IL-4 and IL-13 induce a distinct activation program, referred to as "alternative activation" (M2) (Gordon 2003; Stein et al. 1992). The term "alternatively activated macrophage" has also been applied to mononuclear phagocytes exposed to IL-10, glucocorticoid, or secosteroid (vitamin D3) hormones (Goerdt et al. 1999). Macrophages exposed to immune complexes (IC) and LPS are characterized by an IL-10^{high} and IL-12^{low} phenotype and promote type II responses; they have been called type II activated macrophages (Mosser 2003). Finally, human monocytes dif-

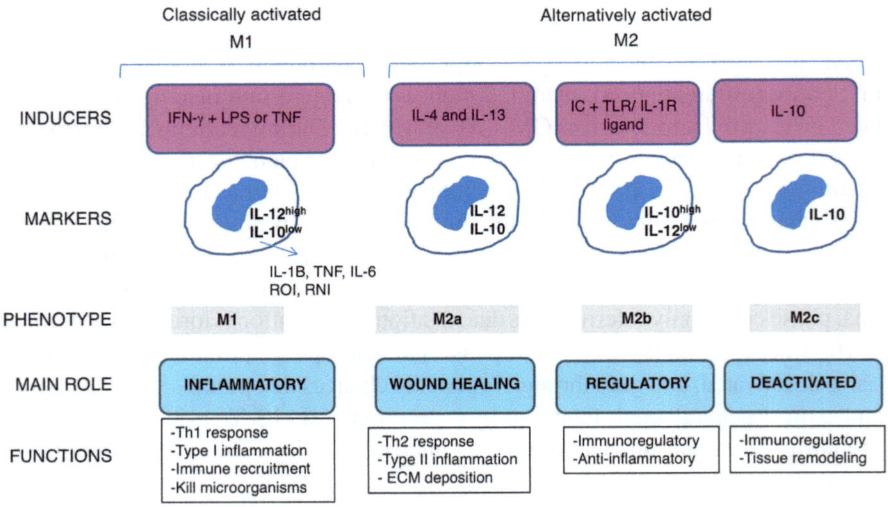

Fig. 1.12 Inducers, markers, and functional features of polarized macrophage populations. Macrophages exposed to IFN-γ in concert with microbial stimuli (e.g., LPS) or cytokines (e.g., TNF or GM-CSF) differentiate into M1 inflammatory macrophages that express high levels of IL-12 and modest levels of IL-10 and are efficient producers of effector molecules (reactive oxygen and nitrogen intermediates) and inflammatory cytokines (IL-1β, TNF, IL-6). M2 macrophages tend to immunoregulatory and tissue remodeling functions. M2a (induced by exposure to IL-4 and IL-13) expresses both IL12 and IL-10 and M2b (induced by combined exposure to immune complexes and TLR or IL-1R ligands) expresses high levels of IL-10 and low levels of IL-12. Finally, M2c, induced by IL-10, expresses IL-10 but not IL-12 (*IFN-γ* interferon-γ, *LPS* lipopolysaccharide, *MR* mannose receptor, *TNF* tumor necrosis factor, *TLR* Toll-like receptor, *RNI* reactive nitrogen intermediates, *ROI* reactive oxygen intermediates). Adapted from Mosser and Edwards (2008), Mantovani et al. (2004), and Gordon and Martinez (2010)

ferentiated with GM-CSF or M-CSF have M1 and M2 properties, respectively, and have been referred to as Mø1 and Mø2 (Mantovani et al. 2004; Verreck et al. 2006).

Although alternatively activated macrophages share some phenotypic and functional properties, three well-defined forms of M2 have been described (Mantovani et al. 2002, 2004; Verreck et al. 2006): M2a, induced by IL-4 or IL-13 and also referred as wound-healing macrophages (Mosser and Edwards 2008); M2b, induced by exposure to IgG-containing immune complexes (IC) and agonists of Toll-like receptors (TLRs) or IL-1R and also referred to as regulatory macrophages (Mosser and Edwards 2008); and M2c, induced by IL-10 and glucocorticoid hormones (Fig. 1.12, Mosser and Edwards 2008; Mantovani et al. 2004; Gordon and Martinez 2010). In fact, Mosser and Edwards suggested a classification based on the fundamental macrophage functions that are involved in maintaining homeostasis: host defense (M1), wound healing (M2a), and immune regulation (M2b) (Mosser and Edwards 2008).

Differential cytokine production is a key feature of polarized macrophages (Jung et al. 2004; Lang et al. 2002; Locati et al. 2002; Perrier et al. 2004; Scotton et al. 2005). M1 phenotype is typically IL-10^{low} and IL-12^{high}, whereas M2 macrophages are typically IL-10^{high} and IL-12^{low}. Human M1 macrophages also produce high levels of IL-23 (Verreck et al. 2004). Components of the IL-1 system are differentially regulated in polarized macrophage populations and it is coordinately regulated by signals that polarize macrophages in an M1 or M2 direction. M2 cells are generally characterized by low production of proinflammatory cytokines (IL-1, TNF, and IL-6). However, macrophages exposed to IC and LPS (M2b) are an exception, in that they retain high levels of inflammatory cytokine production with concomitant high IL-10 and low IL-12 (Mosser 2003), protecting mice against LPS toxicity and promoting Th2 differentiation and antibody production (Mosser 2003; Mosser and Karp 1999).

Functionally, activated M1 macrophages are potent effector cells integrated in Th1 responses, which kill microorganisms and tumor cells and produce copious amounts of proinflammatory cytokines. By contrast, M2 macrophages tune inflammatory responses and adaptive type I immunity, scavenge debris, and promote angiogenesis, tissue remodeling, and repair. More specifically, integration with and promotion of type II responses prevail for IL-4- or IL-13-stimulated M2a macrophages, whereas suppression and regulation of inflammation and immunity are predominant in IL-10-stimulated M2b cells.

IL-4 stimulates arginase activity in macrophages, allowing them to convert arginine to ornithine, a precursor of polyamines and collagen, thereby contributing to the production of the extracellular matrix (Kreider et al. 2007). Adaptive immune responses can also lead to the production of IL-4, and it is thought that this is the primary pathway for the development and maintenance of wound-healing macrophages. TH2-type immune responses are primarily induced in response to disturbances at mucosal surfaces (Reese et al. 2007), and they are particularly important in the lung and intestines.

M2b or regulatory macrophages can arise following innate or adaptive immune responses. They need two stimuli to induce their anti-inflammatory activity. The first signal (for example, immune complexes, prostaglandins, adenosine, or apoptotic cells) generally has little or no stimulatory function on its own. However, when combined with a second stimulus, such as a TLR ligand, the two signals reprogram macrophages to produce IL-10 (Edwards et al. 2006), the production of which is the most important and reliable characteristic of regulatory macrophages. These regulatory M2b macrophages also downregulate IL-12 production (Gerber and Mosser 2001) and therefore the ratio of IL-10 to IL-12 could be used to define regulatory macrophages (Mantovani et al. 2004). Because IL-10 can inhibit the production and activity of various proinflammatory cytokines, these regulatory macrophages are potent inhibitors of inflammation, despite the fact that they retain the ability to produce many proinflammatory cytokines. Unlike wound-healing macrophages, these regulatory macrophages do not contribute to the production of the extracellular matrix, and many of these regulatory cells express high levels of co-stimulatory molecules (CD80 and CD86) and therefore can present antigens to T cells (Mosser and Edwards 2008; Edwards et al. 2006).

Both innate and adaptive signals can influence macrophage physiology, and these alterations allow macrophages to participate in homeostatic processes, such as tissue remodeling and wound healing, as well as in host defense. However, each of these alterations can have potentially dangerous consequences if not appropriately regulated. For example, classically activated macrophages can cause damage to host tissues, predispose surrounding tissue to neoplastic transformation, and influence glucose metabolism by promoting insulin resistance. Macrophages that are normally involved in wound healing can promote fibrosis, exacerbate allergic responses, and be exploited by pathogens for intracellular survival. Regulatory macrophages can contribute to the progression of neoplasia and the high levels of IL-10 that these cells produce can predispose the host to infection (Mosser and Edwards 2008).

Finally, it is important to understand that discrete clustering of M2 macrophages into two or three categories does not fulfill the phenotypes observed in vivo and it is vital that polarization is understood as a continuum that changes depending on many factors as the type of wound (infectious, trauma/hemorrhage, severe burn), involvement or not of adaptive immune responses, etc. In this sense, Daley et al. in a murine model of sterile dermal incision examined the role of IL-4 and IL-13, the prototypic type 2 cytokines defining "alternative macrophage phenotype" in the process of dermal wound healing, and concluded that macrophages participating in sterile dermal wound repair exhibit a complex progression of phenotypes that did not fit the signature of alternative, classical, M1 or M2 categories (Daley et al. 2010).

The Role of Macrophages in Wound Healing

The outcome of the tissue repair response changes dramatically during lifetime ranging from prenatal regeneration toward the formation of function-impairing scar tissue and pathological healing.

Early fetal wounds heal with minimal scarring and this fact has been related to little inflammation and reduced levels of TGF-β1 (Bullard et al. 2003; Ferguson and O'Kane 2004; Katsuyama and Paro 2013; Stramer and Martin 2005). Wounds in newborn PU.1 null mice, which lack macrophages, also heal without scar, although it is worth noting that this model lacks also B cells, mast cells, and eosinophils (Martin et al. 2003). However, physiological repair in adults requires macrophages and in fact macrophages play a key role in tissue regeneration and repair, driving diverse and specific functions during the different and consecutive phases of wound healing. Several mouse models have revealed that macrophages are indispensable for the healing response (Lucas et al. 2010; Mirza et al. 2009; Willenborg et al. 2012; Duffield et al. 2005; Goren et al. 2009) and salamander limb regeneration is also dependent on macrophages (Godwin et al. 2013; Willenborg and Eming 2014).

Functional distinct subpopulations of macrophages exist in the same tissue and they play specific roles during different stages of wound healing, highlighting that macrophages may be both pathogenic and beneficial. In a well-characterized liver fibrosis model induced by carbon tetrachloride (CCl_4) and using a transgenic model (CD11b-DTR), where macrophages could be selectively depleted, depletion of macrophages when liver fibrosis was advanced resulted in reduced scarring and fewer myofibroblasts. However, depletion during recovery phase led to a failure of matrix degradation (Duffield et al. 2005).

Using the same transgenic model, Mirza et al. studied the effects of macrophage depletion in excisional skin wounds (Mirza et al. 2009). They depleted macrophages immediately before and 48 h after injury and observed a delayed reepithelialization, reduced collagen deposition, and impaired angiogenesis. These effects were related to an increased production of TNF-α and reduced production of VEGF and TGF-β1.

Goren et al. reported another model of macrophage depletion, using transgenic mice that expressed the DT receptor under the myeloid cell-specific lysozyme M promoter (LysM-DTR) and administrating DT before and through all the phases of the wound-healing process (Goren et al. 2009). Macrophage depletion resulted in delayed closure and impaired angiogenesis, increased expression of proinflammatory cytokines such as IL-1β and a decrease of TGFβ1, as well as deregulated pattern of VEGF, consistent with the results reported by Mirza et al. (2009).

Using the same model, Lucas et al. depleted macrophages at different phases of the healing process and reported that when macrophages were depleted during the inflammatory phase (0–5 days, early stage), a significant delay of the early repair response was observed, compared to control mice. At later time points, although wounds presented a similar closure rate, the overall amount of granulation tissue, as well as vascularization, cellularity, contractile force, and scar tissue, remained reduced when compared with controls. Ablation of macrophages in the mid-stage (5–9 days, tissue formation stage) of repair response showed that the role of macrophages through that phase consists of stabilizing the vascular structures and transition of granulation tissue into scar tissue, while macrophages at the late stage (9–14 days, tissue-remodeling stage) do not affect tissue maturation and scar development (Lucas et al. 2010).

In conclusion, macrophages are a necessary requisite for wound healing in the adult. However, macrophages present different phenotypes and exert different functions in different stages of the wound healing. For instance, it has been shown recently that delivery of CXCL12 accelerates wound closure through induction of TGFβ-producing macrophages, which further instruct other cells involved in the wound-healing process (Vagesjo et al. 2018). Similarly, Barreiro et al. have suggested that a novel subset of murine dermal perivascular macrophages, that they called STREAM, extend protrusions across the endothelial junctions and possess a distinctive anti-inflammatory transcriptional profile (Barreiro et al. 2016).

Macrophage Polarization/Function Through Different Healing Stages

The wound-healing process can be divided into three functional stages: inflammatory, proliferative, and tissue-remodeling/maturation phases. The kinetics of each phase depends on the severity of injury, presence of infection, age, or pathological situations (Mirza and Koh 2011; Mirza et al. 2013; Brubaker et al. 2011).

M1 polarized macrophages mediate tissue damage and initiate inflammatory responses (Gordon and Martinez 2010; Biswas and Mantovani 2010). They clear cellular debris and necrotic tissue and combat invading pathogens, expressing highly active proteases and proinflammatory mediators to accomplish these functions. During the early stages of the repair response after skin wounding, infiltrating macrophages express an M2 phenotype and present a trophic function (VEGF, PDGF). Their depletion inhibited the formation of a highly vascularized, cellular granulation tissue, and of scar tissues (Lucas et al. 2010). In later stages, macrophages exert an anti-inflammatory function (IL-1Rα, IL-10, TGFβ1) (Willenborg and Eming 2014; Mantovani et al. 2013).

These dynamic changes in macrophage polarization have been studied in models of acute ischemic heart and kidney pathology. Monocytes are recruited into the tissue and their activation state undergoes dynamic changes from a predominantly M1 to a predominantly M2 phenotype (Lambert et al. 2008; Ricardo et al. 2008; Swaminathan and Griffin 2008; Troidl et al. 2009).

Regulation of macrophage polarization plays a vital role in wound healing. In humans, chronic venous leg ulcers represent a failure to switch from an M1 to an M2 phenotype, and this fact inhibits resolution of the inflammatory condition (Sindrilaru et al. 2011).

Inflammatory Phase

During the inflammatory phase of the wound-healing process, that lasts a few days after injury, one of the major functions of monocytes/macrophages is the removal of damaged tissue (Leibovich and Ross 1975) and, in case of pathogen spreading in the wound bed, macrophages phagocytose these pathogens and present antigens to T cells. Moreover, macrophages induce apoptosis and phagocyte neutrophils (Meszaros et al. 1999, 2000), contributing to the transition from the inflammatory phase to the proliferative phase of the healing process. In skin wounds, the effect of macrophage depletion on neutrophil accumulation is dependent on the experimental protocol used (Mirza et al. 2009; Goren et al. 2009; Leibovich and Ross 1975).

In addition to debridement, macrophages of the inflammatory phase display an M1 phenotype (Daley et al. 2010) and consequently secrete inflammatory cytokines, such as IL-1, TNFα, and IL-6. IL-6 and TNFα are essential for the efficient healing in the early days after injury to rodent skin (Gallucci et al. 2000; Lee et al. 2000) and regulate proliferation of keratinocytes and fibroblasts (Hernandez-Quintero et al. 2006; Mateo et al. 1994). IL-1 however reduces fibrosis without affecting the tensile strength of the repaired skin (Thomay et al. 2009). Early skin wound monocytes/macrophages also produce vascular endothelial growth factor (VEGF) and thereby promote granulation tissue formation and angiogenesis (Willenborg et al. 2012; Baum and Arpey 2005). Depletion of monocytes/macrophages during the inflammatory phase reduces granulation tissue formation and cell proliferation in mouse skin wounds (Lucas et al. 2010; Mirza et al. 2009).

Proliferative/Granulation Phase

Cell proliferation begins in the early days after injury to mouse skin and skeletal muscle, peaks around day 5, and persists at low levels until at least days 10–12 (Jun and Lau 2010), a similar time course to that of macrophages. The proliferative phase of cutaneous wound healing, also called granulation phase, is characterized by active fibroplasia, epidermal regeneration, wound contraction, and angiogenic sprouting.

Keratinocyte proliferation and migration allow reepithelialization of the wound and restoration of the barrier function of the skin (Werner and Grose 2003). Proliferation of fibroblasts and endothelial cells allows matrix deposition and angiogenesis, respectively (Werner and Grose 2003; DiPietro 2013), and Schwann cells participate in the regeneration of peripheral nerves as well as contribute to granulation tissue formation (Parfejevs et al. 2018a).

It has been reported that M1 macrophages shift to M2 macrophages (Gordon 2003; Duffield 2003) that contribute to the resolution of inflammation and the wound-healing process (Goerdt et al. 1999; Porcheray et al. 2005). In fact, depletion of macrophages during the proliferative phase significantly disturbed the transition of the mid-stage to the late stage of repair response (Lucas et al. 2010).

As the inflammatory phase declines and the proliferative or granulation phase progresses, macrophages decrease proinflammatory cytokine expression and increase expression of anti-inflammatory cytokines and growth factors, such as IL-10, TGFβ, and IGF-1, supporting the growth of new tissue and promoting collagen deposition and tissue maturation (Novak and Koh 2013a; Mirza and Koh 2011; Willenborg et al. 2012; Daley et al. 2010; Perdiguero et al. 2011). This phenotypic shift may be controlled, in part, by phagocytosis of tissue and cell debris (Stout 2010; Fadok et al. 1998), release of soluble factors from other immune cells (Daley et al. 2005), autocrine anti-inflammatory feedback mechanisms (Perdiguero et al. 2011), and/or composition of the granulation tissue (Blakney et al. 2012; Wehner et al. 2010).

Reepithelialization and Contraction

Macrophage-derived TGFβ plays a crucial role in reepithelialization (Abbott et al. 2010; Mirza et al. 2009). This reepithelialization process is made easier by contraction of the underlying connective tissue, which brings the wound margins toward each other. This contraction process is performed by myofibroblasts, activated by TGFβ and PDGF (Frank et al. 1995). Once the wound surface is covered by a monolayer of keratinocytes, epidermal migration ceases and a new stratified epidermis with underlying basal lamina is re-established from the margins of the wound inwards. At this moment, the defect is filled with granulation tissue and covered by a newly formed epidermal layer. Nevertheless, the wound-healing process, particularly the remodeling phase, can still go on for months.

ECM Deposition

In this phase, macrophage-derived TGFβ stimulates collagen production and also reduces degradation of the wound matrix by collagenase and through increased inhibition of MMPs. TGFβ has three isotypes (TGF-β1, -β2, and -β3), which all stimulate infiltration of inflammatory cells and fibroblasts. However, at gestational ages associated with scarless repair, low levels of TGF-β1 and high levels of TGF-β3 are expressed (Bullard et al. 2003), suggesting that the relative proportion of each subtype may be crucial for scarring.

TGFβ, in combination with other factors as PDGF, FGF2, and IGF-1 which are mainly produced by macrophages, can mediate differentiation of mesenchymal stem cells into myofibroblasts and induce collagen and other ECM component production (Werner and Grose 2003; Ishida et al. 2008; Vogler et al. 2003).

Angiogenesis

Macrophages contribute to angiogenesis during wound healing, by induction of angiogenic sprouting in the wound bed (Greenburg and Hunt 1978; Thakral et al. 1979). When macrophages are depleted during the wound-healing process, VEGF is reduced, resulting in a decreased vascularization (Lucas et al. 2010; Mirza et al. 2009; Goren et al. 2009). Vascularization is crucial for most physiological and pathological processes of tissue growth and several studies have underlined the role of macrophages in angiogenesis (Lucas et al. 2010; Mirza et al. 2009; Murdoch et al. 2008; Okuno et al. 2011).

In order to facilitate migration of endothelial cells, the ECM and basement membrane are degraded by mediation of the MMPs, produced by several cells including macrophages and endothelial cells. Then, endothelial cells migrate into the wound by adherence to integrin cell surface receptors on the ECM. This is predominantly exerted by TNFα and VEGF release by macrophages, and TNFα may in turn induce VEGF expression in keratinocytes and fibroblasts (Frank et al. 1995).

Blood vessel growth and organization depend on macrophage and pericyte interplay (Armulik et al. 2011; Fantin et al. 2010; Rymo et al. 2011; Stefater et al. 2011). The role of macrophages in the vasculature has been studied extensively in the context of tumorigenesis (Schmid and Varner 2012; Wynn et al. 2013), but they have also been implicated in physiological angiogenesis, directly, through physical involvement in the process and indirectly by the secretion of angiogenic factors (Fantin et al. 2010; Rymo et al. 2011).

It has been reported that these macrophages participate in vessel sprouting driven by specialized endothelial tip cells. The formation of tip cells is predominantly stimulated by M2 macrophages, especially through the secretion of VEGF (Tammela et al. 2008), and the macrophages seem to serve as guidance and bridge cells in this fusion process of the tip cells, as well (Fantin et al. 2010). Macrophages can also increase vascular permeability by releasing vasoactive substances such as vascular permeability factor (Berse et al. 1992), substance P (Pascual and Bost 1990), platelet-activating factor, and prostaglandins (Middleton and Thatcher 1998).

There is controversy about what macrophage profile, proinflammatory (M1) versus alternatively activated (M2) macrophages, contributes to angiogenesis. In a recent in vitro study, macrophages with a proinflammatory profile enhanced angiogenesis, increasing the number and length of endothelial sprouts, in a Notch signaling-dependent manner (Tattersall et al. 2016). In contrast, treatment with alternative activator IL-4 did not enhance macrophage angiotropism (Tattersall et al. 2016). In vivo studies, however, have suggested that inflammatory polarized macrophages do not contribute to angiogenesis (Jetten et al. 2014) or that alternatively activated macrophages show additional angiotrophic character due to increased metalloproteinase activity, while inflammatory macrophages provided less proangiogenic stimulus (Zajac et al. 2013). Probably, the role of macrophages in angiogenesis varies with injury site and phase and it is more likely to secrete proangiogenic and anti-angiogenic signals depending on the context (Brancato and

Albina 2011). Perivascular M2 macrophages also seem to regulate vascular permeability (He et al. 2016).

Notch signaling functions in several angiogenic mechanisms, most notably controlling the differentiation between endothelial tip- and stalk-cell identities (Tung et al. 2012). More recently, Notch has been implicated in the interaction between endothelial cells and both macrophages and perivascular cells. In macrophages, Notch signaling has been found to be important for recruitment to sites of active angiogenesis in both developmental and pathological settings, and Notch signaling has been detected in macrophages at the sites of imminent or recent vessel anastomosis, suggesting a role in this process (Outtz et al. 2010, 2011). In fact, inhibition of macrophage Notch signaling was sufficient to abrogate the angiotrophic advantage of inflammatory polarization. This is consistent with the role of Notch as a mediator of macrophage inflammatory polarization (Wang et al. 2010).

Peripheral Nerve Regeneration

The subepidermal nerve plexus is the most peripheral part of the nervous system. Following injury, Schwann cells are responsible for the regenerative capacity of the peripheral nervous system (PNS) and act directly to aid axonal outgrowth and remyelinate the regenerating axon (Gaudet et al. 2011; Zochodne 2012). Schwann cells also participate in regulating the immune response (Stratton and Shah 2016; Stratton et al. 2016) and express several ligands that are known to interact with receptors expressed by macrophages, yet the effects of Schwann cells in regulating macrophage phenotype are not fully understood.

In response to peripheral nerve injury, macrophages are recruited to the injured nerve, and play a vital function in clearing debris, including inhibitory myelin debris (Kang and Lichtman 2013), a process that is necessary for a competent axonal regeneration (Gaudet et al. 2011; Barrette et al. 2008). In addition, macrophages regulate neurotrophin production and angiogenesis (Cattin et al. 2015; Gaudet et al. 2011; Barrette et al. 2008; Bastien and Lacroix 2014).

The interaction between macrophages and Schwann cells has been demonstrated in an inducible in vivo model, where activation of tamoxifen-inducible Raf-kinase transgene causes Schwann cell dedifferentiation in the absence of axonal injury. Consequently, CCL2 is upregulated by Schwann cells, attracting and activating macrophages, that participate in demyelination (Napoli et al. 2012; Klein and Martini 2016).

Schwann cells in the distal stump upregulate the expression of cytokines involved in the activation of an innate immune response, such as TNFα, IL-1α, IL-1β, LIF, and MCP-1 (Gaudet et al. 2011; Martini et al. 2008; Rotshenker 2011). This allows repair Schwann cells to interact with immune cells and to recruit macrophages to the nerve. Cytokines such as IL-6 and LIF not only attract macrophages to the injured nerve but can also act on neurons to promote axonal regeneration (Bauer et al. 2007; Cafferty et al. 2001; Hirota et al. 1996). In addition, macrophages promote vascularization of the distal nerve, via VEGF secretion (Cattin et al. 2015;

Barrette et al. 2008; Niemi et al. 2013). In peripheral nerve injury, macrophages respond selectively to the hypoxia within the Schwann cell regeneration bridge, and via VEGF-A secretion they trigger the polarized vascularization of the bridge region. These newly formed blood vessels are used as a guiding path by Schwann cells to cross the bridge, guiding the regrowing axons with them.

Macrophages also cooperate with Schwann cells to degrade myelin debris that potentially inhibits axon growth during the second phase of myelin clearance (Rotshenker 2011; Hirata and Kawabuchi 2002). Schwann cells themselves take a major part in breaking down their own redundant myelin sheaths during the first 5–7 days after injury (Niemi et al. 2013; Jessen and Mirsky 2016; Perry et al. 1995). Then, the second phase of myelin clearance is dominated by macrophages (Hirata and Kawabuchi 2002; Dubovy et al. 2013; Ramaglia et al. 2008; Vargas et al. 2010).

In peripheral nerve injury, monocytes infiltrated within 6–12 h present an increased expression of proinflammatory-associated factors, including TNFα, IL-1β, IL-6, PPBP, CXCL2, CCL8, SAA3, SLPI, and CD300 at 1 day postinjury (Jha et al. 2015; Nadeau et al. 2011; Painter et al. 2014; Ydens et al. 2012), similar to other tissues under acute injury (Arnold et al. 2007; Aurora and Olson 2014; Gensel and Zhang 2015; Kroner et al. 2014). Interestingly, characteristic proinflammatory-associated genes such as interferon-γ receptor (Ifngr), and in some cases inducible nitric oxide synthase/iNOS (Nos2), are not usually detected at early or late stages following peripheral nerve injury (Painter et al. 2014; Ydens et al. 2012; Peluffo et al. 2015).

There is controversy on whether these proinflammatory factors are detrimental or not (Arnold et al. 2007; Aurora and Olson 2014; Gensel and Zhang 2015; Kroner et al. 2014). In vitro, the exposure of Schwann cells and neurons to M1-primed macrophage-conditioned media enhances Schwann cell proliferation, reduces axonal outgrowth, and compromises neuronal survival (Kigerl et al. 2009; Mokarram et al. 2012). In vivo, however, when proinflammatory macrophage function is interfered, nerve healing is compromised and a reduction in axonal regrowth is observed (Peluffo et al. 2015).

It is worth noting that although the M1-associated macrophage is the predominant macrophage phenotype present at 1 day postinjury (Bastien and Lacroix 2014), the most highly upregulated genes are those that encode for enzymes, such as Arg1 and Chil3, associated with an anti-inflammatory response (Novak and Koh 2013a; Painter et al. 2014; Gensel and Zhang 2015; Novak and Koh 2013b). Then, between 7 and 14 days postinjury, a different anti-inflammatory macrophage subtype wave dominates the nerve, characterized by the expression of interleukin-10 (IL-10) and receptors such as interleukin-10 receptor (IL-10r), interleukin-4 receptor (IL-4r), interleukin-13 receptor alpha 1 (IL-13ra1), and triggering receptor expressed on myeloid cells 2 (Trem2) (Bastien and Lacroix 2014; Ydens et al. 2012; Be'eri et al. 1998). The kinetics of macrophage polarization at later stages of nerve injury is similar to what is observed at sites of injury in tissues such as skin and skeletal muscle (Novak and Koh 2013a; Arnold et al. 2007; Daley et al. 2010), but different from what occurs in spinal cord injury where the expression of M2-associated genes and proteins is only transient.

Remodeling or Maturation

The last phase of wound healing, remodeling, is characterized by maturation of the repaired tissue that includes reorganization of the vasculature, as capillary density declines toward the level of normal tissue (DiPietro 2013), extracellular matrix remodeling from granulation to scar tissue, and senescence of myofibroblasts (Jun and Lau 2010; Desmouliere 1995). Most endothelial cells, macrophages, and myofibroblasts undergo apoptosis, or exit the wound (Gurtner et al. 2008). In the mouse skin, wound healing is completed within 1–2 weeks, depending on the experimental model (Lucas et al. 2010; Mirza et al. 2009; Mirza and Koh 2011). However, scar maturation may take 6 months or more (Robins et al. 2003).

During this phase, macrophages sustain expression of TGFβ; decrease expression of VEGF, arginase-1, and insulin-like growth factor-1 (Mirza and Koh 2011; Willenborg et al. 2012); and downregulate expression of both proinflammatory and anti-inflammatory cytokines (Perdiguero et al. 2011).

Macrophages play an important role in ECM breakdown, which is fundamental for wound healing and tissue remodeling. They secrete MMPs and serine proteases. MMP-12, for example, is a macrophage-specific metallo-elastase (Shapiro et al. 1993), which has been suggested to play a role in capillary regression in skin wounds, resulting in a decrease of erythema (Madlener et al. 1998). Macrophages also promote tissue integrity by producing type VIII collagen (Weitkamp et al. 1999) and some matrix proteins (Gratchev et al. 2001).

However, relevance of macrophages during the remodeling phase of tissue repair remains unclear. In mouse models of chronic liver fibrosis, macrophages of an anti-inflammatory phenotype are required for resolution of established fibrosis (Duffield et al. 2005) but there is no evidence that remodeling-phase macrophages have a similar fibrinolytic role after acute skin injury. In fact, depletion of macrophages from mouse skin wounds during the remodeling phase has no effect on the amount or organization of scar tissue at 14 days after injury (Lucas et al. 2010). Interestingly, macrophage depletion studies suggest that collagen production and scar maturation in skin wounds are regulated by monocytes/macrophages of the inflammatory and proliferative phases, respectively, rather than by macrophages of the remodeling phase (Lucas et al. 2010; Rodero et al. 2013).

Macrophages and Impaired Wound Healing

Activity of macrophages is regulated by cytokines acting in both autocrine and paracrine manners. Proinflammatory cytokines of the inflammatory or early stage include IL-1β, TNF-α, and IL-6. At the proliferative phase, macrophages secrete the anti-inflammatory cytokine IL-10, as well as pro-healing factors IGF-1 and TGFβ1. VEGF is released mainly by macrophages in the early phase of the healing process, while at later stages keratinocyte/fibroblasts/endothelial cells are the main releasers (Mirza and Koh 2015).

Many pathological situations associated with impaired healing are characterized by persistent inflammation, reduced vascularization and granulation tissue, and incomplete wound closure. This impairment is associated with persistent proinflammatory cytokine production and decrease of proangiogenic and pro-healing factors (Loots et al. 1998; Mirza and Koh 2011; Blakytny and Jude 2006; Jeffcoate and Harding 2003; Mirza et al. 2014).

Long-term macrophage accumulation and/or deregulation of macrophage phenotype and function can lead to tissue damage, failure to heal, and/or fibrosis. Skin wounds in diabetic mice and humans present impaired healing and a prolonged proinflammatory macrophage phenotype supported by IL-1β and TNFα (Mirza and Koh 2011; Mirza et al. 2013; Goren et al. 2007). However, impaired healing is not always associated with proinflammatory macrophage activation.

Each tissue injury has a unique set of damage, recruitment, and molecular signals coordinating the repair response. For example, unique microenvironmental signals are created by sterile inflammation versus pathogen-mediated inflammation due to the damage-associated molecular patterns recognized by inflammatory cells rather than pathogen-associated molecular patterns. When wound healing is complicated by infection or other activity that activates the adaptive leukocyte system, macrophages and T cells cross-talk via cytokines and co-stimulators (Doherty 1995; Tredget et al. 2006) and macrophages stimulate T-cell expansion and differentiation to Th1 and Th2 cells. In this case, T cells make wound healing more complex, and can lead to necrosis, fibrosis, ulcer formation, or granuloma formation. Th2 response results in the production of ECM, while Th1 cells mainly produce IFN-γ which results in the differentiation of macrophages into M1 macrophages. In another situation, ischemic injury can activate signaling divergent from traumatic injury based on induction of hypoxia signaling and reactive oxygen species. Physiology/biology of the particular tissue also significantly impacts the way the injury niche orchestrates repair. The local stromal cell and tissue-resident macrophage populations, mechanical properties and organization of the tissue, and extent of vascularization and oxygenation could all contribute to differential engagement of immune cell populations in situ. These factors contribute to the diversity of macrophage phenotypes and repair programs produced by the innate immune system.

Macrophage to Pericyte Transition

An elegant in vivo fate-mapping work by Yamazaki et al. has recently shown that, in the embryonic skin vasculature, tissue-localized F4/80+ myeloid progenitors differentiate into pericytes in a process mediated by TGFβ signaling (Yamazaki et al. 2017). Type 2 TGF-β receptor (Tgfbr2) mutant mice exhibited deficient pericyte development in skin vasculature. The myeloid-derived pericytes constituted a substantial proportion (about 20–30%) of all skin pericytes at E15.5 (Dias Moura Prazeres et al. 2017). Unfortunately, the authors did not test whether such transitions are also happening in adult skin upon wounding, but this certainly remains an enticing possibility.

Another interesting possibility is that macrophages act as signaling hubs to attract pericyte-like cells to the wounds. One of the important macrophage-derived signals seems to be glycoprotein nonmetastatic melanoma protein B (GPNMB) (Yu et al. 2018). Macrophage-derived GPNMB promotes mobilization of CD11b-CD45-CD31-Sca-1+CD29+CD146+ cells to the wound bed (Silva et al. 2018b).

Pericyte Relationship with Other Stromal Cell Types

As introduced earlier, careful reading of the literature is important to extract all available information about these enigmatic cells. First of all, perivascular-localized cells residing both in human scalp (Yamanishi et al. 2012) and glabrous skin (Ruetze et al. 2013) have been classified as putative dermal stem cells (Vapniarsky et al. 2015). However, we and others have reported that dermal stemness can be safely attributed to dermal Schwann cells (Etxaniz et al. 2014; Iribar et al. 2017; Gresset et al. 2015) that become dedifferentiated upon wounding, and that these are closely related to pericytes. This is at least true if we define dermal stem cells as those capable of differentiating to several lineages, including the neural progeny (Toma et al. 2001). Therefore, it seems likely that, upon wounding, we might witness in the skin a similar situation to what was reported in the mouse incisor tooth (Kaukua et al. 2014), which is a permanently regenerating organ: Schwann cells may give rise to pericytes that may give rise to the terminally differentiated cells that repair the tissue. This hypothesis remains to be experimentally proven in the skin, although the fact that a subset of Schwann cells and pericytes derive from the boundary cap, and the fact that both derivatives seem to be interconvertible (at least in development), adds support to this idea. Alternatively or as an addition to this enticing possibility, it has already been demonstrated that paracrine signaling from dedifferentiated Schwann cells may be key in inducing a blastema-like state in stromal cells that are present within the wound (Johnston et al. 2013, 2016; Parfejevs et al. 2018b). Whether the cells responding to the signals are pericytes remains to be determined. For instance, pericytes proliferate upon wounding in the axolotl digit tip regeneration setting and migrate into the blastema but give rise solely to pericytes (Currie et al. 2016).

Some research groups use extensively the "fibroblast subpopulation" terminology, their original studies indicating that at least two distinct such subpopulations populate the dermis (Driskell et al. 2013; Driskell and Watt 2015). We believe that this terminology is confusing since there is no definition for fibroblast other than a dermal cell growing in attachment culture. This is obviously a highly heterogenous culture (Hu et al. 2018), and it includes plenty of different cell types such as pericytes, at least in the lower passages (Paquet-Fifield et al. 2009). Upon wounding, Watt et al. defined two waves of cells populating the wound bed: the first wave derives from cells of the reticular dermis and hypodermis (traced by Dlk1 expression), and which elaborate the collagenous ECM characteristic of fibrosis. In the second wave, the upper dermal lineage (papillary dermal cells defined as CD26+ Sca1−) contributed exclusively to the papillary dermis upon reepithelialization. Interestingly, Rinkevich et al. identified CD26/DPP4 as a surface marker that allows isolation of a pro-fibrotic fibroblast lineage (Rinkevich et al. 2015).

More recent studies, including some where single-cell RNA-sequencing permitted the discrimination of cell subsets in a more defined way (Philippeos et al. 2018; Tabib et al. 2018), have shown a much more complex picture, as it was entirely predictable. Tabib et al. defined 19 distinct clusters of cells in human dermis, of which 3 were defined as "fibroblasts," 2 as "pericytes," 1 as "macrophage/DC," and 1 as "smooth muscle" (Tabib et al. 2018). In a reanalysis of the "fibroblast" clusters (2742 cells), they found two major and five minor subpopulations, defined as follows: (i) SFRP2/DPP4 (1671 cells) that could be pro-fibrotic fibroblasts, and included a WIF1/COMP/NKD2 subset that expressed highest levels of collagen type I; (ii) FMO1/LSP1 (536 cells) that had a perivascular location and expressed CXCL12; (iii) CRABP1/TNN/ASPN, possibly dermal papilla cells; (iv) COL11A1/DPEP1 cells; (v) SFRP4 cells; (vi) PRG4 (59 cells); and (vii) ANGPTL7/C2orf40 (33 cells). In an analysis of 184 cells, Philippeos et al. identified five fibroblast subpopulations, but it is unclear if these groups included perivascular cells and macrophages as well (Philippeos et al. 2018). To add to the complexity, Korosec et al. have separated human papillary and reticular fibroblasts through cell sorting as FAP+CD90− (papillary) and FAP−CD90+ (reticular) fibroblasts, but they have shown that these subsets are not spatially restricted, and that they respond dynamically to the microenvironment (Korosec et al. 2018).

Obviously, careful analysis of these novel single-cell data and comparison with the available literature are needed to obtain some more relevant information out of these experiments. For instance, one key point is which "fibroblast" will give rise to the granulation tissue upon wounding. In mice transiently overexpressing TGF-β1 and PDGF-B growth factors in the skin, a marked macrophage influx and an expansion of the connective tissue cell population were detected, originating from microvascular pericytes (Rodriguez et al. 2013). Of interest, TGF-β1 produced a more stiff, tense ECM consistent with a pro-fibrotic role, while PDGF-B produced immature granulation tissue. Interestingly, a pericyte subpopulation characterized as PECAM1+Sca1+CD38+ cells enters the cell cycle upon wounding and gives rise to myofibroblast-like cells in granulation tissue (Etich et al. 2013). In fact, debrided wound material in burn wounds has been suggested as a source of autologous PDGFRβ+ stem cells for wound repair (Natesan et al. 2011). Thus, it seems likely that some pericyte subset is responsible for this transformation. While hair follicle dermal stem cells (Rahmani et al. 2014) also may seem to contribute to this phenomenon, the relationship of these cells to the cells in other compartments of the dermis is still unclear and needs further clarification. The striking similarities between myofibroblasts, pericytes, and perifollicular dermal sheath cells have previously been noted (Juniantito et al. 2012).

The putative relationship between pericytes (PC) and Schwann cells (SCs) was demonstrated in cultured spheres generated from dermal tissue, which contained

SCs identified as p75+/CD56+ cells, which in vivo are associated to nerve terminal axons innervating the skin, and p75+/CD56− PCs which are found in perivascular location of dermal vascular plexus (Etxaniz et al. 2014). We proposed that the neural potential observed in skin-derived precursor cell cultures is produced as a result of p75+/CD56+ cell differentiation. The transcriptomic analysis performed on those populations corroborated the identity of both populations; thus, p75+/cd56+ fraction expressed Schwann cell precursor genes (according to their immature state), and p75+/CD56− population expressed pericytic markers. The comparison between both populations revealed an extraordinary similarity that suggests a developmental relation between them. Additionally, the correlation between Sox2 levels that was mostly expressed in SCs but also in a high level in PCs, and the in vitro-measured potential on these populations, indicated that Sox2 could be acting as a regulator of the differentiation stage, and these human data have later been confirmed in different mouse models (Iribar et al. 2017; Gresset et al. 2015). Hence we proposed that a switch between SCs and PCs is possible, in which both populations are mutually convertible in base of their Sox2 expression levels (Etxaniz et al. 2014). During pericyte activation in angiogenesis, in normal or damaged conditions also a detach from the vessel wall is implicated with a phenotype change and the acquisition of new functional capacities (Morikawa and Ezaki 2011; Cheng et al. 2011).

Conclusions

In this chapter, we have discussed the different aspects of vascular formation, peripheral innervation, and role of macrophages in cutaneous wound healing and we have hypothesized how these cells may often convert into each other, or influence the conversion of their neighbors. We proposed that vascular formation is regulated by different mechanisms between development and postnatal life, and that sprouting angiogenesis in skin wound healing is driven by pericytes. We have also suggested that nerve-associated cells, specifically the terminal Schwann cells that ensheath the nerve endings and derive from the boundary cap, may have a prominent role as pro-regenerative actors and may do so by transitioning themselves into pericytes, among other things. We have also proposed that macrophages and pericytes might interconvert, and that EPCs may give rise to perivascular cells under certain circumstances. Thus, it may seem that upon wounding, lineage boundaries become frail and virtually any cell can be transitioned into the next relevant cell lineage for repair to occur efficiently. Obviously, we now believe that the cell subsets that we scientists like so much, because they adapt to our classifying mission, may be considered as mere reference points in the continuum of cell states that will be available in a cutaneous wound at any given time. This would apply not only to macrophages, but also to pericytes and Schwann cells. We hope that our vision on the complexity of the role of pericytes in wound healing is now clearer, and that the different aspects that we touched upon will be of use for interested readers. Finally, we would like to mention that the putative role of resident pericyte subtypes in developing cutaneous scarring is gaining momentum (Bodnar et al. 2016; Greenhalgh et al. 2015; Leavitt et al. 2016; Prazeres et al. 2018), but covering this issue at any depth would be outside of the already too wide scope of this chapter.

Acknowledgments Writing of this chapter was supported by grants from Instituto de Salud Carlos III (PI13/02172; PI16/01430; AC17/00012), co-funded by the European Union (ERDF/ESF, "Investing in your future"; and Eracosysmed/H2020 Grant Agreement No. 643271). HI received a studentship from the Department of Education, University and Research of the Basque Government (PRE2013-1-1068).

A tribute is given to Dr. Shunichi Morikawa. From the right to the left, the two co-first authors of this chapter, Dr. Haizea Iribar and Dr. Shunichi Morikawa, and his wife, celebrating Japanese culture

References

Abbott, N. J., Patabendige, A. A., Dolman, D. E., Yusof, S. R., & Begley, D. J. (2010). Structure and function of the blood-brain barrier. *Neurobiology of Disease, 37*, 13–2510.

Adams, R. H., & Alitalo, K. (2007). Molecular regulation of angiogenesis and lymphangiogenesis. *Nature Reviews Molecular Cell Biology, 8*, 464–478.

Aki, R., Amoh, Y., Li, L., Katsuoka, K., & Hoffman, R. M. (2010). Nestin-expressing interfollicular blood vessel network contributes to skin transplant survival and wound healing. *Journal of Cellular Biochemistry, 110*, 80–86.

Albert, P., & Boilly, B. (1988). Effect of transferrin on amphibian limb regeneration: A blastema cell culture study. *Roux's Archives of Developmental Biology, 197*, 193–196.

van Amerongen, M. J., Harmsen, M. C., van Rooijen, N., Petersen, A. H., & van Luyn, M. J. A. (2007). Macrophage depletion impairs wound healing and increases left ventricular remodeling after myocardial injury in mice. *The American Journal of Pathology, 170*, 818–829.

Amoh, Y., et al. (2005). Hair follicle-derived blood vessels vascularize tumors in skin and are inhibited by doxorubicin. *Cancer Research, 65*, 2337–2343.

Amselgruber, W. M., Schafer, M., & Sinowatz, F. (1999). Angiogenesis in the bovine corpus luteum: An immunocytochemical and ultrastructural study. *Anatomia, Histologia, Embryologia, 28*, 157–166.

Anand, P. (1996). Neurotrophins and peripheral neuropathy. *Philosophical Transactions of the Royal Society of London. Series B, Biological Sciences, 351*, 449–454.

Anand, P., et al. (1987). Increase of substance P-like immunoreactivity in the peripheral nerve of the axolotl after injury. *Neuroscience Letters, 82*, 241–245.

Armulik, A., Genové, G., & Betsholtz, C. (2011). Pericytes: Developmental, physiological, and pathological perspectives, problems, and promises. *Developmental Cell, 21*, 193–215.

Arnold, L., et al. (2007). Inflammatory monocytes recruited after skeletal muscle injury switch into anti-inflammatory macrophages to support myogenesis. *The Journal of Experimental Medicine, 204*, 1057–1069.

Arthur, R. P., & Shelley, W. B. (1959). The innervation of human epidermis. *The Journal of Investigative Dermatology, 32*, 397–411.

Asahara, T., et al. (1997). Isolation of putative progenitor endothelial cells for angiogenesis. *Science, 275*, 964–967.

Asahara, T., et al. (1999). Bone marrow origin of endothelial progenitor cells responsible for postnatal vasculogenesis in physiological and pathological neovascularization. *Circulation Research, 85*, 221–228.

Ashrafi, M., Baguneid, M., & Bayat, A. (2016). The role of neuromediators and innervation in cutaneous wound healing. *Acta Dermato-Venereologica, 96*, 587–594.

Ashton, N., & de Oliveira, F. (1966). Nomenclature of pericytes. Intramural and extramural. *The British Journal of Ophthalmology, 50*, 119–123.

Aurora, A. B., & Olson, E. N. (2014). Immune modulation of stem cells and regeneration. *Cell Stem Cell, 15*, 14–25.

Ausprunk, D. H., & Folkman, J. (1977). Migration and proliferation of endothelial cells in preformed and newly formed blood vessels during tumor angiogenesis. *Microvascular Research, 14*, 53–65.

Bagley, R. G., Weber, W., Rouleau, C., & Teicher, B. A. (2005). Pericytes and endothelial precursor cells: Cellular interactions and contributions to malignancy. *Cancer Research, 65*, 9741–9750.

Bai, L., et al. (2015). Genetic identification of an expansive mechanoreceptor sensitive to skin stroking. *Cell, 163*, 1783–1795.

Bain, C. C., et al. (2014). Constant replenishment from circulating monocytes maintains the macrophage pool in the intestine of adult mice. *Nature Immunology, 15*, 929–937.

Balaji, S., King, A., Crombleholme, T. M., & Keswani, S. G. (2013). The role of endothelial progenitor cells in postnatal vasculogenesis: Implications for therapeutic neovascularization and wound healing. *Advances in Wound Care, 2*, 283–295.

Baluk, P., Morikawa, S., Haskell, A., Mancuso, M., & McDonald, D. M. (2003). Abnormalities of basement membrane on blood vessels and endothelial sprouts in tumors. *The American Journal of Pathology, 163*, 1801–1815.

Baron, M., & Gallego, A. (1972). The relation of the microglia with the pericytes in the cat cerebral cortex. *Zeitschrift für Zellforschung und Mikroskopische Anatomie, 128*, 42–57.

Barreiro, O., et al. (2016). Pivotal role for skin transendothelial radio-resistant anti-inflammatory macrophages in tissue repair. *eLife, 5*, e15251.

Barrette, B., et al. (2008). Requirement of myeloid cells for axon regeneration. *Journal of Neuroscience, 28*, 9363–9376.

Bastien, D., & Lacroix, S. (2014). Cytokine pathways regulating glial and leukocyte function after spinal cord and peripheral nerve injury. *Experimental Neurology, 258*, 62–77.

Bauer, S. M., Bauer, R. J., & Velazquez, O. C. (2005). Angiogenesis, vasculogenesis, and induction of healing in chronic wounds. *Vascular and Endovascular Surgery, 39*, 293–306.

Bauer, S., Kerr, B. J., & Patterson, P. H. (2007). The neuropoietic cytokine family in development, plasticity, disease and injury. *Nature Reviews Neuroscience, 8*, 221–232.

Baum, C. L., & Arpey, C. J. (2005). Normal cutaneous wound healing: Clinical correlation with cellular and molecular events. *Dermatologic Surgery, 31*, 674–686;. discussion 686.

Be'eri, H., Reichert, F., Saada, A., & Rotshenker, S. (1998). The cytokine network of Wallerian degeneration: IL-10 and GM-CSF. *European Journal of Neuroscience, 10*, 2707–2713.

Bergmann, C. E., et al. (2006). Arteriogenesis depends on circulating monocytes and macrophage accumulation and is severely depressed in op/op mice. *Journal of Leukocyte Biology, 80*, 59–65.

Berse, B., Brown, L. F., Van de Water, L., Dvorak, H. F., & Senger, D. R. (1992). Vascular permeability factor (vascular endothelial growth factor) gene is expressed differentially in normal tissues, macrophages, and tumors. *Molecular Biology of the Cell, 3*, 211–220.

Birbrair, A., et al. (2017). How plastic are pericytes? *Stem Cells and Development, 26*, 1013–1019.

Biswas, S. K., & Mantovani, A. (2010). Macrophage plasticity and interaction with lymphocyte subsets: Cancer as a paradigm. *Nature Immunology, 11*, 889–896.

Blakney, A. K., Swartzlander, M. D., & Bryant, S. J. (2012). The effects of substrate stiffness on the in vitro activation of macrophages and in vivo host response to poly(ethylene glycol)-based hydrogels. *Journal of Biomedical Materials Research. Part A, 100*, 1375–1386.

Blakytny, R., & Jude, E. (2006). The molecular biology of chronic wounds and delayed healing in diabetes. *Diabetic Medicine, 23*, 594–608.

Blancas, A. A., Wong, L. E., Glaser, D. E., & McCloskey, K. E. (2013). Specialized tip/stalk-like and phalanx-like endothelial cells from embryonic stem cells. *Stem Cells and Development, 22*, 1398–1407.

Bodnar, R. J., Satish, L., Yates, C. C., & Wells, A. (2016). Pericytes: A newly recognized player in wound healing. *Wound Repair and Regeneration, 24*, 204–214.

Brain, S. D., & Grant, A. D. (2004). Vascular actions of calcitonin gene-related peptide and adrenomedullin. *Physiological Reviews, 84*, 903–934.

Brancato, S. K., & Albina, J. E. (2011). Wound macrophages as key regulators of repair: Origin, phenotype, and function. *The American Journal of Pathology, 178*, 19–25.

Braverman, I. M. (2000). The cutaneous microcirculation. *The Journal of Investigative Dermatology Symposium Proceedings, 5*, 3–9.

Brockes, J. P., & Kumar, A. (2008). Comparative aspects of animal regeneration. *Annual Review of Cell and Developmental Biology, 24*, 525–549.

Brooks, M. D., Sengupta, R., Snyder, S. C., & Rubin, J. B. (2013). Hitting them where they live: Targeting the glioblastoma perivascular stem cell niche. *Current Pathobiology Reports, 1*, 101–110.

Brubaker, A. L., Palmer, J. L., & Kovacs, E. J. (2011). Age-related dysregulation of inflammation and innate immunity: Lessons learned from rodent models. *Aging and Disease, 2*, 346–360.

Buckley, G., Metcalfe, A. D., & Ferguson, M. W. (2011). Peripheral nerve regeneration in the MRL/MpJ ear wound model. *Journal of Anatomy, 218*, 163–172.

Buckley, G., Wong, J., Metcalfe, A. D., & Ferguson, M. W. (2012). Denervation affects regenerative responses in MRL/MpJ and repair in C57BL/6 ear wounds. *Journal of Anatomy, 220*, 3–12.

Bullard, K. M., Longaker, M. T., & Lorenz, H. P. (2003). Fetal wound healing: Current biology. *World Journal of Surgery, 27*, 54–61.

Burri, P. H., & Djonov, V. (2002). Intussusceptive angiogenesis—the alternative to capillary sprouting. *Molecular Aspects of Medicine, 23*, S1–S27.

Cafferty, W. B., et al. (2001). Leukemia inhibitory factor determines the growth status of injured adult sensory neurons. *Journal of Neuroscience, 21*, 7161–7170.

Calderon, B., et al. (2015). The pancreas anatomy conditions the origin and properties of resident macrophages. *The Journal of Experimental Medicine, 212*, 1497–1512.

Campbell, M., & Humphries, P. (2011). Size-selective and in vitro assessment of inner blood retina barrier permeability. *Methods in Molecular Biology, 763*, 355–367.
Carlson, M. E., & Conboy, I. M. (2007). Loss of stem cell regenerative capacity within aged niches. *Aging Cell, 6*, 371–382.
Carmeliet, P., & Tessier-Lavigne, M. (2005). Common mechanisms of nerve and blood vessel wiring. *Nature, 436*, 193–200.
Cattin, A.-L., et al. (2015). Macrophage-induced blood vessels guide Schwann cell-mediated regeneration of peripheral nerves. *Cell, 162*, 1127–1139.
Cavallo, T., Sade, R., Folkman, J., & Cotran, R. S. (1973). Ultrastructural autoradiographic studies of the early vasoproliferative response in tumor angiogenesis. *The American Journal of Pathology, 70*, 345–362.
Caviedes-Bucheli, J., et al. (2008). Calcitonin gene-related peptide receptor expression in alternatively activated monocytes/macrophages during irreversible pulpitis. *Journal of Endodontics, 34*, 945–949.
Chan-Ling, T., et al. (2004). Desmin ensheathment ratio as an indicator of vessel stability: Evidence in normal development and in retinopathy of prematurity. *The American Journal of Pathology, 165*, 1301–1313.
Cheng, G., et al. (2011). Engineered blood vessel networks connect to host vasculature via wrapping-and-tapping anastomosis. *Blood, 118*, 4740–4749.
Cheng, L., et al. (2013). Glioblastoma stem cells generate vascular pericytes to support vessel function and tumor growth. *Cell, 153*, 139–152.
Cheverud, J. M., et al. (2012). Healing quantitative trait loci in a combined cross analysis using related mouse strain crosses. *Heredity, 108*, 441–446.
Chiang, H. Y., Chen, C. T., Chien, H. F., & Hsieh, S. T. (2005). Skin denervation, neuropathology, and neuropathic pain in a laser-induced focal neuropathy. *Neurobiology of Disease, 18*, 40–53.
Chong, D. C., Yu, Z., Brighton, H. E., Bear, J. E., & Bautch, V. L. (2017). Tortuous microvessels contribute to wound healing via sprouting angiogenesis. *Arteriosclerosis, Thrombosis, and Vascular Biology, 37*, 1903–1912.
Choo, A. F., Logan, D. M., & Rathbone, M. P. (1978). Nerve trophic effects: An in vitro assay for factors involved in regulation of protein synthesis in regenerating amphibian limbs. *The Journal of Experimental Zoology, 206*, 347–354.
Clark, L. D., Clark, R. K., & Heber-Katz, E. (1998). A new murine model for mammalian wound repair and regeneration. *Clinical Immunology and Immunopathology, 88*, 35–45.
Cliff, W. (1963). Observations on healing tissues: A combined light and electron microscopic investigation. *Philosophical Transactions of the Royal Society, 246*, 305–325.
Conway, E. M., Collen, D., & Carmeliet, P. (2001). Molecular mechanisms of blood vessel growth. *Cardiovascular Research, 49*, 507–521.
Corselli, M., Chen, C. W., Crisan, M., Lazzari, L., & Peault, B. (2010). Perivascular ancestors of adult multipotent stem cells. *Arteriosclerosis, Thrombosis, and Vascular Biology, 30*, 1104–1109.
Crisan, M., Corselli, M., Chen, W. C. W., Peault, B., & Moldovan, N. I. (2012). Perivascular cells for regenerative medicine. *Journal of Cellular and Molecular Medicine, 16*, 2851–2860.
Crocker, D. J., Murad, T. M., & Geer, J. C. (1970). Role of the pericyte in wound healing. An ultrastructural study. *Experimental and Molecular Pathology, 13*, 51–65.
Currie, J. D., et al. (2016). Live imaging of axolotl digit regeneration reveals spatiotemporal choreography of diverse connective tissue progenitor pools. *Developmental Cell, 39*, 411–423.
Daley, J. M., et al. (2005). Modulation of macrophage phenotype by soluble product(s) released from neutrophils. *Journal of Immunology, 174*, 2265–2272.
Daley, J. M., Brancato, S. K., Thomay, A. A., Reichner, J. S., & Albina, J. E. (2010). The phenotype of murine wound macrophages. *Journal of Leukocyte Biology, 87*, 59–67.
Dallos, A., et al. (2006). Effects of the neuropeptides substance P, calcitonin gene-related peptide, vasoactive intestinal polypeptide and galanin on the production of nerve growth factor and inflammatory cytokines in cultured human keratinocytes. *Neuropeptides, 40*, 251–263.

Darland, D. C., & D'Amore, P. A. (2001). TGF beta is required for the formation of capillary-like structures in three-dimensional cocultures of 10T1/2 and endothelial cells. *Angiogenesis, 4*, 11–20.
Darland, D. C, et al. (2003). Pericyte production of cell-associated VEGF is differentiation-dependent and is associated with endothelial survival. *Developmental Biology, 264*, 275–288.
Davies, A. M. (2000). Neurotrophins: Neurotrophic modulation of neurite growth. *Current Biology, 10*, R198–R200.
De Smet, F., Segura, I., De Bock, K., Hohensinner, P. J., & Carmeliet, P. (2009). Mechanisms of vessel branching: Filopodia on endothelial tip cells lead the way. *Arteriosclerosis, Thrombosis, and Vascular Biology, 29*, 639–649.
De Spiegelaere, W., et al. (2012). Intussusceptive angiogenesis: A biologically relevant form of angiogenesis. *Journal of Vascular Research, 49*, 390–404.
Delgado, A. V., McManus, A. T., & Chambers, J. P. (2005). Exogenous administration of Substance P enhances wound healing in a novel skin-injury model. *Experimental Biology and Medicine, 230*, 271–280.
Desmouliere, A. (1995). Factors influencing myofibroblast differentiation during wound healing and fibrosis. *Cell Biology International, 19*, 471–476.
Dias Moura Prazeres, P. H., et al. (2017). Pericytes are heterogeneous in their origin within the same tissue. *Developmental Biology, 427*, 6–11.
Diaz-Flores, L., et al. (2009). Pericytes. Morphofunction, interactions and pathology in a quiescent and activated mesenchymal cell niche. *Histology and Histopathology, 24*, 909–969.
DiPietro, L. A. (2013). Angiogenesis and scar formation in healing wounds. *Current Opinion in Rheumatology, 25*, 87–91.
Djonov, V., Baum, O., & Burri, P. H. (2003). Vascular remodeling by intussusceptive angiogenesis. *Cell and Tissue Research, 314*, 107–117.
Djouhri, L. (2016). Adelta-fiber low threshold mechanoreceptors innervating mammalian hairy skin: A review of their receptive, electrophysiological and cytochemical properties in relation to Adelta-fiber high threshold mechanoreceptors. *Neuroscience and Biobehavioral Reviews, 61*, 225–238.
Doherty, T. M. (1995). T-cell regulation of macrophage function. *Current Opinion in Immunology, 7*, 400–404.
Drake, C. J. (2003). Embryonic and adult vasculogenesis. *Birth Defects Research. Part C, Embryo Today: Reviews, 69*, 73–82.
Driskell, R. R., & Watt, F. M. (2015). Understanding fibroblast heterogeneity in the skin. *Trends in Cell Biology, 25*, 92–99.
Driskell, R. R., et al. (2013). Distinct fibroblast lineages determine dermal architecture in skin development and repair. *Nature, 504*, 277–281.
Dubovy, P., Jancalek, R., & Kubek, T. (2013). Role of inflammation and cytokines in peripheral nerve regeneration. *International Review of Neurobiology, 108*, 173–206.
Duffield, J. S. (2003). The inflammatory macrophage: A story of Jekyll and Hyde. *Clinical Science, 104*, 27–38.
Duffield, J. S., et al. (2005). Selective depletion of macrophages reveals distinct, opposing roles during liver injury and repair. *The Journal of Clinical Investigation, 115*, 56–65.
Dufraine, J., Funahashi, Y., & Kitajewski, J. (2008). Notch signaling regulates tumor angiogenesis by diverse mechanisms. *Oncogene, 27*, 5132–5137.
Dunleavey, J. M., & Dudley, A. C. (2012). Vascular mimicry: Concepts and implications for anti-Angiogenic therapy. *Current Angiogenesis, 1*, 133–138.
Edwards, J. P., Zhang, X., Frauwirth, K. A., & Mosser, D. M. (2006). Biochemical and functional characterization of three activated macrophage populations. *Journal of Leukocyte Biology, 80*, 1298–1307.
Egginton, S., Zhou, A. L., Brown, M. D., & Hudlicka, O. (2001). Unorthodox angiogenesis in skeletal muscle. *Cardiovascular Research, 49*, 634–646.

Eming, S. A., Krieg, T., & Davidson, J. M. (2007a). Inflammation in wound repair: Molecular and cellular mechanisms. *The Journal of Investigative Dermatology, 127*, 514–525.
Eming, S. A., et al. (2007b). Accelerated wound closure in mice deficient for interleukin-10. *The American Journal of Pathology, 170*, 188–202.
Engin, C. (1998). Effects of calcitonin gene-related peptide on wound contraction in denervated and normal rat skin: A preliminary report. *Plastic and Reconstructive Surgery, 101*, 1887–1890.
Engin, C., Demirkan, F., Ayhan, S., Atabay, K., & Baran, N. K. (1996). Delayed effect of denervation on wound contraction in rat skin. *Plastic and Reconstructive Surgery, 98*, 1063–1067.
Epelman, S., et al. (2014). Embryonic and adult-derived resident cardiac macrophages are maintained through distinct mechanisms at steady state and during inflammation. *Immunity, 40*, 91–104.
Etchevers, H. C., Vincent, C., Le Douarin, N. M., & Couly, G. F. (2001). The cephalic neural crest provides pericytes and smooth muscle cells to all blood vessels of the face and forebrain. *Development, 128*, 1059–1068.
Etich, J., et al. (2013). PECAM1(+)/Sca1(+)/CD38(+) vascular cells transform into myofibroblast-like cells in skin wound repair. *PLoS One, 8*, e53262.
Etxaniz, U., et al. (2014). Neural-competent cells of adult human dermis belong to the Schwann lineage. *Stem Cell Reports, 3*, 774–788.
Fadok, V. A., et al. (1998). Macrophages that have ingested apoptotic cells in vitro inhibit proinflammatory cytokine production through autocrine/paracrine mechanisms involving TGF-beta, PGE2, and PAF. *The Journal of Clinical Investigation, 101*, 890–898.
Fantin, A., et al. (2010). Tissue macrophages act as cellular chaperones for vascular anastomosis downstream of VEGF-mediated endothelial tip cell induction. *Blood, 116*, 829–840.
Ferguson, M. W. J., & O'Kane, S. (2004). Scar-free healing: From embryonic mechanisms to adult therapeutic intervention. *Philosophical Transactions of the Royal Society of London. Series B, Biological Sciences, 359*, 839–850.
Frank, S., et al. (1995). Regulation of vascular endothelial growth factor expression in cultured keratinocytes. Implications for normal and impaired wound healing. *The Journal of Biological Chemistry, 270*, 12607–12613.
Frontczak-Baniewicz, M., & Walski, M. (2002). Non-sprouting angiogenesis in neurohypophysis after traumatic injury of the cerebral cortex. Electron-microscopic studies. *Neuro Endocrinology Letters, 23*, 396–404.
Fukai, T., Takeda, A., & Uchinuma, E. (2005). Wound healing in denervated rat skin. *Wound Repair and Regeneration, 13*, 175–180.
Gallagher, K. A., et al. (2007). Diabetic impairments in NO-mediated endothelial progenitor cell mobilization and homing are reversed by hyperoxia and SDF-1 alpha. *The Journal of Clinical Investigation, 117*, 1249–1259.
Gallucci, R. M., et al. (2000). Impaired cutaneous wound healing in interleukin-6-deficient and immunosuppressed mice. *FASEB Journal, 14*, 2525–2531.
Gamse, R., Saria, A., Lundberg, J. M., & Theodorsson-Norheim, E. (1986). Behavioral and neurochemical changes after intracisternal capsaicin treatment of the Guinea pig. *Neuroscience Letters, 64*, 287–292.
Gaudet, A. D., Popovich, P. G., & Ramer, M. S. (2011). Wallerian degeneration: Gaining perspective on inflammatory events after peripheral nerve injury. *Journal of Neuroinflammation, 8*, 110.
Gautier, E. L., et al. (2012). Gene-expression profiles and transcriptional regulatory pathways that underlie the identity and diversity of mouse tissue macrophages. *Nature Immunology, 13*, 1118–1128.
Geissmann, F., Jung, S., & Littman, D. R. (2003). Blood monocytes consist of two principal subsets with distinct migratory properties. *Immunity, 19*, 71–82.
Gensel, J. C., & Zhang, B. (2015). Macrophage activation and its role in repair and pathology after spinal cord injury. *Brain Research, 1619*, 1–11.

Gerber, J. S., & Mosser, D. M. (2001). Reversing lipopolysaccharide toxicity by ligating the macrophage Fc gamma receptors. *Journal of Immunology, 166*, 6861–6868.

Gerhardt, H., et al. (2003). VEGF guides angiogenic sprouting utilizing endothelial tip cell filopodia. *The Journal of Cell Biology, 161*, 1163–1177.

Gibbings, S. L., et al. (2015). Transcriptome analysis highlights the conserved difference between embryonic and postnatal-derived alveolar macrophages. *Blood, 126*, 1357–1366.

Gibran, N. S., et al. (2002). Diminished neuropeptide levels contribute to the impaired cutaneous healing response associated with diabetes mellitus. *The Journal of Surgical Research, 108*, 122–128.

Ginhoux, F., et al. (2010). Fate mapping analysis reveals that adult microglia derive from primitive macrophages. *Science, 330*, 841–845.

Godwin, J. W., Pinto, A. R., & Rosenthal, N. A. (2013). Macrophages are required for adult salamander limb regeneration. *Proceedings of the National Academy of Sciences of the United States of America, 110*, 9415–9420.

Goerdt, S., et al. (1999). Alternative versus classical activation of macrophages. *Pathobiology: Journal of Immunopathology, Molecular and Cellular Biology, 67*, 222–226.

Gomez Perdiguero, E., et al. (2015). Tissue-resident macrophages originate from yolk-sac-derived erythro-myeloid progenitors. *Nature, 518*, 547–551.

Gordon, S. (2003). Do macrophage innate immune receptors enhance atherogenesis? *Developmental Cell, 5*, 666–668.

Gordon, S., & Martinez, F. O. (2010). Alternative activation of macrophages: Mechanism and functions. *Immunity, 32*, 593–604.

Goren, I., et al. (2007). Systemic anti-TNFalpha treatment restores diabetes-impaired skin repair in Ob/Ob mice by inactivation of macrophages. *The Journal of Investigative Dermatology, 127*, 2259–2267.

Goren, I., et al. (2009). A transgenic mouse model of inducible macrophage depletion: Effects of diphtheria toxin-driven lysozyme M-specific cell lineage ablation on wound inflammatory, angiogenic, and contractive processes. *The American Journal of Pathology, 175*, 132–147.

Gratchev, A., et al. (2001). Alternatively activated macrophages differentially express fibronectin and its splice variants and the extracellular matrix protein betaIG-H3. *Scandinavian Journal of Immunology, 53*, 386–392.

Greenburg, G. B., & Hunt, T. K. (1978). The proliferative response in vitro of vascular endothelial and smooth muscle cells exposed to wound fluids and macrophages. *Journal of Cellular Physiology, 97*, 353–360.

Greenhalgh, S. N., Conroy, K. P., & Henderson, N. C. (2015). Healing scars: Targeting pericytes to treat fibrosis. *QJM, 108*, 3–7.

Gresset, A., et al. (2015). Boundary caps give rise to neurogenic stem cells and terminal glia in the skin. *Stem Cell Reports, 5*, 278–290.

Guichet, P. O., et al. (2015). Notch1 stimulation induces a vascularization switch with pericyte-like cell differentiation of glioblastoma stem cells. *Stem Cells, 33*, 21–34.

Gurtner, G. C., Werner, S., Barrandon, Y., & Longaker, M. T. (2008). Wound repair and regeneration. *Nature, 453*, 314–321.

Haas, T. L., et al. (2000). Matrix metalloproteinase activity is required for activity-induced angiogenesis in rat skeletal muscle. *American Journal of Physiology, 279*, H1540–H1547.

Hagner, S., Stahl, U., Knoblauch, B., McGregor, G. P., & Lang, R. E. (2002). Calcitonin receptor-like receptor: Identification and distribution in human peripheral tissues. *Cell and Tissue Research, 310*, 41–50.

Hamilton, J. A. (2002). GM-CSF in inflammation and autoimmunity. *Trends in Immunology, 23*, 403–408.

Han, M., Yang, X., Lee, J., Allan, C. H., & Muneoka, K. (2008). Development and regeneration of the neonatal digit tip in mice. *Developmental Biology, 315*, 125–135.

He, H., et al. (2016). Perivascular macrophages limit permeability. *Arteriosclerosis, Thrombosis, and Vascular Biology, 36*, 2203–2212.

Herbert, S. P., & Stainier, D. Y. (2011). Molecular control of endothelial cell behaviour during blood vessel morphogenesis. *Nature Reviews. Molecular Cell Biology, 12*, 551–564.

Hernandez-Quintero, M., Kuri-Harcuch, W., Gonzalez Robles, A., & Castro-Muñozledo, F. (2006). Interleukin-6 promotes human epidermal keratinocyte proliferation and keratin cytoskeleton reorganization in culture. *Cell and Tissue Research, 325*, 77–90.

Hetheridge, C., et al. (2012). The formin FMNL3 is a cytoskeletal regulator of angiogenesis. *Journal of Cell Science, 125*, 1420–1428.

Hirata, K., & Kawabuchi, M. (2002). Myelin phagocytosis by macrophages and nonmacrophages during Wallerian degeneration. *Microscopy Research and Technique, 57*, 541–547.

Hirota, H., Kiyama, H., Kishimoto, T., & Taga, T. (1996). Accelerated nerve regeneration in mice by upregulated expression of interleukin (IL) 6 and IL-6 receptor after trauma. *Journal of Experimental Medicine, 183*, 2627–2634.

Hirschi, K. K., & D'Amore, P. A. (1996). Pericytes in the microvasculature. *Cardiovascular Research, 32*, 687–698.

Hirsh, J., & Weitz, J. I. (1999). Thrombosis and anticoagulation. *Seminars in Hematology, 36*, 118–132.

Hordinsky, M., Ericson, M., Snow, D., Boeck, C., & Lee, W. S. (1999). Peribulbar innervation and substance P expression following nonpermanent injury to the human scalp hair follicle. *The Journal of Investigative Dermatology. Symposium Proceedings, 4*, 316–319.

Hosoi, J., et al. (1993). Regulation of Langerhans cell function by nerves containing calcitonin gene-related peptide. *Nature, 363*, 159–163.

Hu, M. S., Moore, A. L., & Longaker, M. T. (2018). A fibroblast is not a fibroblast is not a fibroblast. *Journal of Investigative Dermatology, 138*, 729–730.

Iribar, H., Perez-Lopez, V., Etxaniz, U., Gutierrez-Rivera, A., & Izeta, A. (2017). Schwann cells in the ventral dermis do not derive from Myf5-expressing precursors. *Stem Cell Reports, 9*, 1477–1487.

Ishida, Y., Gao, J.-L., & Murphy, P. M. (2008). Chemokine receptor CX3CR1 mediates skin wound healing by promoting macrophage and fibroblast accumulation and function. *Journal of Immunology, 180*, 569–579.

Ishikawa, S., et al. (2014). Effects of neuropeptides and their local administration to cutaneous wounds in sensory-impaired areas. *Journal of Plastic Surgery and Hand Surgery, 48*, 143–147.

Jancso, N., Jancso-Gabor, A., & Szolcsanyi, J. (1967). Direct evidence for neurogenic inflammation and its prevention by denervation and by pretreatment with capsaicin. *British Journal of Pharmacology and Chemotherapy, 31*, 138–151.

Jeffcoate, W. J., & Harding, K. G. (2003). Diabetic foot ulcers. *Lancet, 361*, 1545–1551.

Jenkins, S. J., et al. (2011). Local macrophage proliferation, rather than recruitment from the blood, is a signature of TH2 inflammation. *Science, 332*, 1284–1288.

Jenkins, S. J., et al. (2013). IL-4 directly signals tissue-resident macrophages to proliferate beyond homeostatic levels controlled by CSF-1. *The Journal of Experimental Medicine, 210*, 2477–2491.

Jeon, H., et al. (1996). Pericytes from microvessel fragment produce type IV collagen and multiple laminin isoforms. *Bioscience, Biotechnology, and Biochemistry, 60*, 856–861.

Jessen, K. n. R., & Mirsky, R. (2008). Negative regulation of myelination: Relevance for development, injury, and demyelinating disease. *Glia, 56*, 1552–1565.

Jessen, K. R., & Mirsky, R. (2016). The repair Schwann cell and its function in regenerating nerves. *Journal of Physiology, 594*, 3521–3531.

Jetten, N., et al. (2014). Anti-inflammatory M2, but not pro-inflammatory M1 macrophages promote angiogenesis in vivo. *Angiogenesis, 17*, 109–118.

Jha, A. K., et al. (2015). Network integration of parallel metabolic and transcriptional data reveals metabolic modules that regulate macrophage polarization. *Immunity, 42*, 419–430.

Johnson, K. E., & Wilgus, T. A. (2014). Vascular endothelial growth factor and angiogenesis in the regulation of cutaneous wound repair. *Advances in Wound Care, 3*, 647–661.

Johnston, A. P., et al. (2013). Sox2-mediated regulation of adult neural crest precursors and skin repair. *Stem Cell Reports, 1*, 38–45.

Johnston, A. P., et al. (2016). Dedifferentiated Schwann cell precursors secreting paracrine factors are required for regeneration of the mammalian digit tip. *Cell Stem Cell, 19*(4), 433–448.

Jun, J.-I., & Lau, L. F. (2010). The matricellular protein CCN1 induces fibroblast senescence and restricts fibrosis in cutaneous wound healing. *Nature Cell Biology, 12*, 676–685.

Jung, S., et al. (2004). Induction of IL-10 in rat peritoneal macrophages and dendritic cells by glatiramer acetate. *Journal of Neuroimmunology, 148*, 63–73.

Juniantito, V., et al. (2012). Immunophenotypical analyses of myofibroblasts in rat excisional wound healing: Possible transdifferentiation of blood vessel pericytes and perifollicular dermal sheath cells into myofibroblasts. *Histology and Histopathology, 27*, 515–527.

Kabara, M., et al. (2014). Immortalized multipotent pericytes derived from the vasa vasorum in the injured vasculature. A cellular tool for studies of vascular remodeling and regeneration. *Laboratory Investigation, 94*, 1340–1354.

Kang, H., & Lichtman, J. W. (2013). Motor axon regeneration and muscle reinnervation in young adult and aged animals. *Journal of Neuroscience, 33*, 19480–19491.

Katsuyama, T., & Paro, R. (2013). Innate immune cells are dispensable for regenerative growth of imaginal discs. *Mechanisms of Development, 130*, 112–121.

Kaukua, N., et al. (2014). Glial origin of mesenchymal stem cells in a tooth model system. *Nature, 513*, 551–554.

Kawamoto, K., & Matsuda, H. (2004). Nerve growth factor and wound healing. *Progress in Brain Research, 146*, 369–384.

Kigerl, K. A., et al. (2009). Identification of two distinct macrophage subsets with divergent effects causing either neurotoxicity or regeneration in the injured mouse spinal cord. *Journal of Neuroscience, 29*, 13435–13444.

Kilarski, W. W., & Gerwins, P. (2009). A new mechanism of blood vessel growth—hope for new treatment strategies. *Discovery Medicine, 8*, 23–27.

Kilicaslan, S. M., Cevher, S. C., & Peker, E. G. (2013). Ultrastructural changes in blood vessels in epidermal growth factor treated experimental cutaneous wound model. *Pathology, Research and Practice, 209*, 710–715.

Kim, L. R., Whelpdale, K., Zurowski, M., & Pomeranz, B. (1998). Sympathetic denervation impairs epidermal healing in cutaneous wounds. *Wound Repair and Regeneration, 6*, 194–201.

King, A., Balaji, S., Keswani, S. G., & Crombleholme, T. M. (2014). The role of stem cells in wound angiogenesis. *Advances in Wound Care, 3*, 614–625.

Klein, D., & Martini, R. (2016). Myelin and macrophages in the PNS: An intimate relationship in trauma and disease. *Brain Research, 1641*, 130–138.

Koh, T. J., Novak, M. L., & Mirza, R. E. (2013). Assessing macrophage phenotype during tissue repair. *Methods in Molecular Biology, 1037*, 507–518.

Korosec, A., et al. (2018). Lineage identity and location within the dermis determine the function of papillary and reticular fibroblasts in human skin. *Journal of Investigative Dermatology, 139*(2), 342–351.

Kreider, T., Anthony, R. M., Urban, J. F., & Gause, W. C. (2007). Alternatively activated macrophages in helminth infections. *Current Opinion in Immunology, 19*, 448–453.

Krishna Priya, S., et al. (2016). Tumour angiogenesis-origin of blood vessels. *International Journal of Cancer, 139*, 729–735.

Kroner, A., et al. (2014). TNF and increased intracellular iron alter macrophage polarization to a detrimental M1 phenotype in the injured spinal cord. *Neuron, 83*, 1098–1116.

Kumar, A., & Brockes, J. P. (2012). Nerve dependence in tissue, organ, and appendage regeneration. *Trends in Neurosciences, 35*, 691–699.

Kumar, A., Nevill, G., Brockes, J. P., & Forge, A. (2010). A comparative study of gland cells implicated in the nerve dependence of salamander limb regeneration. *Journal of Anatomy, 217*, 16–25.

Kuncova, J., Svíglerová, J., Tonar, Z., & Slavíková, J. (2005). Heterogenous changes in neuropeptide Y, norepinephrine and epinephrine concentrations in the hearts of diabetic rats. *Autonomic Neuroscience : Basic & Clinical, 121*, 7–15.

Kurz, H., Burri, P. H., & Djonov, V. G. (2003). Angiogenesis and vascular remodeling by intussusception: From form to function. *News in Physiological Sciences, 18*, 65–70.

Lambert, J. M., Lopez, E. F., & Lindsey, M. L. (2008). Macrophage roles following myocardial infarction. *International Journal of Cardiology, 130*, 147–158.

Lang, R., Patel, D., Morris, J. J., Rutschman, R. L., & Murray, P. J. (2002). Shaping gene expression in activated and resting primary macrophages by IL-10. *Journal of Immunology, 169*, 2253–2263.

Laverdet, B., et al. (2015). Skin innervation: Important roles during normal and pathological cutaneous repair. *Histology and Histopathology, 30*, 875–892.

Lavine, K. J., et al. (2014). Distinct macrophage lineages contribute to disparate patterns of cardiac recovery and remodeling in the neonatal and adult heart. *Proceedings of the National Academy of Sciences of the United States of America, 111*, 16029–16034.

Leavitt, T., et al. (2016). Scarless wound healing: Finding the right cells and signals. *Cell and Tissue Research, 365*, 483–493.

Lee, P. C., et al. (1999). Impaired wound healing and angiogenesis in eNOS-deficient mice. *The American Journal of Physiology, 277*, H1600–H1608.

Lee, R. H., et al. (2000). Inhibition of tumor necrosis factor-alpha attenuates wound breaking strength in rats. *Wound Repair and Regeneration, 8*, 547–553.

Leibovich, S. J., & Ross, R. (1975). The role of the macrophage in wound repair. A study with hydrocortisone and antimacrophage serum. *The American Journal of Pathology, 78*, 71–100.

Lewin, G. R., & Mendell, L. M. (1993). Nerve growth factor and nociception. *Trends in Neurosciences, 16*, 353–359.

Li, L., & Ginty, D. D. (2014). The structure and organization of lanceolate mechanosensory complexes at mouse hair follicles. *eLife, 3*, e01901.

Li, A. K., Koroly, M. J., Schattenkerk, M. E., Malt, R. A., & Young, M. (1980). Nerve growth factor: Acceleration of the rate of wound healing in mice. *Proceedings of the National Academy of Sciences of the United States of America, 77*, 4379–4381.

Li, L., Yan, B., Shi, Y.-Q., Zhang, W.-Q., & Wen, Z.-L. (2012). Live imaging reveals differing roles of macrophages and neutrophils during zebrafish tail fin regeneration. *The Journal of Biological Chemistry, 287*, 25353–25360.

Locati, M., et al. (2002). Analysis of the gene expression profile activated by the CC chemokine ligand 5/RANTES and by lipopolysaccharide in human monocytes. *Journal of Immunology, 168*, 3557–3562.

Loots, M. A., et al. (1998). Differences in cellular infiltrate and extracellular matrix of chronic diabetic and venous ulcers versus acute wounds. *The Journal of Investigative Dermatology, 111*, 850–857.

Lucas, T., et al. (2010). Differential roles of macrophages in diverse phases of skin repair. *Journal of Immunology, 184*, 3964–3977.

Madlener, M., Parks, W. C., & Werner, S. (1998). Matrix metalloproteinases (MMPs) and their physiological inhibitors (TIMPs) are differentially expressed during excisional skin wound repair. *Experimental Cell Research, 242*, 201–210.

Mahdavian Delavary, B., van der Veer, W. M., van Egmond, M., Niessen, F. B., & Beelen, R. H. J. (2011). Macrophages in skin injury and repair. *Immunobiology, 216*, 753–762.

Makanya, A. N., Hlushchuk, R., & Djonov, V. G. (2009). Intussusceptive angiogenesis and its role in vascular morphogenesis, patterning, and remodeling. *Angiogenesis, 12*, 113–123.

Mantovani, A., Sozzani, S., Locati, M., Allavena, P., & Sica, A. (2002). Macrophage polarization: Tumor-associated macrophages as a paradigm for polarized M2 mononuclear phagocytes. *Trends in Immunology, 23*, 549–555.

Mantovani, A., et al. (2004). The chemokine system in diverse forms of macrophage activation and polarization. *Trends in Immunology, 25*, 677–686.

Mantovani, A., Biswas, S. K., Galdiero, M. R., Sica, A., & Locati, M. (2013). Macrophage plasticity and polarization in tissue repair and remodelling. *The Journal of Pathology, 229*, 176–185.

Mapp, P. I., McWilliams, D. F., Turley, M. J., Hargin, E., & Walsh, D. A. (2012). A role for the sensory neuropeptide calcitonin gene-related peptide in endothelial cell proliferation in vivo. *British Journal of Pharmacology, 166*, 1261–1271.

Marcola, M., & Rodrigues, C. E. (2015). Endothelial progenitor cells in tumor angiogenesis: Another brick in the wall. *Stem Cells International, 2015*, 832649.

Martin, P., & Leibovich, S. J. (2005). Inflammatory cells during wound repair: The good, the bad and the ugly. *Trends in Cell Biology, 15*, 599–607.

Martin, P., et al. (2003). Wound healing in the PU.1 null mouse—tissue repair is not dependent on inflammatory cells. *Current Biology, 13*, 1122–1128.

Martinez, F. O., & Gordon, S. (2014). The M1 and M2 paradigm of macrophage activation: Time for reassessment. *F1000Prime Reports, 6*, 13.

Martinez, F. O., Sica, A., Mantovani, A., & Locati, M. (2008). Macrophage activation and polarization. *Frontiers in Bioscience, 13*, 453–461.

Martini, R., Fischer, S., López-Vales, R., & David, S. (2008). Interactions between Schwann cells and macrophages in injury and inherited demyelinating disease. *Glia, 56*, 1566–1577.

Massena, S., et al. (2015). Identification and characterization of VEGF-A-responsive neutrophils expressing CD49d, VEGFR1, and CXCR4 in mice and humans. *Blood, 126*, 2016–2026.

Mateo, R. B., Reichner, J. S., & Albina, J. E. (1994). Interleukin-6 activity in wounds. *The American Journal of Physiology, 266*, R1840–R1844.

McGovern, N., et al. (2014). Human dermal CD14+ cells are a transient population of monocyte-derived macrophages. *Immunity, 41*, 465–477.

Mescher, A. L., & Kiffmeyer, W. R. (1992). Axonal release of transferrin in peripheral nerves of axolotls during regeneration. *Monographs in Developmental Biology, 23*, 100–109.

Mescher, A. L., Connell, E., Hsu, C., Patel, C., & Overton, B. (1997). Transferrin is necessary and sufficient for the neural effect on growth in amphibian limb regeneration blastemas. *Development, Growth & Differentiation, 39*, 677–684.

Meszaros, A. J., Reichner, J. S., & Albina, J. E. (1999). Macrophage phagocytosis of wound neutrophils. *Journal of Leukocyte Biology, 65*, 35–42.

Meszaros, A. J., Reichner, J. S., & Albina, J. E. (2000). Macrophage-induced neutrophil apoptosis. *Journal of Immunology, 165*, 435–441.

Micera, A., et al. (2001). Nerve growth factor displays stimulatory effects on human skin and lung fibroblasts, demonstrating a direct role for this factor in tissue repair. *Proceedings of the National Academy of Sciences of the United States of America, 98*, 6162–6167.

Middleton, M., & Thatcher, N. (1998). G- and GM-CSF. *International Journal of Antimicrobial Agents, 10*, 91–93.

Miller, M. S., Buck, S. H., Sipes, I. G., Yamamura, H. I., & Burks, T. F. (1982). Regulation of substance P by nerve growth factor: Disruption by capsaicin. *Brain Research, 250*, 193–196.

Mirza, R., & Koh, T. J. (2011). Dysregulation of monocyte/macrophage phenotype in wounds of diabetic mice. *Cytokine, 56*, 256–264.

Mirza, R. E., & Koh, T. J. (2015). Contributions of cell subsets to cytokine production during normal and impaired wound healing. *Cytokine, 71*, 409–412.

Mirza, R., DiPietro, L. A., & Koh, T. J. (2009). Selective and specific macrophage ablation is detrimental to wound healing in mice. *The American Journal of Pathology, 175*, 2454–2462.

Mirza, R. E., Fang, M. M., Ennis, W. J., & Koh, T. J. (2013). Blocking interleukin-1beta induces a healing-associated wound macrophage phenotype and improves healing in type 2 diabetes. *Diabetes, 62*, 2579–2587.

Mirza, R. E., Fang, M. M., Weinheimer-Haus, E. M., Ennis, W. J., & Koh, T. J. (2014). Sustained inflammasome activity in macrophages impairs wound healing in type 2 diabetic humans and mice. *Diabetes, 63*, 1103–1114.

Mokarram, N., Merchant, A., Mukhatyar, V., Patel, G., & Bellamkonda, R. V. (2012). Effect of modulating macrophage phenotype on peripheral nerve repair. *Biomaterials, 33*, 8793–8801.

Molawi, K., et al. (2014). Progressive replacement of embryo-derived cardiac macrophages with age. *The Journal of Experimental Medicine, 211*, 2151–2158.
Montagna, W., & Parakkal, P. (1974). In W. Montagna & P. F. Parakkal (Eds.), *The structure and function of skin*. Cambridge, MA: Academic Press.
Morikawa, S., & Ezaki, T. (2011). Phenotypic changes and possible angiogenic roles of pericytes during wound healing in the mouse skin. *Histology and Histopathology, 26*, 979–995.
Morikawa, S., et al. (2002). Abnormalities in pericytes on blood vessels and endothelial sprouts in tumors. *The American Journal of Pathology, 160*, 985–1000.
Mosser, D. M. (2003). The many faces of macrophage activation. *Journal of Leukocyte Biology, 73*, 209–212.
Mosser, D. M., & Edwards, J. P. (2008). Exploring the full spectrum of macrophage activation. *Nature Reviews Immunology, 8*, 958–969.
Mosser, D. M., & Karp, C. L. (1999). Receptor mediated subversion of macrophage cytokine production by intracellular pathogens. *Current Opinion in Immunology, 11*, 406–411.
Motegi, S. I., & Ishikawa, O. (2017). Mesenchymal stem cells: The roles and functions in cutaneous wound healing and tumor growth. *Journal of Dermatological Science, 86*, 83–89.
Muangman, P., et al. (2004). Nerve growth factor accelerates wound healing in diabetic mice. *Wound Repair and Regeneration, 12*, 44–52.
Mukouyama, Y. S., Shin, D., Britsch, S., Taniguchi, M., & Anderson, D. J. (2002). Sensory nerves determine the pattern of arterial differentiation and blood vessel branching in the skin. *Cell, 109*, 693–705.
Murdoch, C., Muthana, M., Coffelt, S. B., & Lewis, C. E. (2008). The role of myeloid cells in the promotion of tumour angiogenesis. *Nature Reviews. Cancer, 8*, 618–631.
Nadeau, S., et al. (2011). Functional recovery after peripheral nerve injury is dependent on the pro-inflammatory cytokines IL-1β and TNF: Implications for neuropathic pain. *Journal of Neuroscience, 31*, 12533–12542.
Nakayasu, K. (1988). Origin of pericytes in neovascularization of rat cornea. *Japanese Journal of Ophthalmology, 32*, 105–112.
Napoli, I., et al. (2012). A central role for the ERK-signaling pathway in controlling Schwann cell plasticity and peripheral nerve regeneration in vivo. *Neuron, 73*, 729–742.
Natesan, S., Wrice, N. L., Baer, D. G., & Christy, R. J. (2011). Debrided skin as a source of autologous stem cells for wound repair. *Stem Cells, 29*, 1219–1230.
Nehls, V., & Drenckhahn, D. (1993). The versatility of microvascular pericytes: From mesenchyme to smooth muscle? *Histochemistry, 99*, 1–12.
Nehls, V., Denzer, K., & Drenckhahn, D. (1992). Pericyte involvement in capillary sprouting during angiogenesis in situ. *Cell and Tissue Research, 270*, 469–474.
Niemi, J. P., et al. (2013). A critical role for macrophages near axotomized neuronal cell bodies in stimulating nerve regeneration. *Journal of Neuroscience, 33*, 16236–16248.
Nishimura, H., & Asahara, T. (2005). Bone marrow-derived endothelial progenitor cells for neovascular formation. *EXS, 94*, 147–154.
Nithya, M., Suguna, L., & Rose, C. (2003). The effect of nerve growth factor on the early responses during the process of wound healing. *Biochimica et Biophysica Acta, 1620*, 25–31.
Novak, M. L., & Koh, T. J. (2013a). Phenotypic transitions of macrophages orchestrate tissue repair. *The American Journal of Pathology, 183*, 1352–1363.
Novak, M. L., & Koh, T. J. (2013b). Macrophage phenotypes during tissue repair. *Journal of Leukocyte Biology, 93*, 875–881.
Okuno, Y., Nakamura-Ishizu, A., Kishi, K., Suda, T., & Kubota, Y. (2011). Bone marrow-derived cells serve as proangiogenic macrophages but not endothelial cells in wound healing. *Blood, 117*, 5264–5272.
Olingy, C. E., et al. (2017). Non-classical monocytes are biased progenitors of wound healing macrophages during soft tissue injury. *Scientific Reports, 7*, 447.
O'Shea, J. J., & Murray, P. J. (2008). Cytokine signaling modules in inflammatory responses. *Immunity, 28*, 477–487.

Outtz, H. H., Wu, J. K., Wang, X., & Kitajewski, J. (2010). Notch1 deficiency results in decreased inflammation during wound healing and regulates vascular endothelial growth factor receptor-1 and inflammatory cytokine expression in macrophages. *Journal of Immunology, 185*, 4363–4373.

Outtz, H. H., Tattersall, I. W., Kofler, N. M., Steinbach, N., & Kitajewski, J. (2011). Notch1 controls macrophage recruitment and notch signaling is activated at sites of endothelial cell anastomosis during retinal angiogenesis in mice. *Blood, 118*, 3436–3439.

Ozerdem, U., & Stallcup, W. B. (2003). Early contribution of pericytes to angiogenic sprouting and tube formation. *Angiogenesis, 6*, 241–249.

Painter, M. W., et al. (2014). Diminished Schwann cell repair responses underlie age-associated impaired axonal regeneration. *Neuron, 83*, 331–343.

Paku, S., et al. (2011). A new mechanism for pillar formation during tumor-induced intussusceptive angiogenesis: Inverse sprouting. *The American Journal of Pathology, 179*, 1573–1585.

Palazzo, E., et al. (2012). Role of neurotrophins on dermal fibroblast survival and differentiation. *Journal of Cellular Physiology, 227*, 1017–1025.

Paquet-Fifield, S., et al. (2009). A role for pericytes as microenvironmental regulators of human skin tissue regeneration. *The Journal of Clinical Investigation, 119*, 2795–2806.

Parfejevs, V., et al. (2018a). Injury-activated glial cells promote wound healing of the adult skin in mice. *Nature Communications, 9*, 236.

Parfejevs, V., Antunes, A. T., & Sommer, L. (2018b). Injury and stress responses of adult neural crest-derived cells. *Developmental Biology, 444*, S356–S365.

Park, K. A., et al. (2010). Signaling pathways that mediate nerve growth factor-induced increase in expression and release of calcitonin gene-related peptide from sensory neurons. *Neuroscience, 171*, 910–923.

Parrinello, S., et al. (2010). EphB signaling directs peripheral nerve regeneration through Sox2-dependent Schwann cell sorting. *Cell, 143*, 145–155.

Pascual, D. W., & Bost, K. L. (1990). Substance P production by P388D1 macrophages: A possible autocrine function for this neuropeptide. *Immunology, 71*, 52–56.

Passlick, B., Flieger, D., & Ziegler-Heitbrock, H. W. (1989). Identification and characterization of a novel monocyte subpopulation in human peripheral blood. *Blood, 74*, 2527–2534.

Patan, S., et al. (2001). Vascular morphogenesis and remodeling in a model of tissue repair: Blood vessel formation and growth in the ovarian pedicle after ovariectomy. *Circulation Research, 89*, 723–731.

Peluffo, H., et al. (2015). CD300f immunoreceptor contributes to peripheral nerve regeneration by the modulation of macrophage inflammatory phenotype. *Journal of Neuroinflammation, 12*, 145.

Perdiguero, E., et al. (2011). p38/MKP-1-regulated AKT coordinates macrophage transitions and resolution of inflammation during tissue repair. *The Journal of Cell Biology, 195*, 307–322.

Perrier, P., et al. (2004). Distinct transcriptional programs activated by interleukin-10 with or without lipopolysaccharide in dendritic cells: Induction of the B cell-activating chemokine, CXC chemokine ligand 13. *Journal of Immunology, 172*, 7031–7042.

Perry, V. H., Tsao, J. W., Fearn, S., & Brown, M. C. (1995). Radiation-induced reductions in macrophage recruitment have only slight effects on myelin degeneration in sectioned peripheral nerves of mice. *European Journal of Neuroscience, 7*, 271–280.

Philippeos, C., et al. (2018). Spatial and single-cell transcriptional profiling identifies functionally distinct human dermal fibroblast subpopulations. *Journal of Investigative Dermatology, 138*, 811–825.

Polverini, P. J. (2011). Angiogenesis and wound healing: Basic discoveries, clinical implications, and therapeutic opportunities. *Endodontic Topics, 24*, 130–145.

Porcheray, F., et al. (2005). Macrophage activation switching: An asset for the resolution of inflammation. *Clinical and Experimental Immunology, 142*, 481–489.

Pradhan, L., et al. (2011). Gene expression of pro-inflammatory cytokines and neuropeptides in diabetic wound healing. *The Journal of Surgical Research, 167*, 336–342.

Prazeres, P., et al. (2018). Perivascular cell alphav integrins as a target to treat skeletal muscle fibrosis. *The International Journal of Biochemistry & Cell Biology, 99*, 109–113.

Provitera, V., et al. (2007). Myelinated nerve endings in human skin. *Muscle & Nerve, 35*, 767–775.

Radomska, K. J., & Topilko, P. (2017). Boundary cap cells in development and disease. *Current Opinion in Neurobiology, 47*, 209–215.

Raes, G., et al. (2002). FIZZ1 and Ym as tools to discriminate between differentially activated macrophages. *Developmental Immunology, 9*, 151–159.

Rahmani, W., et al. (2014). Hair follicle dermal stem cells regenerate the dermal sheath, repopulate the dermal papilla, and modulate hair type. *Developmental Cell, 31*, 543–558.

Ramaglia, V., et al. (2008). Soluble complement receptor 1 protects the peripheral nerve from early axon loss after injury. *American Journal of Pathology, 172*, 1043–1052.

Redmer, D. A., et al. (2001). Evidence for a role of capillary pericytes in vascular growth of the developing ovine corpus luteum. *Biology of Reproduction, 65*, 879–889.

Reese, T. A., et al. (2007). Chitin induces accumulation in tissue of innate immune cells associated with allergy. *Nature, 447*, 92–96.

Rehman, J., Li, J., Orschell, C. M., & March, K. L. (2003). Peripheral blood "endothelial progenitor cells" are derived from monocyte/macrophages and secrete angiogenic growth factors. *Circulation, 107*, 1164–1169.

Reinisch, C. M., & Tschachler, E. (2005). The touch dome in human skin is supplied by different types of nerve fibers. *Annals of Neurology, 58*, 88–95.

Reynolds, L. P., & Redmer, D. A. (1998). Expression of the angiogenic factors, basic fibroblast growth factor and vascular endothelial growth factor, in the ovary. *Journal of Animal Science, 76*, 1671–1681.

Rezaie, P., & Male, D. (2002). Mesoglia & microglia—a historical review of the concept of mononuclear phagocytes within the central nervous system. *Journal of the History of the Neurosciences, 11*, 325–374.

Rhodin, J. A., & Fujita, H. (1989). Capillary growth in the mesentery of normal young rats. Intravital video and electron microscope analyses. *Journal of Submicroscopic Cytology and Pathology, 21*, 1–34.

Ribatti, D., & Crivellato, E. (2012). Sprouting angiogenesis, a reappraisal. *Developmental Biology, 372*, 157–165.

Ribatti, D., Vacca, A., Nico, B., Roncali, L., & Dammacco, F. (2001). Postnatal vasculogenesis. *Mechanisms of Development, 100*, 157–163.

Ricardo, S. D., van Goor, H., & Eddy, A. A. (2008). Macrophage diversity in renal injury and repair. *The Journal of Clinical Investigation, 118*, 3522–3530.

Richards, A. M., Floyd, D. C., Terenghi, G., & McGrouther, D. A. (1999). Cellular changes in denervated tissue during wound healing in a rat model. *The British Journal of Dermatology, 140*, 1093–1099.

Rinkevich, Y., et al. (2015). Skin fibrosis. Identification and isolation of a dermal lineage with intrinsic fibrogenic potential. *Science, 348*, aaa2151.

Risau, W., et al. (1988). Vasculogenesis and angiogenesis in embryonic-stem-cell-derived embryoid bodies. *Development, 102*, 471–478.

Robins, S. P., et al. (2003). Increased skin collagen extractability and proportions of collagen type III are not normalized after 6 months healing of human excisional wounds. *Journal of Investigative Dermatology, 121*, 267–272.

Rodero, M. P., Legrand, J. M. D., Bou-Gharios, G., & Khosrotehrani, K. (2013). Wound-associated macrophages control collagen 1alpha2 transcription during the early stages of skin wound healing. *Experimental Dermatology, 22*, 143–145.

Rodriguez, A., et al. (2013). Phenotypical differences in connective tissue cells emerging from microvascular pericytes in response to overexpression of PDGF-B and TGF-beta1 in normal skin in vivo. *The American Journal of Pathology, 182*, 2132–2146.

Rook, J. M., Hasan, W., & McCarson, K. E. (2009). Morphine-induced early delays in wound closure: Involvement of sensory neuropeptides and modification of neurokinin receptor expression. *Biochemical Pharmacology, 77*, 1747–1755.
Roosterman, D., Goerge, T., Schneider, S. W., Bunnett, N. W., & Steinhoff, M. (2006). Neuronal control of skin function: The skin as a neuroimmunoendocrine organ. *Physiological Reviews, 86*, 1309–1379.
Rotshenker, S. (2011). Wallerian degeneration: The innate-immune response to traumatic nerve injury. *Journal of Neuroinflammation, 8*, 109.
Ruetze, M., et al. (2013). A novel niche for skin derived precursors in non-follicular skin. *Journal of Dermatological Science, 69*, 132–139.
Rundhaug, J. E. (2005). Matrix metalloproteinases and angiogenesis. *Journal of Cellular and Molecular Medicine, 9*, 267–285.
Rutlin, M., et al. (2014). The cellular and molecular basis of direction selectivity of Adelta-LTMRs. *Cell, 159*, 1640–1651.
Rymo, S. F., et al. (2011). A two-way communication between microglial cells and angiogenic sprouts regulates angiogenesis in aortic ring cultures. *PLoS One, 6*, e15846.
Sakuma, R., et al. (2016). Brain pericytes serve as microglia-generating multipotent vascular stem cells following ischemic stroke. *Journal of Neuroinflammation, 13*, 57.
Sasaki, M., et al. (2008). Mesenchymal stem cells are recruited into wounded skin and contribute to wound repair by transdifferentiation into multiple skin cell type. *Journal of Immunology, 180*, 2581–2587.
Sato, Y., & Rifkin, D. B. (1989). Inhibition of endothelial cell movement by pericytes and smooth muscle cells: Activation of a latent transforming growth factor-beta 1-like molecule by plasmin during co-culture. *The Journal of Cell Biology, 109*, 309–315.
Sava, P., Cook, I. O., Mahal, R. S., & Gonzalez, A. L. (2015). Human microvascular pericyte basement membrane remodeling regulates neutrophil recruitment. *Microcirculation, 22*, 54–67.
Schlundt, C., et al. (2018). Macrophages in bone fracture healing: Their essential role in endochondral ossification. *Bone, 106*, 78–89.
Schmid, M. C., & Varner, J. A. (2012). Myeloid cells in tumor inflammation. *Vascular Cell, 4*, 14.
Schoefl, G. I. (1963). Studies on inflammation. Iii. Growing capillaries: Their structure and permeability. *Virchows Archiv für Pathologische Anatomie und Physiologie und für Klinische Medizin, 337*, 97–141.
Schulz, C., et al. (2012). A lineage of myeloid cells independent of Myb and hematopoietic stem cells. *Science, 336*, 86–90.
Scott, C. L., et al. (2016). Bone marrow-derived monocytes give rise to self-renewing and fully differentiated Kupffer cells. *Nature Communications, 7*, 10321.
Scotton, C. J., et al. (2005). Transcriptional profiling reveals complex regulation of the monocyte IL-1 beta system by IL-13. *Journal of Immunology, 174*, 834–845.
Seifert, A. W., et al. (2012a). Skin shedding and tissue regeneration in African spiny mice (Acomys). *Nature, 489*, 561–565.
Seifert, A. W., et al. (2012b). The influence of fundamental traits on mechanisms controlling appendage regeneration. *Biological Reviews of the Cambridge Philosophical Society, 87*, 330–345.
Senapati, A., et al. (1986). Depletion of neuropeptides during wound healing in rat skin. *Neuroscience Letters, 71*, 101–105.
Shapiro, S. D., Kobayashi, D. K., & Ley, T. J. (1993). Cloning and characterization of a unique elastolytic metalloproteinase produced by human alveolar macrophages. *Journal of Biological Chemistry, 268*, 23824–23829.
Shi, C., & Pamer, E. G. (2011). Monocyte recruitment during infection and inflammation. *Nature Reviews Immunology, 11*, 762–774.
Shi, Q., et al. (1994). Proof of fallout endothelialization of impervious Dacron grafts in the aorta and inferior vena cava of the dog. *Journal of Vascular Surgery, 20*, 546–556. discussion 556–557.

Shi, C. M., Qu, J. F., & Cheng, T. M. (2003). Effects of the nerve growth factor on the survival and wound healing in mice with combined radiation and wound injury. *Journal of Radiation Research, 44*, 223–228.
Silva, W. N., et al. (2018a). Role of Schwann cells in cutaneous wound healing. *Wound Repair and Regeneration, 26*(5), 392–397.
Silva, W. N., et al. (2018b). Macrophage-derived GPNMB accelerates skin healing. *Experimental Dermatology, 27*, 630–635.
Simoes, M. G., et al. (2014). Denervation impairs regeneration of amputated zebrafish fins. *BMC Developmental Biology, 14*, 49.
Sims, D. E. (1986). The pericyte—a review. *Tissue & Cell, 18*, 153–174.
Sindrilaru, A., et al. (2011). An unrestrained proinflammatory M1 macrophage population induced by iron impairs wound healing in humans and mice. *The Journal of Clinical Investigation, 121*, 985–997.
Singer, M. (1951). Introduction of regeneration of forelimb of the frog by augmentation of the nerve supply. *Proceedings of the Society for Experimental Biology and Medicine, 76*, 413–416.
Singer, M. (1952). The influence of the nerve in regeneration of the amphibian extremity. *The Quarterly Review of Biology, 27*, 169–200.
Singer, M. (1964). The trophic quality of the neuron: Some theoretical considerations. *Progress in Brain Research, 13*, 228–232.
Smith, P. G., & Liu, M. (2002). Impaired cutaneous wound healing after sensory denervation in developing rats: Effects on cell proliferation and apoptosis. *Cell and Tissue Research, 307*, 281–291.
Sorrell, J. M., & Caplan, A. I. (2004). Fibroblast heterogeneity: More than skin deep. *Journal of Cell Science, 117*, 667–675.
Souza, B. R., Cardoso, J. F., Amadeu, T. P., Desmouliere, A., & Costa, A. M. (2005). Sympathetic denervation accelerates wound contraction but delays reepithelialization in rats. *Wound Repair and Regeneration, 13*, 498–505.
Spanel-Borowski, K., Amselgruber, W., & Sinowatz, F. (1987). Capillary sprouts in ovaries of immature superstimulated golden hamsters: A SEM study of microcorrosion casts. *Anatomy and Embryology, 176*, 387–391.
Spenny, M. L., et al. (2002). Neutral endopeptidase inhibition in diabetic wound repair. *Wound Repair and Regeneration, 10*, 295–301.
Stefater, J. A., et al. (2011). Regulation of angiogenesis by a non-canonical Wnt-Flt1 pathway in myeloid cells. *Nature, 474*, 511–515.
Stein, M., Keshav, S., Harris, N., & Gordon, S. (1992). Interleukin 4 potently enhances murine macrophage mannose receptor activity: A marker of alternative immunologic macrophage activation. *The Journal of Experimental Medicine, 176*, 287–292.
Stocum, D. L. (2011). The role of peripheral nerves in urodele limb regeneration. *The European Journal of Neuroscience, 34*, 908–916.
Stout, R. D. (2010). Editorial: Macrophage functional phenotypes: No alternatives in dermal wound healing? *Journal of Leukocyte Biology, 87*, 19–21.
Stout, R. D., et al. (2005). Macrophages sequentially change their functional phenotype in response to changes in microenvironmental influences. *Journal of Immunology, 175*, 342–349.
Stramer, B., & Martin, P. (2005). Cell biology: Master regulators of sealing and healing. *Current Biology, 15*, R425–R427.
Stratton, J. A., & Shah, P. T. (2016). Macrophage polarization in nerve injury: Do Schwann cells play a role? *Neural Regeneration Research, 11*, 53–57.
Stratton, J. A., et al. (2016). The immunomodulatory properties of adult skin-derived precursor Schwann cells: Implications for peripheral nerve injury therapy. *The European Journal of Neuroscience, 43*, 365–375.
Strauss-Ayali, D., Conrad, S. M., & Mosser, D. M. (2007). Monocyte subpopulations and their differentiation patterns during infection. *Journal of Leukocyte Biology, 82*, 244–252.

Styp-Rekowska, B., Hlushchuk, R., Pries, A. R., & Djonov, V. (2011). Intussusceptive angiogenesis: Pillars against the blood flow. *Acta Physiologica, 202*, 213–223.

Summan, M., et al. (2006). Macrophages and skeletal muscle regeneration: A clodronate-containing liposome depletion study. *American Journal of Physiology. Regulatory, Integrative and Comparative Physiology, 290*, R1488–R1495.

Swaminathan, S., & Griffin, M. D. (2008). First responders: Understanding monocyte-lineage traffic in the acutely injured kidney. *Kidney International, 74*, 1509–1511.

Tabib, T., Morse, C., Wang, T., Chen, W., & Lafyatis, R. (2018). SFRP2/DPP4 and FMO1/LSP1 define major fibroblast populations in human skin. *Journal of Investigative Dermatology, 138*, 802–810.

Tagawa, S., et al. (2015). Determination of early and late endothelial progenitor cells in peripheral circulation and their clinical association with coronary artery disease. *International Journal of Vascular Medicine, 2015*, 674213.

Tammela, T., et al. (2008). Blocking VEGFR-3 suppresses angiogenic sprouting and vascular network formation. *Nature, 454*, 656–660.

Tamoutounour, S., et al. (2013). Origins and functional specialization of macrophages and of conventional and monocyte-derived dendritic cells in mouse skin. *Immunity, 39*, 925–938.

Tattersall, I. W., et al. (2016). In vitro modeling of endothelial interaction with macrophages and pericytes demonstrates notch signaling function in the vascular microenvironment. *Angiogenesis, 19*, 201–215.

Thakral, K. K., Goodson, W. H., & Hunt, T. K. (1979). Stimulation of wound blood vessel growth by wound macrophages. *The Journal of Surgical Research, 26*, 430–436.

Thomay, A. A., et al. (2009). Disruption of interleukin-1 signaling improves the quality of wound healing. *The American Journal of Pathology, 174*, 2129–2136.

Timmermans, F., et al. (2007). Endothelial outgrowth cells are not derived from CD133+ cells or CD45+ hematopoietic precursors. *Arteriosclerosis, Thrombosis, and Vascular Biology, 27*, 1572–1579.

Toda, M., et al. (2008). Neuronal system-dependent facilitation of tumor angiogenesis and tumor growth by calcitonin gene-related peptide. *Proceedings of the National Academy of Sciences of the United States of America, 105*, 13550–13555.

Toma, J. G., et al. (2001). Isolation of multipotent adult stem cells from the dermis of mammalian skin. *Nature Cell Biology, 3*, 778–784.

Tredget, E. E., Yang, L., Delehanty, M., Shankowsky, H., & Scott, P. G. (2006). Polarized Th2 cytokine production in patients with hypertrophic scar following thermal injury. *Journal of Interferon & Cytokine Research, 26*, 179–189.

Troidl, C., et al. (2009). Classically and alternatively activated macrophages contribute to tissue remodelling after myocardial infarction. *Journal of Cellular and Molecular Medicine, 13*, 3485–3496.

Trost, A., et al. (2016). Brain and retinal pericytes: Origin, function and role. *Frontiers in Cellular Neuroscience, 10*, 20. https://doi.org/10.3389/fncel.2016.00020.

Tschachler, E., et al. (2004). Sheet preparations expose the dermal nerve plexus of human skin and render the dermal nerve end organ accessible to extensive analysis. *The Journal of Investigative Dermatology, 122*, 177–182.

Tung, J. J., Tattersall, I. W., & Kitajewski, J. (2012). Tips, stalks, tubes: Notch-mediated cell fate determination and mechanisms of tubulogenesis during angiogenesis. *Cold Spring Harbor Perspectives in Medicine, 2*, a006601.

Uchiyama, A., et al. (2014). MFG-E8 regulates angiogenesis in cutaneous wound healing. *The American Journal of Pathology, 184*, 1981–1990.

Vagesjo, E., et al. (2018). Accelerated wound healing in mice by on-site production and delivery of CXCL12 by transformed lactic acid bacteria. *Proceedings of the National Academy of Sciences of the United States of America, 115*, 1895–1900.

Vapniarsky, N., Arzi, B., Hu, J. C., Nolta, J. A., & Athanasiou, K. A. (2015). Concise review: Human dermis as an autologous source of stem cells for tissue engineering and regenerative medicine. *Stem Cells Translational Medicine, 4*, 1187–1198.
Vargas, M. E., Watanabe, J., Singh, S. J., Robinson, W. H., & Barres, B. A. (2010). Endogenous antibodies promote rapid myelin clearance and effective axon regeneration after nerve injury. *Proceedings of the National Academy of Sciences of the United States of America, 107*, 11993–11998.
Velazquez, O. C. (2007). Angiogenesis and vasculogenesis: Inducing the growth of new blood vessels and wound healing by stimulation of bone marrow-derived progenitor cell mobilization and homing. *Journal of Vascular Surgery, 45*(Suppl A), A39–A47.
Verreck, F. A. W., et al. (2004). Human IL-23-producing type 1 macrophages promote but IL-10-producing type 2 macrophages subvert immunity to (myco)bacteria. *Proceedings of the National Academy of Sciences of the United States of America, 101*, 4560–4565.
Verreck, F. A. W., de Boer, T., Langenberg, D. M. L., van der Zanden, L., & Ottenhoff, T. H. M. (2006). Phenotypic and functional profiling of human proinflammatory type-1 and anti-inflammatory type-2 macrophages in response to microbial antigens and IFN-gamma- and CD40L-mediated costimulation. *Journal of Leukocyte Biology, 79*, 285–293.
Villalta, S. A., et al. (2011). Interleukin-10 reduces the pathology of mdx muscular dystrophy by deactivating M1 macrophages and modulating macrophage phenotype. *Human Molecular Genetics, 20*, 790–805.
Virgintino, D., et al. (2007). An intimate interplay between precocious, migrating pericytes and endothelial cells governs human fetal brain angiogenesis. *Angiogenesis, 10*, 35–45.
Vogler, R., Sauer, B., Kim, D. S., Schäfer-Korting, M., & Kleuser, B. (2003). Sphingosine-1-phosphate and its potentially paradoxical effects on critical parameters of cutaneous wound healing. *The Journal of Investigative Dermatology, 120*, 693–700.
Wang, N., & Gibbons, C. H. (2013). Skin biopsies in the assessment of the autonomic nervous system. *Handb Clin Neurol, 117*, 371–378.
Wang, Y.-C., et al. (2010). Notch signaling determines the M1 versus M2 polarization of macrophages in antitumor immune responses. *Cancer Research, 70*, 4840–4849.
Watanabe, T., & Raff, M. C. (1988). Retinal astrocytes are immigrants from the optic nerve. *Nature, 332*, 834–837.
Wehner, S., et al. (2010). Mechanical strain and TLR4 synergistically induce cell-specific inflammatory gene expression in intestinal smooth muscle cells and peritoneal macrophages. *American Journal of Physiology. Gastrointestinal and Liver Physiology, 299*, G1187–G1197.
Wei, T., et al. (2012). Keratinocyte expression of inflammatory mediators plays a crucial role in substance P-induced acute and chronic pain. *Journal of Neuroinflammation, 9*, 181.
Weitkamp, B., Cullen, P., Plenz, G., Robenek, H., & Rauterberg, J. (1999). Human macrophages synthesize type VIII collagen in vitro and in the atherosclerotic plaque. *FASEB Journal, 13*, 1445–1457.
Werner, S., & Grose, R. (2003). Regulation of wound healing by growth factors and cytokines. *Physiological Reviews, 83*, 835–870.
West, H., Richardson, W. D., & Fruttiger, M. (2005). Stabilization of the retinal vascular network by reciprocal feedback between blood vessels and astrocytes. *Development, 132*, 1855–1862.
Willenborg, S., & Eming, S. A. (2014). Macrophages—sensors and effectors coordinating skin damage and repair. *Journal of the German Society of Dermatology, 12*, 214–221,. 214–223.
Willenborg, S., et al. (2012). CCR2 recruits an inflammatory macrophage subpopulation critical for angiogenesis in tissue repair. *Blood, 120*, 613–625.
Wynn, T. A., et al. (2011). Quantitative assessment of macrophage functions in repair and fibrosis. *Current Protocols in Immunology*,. Chapter 14, Unit14.22.
Wynn, T. A., Chawla, A., & Pollard, J. W. (2013). Macrophage biology in development, homeostasis and disease. *Nature, 496*, 445–455.
Xiao, Y., et al. (2013). Perivascular hair follicle stem cells associate with a venule annulus. *The Journal of Investigative Dermatology, 133*, 2324–2331.

Xueyong, L., et al. (2008). Differentiation of the pericyte in wound healing: The precursor, the process, and the role of the vascular endothelial cell. *Wound Repair and Regeneration, 16*, 346–355.
Yamanishi, H., Fujiwara, S., & Soma, T. (2012). Perivascular localization of dermal stem cells in human scalp. *Experimental Dermatology, 21*, 78–80.
Yamazaki, T., et al. (2017). Tissue myeloid progenitors differentiate into pericytes through TGF-beta Signaling in developing skin vasculature. *Cell Reports, 18*, 2991–3004.
Ydens, E., et al. (2012). Acute injury in the peripheral nervous system triggers an alternative macrophage response. *Journal of Neuroinflammation, 9*, 176.
Yona, S., et al. (2013). Fate mapping reveals origins and dynamics of monocytes and tissue macrophages under homeostasis. *Immunity, 38*, 79–91.
Young, B., Woodford, P., & O'Dowd, G. (2014). *Wheater's functional histology: A text and colour atlas* (6th ed.). London: Elsevier Churchill Livingstone.
Yu, B., Alboslemy, T., Safadi, F., & Kim, M. H. (2018). Glycoprotein nonmelanoma clone B regulates the crosstalk between macrophages and mesenchymal stem cells toward wound repair. *Journal of Investigative Dermatology, 138*, 219–227.
Zajac, E., et al. (2013). Angiogenic capacity of M1- and M2-polarized macrophages is determined by the levels of TIMP-1 complexed with their secreted proMMP-9. *Blood, 122*, 4054–4067.
Zimmerman, A., Bai, L., & Ginty, D. D. (2014). The gentle touch receptors of mammalian skin. *Science, 346*, 950–954
Ziyad, S., & Iruela-Arispe, M. L. (2011). Molecular mechanisms of tumor angiogenesis. *Genes & Cancer, 2*, 1085–1096.
Zochodne, D. W. (2012). The challenges and beauty of peripheral nerve regrowth. *Journal of the Peripheral Nervous System, 17*, 1–18.

Chapter 2
Pericytes in Glioblastomas: Multifaceted Role Within Tumor Microenvironments and Potential for Therapeutic Interventions

Anirudh Sattiraju and Akiva Mintz

Abstract Glioblastoma (GBM) is an aggressive and lethal disease that often results in a poor prognosis. Unlike most solid tumors, GBM is characterized by diffuse infiltrating margins, extensive angiogenesis, hypoxia, necrosis, and clonal heterogeneity. Recurrent disease is an unavoidable consequence for many patients as standard treatment options such as surgery, radiotherapy, and chemotherapy have proven to be insufficient in causing long-term survival benefits. Systemic delivery of promising drugs is hindered due to the blood–brain barrier and non-uniform perfusion within GBM tissue. In recent years, many investigations have highlighted the role of GBM stem cells (GSCs) and their microenvironment in the initiation and maintenance of tumor tissue. Preclinical and early clinical studies to target GSCs and microenvironmental components are currently underway. Of these strategies, immunotherapy using checkpoint inhibitors and redirected cytotoxic T cells have shown promising results in early investigations. But, GBM microenvironment is heterogenous and recent investigations have shown cell populations within this microenvironment to be plastic. These studies underline the importance of identifying the role of and targeting multiple cell populations within the GBM microenvironment which could have a synergistic effect when combined with novel therapies. Pericytes are multipotent perivascular cells that play a vital role within the GBM microenvironment by assisting in tumor initiation, survival, and progression. Due to their role in regulating the blood–brain barrier permeability, promoting angiogenesis, tumor growth, clearing extracellular matrix for infiltrating GBM cells and in helping GBM cells evade immune surveillance, pericytes could be ideal therapeutic targets for stymieing or exploiting their role within the GBM microenvironment. This chapter will introduce hallmarks of GBM and elaborate on the contributions of

A. Sattiraju, PhD · A. Mintz (✉)
Department of Radiology, Columbia University Irving Medical Center, New York, NY, USA
e-mail: anirudh.sattiraju@mssm.edu; am4754@cumc.columbia.edu

A. Birbrair (ed.), *Pericyte Biology in Disease*, Advances in Experimental Medicine and Biology 1147, https://doi.org/10.1007/978-3-030-16908-4_2

pericytes to these hallmarks by examining recent findings. In addition, the chapter also highlights the therapeutic value of targeting pericytes, while discussing conventional and novel GBM therapies and obstacles to their efficacy.

Keywords Glioblastoma · GBM · Pericytes · Microenvironment

Introduction

Glioblastoma: Incidence and Histological Characteristics

GBM is the most common and aggressive primary malignant brain cancer with a dismal prognosis (Sattiraju et al. 2017a; Van Meir et al. 2010). The median overall survival rate is currently 20.8 months for patients undergoing treatment with adjuvant chemotherapy (temozolomide) following maximum safe resection (Roh et al. 2017). It is estimated that about 13,000 patients are newly diagnosed with GBM annually with an estimated 5-year survival rate of about 10% for adults and about 40% for children (Rostomily et al. 2005; Song et al. 2010; Stupp et al. 2009).

Histopathological characteristics of GBM include anaplasia, macrophage and microglial infiltration, extensive angiogenesis and regions of severe hypoxia, resulting in pseudopalisading structures around necrosing neural tissue (Bissell and Radisky 2001; Van Meir et al. 2010). At a cellular level, GBM is characterized by the rapid proliferation of malignant cells which invade diffusely into the surrounding normal brain parenchyma accompanied by extensive proliferation of endothelial cells which assemble into highly torturous, disorganized and leaky blood vessels which often result in hypoxic microenvironments within tumor tissues (Van Meir et al. 2010; Zong et al. 2012). Invading GBM cells tend to displace preexisting astrocytes and pericytes that are otherwise tightly wrapped around endothelial cells, thereby disrupting the blood–brain barrier (BBB), resulting in leaky blood vessels (Dubois et al. 2014; Watkins et al. 2014). The cell of origin (COI) which is hypothesized to be the first malignant cell that sets the eventual formation of aberrant tissue into motion is thought to exist within the tumor mass in areas of severe hypoxia. GBM is a highly heterogenous disease indicated by the high degree of genetic variations and existence of multiple subclones within tumor tissues. This high degree of heterogeneity has also prompted some researchers to believe that there may be multiple cells of origin, giving rise to a mosaic of aberrant cells which we consider as the tumor tissue (Bradshaw et al. 2016; Cabrera et al. 2015; Dalerba et al. 2007; Fidoamore et al. 2016; Friedmann-Morvinski and Verma 2014; Heddleston et al. 2011).

Hypoxic Microenvironment Within GBM

GBM is an aggressive disease which requires abundant supply of oxygen and nutrients for the survival of highly proliferating cells. The high rate of cellular proliferation within the tumor tissue causes severe hypoxia (0.1–0.5%) within the tumor core and mild hypoxia (0.5–2.5%) in peripheral regions of the tumor (Evans et al. 2004).

Within these hypoxic regions, cells residing far away from preexisting blood vessels that cannot adapt to their hypoxic microenvironment undergo apoptosis or coagulative necrosis (Brat and Van Meir 2004; Martínez-González et al. 2012). This results in the outward movement of cells in the surrounding areas towards the periphery of the hypoxic region. These outward moving cells form palisading structures called "Pseudopalisades" and these events are thought to enhance the invasiveness of GBM cells, which in turn stimulate endothelial proliferation and angiogenesis through secretion of VEGF and other factors. Pseudopalisades are a distinct feature of GBMs and a marker of aggressive disease, often distinguishing them from low-grade astrocytomas (Brat et al. 2004; Heddleston et al. 2009; Rong et al. 2006; Sattiraju et al. 2017a).

Hallmarks of GBM

Glioblastoma Initiation and Maintenance

Recent investigations have pointed towards progenitor cells present within the sub-ventricular zone (SVZ) and in peripheral white matter of the cortex as potential COIs for GBM (Alcantara Llaguno et al. 2015). These investigations had employed cell population-specific knockdown of proto-oncogenes and in vivo transduction of progenitor cells expressing fluorescent labels in the cortex of the brain of transgenic mice to show that primitive progenitor cells give rise to aberrant populations of rapidly proliferating cells in a hierarchical fashion. *Llaguno et al.* pointed to the mutation of proto-oncogenes as a prerequisite for the transformation of otherwise normal progenitor cells and subsequent altered migration and production of aberrant glial cells. *Assanah et al.*, on the other hand, showed that the overexpression of growth factors receptors without mutations to proto-oncogenes was sufficient to alter the function of progenitor cells and cause the growth of tumors within brains of mice (Assanah et al. 2006).

Even though the question of which cell population(s) initiate GBM is still being hotly debated (Safa et al. 2015; Soda et al. 2011), studies examining the effect of therapeutics on established GBM models have reported the existence of stem-like, plastic cells present within GBM tissue that tend to survive therapeutic exposure and later cause disease relapse (Bao et al. 2006; Baskar et al. 2012; Chen et al. 2012; Jackson et al. 2015; Mannino and Chalmers 2011; Murat et al. 2008; Weller et al. 2012). These tumor cells which show the genetic expression and functional characteristics of stem cells have been termed as glioblastoma stem cells (GSCs) (Bradshaw et al. 2016; Cabrera et al. 2015; Calabrese et al. 2007; Dalerba et al. 2007; Gilbertson and Rich 2007; Hanahan and Weinberg 2011; Heddleston et al. 2009; Lathia et al. 2015; Liebelt et al. 2016; Singh et al. 2004a, b). In recent years, further studies into the role of GSCs have indicated their importance in the maintenance of GBMs and reconstitution of tumors post therapy (Chou et al. 2012; Dewhirst et al. 2008; El Hallani et al. 2010; Heddleston et al. 2009; Jhaveri et al. 2016; Lathia et al. 2010;

Nakada et al. 2013; Ogden et al. 2008; Ricci-Vitiani et al. 2010). Although several investigations have elucidated the role of GSCs and explored potential ways to target them to enhance the efficacy of current and future therapies, questions regarding the ontology of GSCs have not been conclusively answered (Brooks et al. 2013; Chen et al. 2014; Fan et al. 2010; Huang et al. 2012; Mendez et al. 2010; Persano et al. 2012; Sattiraju et al. 2017a). In addition, *Segerman et al. Chaffer et al.* and others showed that cancer cells are highly plastic and that terminally differentiated cancer cells dedifferentiate into cancer stem cells (CSCs) in response to stressors and therapy (Chaffer et al. 2013; Niklasson et al. 2017; Segerman et al. 2016). These studies highlighted the importance of tissue microenvironment in the regulation of cancer cell state, but further investigations are necessary to understand the importance of cellular plasticity in tumor maintenance and therapeutic resistance.

Angiogenesis and Perfusion with GBM

GBM is characterized by extensive angiogenesis to allow the growth and survival of rapidly proliferating cells (Das and Marsden 2013; Jain et al. 2007). Hypoxic microenvironments cause GBM cell invasion and stimulate vascular and perivascular cells to produce pro-angiogenic factors (Brat et al. 2004; McCord et al. 2009; Rong et al. 2006). Vascular endothelial growth factor (VEGF) and its receptors, platelet-derived growth factor (PDGF), PDGF receptor-beta (PDGFRβ), angiopoietins (Ang1 and Ang2), Tie2, matrix metalloproteinases (MMP-2 and MMP-9), bone morphogenic proteins (BMPs), etc. have been shown to be involved in this process (Jackson et al. 2017; Ribeiro and Okamoto 2015). The central nervous system (CNS) vasculature consists of tightly packed endothelial cells that are wrapped around by pericytes which provide structural support. Additionally, these vessels are further wrapped around by astrocytes that extend across endothelial cell tight junctions and by interneurons which together form the BBB. BBB is a protective vascular barrier that only allows the passive diffusion of water, oxygen, carbon dioxide, and highly lipophilic molecules. Glucose, amino acids, hormones, and larger molecules are actively transported across the BBB (Abbott 2002; Abbott et al. 2010). Co-opted vasculature and the newly formed angiogenic blood vessels within the GBM tissue show greater vascular permeability than normal CNS vasculature due to their disorganized architecture. This altered, often leaky vascular barrier within the tumor tissue is termed as the blood–tumor barrier (BTB) and results in regions of edema which hinder effective drug delivery (Agarwal et al. 2013; Dubois et al. 2014; Sattiraju et al. 2017b). The presence of BBB in peritumoral regions and a leaky BTB within the tumor results in ineffective perfusion of GBM tissue which further contributes to necrosis and hypoxia while certain parts of a tumor might escape exposure to systemic therapies, thereby resulting in recurrent disease (Pardridge 2005, 2012).

Immune Microenvironment with GBM

A bulk of GBM tissue consists of infiltrated microglia and macrophages, but their response to tumor cells is often suppressed. Immune suppressive factors such as interleukin-10 (IL-10), IL6, transforming growth factor-beta (TGF-β), prostaglandin E2 (PGE2) suppress immune response against GBM cells, promote transformation of dendritic cells (DCs) into a regulatory phenotype and promote the activation of FOXP3+ regulatory T cells (Tregs). Hypoxia and subsequent expression of HIF-1α and VEGF production have also been reported to cause Treg activation and immune suppression. Macrophages isolated from GBMs tend to polarize towards M1 (pro-inflammatory) and M2 (anti-inflammatory) phenotypes. Tumor-activated macrophages (TAMs) which are similar to M2 macrophages in function and cell surface marker presentation have been shown to play a vital role within the GBM microenvironment by promoting invasion and growth of tumor cells. TAMs have been reported to cause matrix degradation and enhance survival of GBM cells and GSCs. Hypoxic microenvironments within GBM have been reported to stimulate activation of microglia into TAMs (Jackson et al. 2017; Nduom et al. 2015; Razavi et al. 2016).

Activation of naïve T cells within the GBM microenvironment not only requires contact with major histocompatibility complexes (MHCs) on antigen-presenting cells (APCs) but also activation of co-stimulatory receptors (Driessens et al. 2009; Sharpe and Abbas 2006). Based on significant findings in the past decade that shined the spotlight on the role of immune cells and immune-related mechanisms within tumor microenvironment that effect tumor evasion and clearance, a lot of efforts are currently being made to exploit these mechanisms to deliver therapeutics which could suppress inhibitory signals and allow immune cells to identify and clear tumor cells (Rosenberg and Restifo 2015; Wang et al. 2014; Yang 2015). Inhibiting immune checkpoints is currently the most commonly exploited mechanism for therapeutic purposes. Immune checkpoints prevent naïve T cells from causing autoimmune responses during infections by producing inhibitory signals. Programmed death-1 (PD-1) is an immune checkpoint co-stimulatory receptor expressed on the cell surface of T cells. Normal cells of the body express ligands for PD-1, namely PDL-L1 and PDL-L2, to prevent activation of naïve T cells and subsequent cytotoxicity. GBM cells exploit this mechanism and overexpress PDL-L1 on their cell surface to evade immune recognition and attack. Monoclonal antibodies that inhibit PD-1-PDL-L1 binding have shown to boost immune response against tumor cells and enhance survival in pre-clinical studies (Iwai et al. 2017). Promising results in clinical trials resulted in the approval of immune checkpoint inhibitors (ICIs) such as Nivolumab (approved for NSCLC, melanoma and renal cell carcinoma), Pembrolizumab (approved for melanoma, lung cancer and head and neck cancer), and Azetolizumab (approved for Urothelial and lung cancers) for the treatment of cancer patients (Alsaab et al. 2017; Hamanishi et al. 2016; Kang et al. 2016).

Cytotoxic T cell antigen-4 (CTLA4) is a co-stimulatory receptor, similar to PD-1, which negatively regulates T-cell activation. Inhibiting the binding of CTLA4 to its ligand had been shown to cause increased immune response against tumors. Monoclonal antibodies that inhibit CTLA4 binding such as Ipilimumab were approved for use in the clinic for melanoma patients after they showed promising results in preclinical and clinical trials (Alsaab et al. 2017; Hamanishi et al. 2016; Kang et al. 2016). The delivery of ICIs to GBMs is hindered, however, by the presence of BBB and the ineffective perfusion within tumor tissue due to angiogenic blood vessels (Hodges et al. 2016; Lyon et al. 2017; Rolle et al. 2010).

A recent phase-I study showed the potential for using interleukin-13 receptor alpha 2 (IL13RA2) directed chimeric antigen receptor (CAR) T cells against recurrent GBM. The IL13RA2 redirected CAR T cells in this study were delivered into the resection cavity of a 50-year-old patient using a neurosurgical technique called convection-enhanced delivery (CED) (Brown et al. 2016). CED provides the opportunity to bypass the BBB and enhance the efficacy of drugs by delivering them directly into the bed of the tumor (Bobo et al. 1994; Pardridge 2005; Lidar et al. 2004). Even though results from the study were promising, more efforts are needed to enhance the efficacy of this strategy.

Failure of Conventional Therapies Against GBM and Trends for the Future

Recurrent disease is a major contributor to GBM patient mortality and is almost an eventual outcome for newly diagnosed patients whose tumors regress following adjuvant chemotherapy using temozolomide. Invasive GBM cells residing outside the area of resection that escape surgical debulking, therapy-resistant GBM cells that either attain stem cell-like characteristics through de-differentiation or retain their stem cell state after tumor initiation (COIs), insufficient diffusion due to the BBB, and resulting accumulation of ineffective concentrations of systemic therapies within GBM tissue are thought to be major reasons for therapeutic failure and disease relapse (Chaichana 2014; Eyupoglu et al. 2013; Lathia et al. 2015; Pardridge 2005; Persidsky et al. 2006; Wolburg and Lippoldt 2002; Wolburg et al. 2012).

In recent years, a lot of attention has been placed on stymieing angiogenic processes using anti-angiogenic therapies such as Bevacizumab (which binds to and inhibits VEGF function) and to target GBM-specific cell surface receptors, often overexpressed by tumors such as the mutated EGFRvIII (Gilbert et al. 2014). These therapies have proven to be ineffective in causing a long-term effect on the progression of tumors and have not dramatically increased the median survival rate. Tumor heterogeneity, barriers to effective drug delivery, and additional factors involved in the promotion and maintenance of angiogenic vessels are thought to have caused the failure of these therapies in the clinic. Latest strategies to modulate the immune activity within GBM microenvironment, redirecting T cells using chimeric antigen receptors (CARs) towards GBM cells and oncolytic viruses have shown promise in

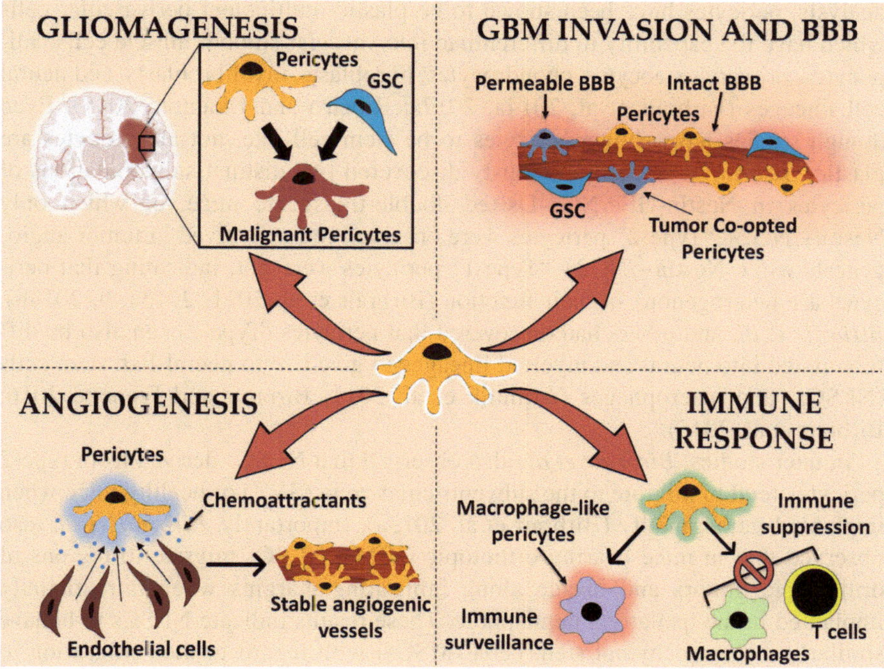

Fig. 2.1 Diagram showing the multifaceted role of pericytes in various critical events of GBM initiation, establishment, maintenance, and progression. *GSC* glioblastoma stem cells, *BBB* blood–brain barrier

preclinical investigations and phase-I clinical trials, but further research is needed to increase their effectiveness in the future. Additionally, preclinical investigations evaluating the efficacy of targeting GSCs, strategies to transiently enhance BBB permeability, therapeutic stem cells-based drug delivery and targeted molecular irradiation are currently underway to deliver systemic therapies more effectively to GBM tissue (Sattiraju et al. 2017b, c) (Fig. 2.1).

Multifaceted Role of Pericytes Within GBM

Stemness and Tumor Initiation

Pericytes were previously thought to mainly play a role in supporting vascular architectures within the brain as part of the neurovascular unit (NVU) and to regulate blood flow within capillaries, but recent investigations have shed light on their role in tissue homeostasis and disease pathologies (Jackson et al. 2017; Sweeney et al. 2016). Using transgenic mouse models and cell surface receptor expression analysis, pericytes have been shown to be plastic multipotent perivascular cells which have the capability to differentiate into vascular smooth muscle cells, adipocytes, primary osteocytes, chondrocytes, fibroblasts, myofibroblasts, and neural cell lineages (Birbrair et al. 2014a, 2017a; Ribeiro and Okamoto 2015). Even though studies have shown pericytes to be stem cell-like, not all pericytes are plastic. *Birbrair et al.* had previously discovered two distinct subpopulations of pericytes in Nestin-GFP/NG2-DsRed double-transgenic mice, of which only Nestin+/NG2+ "Type-2" pericytes were observed to be involved in tumor angiogenesis while Nestin−/NG2+ "Type-1" pericytes were not, indicating that pericytes are heterogenous in their function (Birbrair et al. 2011, 2013a, b, 2014b). *Birbrair et al.* and others had discovered that pericytes (Type-2) can also be differentiated into neural and myeloid lineages to give rise to neural-like stem cells (NLSCs) and macrophages (Armulik et al. 2011; Birbrair and Frenette 2016; Birbrair et al. 2015).

In later studies, *Birbrair et al.* also observed that NLSCs derived from type-2 pericytes tend to migrate to the subventricular zone (SVZ) of healthy mice when implanted intracranially (Birbrair et al. 2013a). Importantly, *Birbrair et al.* also observed that in mice bearing orthotopic GBMs, NLSCs migrate to regions of orthotropic tumors and invade along infiltrating margins when intracranially implanted in the ipsilateral hemisphere. These results indicate NLSCs to behave similarly to mesenchymal stem cells (MSCs) with regard to their migration to GBM sites when implanted intracranially but unlike MSCs which are co-opted by the tumor cells and transform into tumor-associated fibroblasts (TAFs), which assist in tumor growth and expansion, NLSCs did not differentiate into a TAF-like phenotype when co-cultured with GBM cells. In addition, unlike MSCs, NLSCs did not promote angiogenesis when in contact with GBM cells (Birbrair et al. 2017b). These results further contribute to the hypothesis that pericytes could be closely related to MSCs either as their precursors or as a specialized subpopulation of MSCs within the brain. Further adding fuel to this line of thought are reports showing that pericytes and MSCs share cell surface marker expression such as NG2, CD44, αSMA, PDGFRβ, CD90, CD73, CD105, and Sca-1 (Crisan et al. 2008; Ribeiro and Okamoto 2015). The ability of pericytes to differentiate into neural and myeloid lineages could point to their role of pericytes in responding to injury and neurological diseases (Dore-Duffy et al. 2000; ElAli et al. 2014).

In GBM, GSCs have been reported to transdifferentiate into tumor pericytes, which form the majority of pericytes found within tumor tissue and assist in GBM cell proliferation and GSC self-renewal (Caspani et al. 2014; Cheng et al. 2013; Jackson et al. 2017). Neural stem cells (NSCs) have previously been shown to harbor the capacity to transdifferentiate into pericytes in normal brain tissue, indicating that GSCs could exploit such mechanisms to give rise to perivascular niche components (Cheng et al. 2013; Goldberg and Hirschi 2009). Due to the tendency of GSCs to localize near vasculature and due to their ability to give rise to malignant multipotent pericytes, researchers have suggested the possibility for such GSC-derived

CD133$^+$ "malignant pericytes" to drive tumor progression. *Appaix et al.* suggested an alternative "Cancer Pericytes" model for GBM initiation and progression where they hypothesized malignant CD133$^+$ pericytes to act as COIs (Appaix et al. 2014). According to this hypothesis, existing pericytes behaving as MSCs could attain a neural stem cell-like phenotype by detaching from basement membrane, thereby forming a cancer stem-cell pool. These malignant pericyte-derived CSCs could then proliferate extensively, giving rise to a tumor mass which results in a hypoxic microenvironment. Subsequently, endothelial cells recruited through the CXCR4/CXCL12 pathway could initiate angiogenesis. Malignant pericyte-derived CSCs could then differentiate into an aggressive mesenchymal phenotype, transdifferentiate into other cell types and drive tumor heterogeneity within the tumor mass or migrate to co-opted blood vessels to initiate another cycle of tumor formation. The authors speculated that such malignant pericytes could gain and lose CD133 expression during different stages of tumor formation and could thus explain the existence of CD133$^-$ stem cells within GBMs.

Additionally, *Zhang et al.* have reported that overexpression of cytoplasmic GT198 (a DNA repair gene that activates VEGF) within pericytes gives rise to tumors. The authors suggested that malignant pericytes could be derived from GT198 expressing GSCs or from normal pericytes that undergo somatic mutations upon microenvironmental stimuli. GT198$^+$ malignant pericytes could also be resistant to radiotherapy and cause the failure of anti-VEGF therapies (Zhang et al. 2017).

Endothelial cells on co-opted blood vessels within the brain are thought to stimulate the migration of GSCs towards them through SDF-1/CXCR4 pathway and later induce their transdifferentiation into tumor pericytes by secreting TGF-β (Cheng et al. 2013). This has been suggested as a mechanism to allow for the proliferation of endothelial cells, as tumor pericytes present within the perivascular niche secrete VEGF and other paracrine factors. In addition, enhanced pericyte coverage of co-opted and angiogenic blood vessels is thought to render resistance to anti-angiogenic therapies (Gabriele Bergers and Hanahan 2008; Ribeiro and Okamoto 2015). The ability of GSCs to give rise to tumor pericytes would also indicate their independence from relying on perivascular cells and their progenitors within the peri-tumoral region for engineering a suitable microenvironment.

Caspani et al. have suggested a "Dual Cell of Origin" hypothesis where pericytes drive tumor diversification upon direct contact with GSCs. In their study, the authors reported that pericytes attain a stem cell-like state upon transfer of cytoplasm from GSCs giving rise to GSC–pericyte cell fusions. The authors reported the existence of GBM cells that expressed labels for both pericytes and GBM cells within orthotopically implanted xenografts, suggesting that aneuploid cells derived from multipolar division of these GSC–pericyte cell fusions could drive GBM diversification (Caspani et al. 2014). The findings mentioned in the above section indicate that pericytes within NVU are plastic and play a critical role in GBM initiation and progression.

Pericytes Are Involved in GBM Invasion

Pericytes are thought to play a protective role against tumor invasion by acting as physical barriers but poor and disorganized pericyte coverage is often observed within the tumor microenvironments which allows tumor to spread (Xian et al. 2006). Invading GBM cells tend to incorporate existing blood vessels into tumor tissue by a process termed "Vascular Co-option." These co-opted blood vessels undergo necrosis upon angiopoietin-2 secretion by GBM cells, repeating events of hypoxia and subsequent invasion into surrounding areas of normal brain parenchyma (Liebelt et al. 2016; Reiss et al. 2005). Alternatively, edema caused due to incomplete coverage of angiogenic blood vessels by pericytes within tumor tissue has also been hypothesized to enhance tumor invasion by increasing intratumoral fluid pressure. This increased intratumoral fluid pressure is further thought to suppress surrounding blood circulation, resulting in hypoxic microenvironments, thereby stimulating the invasion of GBM cells into surrounding normal brain parenchyma (Cooke et al. 2012).

Cytoplasmic extensions of invading GBM cells around pericytes that they encounter, termed "Flectopodia," have been suggested to facilitate the co-option of existing blood vessels by trafficking GBM cell cytoplasm into the cellular cortex of pericytes. GTPase Cdc42 was shown to play a critical role in the formation of such cytoplasmic extensions and pericyte activation (Caspani et al. 2014).

Poor pericyte coverage allows spreading of GBM, as incomplete coverage of angiogenic vessels allows vascular invasion by GBM cells. In addition, poor pericyte coverage also enhances the metastatic potential of other solid tumors that tend to migrate to the brain. Brain metastatic lung and melanoma cells that extravasate through capillaries were shown to survive and proliferate for long periods of time only when in contact with the abluminal endothelial cells of capillaries in a pericyte-like position. This mechanism where tumor cells position themselves similar to pericytes when in contact with the abluminal endothelial cells has been extensively studied in melanomas and is termed as pericyte mimicry or angiotropism (Bentolila et al. 2016; Lugassy et al. 2014; Scott et al. 2015). Pericyte mimicry not only allows for the survival and proliferation of tumor cells but has also been reported to be exploited by tumor cells for extravascular migratory metastasis. This mechanism allows tumor cells to spread to local and distant sites by avoiding vascular invasion. GBM cells have been well documented in in vivo and in vitro studies to invade surrounding normal brain parenchyma along abluminal side of capillaries through the mechanism of pericyte mimicry (Scott et al. 2015).

As mentioned in the previous section, the ability of GSCs to attain a pericyte-like phenotype and their similarities to pericytes within the GBM microenvironment, with regard to cell surface marker expression, could be a result of pericyte mimicry of tumor cells and could support the hypothesis that pericyte-like cells within the GBM microenvironment could also act as initiating cells.

Pericytes and the BBB

Pericytes have been shown to play a vital role in regulating BBB permeability by controlling the expression and alignment of tight and adherens junction proteins along with the transcytosis of molecules across the BBB (Sweeney et al. 2016). The importance of pericytes for the structural stability of CNS vessels and for maintaining the BBB was observed in PDGFβ knockout mice where pericyte depletion resulted in enhanced CNS vascular permeability due to BBB disruption (Armulik et al. 2011; Bell et al. 2010; Daneman et al. 2010; Sweeney et al. 2016). PDGF-BB(ligand)-PDGFRβ signalling pathway is critical for pericyte–endothelial cell interactions to maintain BBB stability and to regulate extravasation across the barrier into brain parenchyma. Transgenic mice with both PDGF and PDGFRβ null mutations have shown embryonic lethality due to the development of microvascular instability and hyperplasia, aneurysms, and microhemorrhages (Sweeney et al. 2016). In studies where PDGF–BB–PDGFRβ signalling pathway was partially disrupted, age-dependent BBB disruption was observed (Armulik et al. 2011; Bell et al. 2010; Daneman et al. 2010; Sweeney et al. 2016). Similarly, *Abramsson et al.* showed that the lack of PDGFRβ expression on pericytes resulted in their poor vascular coverage and enlargement of blood vessel diameter, resulting in increased leakiness (Abramsson et al. 2002). In addition, TGFβ, Ang1, and Notch signaling have been reported to play an important role in maintaining the integrity of the BBB (Bruna et al. 2007; Cheng et al. 2013; Liebner and Plate 2010; Reis and Liebner 2013; Sweeney et al. 2016). TGFβ-TGFβR2 signaling promotes pericyte maturation, proliferation, and attachment to endothelial cells. Aberrant signaling of downstream effectors of TGFβ-TGFβR2 pathway such as Smad1, Smad2, and Smad4 results in vascular destabilization and brain hemorrhages (Goumans and Mummery 2000; Li et al. 2011; Maddaluno et al. 2013; Sweeney et al. 2016). Additionally, forkhead transcription factor (Foxf2) which affects TGFβ signaling also plays a role in maintaining BBB integrity as Foxf2 knockout transgenic mice show BBB disruption, hemorrhages, perivascular edema, increase in luminal endothelial caveolae, and thinning of basal lamina of capillaries (Reyahi et al. 2015).

Pericytes have been observed to attach loosely and improperly cover angiogenic vessels within GBMs, yet proper pericyte coverage has been shown to result in stabilization of tumor blood vessels, thereby accelerating GBM cell proliferation and invasion. High TGFβ-Smad activity was shown to result in highly aggressive and proliferative GBMs in patients which conferred a poor prognosis. TGFβ-Smad pathway has also been shown to activate PDGFB gene expression in primary GBM cells in vitro (Bruna et al. 2007). Majority of pericytes found in GBMs are thought to arise from GSCs that transdifferentiate upon migration to endothelial cells through SDF-1-CXCR4 signaling pathway. TGFβ has been shown to play a vital role in inducing this transdifferentiation of GSCs into pericyte-like cells which help promote angiogenesis, stabilize tumor vessels, and contribute to GBM growth (Cheng et al. 2013).

Notch3, expressed by brain pericytes, has been shown to play an important role in interactions with endothelial cells and in maintaining the integrity of BBB. Abnormal TGFβ-Notch signaling has been reported to cause cerebral cavernous malformation. Transgenic mice with dysfunctional Notch signaling caused by knockout of RBP-Jκ transcriptional factor, showed brain hemorrhages (Li et al. 2011; Maddaluno et al. 2013; Sweeney et al. 2016). Like TGFβ, Notch1 signaling has also been shown to induce transdifferentiation of GSCs into pericyte-like cells in vitro which promote angiogenesis (Guichet et al. 2015). Angiopoietin-1 (Ang1) expressed by pericytes, which binds to Tie2 receptor tyrosine kinase on endothelial cells, maintains BBB integrity. Like TGFβ and Notch1, knockout of Ang1 in transgenic mice resulted in BBB disruption and promotion of angiogenesis (Suri et al. 1996). The major facilitator superfamily domain-containing 2a (MFSD2A), a BBB transporter whose expression is thought to depend on presence of pericytes, has also been reported to play a role in the formation and maintenance of BBB integrity (Ben-Zvi et al. 2014).

Pericytes interact with astrocytes through apolipoprotein E (APOE4)–LRP1 interaction, resulting in the activation of MMP-9 activity by signaling through cyclophilin A (CypA)-NFκB pathway, which promotes inflammation. Increased APOE4-LRP1 signaling and resulting increase in the activity of MMP-9 causes BBB disruption due to degradation of endothelial tight junctions and basement membrane (Bell et al. 2012; Sweeney et al. 2016). Pericytes also regulate perfusion through brain capillaries and regulate BBB permeability through their contractile nature which is facilitated by synthesis of vimentin, actin and myosin microfilaments, tropomyosin and desmin. As part of the NVU, pericytes are also in contact with interneurons and receive communications from the nervous system as the average distance between a neuron and a brain capillary is 8–23 μm. Conditioned media from cultured human brain pericytes has shown the presence of neurotrophic factors and pericytes have been shown to regulate capillary diameter and cerebral blood flow upon signaling from neurons (Hawkes et al. 2011; Lovick et al. 1999). Studies have shown that norepinephrine leads to pericyte contraction (reduction of capillary diameter) while GABA, dopamine, glutamate, and adenosine cause pericyte relaxation (increase in capillary diameter) (Sweeney et al. 2016).

The BBB provides a formidable challenge for delivering drugs into the CNS, especially to brain tumors. In addition to intact BBB in areas around brain tumors, leaky angiogenic vessels which cause edema in certain areas of tumors result in the ineffective and non-uniform delivery of systemically delivered therapies (Sattiraju et al. 2017a, b). Ineffective drug delivery is thought to be one of the major reasons for GBM relapse, as cells that are not exposed to effective concentrations of systemic therapies reduce therapeutic efficacy. Therefore, efforts are being made to either transiently disrupt the BBB or to target cells involved in regulating BBB permeability, in order to enhance the extravasation of systemic therapies into GBM. As pericyte recruitment and stabilization of blood vessels within brain tumors have been shown to regulate vascular permeability, strategies to disrupt pericyte recruitment by angiogenic endothelial cells are currently being investigated.

In a recent study by *Behling et al.* by targeting monomeric vascular endothelial cadherin, which is expressed on angiogenic vessels and endothelial progenitor cells (EPCs), using a monoclonal antibody E4G10 that was labeled to α particles, the authors reported enhanced survival in Ntva transgenic mice bearing GBMs. As extensive proliferation of pericytes is usually observed in this GBM model, their higher density was suspected to protect endothelial cells from radiotherapy. In addition to mitigating tumor growth, the authors also observed normalization of the morphology of angiogenic vessels, reduction of edema, and a decrease in pericyte coverage of these vessels. In addition, depletion of regulatory T cells (Tregs) and EPCs was also observed (Behling et al. 2016). In another recent study, *Sattiraju et al.* showed enhancement in BBB permeability in brains of mice bearing orthotopic GBMs by targeting integrin alpha-V beta-3 ($\alpha_v\beta_3$) using an $\alpha_v\beta_3$ targeted antagonist conjugated to partially polymerized liposomes that was labeled to α particles. Apart from observing enhanced BBB permeability, the authors also reported tumor cytotoxicity evidenced by nuclear accumulations of γH2Ax double-strand DNA break repair protein within tumor mass. Overexpression of $\alpha_v\beta_3$ by GBM cells, especially at invasive ends and the presence of proliferating malignant cells in perivascular regions could explain this effect. The authors in this study also observed enhanced BBB permeability in peritumoral and normal regions of the brain surrounding tumors, indicating that enhanced vascular permeability in these distal regions might not have been caused due to direct α particle-induced cellular effects (Sattiraju et al. 2017c).

Studies by both *Behling et al.* and *Sattiraju et al.* show the feasibility of targeting vascular and perivascular components within the NVU using short-ranged, targeted molecular irradiation to either enhance vascular perfusion or BBB permeability, to better deliver systemic drugs to GBMs and to reduce radio-resistance conferred by perivascular cells within GBM microenvironment. But, as evidenced in previous studies where pericyte ablation caused enhanced invasion and metastasis, further long-term evaluations to assess tumor resistance and remission in abovementioned strategies are necessary (Bentolila et al. 2016; Lugassy et al. 2014; Scott et al. 2015). *Xiong et al.* reported an alternative strategy of enhancing BBB permeability by remotely activating intracranially implanted genetically engineered MSCs using high-frequency focused ultrasound (HIFU) to secrete TNF-α. This study highlights the ability to locally deliver factors which can transiently influence vascular and perivascular cells that regulate BBB integrity and perfusion within normal brain parenchyma and GBMs, thus enhancing the delivery of systemic drugs to GBMs and CNS (Sattiraju et al. 2017b; Xiong et al. 2015).

Pericytes Drive Angiogenesis

Angiogenesis is a crucial multi-stage process resulting in the transformation of GBM into an aggressive disease (Jain et al. 2007). Highly proliferating cells within GBM tissue drive up nutrient and oxygen demand and the resulting hypoxic microenvironment stimulates the production of new and often-disorganized angiogenic blood vessels. Pericytes play a major role in this process by forming a scaffold for newly proliferating endothelial cells to form blood vessels and secrete factors that stabilize these newly formed angiogenic vessels (Mancuso et al. 2006; Ribeiro and Okamoto 2015). The process of angiogenesis is tightly regulated and requires direct contact and crosstalk between endothelial cells and pericytes. Pericytes are involved in secreting factors that stimulate endothelial tip sprouting from existing blood vessels in their vicinity, recruit endothelial cells, and aid in their proliferation. Upon endothelial recruitment and formation of a vessel structure, pericytes within GBM microenvironment wrap endothelial cells which results in vessel stabilization and maturation (Caspani et al. 2014).

Endothelial cells are thought to produce PDGFβ which recruits PDGFRβ expressing pericytes, which in turn secrete VEGF and Ang-1 to stabilize angiogenic blood vessels (Hellstrom et al. 1999). Studies examining the effect of a pan tyrosine kinase inhibitor (SU6668), which preferentially targets PDGFRβ in RIP1-Tag2 transgenic mice that generate pancreatic islet carcinomas showed that ablating pericytes from tumor blood vessels caused vascular regression (Bergers et al. 2003). GSC-derived pericyte-like cells within the GBM microenvironment have also been reported to express PDGFRβ and aid in vessel stabilization and maturation (Cheng et al. 2013). Inhibiting PDGF–BB–PDGFRβ signaling resulted in regression of tumor vessels, indicating that it affects the survival of endothelial cells, GSC-derived pericyte-like cells and co-opted pericytes (Hellstrom et al. 2001). Overexpression of PDGFβ by endothelial cells has been reported to increase pericyte coverage and accelerate tumor growth (Furuhashi et al. 2004).

TGF-β, expressed by pericytes in a latent form, also plays a critical role in angiogenesis and GBM progression. TGF-β-TGFβR2 signaling has been reported to result in the stabilization of angiogenic vessels, and high expression of TGF-β-Smad has been reported to result in a poor prognosis for GBM patients. TGF-β-Smad pathway has been shown to induce expression of PDGF-B in GBM cells and the transdifferentiation of GSCs into pericyte-like cells, which aid in angiogenic vessels stabilization and maturation (Goumans and Mummery 2000; Maddaluno et al. 2013). TGF-β signaling alone has been reported to be able to induce genetic and phenotypic changes, often seen in altered vasculature within GBMs.

Notch1-Dll4 signaling has been reported to be critical in regulating sprouting angiogenesis. Inactivation of Notch1 or Dll4 genes have shown to cause increased endothelial tip sprouting and tip-cell numbers, indicating that Notch1-Dll4 signaling at endothelial sprout restricts tip sprouting towards VEGF-A gradients, thereby ensuring proper sprouting and branching patterns (Gerhardt et al. 2003). Pericytes have been reported to enhance the survival of endothelial cells and promote angiogenesis by secreting VEGF-A, which binds to VEGFR2 expressed on endothelial cells. VEGFR2 stimulates the upregulation of survival genes such as Bcl-2, survivin and X-linked inhibitor of apoptosis protein (XIAP) in endothelial cells (Franco et al. 2011). VEFG-A–VEGFR2 autocrine signaling within endothelial cells also promotes their survival. Vitronectin secreted by pericytes also causes an increase of VEGF-A expression in endothelial cells through integrin α_v-NFκB signaling, which in turn causes intracellular stimulation of VEGFR2 to promote endothelial cell survival (Franco et al. 2011; Sweeney et al. 2016). Inhibiting angiogenesis by interfering with VEGF-A–VEGFR2 signaling was thought to provide significant benefit to GBM patients as it would result in the collapse of existing angiogenic vessels and prevent further tumor vascularization and growth. Bevacizumab (Avastin), a monoclonal antibody against VEGF-A showed very promising results in preclinical studies and was tested in two phase-III clinical trials (RTOG and AVAGlio) as a stand-alone and combinational therapeutic for patients with recurrent GBM (Friedman et al. 2009; Gilbert et al. 2014). Although patients initially showed decreased tumor growth, their tumors eventually grew resistant to the drug, resulting in a progression-free survival of about 10.6 months in Phase-III clinical trials with no added benefit to overall survival when compared to placebo arm (Friedman et al. 2009; Hamza et al. 2014).

Ephrin receptor EphB4, which controls vascular morphogenesis within the NVU during developmental angiogenesis, and its ligand ephrin-B2, which is expressed on brain pericytes and endothelial cells, play a vital role in pericyte–endothelial cell interactions. Ephrin-B2-EphB4 signaling has been reported to regulate pericyte migration and interaction with maturing blood vessels and could be involved in the vascular remodeling within the GBM microenvironment (Augustin and Reiss 2003). Vascular cell adhesion molecule-1 (VCAM-1), N-cadherin and integrin $\alpha_4\beta_1$ play critical roles in regulating pericyte migration, vessel maturation, and survival of endothelial and mural cells during angiogenesis (Gerhardt et al. 2000).

Overexpression of endosialin (CD248) has also been linked to pro-angiogenic role of pericytes within the GBM microenvironment (Brady et al. 2004). Angiogenic vessels within GBMs were reported to be composed of two layers of pericytes, an abluminal layer of proliferating cells and an adluminal layer of cells surrounded by basal lamina. By restricting the recruitment of pericytes and their signaling with endothelial cells during angiogenesis, newly formed GBM blood vessels could be disrupted causing reduced GBM invasion and growth.

In a study by *Svensson et al.* using a RGS5-GFP transgenic mouse model, the authors showed that PDGFRβ and neuroregulin-2 (NG2)-expressing pericytes were activated and recruited to blood vessels within orthotopically implanted GBMs from peritumoral and distal regions within ipsilateral and contralateral hemispheres, including from the rostral SVZ. Activation and recruitment of pericytes from distant regions of the brain towards tumor tissue has been attributed to parenchymal diffusion of paracrine factors such as hypoxia inducible factor-1α (HIF-1α) through cerebrospinal fluid or cerebral edema. These tumor trophic pericytes showed CD13 MSC cell surface marker expression in areas surrounding tumor tissue, but did not show CD13 expression within tumor tissues, indicating a phenotypic shift as pericytes enter paracrine signaling networks within the GBM microenvironment. The results from this study indicate that pericytes migrate and integrate into the vasculature within GBMs, stabilize angiogenic blood vessels, and promote further angiogenesis (Svensson et al. 2015). In another study by *Huang* et al., NG2 was identified to be critical in integrin β1-dependent pericyte–endothelial cell interactions and ablation of NG2 resulted in two-fold reduction in vessel ensheathment by pericytes. NG2 ablation was also reported to cause reduced collagen IV deposition due to loss of collagen VI anchorage. Impaired vessel maturation and stabilization within GBM tissues of mice lacking NG2 expression resulted in decreased tumor progression, highlighting the importance of NG2 for successful angiogenesis (Huang et al. 2010).

Pericytes Contribute to Immune Microenvironment

GBM cell survival, proliferation, and invasion into peritumoral regions are thought to be facilitated by their evasion of immune surveillance by using multiple immunosuppressive mechanisms. Pericytes have been shown to regulate immune cell activity by secreting chemokines and other factors in addition to differentiating into macrophages, thus playing a critical role in immune surveillance and tumor clearance within the GBM microenvironment (Valdor et al. 2017). Brain pericytes have been reported to exhibit phagocytic properties and express macrophages cell surface markers such as CD11b, CD68, and MHC class II. Stimulation by IL-1β has been reported to result in the upregulation of iNOS and COX-2 within porcine brain pericytes (Balabanov et al. 1996). In addition, Pieper et al. reported that stimulation by IFN-γ or TNF-α resulted in an antigen-presenting activity within brain pericytes (Pieper et al. 2014). Pieper et al. also showed that stimulation by LPS, TNF-α, and IL-1β resulted in chemotactic recruitment and transmigration of neutrophils into the brain parenchyma (Pieper et al. 2013).

During GBM establishment and expansion, pericytes have also been reported to be co-opted by tumor cells through direct cell–cell contact to facilitate immunosuppression and eventual immune evasion of GBM cells (Caspani et al. 2014). Upregulation of PD-L1, CD90, PDGFRβ, CD248, and *Rgs5* which inhibit CD4+ and CD8+ cytotoxic T cell activity was reported in pericytes derived from within-tumor microenvironments (Bose et al. 2013; Ochs et al. 2013; Ribeiro and Okamoto 2015). *Valdor et al.* recently reported that pericytes conditioned in vitro by GBM cells are characterized by high levels of anti-inflammatory cytokines, suppress the function of cytotoxic T cells and reduce antigen presentation within perivascular regions of orthotopically implanted GBMs. GBM conditioned pericytes (GBM-PCs) showed high levels of IL-10 and TGFβ anti-inflammatory cytokine expression (100–400 pg/mL) and low levels of TNFα expression (45 pg/mL) in vitro when compared to control native pericytes, indicating an immunosuppressive phenotype. GBM-PCs also showed upregulation of *Il4ra* (encoding interleukin-4 receptor alpha), *Il1rn* (encoding interleukin-1 receptor inhibitor), and increased expression of angiogenic cytokine IL-6. However, expression of CD80 and CD86 co-stimulatory molecules was reported to be reduced (Valdor et al. 2017).

The exact mechanisms through which pericytes influence immune activity within the tumor microenvironment are not yet clearly understood and further research is required to gain a pristine understanding of the multifaceted role that pericytes play

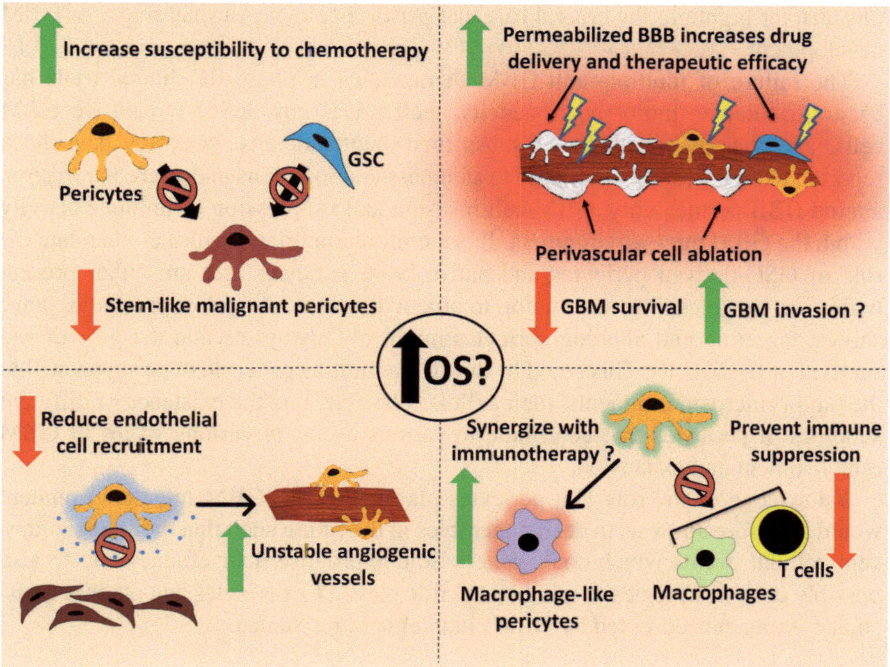

Fig. 2.2 Diagram showing potential therapeutic interventions which could limit the contribution of pericytes to GBM overall survival and progression. *OS* overall survival, *GSC* glioblastoma stem cells, *GBM glioblastoma*, *BBB* blood–brain barrier

within tumor microenvironments and the way in which their role changes during tumor progression and in response to therapy (Fig. 2.2).

Conclusions: Opportunities for Therapeutic Intervention

The median overall survival of GBM patients has been improved to ~16 months using surgical resection followed by adjuvant temozolomide, but recurrent disease remains a major contributor to patient mortality (Roh et al. 2017). Novel therapeutic strategies such as immune checkpoint inhibitors, alternating electric fields, CAR T cells, stem cell-based drug and oncolytic viral delivery and strategies to enhance delivery of systemic therapies show the potential to increase patient survival and to reduce GBM recurrence in the future. As highlighted in this chapter, microenvironmental components play a vital role in the survival and proliferation of GBM cells. It is therefore critical to appreciate the complex relationship between GBM and their microenvironment and to design therapeutic strategies that would not just target one component of their microenvironment, as it would only serve to partially impede GBM survival and progression. The studies and their finding mentioned in this chapter highlight the critical role that pericytes play at various stages of GBM development.

The failure of multiple anti-GBM therapeutics in phase-III clinical trials has been attributed to tumor heterogeneity, ineffective drug delivery, invasive GBM cells, and therapy-resistant GSCs. As this chapter details, pericytes have been reported to significantly contribute to stemness, angiogenesis and altered perfusion, altered BBB permeability, GBM cell invasion and suppression of immune activity within the GBM microenvironment. It is therefore important to further elucidate the role of GSC-derived pericytes and native brain pericytes that are either present within or recruited into the tumor microenvironment. As previous studies have shown, targeting and ablating pericytes might not always stymie the growth and survival of GBM cells. This could be due to the multifaceted role of pericytes within the tumor microenvironments, their cellular plasticity and the existence of different subtypes of pericytes that contribute to various events at various stages of GBM establishment and expansion.

Knowledge of the role that pericytes play within the GBM microenvironment would therefore allow us to design therapies in the future such that they can circumvent cellular events which could otherwise compromise their efficacy and to also possibly design strategies to mitigate the role played by pericytes and other perivascular components co-opted by GBMs in treatment resistance.

References

Abbott, N. J. (2002). Astrocyte-endothelial interactions and blood-brain barrier permeability. *Journal of Anatomy, 200*(6), 629–638.

Abbott, N. J., Patabendige, A. A. K., Dolman, D. E. M., Yusof, S. R., & Begley, D. J. (2010). Structure and function of the blood–brain barrier. *Neurobiology of Disease, 37*(1), 13–25. https://doi.org/10.1016/j.nbd.2009.07.030.

Abramsson, A., Berlin, Ö., Papayan, H., Paulin, D., Shani, M., & Betsholtz, C. (2002). Analysis of mural cell recruitment to tumor vessels. *Circulation, 105*(1), 112.

Agarwal, S., Manchanda, P., Vogelbaum, M. A., Ohlfest, J. R., & Elmquist, W. F. (2013). Function of the blood-brain barrier and restriction of drug delivery to invasive glioma cells: Findings in an orthotopic rat xenograft model of glioma. *Drug Metabolism and Disposition, 41*(1), 33–39. https://doi.org/10.1124/dmd.112.048322.

Alcantara Llaguno, S. R., Wang, Z., Sun, D., Chen, J., Xu, J., Kim, E., et al. (2015). Adult lineage restricted CNS progenitors specify distinct glioblastoma subtypes. *Cancer Cell, 28*(4), 429–440. https://doi.org/10.1016/j.ccell.2015.09.007.

Alsaab, H. O., Sau, S., Alzhrani, R., Tatiparti, K., Bhise, K., Kashaw, S. K., & Iyer, A. K. (2017). PD-1 and PD-L1 checkpoint signaling inhibition for cancer immunotherapy: mechanism, combinations, and clinical outcome. *Frontiers in Pharmacology, 8*, 561. https://doi.org/10.3389/fphar.2017.00561.

Appaix, F., Nissou, M. F., van der Sanden, B., Dreyfus, M., Berger, F., Issartel, J. P., & Wion, D. (2014). Brain mesenchymal stem cells: The other stem cells of the brain? *World Journal of Stem Cells, 6*(2), 134–143. https://doi.org/10.4252/wjsc.v6.i2.134.

Armulik, A., Genové, G., & Betsholtz, C. (2011). Pericytes: Developmental, physiological, and pathological perspectives, problems, and promises. *Developmental Cell, 21*(2), 193–215. https://doi.org/10.1016/j.devcel.2011.07.001.

Assanah, M., Lochhead, R., Ogden, A., Bruce, J., Goldman, J., & Canoll, P. (2006). Glial progenitors in adult white matter are driven to form malignant gliomas by platelet-derived growth factor-expressing retroviruses. *The Journal of Neuroscience, 26*(25), 6781.

Augustin, H. G., & Reiss, Y. (2003). EphB receptors and ephrinB ligands: Regulators of vascular assembly and homeostasis. *Cell and Tissue Research, 314*(1), 25–31. https://doi.org/10.1007/s00441-003-0770-9.

Balabanov, R., Washington, R., Wagnerova, J., & Dore-Duffy, P. (1996). CNS microvascular pericytes express macrophage-like function, cell surface integrin alpha M, and macrophage marker ED-2. *Microvascular Research, 52*(2), 127–142. https://doi.org/10.1006/mvre.1996.0049.

Bao, S., Wu, Q., McLendon, R. E., Hao, Y., Shi, Q., Hjelmeland, A. B., et al. (2006). Glioma stem cells promote radioresistance by preferential activation of the DNA damage response. *Nature, 444*(7120), 756–760. https://doi.org/10.1038/nature05236.

Baskar, R., Lee, K. A., Yeo, R., & Yeoh, K.-W. (2012). Cancer and radiation therapy: Current advances and future directions. *International Journal of Medical Sciences, 9*(3), 193–199. https://doi.org/10.7150/ijms.3635.

Behling, K., Maguire, W. F., Di Gialleonardo, V., Heeb, L. E. M., Hassan, I. F., Veach, D. R., et al. (2016). Remodeling the vascular microenvironment of glioblastoma with α-particles. *Journal of Nuclear Medicine, 57*(11), 1771–1777. https://doi.org/10.2967/jnumed.116.173559.

Bell, R. D., Winkler, E. A., Sagare, A. P., Singh, I., LaRue, B., Deane, R., & Zlokovic, B. V. (2010). Pericytes control key neurovascular functions and neuronal phenotype in the adult brain and during brain aging. *Neuron, 68*(3), 409–427. https://doi.org/10.1016/j.neuron.2010.09.043.

Bell, R. D., Winkler, E. A., Singh, I., Sagare, A. P., Deane, R., Wu, Z., et al. (2012). Apolipoprotein E controls cerebrovascular integrity via cyclophilin A. *Nature, 485*(7399), 512–516. https://doi.org/10.1038/nature11087.

Ben-Zvi, A., Lacoste, B., Kur, E., Andreone, B. J., Mayshar, Y., Yan, H., & Gu, C. (2014). Mfsd2a is critical for the formation and function of the blood–brain barrier. *Nature, 509*, 507. https://doi.org/10.1038/nature13324. https://www.nature.com/articles/nature13324#supplementary-information.

Bentolila, L. A., Prakash, R., Mihic-Probst, D., Wadehra, M., Kleinman, H. K., Carmichael, T. S., et al. (2016). Imaging of angiotropism/vascular co-option in a murine model of brain melanoma: Implications for melanoma progression along extravascular pathways. *Scientific Reports, 6*, 23834. https://doi.org/10.1038/srep23834. https://www.nature.com/articles/srep23834#supplementary-information.

Bergers, G., & Hanahan, D. (2008). Modes of resistance to anti-angiogenic therapy. *Nature Reviews Cancer, 8*(8), 592–603. https://doi.org/10.1038/nrc2442.

Bergers, G., Song, S., Meyer-Morse, N., Bergsland, E., & Hanahan, D. (2003). Benefits of targeting both pericytes and endothelial cells in the tumor vasculature with kinase inhibitors. *The Journal of Clinical Investigation, 111*(9), 1287–1295. https://doi.org/10.1172/jci17929.

Birbrair, A., Borges, I. D. T., Gilson Sena, I. F., Almeida, G. G., da Silva Meirelles, L., Goncalves, R., et al. (2017a). How plastic are pericytes? *Stem Cells and Development, 26*(14), 1013–1019. https://doi.org/10.1089/scd.2017.0044.

Birbrair, A., & Frenette, P. S. (2016). Niche heterogeneity in the bone marrow. *Annals of the New York Academy of Sciences, 1370*(1), 82–96. https://doi.org/10.1111/nyas.13016.

Birbrair, A., Sattiraju, A., Zhu, D., Zulato, G., Batista, I., Nguyen, V. T., et al. (2017b). Novel peripherally derived neural-like stem cells as therapeutic carriers for treating glioblastomas. *Stem Cells Translational Medicine, 6*(2), 471–481. https://doi.org/10.5966/sctm.2016-0007.

Birbrair, A., Wang, Z.-M., Messi, M. L., Enikolopov, G. N., & Delbono, O. (2011). Nestin-GFP transgene reveals neural precursor cells in adult skeletal muscle. *PLoS One, 6*(2), e16816. https://doi.org/10.1371/journal.pone.0016816.

Birbrair, A., Zhang, T., Wang, Z.-M., Messi, M. L., Enikolopov, G. N., Mintz, A., & Delbono, O. (2013a). Skeletal muscle neural progenitor cells exhibit properties of NG2-glia. *Experimental Cell Research, 319*(1), 45–63. https://doi.org/10.1016/j.yexcr.2012.09.008.

Birbrair, A., Zhang, T., Wang, Z.-M., Messi, M. L., Mintz, A., & Delbono, O. (2013b). Type-1 pericytes participate in fibrous tissue deposition in aged skeletal muscle. *American Journal of Physiology-Cell Physiology, 305*(11), C1098–C1113. https://doi.org/10.1152/ajpcell.00171.2013.

Birbrair, A., Zhang, T., Wang, Z.-M., Messi, M. L., Mintz, A., & Delbono, O. (2014a). Pericytes: Multitasking cells in the regeneration of injured, diseased, and aged skeletal muscle. *Frontiers in Aging Neuroscience, 6*, 245. https://doi.org/10.3389/fnagi.2014.00245.

Birbrair, A., Zhang, T., Wang, Z.-M., Messi, M. L., Olson, J. D., Mintz, A., & Delbono, O. (2014b). Type-2 pericytes participate in normal and tumoral angiogenesis. *American Journal of Physiology-Cell Physiology, 307*(1), C25–C38. https://doi.org/10.1152/ajpcell.00084.2014.

Birbrair, A., Zhang, T., Wang, Z. M., Messi, M. L., Mintz, A., & Delbono, O. (2015). Pericytes at the intersection between tissue regeneration and pathology. *Clinical Science (London, England), 128*(2), 81–93. https://doi.org/10.1042/cs20140278.

Bissell, M. J., & Radisky, D. (2001). Putting tumours in context. *Nature Reviews Cancer, 1*(1), 46–54. https://doi.org/10.1038/35094059.

Bobo, R. H., Laske, D. W., Akbasak, A., Morrison, P. F., Dedrick, R. L., & Oldfield, E. H. (1994). Convection-enhanced delivery of macromolecules in the brain. *Proceedings of the National Academy of Sciences, 91*(6), 2076–2080.

Bose, A., Barik, S., Banerjee, S., Ghosh, T., Mallick, A., Bhattacharyya Majumdar, S., et al. (2013). Tumor-derived vascular pericytes anergize Th cells. *The Journal of Immunology, 191*, 971–981.

Bradshaw, A., Wickremsekera, A., Tan, S. T., Peng, L., Davis, P. F., & Itinteang, T. (2016). Cancer stem cell hierarchy in glioblastoma multiforme. *Frontiers in Surgery, 3*, 21. https://doi.org/10.3389/fsurg.2016.00021.

Brady, J., Neal, J., Sadakar, N., & Gasque, P. (2004). Human endosialin (tumor endothelial marker 1) is abundantly expressed in highly malignant and invasive brain tumors. *Journal of Neuropathology and Experimental Neurology, 63*(12), 1274–1283.

Brat, D. J., Castellano-Sanchez, A. A., Hunter, S. B., Pecot, M., Cohen, C., Hammond, E. H., et al. (2004). Pseudopalisades in glioblastoma are hypoxic, express extracellular matrix proteases, and are formed by an actively migrating cell population. *Cancer Research, 64*(3), 920–927. https://doi.org/10.1158/0008-5472.can-03-2073.

Brat, D. J., & Van Meir, E. G. (2004). Vaso-occlusive and prothrombotic mechanisms associated with tumor hypoxia, necrosis, and accelerated growth in glioblastoma. *Laboratory Investigation, 84*, 397. https://doi.org/10.1038/labinvest.3700070.

Brooks, M. D., Sengupta, R., Snyder, S. C., & Rubin, J. B. (2013). Hitting them where they live: Targeting the glioblastoma perivascular stem cell niche. *Current Pathobiology Reports, 1*(2), 101–110. https://doi.org/10.1007/s40139-013-0012-0.

Brown, C. E., Alizadeh, D., Starr, R., Weng, L., Wagner, J. R., Naranjo, A., et al. (2016). Regression of glioblastoma after chimeric antigen receptor T-cell therapy. *The New England Journal of Medicine, 375*(26), 2561–2569. https://doi.org/10.1056/NEJMoa1610497.

Bruna, A., Darken, R. S., Rojo, F., Ocana, A., Penuelas, S., Arias, A., et al. (2007). High TGFbeta-Smad activity confers poor prognosis in glioma patients and promotes cell proliferation depending on the methylation of the PDGF-B gene. *Cancer Cell, 11*(2), 147–160. https://doi.org/10.1016/j.ccr.2006.11.023.

Cabrera, M. C., Hollingsworth, R. E., & Hurt, E. M. (2015). Cancer stem cell plasticity and tumor hierarchy. *World Journal of Stem Cells, 7*(1), 27–36. https://doi.org/10.4252/wjsc.v7.i1.27.

Calabrese, C., Poppleton, H., Kocak, M., Hogg, T. L., Fuller, C., Hamner, B., et al. (2007). A perivascular niche for brain tumor stem cells. *Cancer Cell, 11*(1), 69–82. https://doi.org/10.1016/j.ccr.2006.11.020.

Caspani, E. M., Crossley, P. H., Redondo-Garcia, C., & Martinez, S. (2014). Glioblastoma: A pathogenic crosstalk between tumor cells and pericytes. *PLoS One, 9*(7), e101402. https://doi.org/10.1371/journal.pone.0101402.

Chaffer, C. L., Marjanovic, N. D., Lee, T., Bell, G., Kleer, C. G., Reinhardt, F., et al. (2013). Poised chromatin at the ZEB1 promoter enables breast cancer cell plasticity and enhances tumorigenicity. *Cell, 154*(1), 61–74. https://doi.org/10.1016/j.cell.2013.06.005.

Chaichana, K. L. (2014). The need to continually redefine the goals of surgery for glioblastoma. *Neuro-Oncology, 16*, 611–612. https://doi.org/10.1093/neuonc/not326.

Chen, J., Ding, Z., Peng, Y., Pan, F., Li, J., Zou, L., et al. (2014). HIF-1α inhibition reverses multidrug resistance in colon cancer cells via downregulation of MDR1/P-glycoprotein. *PLoS One, 9*(6), e98882. https://doi.org/10.1371/journal.pone.0098882.

Chen, J., Li, Y., Yu, T. S., McKay, R. M., Burns, D. K., Kernie, S. G., & Parada, L. F. (2012). A restricted cell population propagates glioblastoma growth after chemotherapy. *Nature, 488*(7412), 522–526. https://doi.org/10.1038/nature11287.

Cheng, L., Huang, Z., Zhou, W., Wu, Q., Donnola, S., Liu, J. K., et al. (2013). Glioblastoma stem cells generate vascular pericytes to support vessel function and tumor growth. *Cell, 153*(1), 139–152. https://doi.org/10.1016/j.cell.2013.02.021.

Chou, C.-W., Wang, C.-C., Wu, C.-P., Lin, Y.-J., Lee, Y.-C., Cheng, Y.-W., & Hsieh, C.-H. (2012). Tumor cycling hypoxia induces chemoresistance in glioblastoma multiforme by upregulating the expression and function of ABCB1. *Neuro-Oncology, 14*(10), 1227–1238. https://doi.org/10.1093/neuonc/nos195.

Cooke, V. G., LeBleu, V. S., Keskin, D., Khan, Z., O'Connell, J. T., Teng, Y., et al. (2012). Pericyte depletion results in hypoxia-associated epithelial-to-mesenchymal transition and metastasis mediated by met signaling pathway. *Cancer Cell, 21*(1), 66–81. https://doi.org/10.1016/j.ccr.2011.11.024.

Crisan, M., Yap, S., Casteilla, L., Chen, C. W., Corselli, M., Park, T. S., et al. (2008). A perivascular origin for mesenchymal stem cells in multiple human organs. *Cell Stem Cell, 3*(3), 301–313. https://doi.org/10.1016/j.stem.2008.07.003.

Dalerba, P., Cho, R. W., & Clarke, M. F. (2007). Cancer stem cells: Models and concepts. *Annual Review of Medicine, 58*, 267–284. https://doi.org/10.1146/annurev.med.58.062105.204854.

Daneman, R., Zhou, L., Kebede, A. A., & Barres, B. A. (2010). Pericytes are required for blood-brain barrier integrity during embryogenesis. *Nature, 468*(7323), 562–566. https://doi.org/10.1038/nature09513.

Das, S., & Marsden, P. A. (2013). Angiogenesis in Glioblastoma. *New England Journal of Medicine, 369*(16), 1561–1563. https://doi.org/10.1056/NEJMcibr1309402.

Dewhirst, M. W., Cao, Y., & Moeller, B. (2008). Cycling hypoxia and free radicals regulate angiogenesis and radiotherapy response. *Nature Reviews Cancer, 8*(6), 425–437. http://www.nature.com/nrc/journal/v8/n6/suppinfo/nrc2397_S1.html.

Dore-Duffy, P., Owen, C., Balabanov, R., Murphy, S., Beaumont, T., & Rafols, J. A. (2000). Pericyte migration from the vascular wall in response to traumatic brain injury. *Microvascular Research, 60*(1), 55–69. https://doi.org/10.1006/mvre.2000.2244.

Driessens, G., Kline, J., & Gajewski, T. F. (2009). Costimulatory and coinhibitory receptors in anti-tumor immunity. *Immunological Reviews, 229*(1), 126–144. https://doi.org/10.1111/j.1600-065X.2009.00771.x.

Dubois, L. G., Campanati, L., Righy, C., D'Andrea-Meira, I., Spohr, T. C. L. d. S. e., Porto-Carreiro, I., et al. (2014). Gliomas and the vascular fragility of the blood brain barrier. *Frontiers in Cellular Neuroscience, 8*, 418. https://doi.org/10.3389/fncel.2014.00418.

El Hallani, S., Boisselier, B., Peglion, F., Rousseau, A., Colin, C., Idbaih, A., et al. (2010). A new alternative mechanism in glioblastoma vascularization: Tubular vasculogenic mimicry. *Brain, 133*(Pt 4), 973–982. https://doi.org/10.1093/brain/awq044.

ElAli, A., Thériault, P., & Rivest, S. (2014). The role of pericytes in neurovascular unit remodeling in brain disorders. *International Journal of Molecular Sciences, 15*(4), 6453–6474. https://doi.org/10.3390/ijms15046453.

Evans, S. M., Judy, K. D., Dunphy, I., Jenkins, W. T., Nelson, P. T., Collins, R., et al. (2004). Comparative measurements of hypoxia in human brain tumors using needle electrodes and EF5 binding. *Cancer Research, 64*(5), 1886–1892.

Eyupoglu, I. Y., Buchfelder, M., & Savaskan, N. E. (2013). Surgical resection of malignant gliomas-role in optimizing patient outcome. *Nature Reviews Neurology, 9*(3), 141–151.

Fan, X., Khaki, L., Zhu, T. S., Soules, M. E., Talsma, C. E., Gul, N., et al. (2010). NOTCH pathway blockade depletes CD133-positive glioblastoma cells and inhibits growth of tumor neurospheres and xenografts. *Stem Cells, 28*(1), 5–16. https://doi.org/10.1002/stem.254.

Fidoamore, A., Cristiano, L., Antonosante, A., d'Angelo, M., Di Giacomo, E., Astarita, C., et al. (2016). Glioblastoma stem cells microenvironment: The paracrine roles of the niche in drug and radioresistance. *Stem Cells International, 2016*, 6809105. https://doi.org/10.1155/2016/6809105.

Franco, M., Roswall, P., Cortez, E., Hanahan, D., & Pietras, K. (2011). Pericytes promote endothelial cell survival through induction of autocrine VEGF-A signaling and Bcl-w expression. *Blood, 118*(10), 2906–2917. https://doi.org/10.1182/blood-2011-01-331694.

Friedman, H. S., Prados, M. D., Wen, P. Y., Mikkelsen, T., Schiff, D., Abrey, L. E., et al. (2009). Bevacizumab alone and in combination with irinotecan in recurrent glioblastoma. *Journal of Clinical Oncology, 27*(28), 4733–4740. https://doi.org/10.1200/jco.2008.19.8721.

Friedmann-Morvinski, D., & Verma, I. M. (2014). Dedifferentiation and reprogramming: Origins of cancer stem cells. *EMBO Reports, 15*(3), 244–253. https://doi.org/10.1002/embr.201338254.

Furuhashi, M., Sjoblom, T., Abramsson, A., Ellingsen, J., Micke, P., Li, H., et al. (2004). Platelet-derived growth factor production by B16 melanoma cells leads to increased pericyte abundance in tumors and an associated increase in tumor growth rate. *Cancer Research, 64*(8), 2725–2733.

Gerhardt, H., Golding, M., Fruttiger, M., Ruhrberg, C., Lundkvist, A., Abramsson, A., et al. (2003). VEGF guides angiogenic sprouting utilizing endothelial tip cell filopodia. *The Journal of Cell Biology, 161*(6), 1163–1177. https://doi.org/10.1083/jcb.200302047.

Gerhardt, H., Wolburg, H., & Redies, C. (2000). N-cadherin mediates pericytic-endothelial interaction during brain angiogenesis in the chicken. *Developmental Dynamics, 218*(3), 472–479. https://doi.org/10.1002/1097-0177(200007)218:3<472::aid-dvdy1008>3.0.co;2-#.

Gilbert, M. R., Dignam, J. J., Armstrong, T. S., Wefel, J. S., Blumenthal, D. T., Vogelbaum, M. A., et al. (2014). A randomized trial of bevacizumab for newly diagnosed glioblastoma. *New England Journal of Medicine, 370*(8), 699–708. https://doi.org/10.1056/NEJMoa1308573.

Gilbertson, R. J., & Rich, J. N. (2007). Making a tumour's bed: Glioblastoma stem cells and the vascular niche. *Nature Reviews Cancer, 7*(10), 733–736. https://doi.org/10.1038/nrc2246.

Goldberg, J. S., & Hirschi, K. K. (2009). Diverse roles of the vasculature within the neural stem cell niche. *Regenerative Medicine, 4*(6), 879–897. https://doi.org/10.2217/rme.09.61.

Goumans, M. J., & Mummery, C. (2000). Functional analysis of the TGFbeta receptor/Smad pathway through gene ablation in mice. *The International Journal of Developmental Biology, 44*(3), 253–265.

Guichet, P. O., Guelfi, S., Teigell, M., Hoppe, L., Bakalara, N., Bauchet, L., et al. (2015). Notch1 stimulation induces a vascularization switch with pericyte-like cell differentiation of glioblastoma stem cells. *Stem Cells, 33*(1), 21–34. https://doi.org/10.1002/stem.1767.

Hamanishi, J., Mandai, M., Matsumura, N., Abiko, K., Baba, T., & Konishi, I. (2016). PD-1/PD-L1 blockade in cancer treatment: Perspectives and issues. *International Journal of Clinical Oncology, 21*(3), 462–473. https://doi.org/10.1007/s10147-016-0959-z.

Hamza, M. A., Mandel, J. J., Conrad, C. A., Gilbert, M. R., Yung, W. K., Puduvalli, V. K., & DeGroot, J. F. (2014). Survival outcome of early versus delayed bevacizumab treatment in patients with recurrent glioblastoma. *Journal of Neuro-Oncology, 119*(1), 135–140. https://doi.org/10.1007/s11060-014-1460-z.

Hanahan, D., & Weinberg, R. A. (2011). Hallmarks of cancer: The next generation. *Cell, 144*(5), 646–674. https://doi.org/10.1016/j.cell.2011.02.013.

Hawkes, C. A., Hartig, W., Kacza, J., Schliebs, R., Weller, R. O., Nicoll, J. A., & Carare, R. O. (2011). Perivascular drainage of solutes is impaired in the ageing mouse brain and in the presence of cerebral amyloid angiopathy. *Acta Neuropathologica, 121*(4), 431–443. https://doi.org/10.1007/s00401-011-0801-7.

Heddleston, J. M., Hitomi, M., Venere, M., Flavahan, W. A., Yang, K., Kim, Y., et al. (2011). Glioma stem cell maintenance: The role of the microenvironment. *Current Pharmaceutical Design, 17*(23), 2386–2401.

Heddleston, J. M., Li, Z., Hjelmeland, A. B., & Rich, J. N. (2009). The hypoxic microenvironment maintains glioblastoma stem cells and promotes reprogramming towards a cancer stem cell phenotype. *Cell Cycle, 8*(20), 3274–3284.

Hellstrom, M., Gerhardt, H., Kalen, M., Li, X., Eriksson, U., Wolburg, H., & Betsholtz, C. (2001). Lack of pericytes leads to endothelial hyperplasia and abnormal vascular morphogenesis. *The Journal of Cell Biology, 153*(3), 543–553.

Hellstrom, M., Kalen, M., Lindahl, P., Abramsson, A., & Betsholtz, C. (1999). Role of PDGF-B and PDGFR-beta in recruitment of vascular smooth muscle cells and pericytes during embryonic blood vessel formation in the mouse. *Development, 126*(14), 3047–3055.

Hodges, T. R., Ferguson, S. D., & Heimberger, A. B. (2016). Immunotherapy in glioblastoma: Emerging options in precision medicine. *CNS Oncology, 5*(3), 175–186. https://doi.org/10.2217/cns-2016-0009.

Huang, F. J., You, W. K., Bonaldo, P., Seyfried, T. N., Pasquale, E. B., & Stallcup, W. B. (2010). Pericyte deficiencies lead to aberrant tumor vascularizaton in the brain of the NG2 null mouse. *Developmental Biology, 344*(2), 1035–1046. https://doi.org/10.1016/j.ydbio.2010.06.023.

Huang, P., Rani, M. R., Ahluwalia, M. S., Bae, E., Prayson, R. A., Weil, R. J., et al. (2012). Endothelial expression of TNF receptor-1 generates a proapoptotic signal inhibited by integrin alpha6beta1 in glioblastoma. *Cancer Research, 72*(6), 1428–1437. https://doi.org/10.1158/0008-5472.can-11-2621.

Iwai, Y., Hamanishi, J., Chamoto, K., & Honjo, T. (2017). Cancer immunotherapies targeting the PD-1 signaling pathway. *Journal of Biomedical Science, 24*(1), 26. https://doi.org/10.1186/s12929-017-0329-9.

Jackson, M., Hassiotou, F., & Nowak, A. (2015). Glioblastoma stem-like cells: At the root of tumor recurrence and a therapeutic target. *Carcinogenesis, 36*(2), 177–185. https://doi.org/10.1093/carcin/bgu243.

Jackson, S., ElAli, A., Virgintino, D., & Gilbert, M. R. (2017). Blood-brain barrier pericyte importance in malignant gliomas: What we can learn from stroke and Alzheimer's disease. *Neuro-Oncology, 19*(9), 1173–1182. https://doi.org/10.1093/neuonc/nox058.

Jain, R. K., di Tomaso, E., Duda, D. G., Loeffler, J. S., Sorensen, A. G., & Batchelor, T. T. (2007). Angiogenesis in brain tumours. *Nature Reviews Neuroscience, 8*(8), 610–622. http://www.nature.com/nrn/journal/v8/n8/suppinfo/nrn2175_S1.html.

Jhaveri, N., Chen, T. C., & Hofman, F. M. (2016). Tumor vasculature and glioma stem cells: Contributions to glioma progression. *Cancer Letters, 380*(2), 545–551. https://doi.org/10.1016/j.canlet.2014.12.028.

Kang, J., Demaria, S., & Formenti, S. (2016). Current clinical trials testing the combination of immunotherapy with radiotherapy. *Journal for Immunotherapy of Cancer, 4*, 51. https://doi.org/10.1186/s40425-016-0156-7.

Lathia, J. D., Gallagher, J., Heddleston, J. M., Wang, J., Eyler, C. E., Macswords, J., et al. (2010). Integrin alpha 6 regulates glioblastoma stem cells. *Cell Stem Cell, 6*(5), 421–432. https://doi.org/10.1016/j.stem.2010.02.018.

Lathia, J. D., Mack, S. C., Mulkearns-Hubert, E. E., Valentim, C. L. L., & Rich, J. N. (2015). Cancer stem cells in glioblastoma. *Genes & Development, 29*(12), 1203–1217. https://doi.org/10.1101/gad.261982.115.

Li, F., Lan, Y., Wang, Y., Wang, J., Yang, G., Meng, F., et al. (2011). Endothelial Smad4 maintains cerebrovascular integrity by activating N-cadherin through cooperation with Notch. *Developmental Cell, 20*(3), 291–302. https://doi.org/10.1016/j.devcel.2011.01.011.

Liebelt, B. D., Shingu, T., Zhou, X., Ren, J., Shin, S. A., & Hu, J. (2016). Glioma stem cells: Signaling, microenvironment, and therapy. *Stem Cells International, 2016*, 7849890. https://doi.org/10.1155/2016/7849890.

Liebner, S., & Plate, K. H. (2010). Differentiation of the brain vasculature: The answer came blowing by the Wnt. *Journal of Angiogenesis Research, 2*(1), 1. https://doi.org/10.1186/2040-2384-2-1.
Lovick, T. A., Brown, L. A., & Key, B. J. (1999). Neurovascular relationships in hippocampal slices: Physiological and anatomical studies of mechanisms underlying flow-metabolism coupling in intraparenchymal microvessels. *Neuroscience, 92*(1), 47–60.
Lugassy, C., Zadran, S., Bentolila, L. A., Wadehra, M., Prakash, R., Carmichael, S. T., et al. (2014). Angiotropism, pericytic mimicry and extravascular migratory metastasis in melanoma: An alternative to intravascular cancer dissemination. *Cancer Microenvironment, 7*(3), 139–152. https://doi.org/10.1007/s12307-014-0156-4.
Lyon, J. G., Mokarram, N., Saxena, T., Carroll, S. L., & Bellamkonda, R. V. (2017). Engineering challenges for brain tumor immunotherapy. *Advanced Drug Delivery Reviews, 114*, 19–32. https://doi.org/10.1016/j.addr.2017.06.006.
Maddaluno, L., Rudini, N., Cuttano, R., Bravi, L., Giampietro, C., Corada, M., et al. (2013). EndMT contributes to the onset and progression of cerebral cavernous malformations. *Nature, 498*(7455), 492–496. https://doi.org/10.1038/nature12207.
Mancuso, M. R., Davis, R., Norberg, S. M., O'Brien, S., Sennino, B., Nakahara, T., et al. (2006). Rapid vascular regrowth in tumors after reversal of VEGF inhibition. *The Journal of Clinical Investigation, 116*(10), 2610–2621. https://doi.org/10.1172/jci24612.
Mannino, M., & Chalmers, A. J. (2011). Radioresistance of glioma stem cells: Intrinsic characteristic or property of the 'microenvironment-stem cell unit'? *Molecular Oncology, 5*(4), 374–386. https://doi.org/10.1016/j.molonc.2011.05.001.
Martínez-González, A., Calvo, G. F., Pérez Romasanta, L. A., & Pérez-García, V. M. (2012). Hypoxic cell waves around necrotic cores in glioblastoma: A biomathematical model and its therapeutic implications. *Bulletin of Mathematical Biology, 74*(12), 2875–2896. https://doi.org/10.1007/s11538-012-9786-1.
McCord, A. M., Jamal, M., Shankavaram, U. T., Lang, F. F., Camphausen, K., & Tofilon, P. J. (2009). Physiologic oxygen concentration enhances the stem-like properties of CD133+ human glioblastoma cells in vitro. *Molecular Cancer Research, 7*(4), 489–497. https://doi.org/10.1158/1541-7786.mcr-08-0360.
Mendez, O., Zavadil, J., Esencay, M., Lukyanov, Y., Santovasi, D., Wang, S. C., et al. (2010). Knock down of HIF-1alpha in glioma cells reduces migration in vitro and invasion in vivo and impairs their ability to form tumor spheres. *Molecular Cancer, 9*, 133. https://doi.org/10.1186/1476-4598-9-133.
Murat, A., Migliavacca, E., Gorlia, T., Lambiv, W. L., Shay, T., Hamou, M. F., et al. (2008). Stem cell-related "self-renewal" signature and high epidermal growth factor receptor expression associated with resistance to concomitant chemoradiotherapy in glioblastoma. *Journal of Clinical Oncology, 26*(18), 3015–3024. https://doi.org/10.1200/jco.2007.15.7164.
Nakada, M., Nambu, E., Furuyama, N., Yoshida, Y., Takino, T., Hayashi, Y., et al. (2013). Integrin α3 is overexpressed in glioma stem-like cells and promotes invasion. *British Journal of Cancer, 108*(12), 2516–2524. https://doi.org/10.1038/bjc.2013.218.
Nduom, E. K., Weller, M., & Heimberger, A. B. (2015). Immunosuppressive mechanisms in glioblastoma. *Neuro-Oncology, 17*(suppl_7), vii9–vii14. https://doi.org/10.1093/neuonc/nov151.
Niklasson, M., Jarvius, M., Haglund, C., Chantzi, E., Bergström, T., Nyberg, F., et al. (2017). Abstract 4175: Targeting of a mesenchymal profile in order to sensitize multitherapy resistant glioblastoma clones. *Cancer Research, 77*(13 Supplement), 4175.
Ochs, K., Sahm, F., Opitz, C. A., Lanz, T. V., Oezen, I., Couraud, P. O., et al. (2013). Immature mesenchymal stem cell-like pericytes as mediators of immunosuppression in human malignant glioma. *Journal of Neuroimmunology, 265*(1-2), 106–116. https://doi.org/10.1016/j.jneuroim.2013.09.011.
Ogden, A. T., Waziri, A. E., Lochhead, R. A., Fusco, D., Lopez, K., Ellis, J. A., et al. (2008). Identification of A2B5+CD133- tumor-initiating cells in adult human gliomas. *Neurosurgery, 62*(2), 505–514.; discussion 514–505. https://doi.org/10.1227/01.neu.0000316019.28421.95.

Pardridge, W. M. (2005). The blood-brain barrier: Bottleneck in brain drug development. *NeuroRx, 2*(1), 3–14.

Pardridge, W. M. (2012). Drug transport across the blood–brain barrier. *Journal of Cerebral Blood Flow & Metabolism, 32*(11), 1959–1972. https://doi.org/10.1038/jcbfm.2012.126.

Persano, L., Pistollato, F., Rampazzo, E., Della Puppa, A., Abbadi, S., Frasson, C., et al. (2012). BMP2 sensitizes glioblastoma stem-like cells to Temozolomide by affecting HIF-1α stability and MGMT expression. *Cell Death & Disease, 3*, e412. http://www.nature.com/cddis/journal/v3/n10/suppinfo/cddis2012153s1.html.

Persidsky, Y., Ramirez, S. H., Haorah, J., & Kanmogne, G. D. (2006). Blood–brain barrier: Structural components and function under physiologic and pathologic conditions. *Journal of Neuroimmune Pharmacology, 1*(3), 223–236. https://doi.org/10.1007/s11481-006-9025-3.

Pieper, C., Marek, J. J., Unterberg, M., Schwerdtle, T., & Galla, H. J. (2014). Brain capillary pericytes contribute to the immune defense in response to cytokines or LPS in vitro. *Brain Research, 1550*, 1–8. https://doi.org/10.1016/j.brainres.2014.01.004.

Pieper, C., Pieloch, P., & Galla, H. J. (2013). Pericytes support neutrophil transmigration via interleukin-8 across a porcine co-culture model of the blood-brain barrier. *Brain Research, 1524*, 1–11. https://doi.org/10.1016/j.brainres.2013.05.047.

Razavi, S.-M., Lee, K. E., Jin, B. E., Aujla, P. S., Gholamin, S., & Li, G. (2016). Immune evasion strategies of glioblastoma. *Frontiers in Surgery, 3*, 11. https://doi.org/10.3389/fsurg.2016.00011.

Reis, M., & Liebner, S. (2013). Wnt signaling in the vasculature. *Experimental Cell Research, 319*(9), 1317–1323. https://doi.org/10.1016/j.yexcr.2012.12.023.

Reiss, Y., Machein, M. R., & Plate, K. H. (2005). The role of angiopoietins during angiogenesis in gliomas. *Brain Pathology, 15*(4), 311–317. https://doi.org/10.1111/j.1750-3639.2005.tb00116.x.

Reyahi, A., Nik, A. M., Ghiami, M., Gritli-Linde, A., Pontén, F., Johansson, B. R., & Carlsson, P. (2015). Foxf2 is required for brain pericyte differentiation and development and maintenance of the blood-brain barrier. *Developmental Cell, 34*(1), 19–32. https://doi.org/10.1016/j.devcel.2015.05.008.

Ribeiro, A. L., & Okamoto, O. K. (2015). Combined effects of pericytes in the tumor microenvironment. *Stem Cells International, 2015*, 8. https://doi.org/10.1155/2015/868475.

Ricci-Vitiani, L., Pallini, R., Biffoni, M., Todaro, M., Invernici, G., Cenci, T., et al. (2010). Tumour vascularization via endothelial differentiation of glioblastoma stem-like cells. *Nature, 468*(7325), 824–828. https://doi.org/10.1038/nature09557.

Roh, T. H., Park, H. H., Kang, S.-G., Moon, J. H., Kim, E. H., Hong, C.-K., et al. (2017). Long-term outcomes of concomitant chemoradiotherapy with temozolomide for newly diagnosed glioblastoma patients: A single-center analysis. *Medicine, 96*(27), e7422. https://doi.org/10.1097/md.0000000000007422.

Rolle, C. E., Sengupta, S., & Lesniak, M. S. (2010). Challenges in clinical design of immunotherapy trials for malignant glioma. *Neurosurgery Clinics of North America, 21*(1), 201–214. https://doi.org/10.1016/j.nec.2009.08.002.

Rong, Y., Durden, D. L., Van Meir, E. G., & Brat, D. J. (2006). 'Pseudopalisading' necrosis in glioblastoma: A familiar morphologic feature that links vascular pathology, hypoxia, and angiogenesis. *Journal of Neuropathology & Experimental Neurology, 65*(6), 529.

Rosenberg, S. A., & Restifo, N. P. (2015). Adoptive cell transfer as personalized immunotherapy for human cancer. *Science, 348*(6230), 62–68. https://doi.org/10.1126/science.aaa4967.

Rostomily, R. C., Spence, A. M., & Silbergeld, D. L. (2005). Neurosurgical management of high-grade gliomas. In A. J. Moore & D. W. Newell (Eds.), *Neurosurgery: Principles and practice* (pp. 167–186). London: Springer London.

Safa, A. R., Saadatzadeh, M. R., Cohen-Gadol, A. A., Pollok, K. E., & Bijangi-Vishehsaraei, K. (2015). Glioblastoma stem cells (GSCs) epigenetic plasticity and interconversion between differentiated non-GSCs and GSCs. *Genes & Diseases, 2*(2), 152–163. https://doi.org/10.1016/j.gendis.2015.02.001.

Sattiraju, A., Sai, K. K. S., & Mintz, A. (2017a). Glioblastoma stem cells and their microenvironment. *Advances in Experimental Medicine and Biology, 1041*, 119–140. https://doi.org/10.1007/978-3-319-69194-7_7.

Sattiraju, A., Sun, Y., Solingapuram Sai, K. K., Li, K. C. P., & Mintz, A. (2017b). Maximizing local access to therapeutic deliveries in glioblastoma. Part IV: Image-guided, remote-controlled opening of the blood-brain barrier for systemic brain tumor therapy. In S. De Vleeschouwer (Ed.), *Glioblastoma*. Brisbane (AU): Codon Publications. Copyright: The Authors.

Sattiraju, A., Xiong, X., Pandya, D. N., Wadas, T. J., Xuan, A., Sun, Y., et al. (2017c). Alpha particle enhanced blood brain/tumor barrier permeabilization in glioblastomas using integrin alpha-v beta-3-targeted liposomes. *Molecular Cancer Therapeutics, 16*(10), 2191–2200. https://doi.org/10.1158/1535-7163.mct-16-0907.

Scott, M. A., Shen, J., Lam, K., Yen, Y. H., Shrestha, S., Mravic, M., Asatrian, G., Chung, C. G., Lugassy, C., Barnhill, R. L., Dry, S. M., Peault, B., & James, A. W. (2015). Review of pericytes in tumor biology. *International Journal of Orthopaedics, 2*(3), 301–306.

Segerman, A., Niklasson, M., Haglund, C., Bergström, T., Jarvius, M., Xie, Y., et al. (2016). Clonal variation in drug and radiation response among glioma-initiating cells is linked to proneural-mesenchymal transition. *Cell Reports, 17*(11), 2994–3009. https://doi.org/10.1016/j.celrep.2016.11.056.

Sharpe, A. H., & Abbas, A. K. (2006). T-cell costimulation—biology, therapeutic potential, and challenges. *New England Journal of Medicine, 355*(10), 973–975. https://doi.org/10.1056/NEJMp068087.

Singh, S. K., Clarke, I. D., Hide, T., & Dirks, P. B. (2004a). Cancer stem cells in nervous system tumors. *Oncogene, 23*(43), 7267–7273. https://doi.org/10.1038/sj.onc.1207946.

Singh, S. K., Hawkins, C., Clarke, I. D., Squire, J. A., Bayani, J., Hide, T., et al. (2004b). Identification of human brain tumour initiating cells. *Nature, 432*, 396–401. https://doi.org/10.1038/nature03128.

Soda, Y., Marumoto, T., Friedmann-Morvinski, D., Soda, M., Liu, F., Michiue, H., et al. (2011). Transdifferentiation of glioblastoma cells into vascular endothelial cells. *Proceedings of the National Academy of Sciences of the United States of America, 108*(11), 4274–4280. https://doi.org/10.1073/pnas.1016030108.

Song, K. S., Phi, J. H., Cho, B.-K., Wang, K.-C., Lee, J. Y., Kim, D. G., et al. (2010). Long-term outcomes in children with glioblastoma. *Journal of Neurosurgery: Pediatrics, 6*(2), 145–149. https://doi.org/10.3171/2010.5.PEDS09558.

Stupp, R., Hegi, M. E., Mason, W. P., van den Bent, M. J., Taphoorn, M. J. B., Janzer, R. C., et al. (2009). Effects of radiotherapy with concomitant and adjuvant temozolomide versus radiotherapy alone on survival in glioblastoma in a randomised phase III study: 5-year analysis of the EORTC-NCIC trial. *The Lancet Oncology, 10*(5), 459–466. https://doi.org/10.1016/S1470-2045(09)70025-7.

Suri, C., Jones, P. F., Patan, S., Bartunkova, S., Maisonpierre, P. C., Davis, S., et al. (1996). Requisite role of angiopoietin-1, a ligand for the TIE2 receptor, during embryonic angiogenesis. *Cell, 87*(7), 1171–1180.

Svensson, A., Özen, I., Genové, G., Paul, G., & Bengzon, J. (2015). Endogenous brain pericytes are widely activated and contribute to mouse glioma microvasculature. *PLoS One, 10*(4), e0123553. https://doi.org/10.1371/journal.pone.0123553.

Sweeney, M. D., Ayyadurai, S., & Zlokovic, B. V. (2016). Pericytes of the neurovascular unit: Key functions and signaling pathways. *Nature Neuroscience, 19*, 771. https://doi.org/10.1038/nn.4288. https://www.nature.com/articles/nn.4288#supplementary-information.

Valdor, R., Garcia-Bernal, D., Bueno, C., Rodenas, M., Moraleda, J. M., Macian, F., & Martinez, S. (2017). Glioblastoma progression is assisted by induction of immunosuppressive function of pericytes through interaction with tumor cells. *Oncotarget, 8*(40), 68614–68626. https://doi.org/10.18632/oncotarget.19804.

Van Meir, E. G., Hadjipanayis, C. G., Norden, A. D., Shu, H. K., Wen, P. Y., & Olson, J. J. (2010). Exciting new advances in neuro-oncology: The avenue to a cure for malignant glioma. *CA: A Cancer Journal for Clinicians, 60*(3), 166–193. https://doi.org/10.3322/caac.20069.

Wang, M., Yin, B., Wang, H. Y., & Wang, R. F. (2014). Current advances in T-cell-based cancer immunotherapy. *Immunotherapy, 6*(12), 1265–1278. https://doi.org/10.2217/imt.14.86.

Watkins, S., Robel, S., Kimbrough, I. F., Robert, S. M., Ellis-Davies, G., & Sontheimer, H. (2014). Disruption of astrocyte-vascular coupling and the blood-brain barrier by invading glioma cells. *Nature Communications, 5*, 4196–4196. https://doi.org/10.1038/ncomms5196.

Weller, M., Cloughesy, T., Perry, J. R., & Wick, W. (2012). Standards of care for treatment of recurrent glioblastoma—Are we there yet? *Neuro-Oncology, 15*, 4–27. https://doi.org/10.1093/neuonc/nos273.

Wolburg, H., & Lippoldt, A. (2002). Tight junctions of the blood–brain barrier: Development, composition and regulation. *Vascular Pharmacology, 38*(6), 323–337. https://doi.org/10.1016/S1537-1891(02)00200-8.

Wolburg, H., Noell, S., Fallier-Becker, P., Mack, A. F., & Wolburg-Buchholz, K. (2012). The disturbed blood-brain barrier in human glioblastoma. *Molecular Aspects of Medicine, 33*(5-6), 579–589. https://doi.org/10.1016/j.mam.2012.02.003.

Xian, X., Håkansson, J., Ståhlberg, A., Lindblom, P., Betsholtz, C., Gerhardt, H., & Semb, H. (2006). Pericytes limit tumor cell metastasis. *The Journal of Clinical Investigation, 116*(3), 642–651. https://doi.org/10.1172/JCI25705.

Xiong, X., Sun, Y., Sattiraju, A., Jung, Y., Mintz, A., Hayasaka, S., & Li, K. C. P. (2015). Remote spatiotemporally controlled and biologically selective permeabilization of blood-brain barrier. *Journal of Controlled Release: Official Journal of the Controlled Release Society, 217*, 113–120. https://doi.org/10.1016/j.jconrel.2015.08.044.

Yang, Y. (2015). Cancer immunotherapy: Harnessing the immune system to battle cancer. *The Journal of Clinical Investigation, 125*(9), 3335–3337. https://doi.org/10.1172/jci83871.

Zhang, L., Wang, Y., Rashid, M. H., Liu, M., Angara, K., Mivechi, N. F., et al. (2017). Malignant pericytes expressing GT198 give rise to tumor cells through angiogenesis. *Oncotarget, 8*(31), 51591–51607. https://doi.org/10.18632/oncotarget.18196.

Zong, H., Verhaak, R. G. W., & Canoll, P. (2012). The cellular origin for malignant glioma and prospects for clinical advancements. *Expert Review of Molecular Diagnostics, 12*(4), 383–394. https://doi.org/10.1586/erm.12.30.

Lidar, Z., Mardor, Y., Jonas, T., Pfeffer, R., Faibel, M., Nass, D., et al. (2004). Convection-enhanced delivery of paclitaxel for the treatment of recurrent malignant glioma: A phase I/II clinical study. *Journal of Neurosurgery, 100*(3), 472–479. https://doi.org/10.3171/jns.2004.100.3.0472.

Chapter 3
Pericytes in Breast Cancer

Jiha Kim

Abstract Breast cancer is a heterogeneous disease driven not only by evolutionally diverse cancer cell themselves but also by highly dynamic microenvironment. At the center of the tumor microenvironment, tumor vasculature plays multiple roles from supporting tumor growth to providing a route for metastasis to the distant organ sites. Blood vessels in breast cancer present with perfusion defects associated with vessel dilation, tortuosity, and poor perivascular coverage (Li et al., Ultrasound Med 32:1145–1155, 2013; Eberhard et al., Cancer Res 60:1388–1393, 2000; Cooke et al., Cancer Cell 21:66–81, 2012). Such abnormal vascular system is partly due to the morphological and molecular alteration of pericytes that is accompanied by a significant heterogeneity within the populations (Kim et al., JCI Insight 1:e90733, 2016). While pericytes are implicated for their controversial roles in breast cancer metastasis (Cooke et al., Cancer Cell 21:66–81, 2012; Gerhardt and Semb, J Mol Med (Berl) 86:135–144, 2008; Keskin et al., Cell Rep 10:1066–1081, 2015; Meng et al., Future Oncol 11:169–179, 2015; Xian et al., J Clin Invest 116:642–651, 2006), the impact of their heterogeneity on breast cancer progression, metastasis, intratumoral immunity, and response to chemotherapy are largely unknown. Due to the complexity of angiogenic programs of breast cancer, the anti-angiogenic or anti-vascular treatment has been mostly unsuccessful (Tolaney et al., Proc Natl Acad Sci U S A 112:14325–14330, 2015; Mackey et al., Cancer Treat Rev 38:673–688, 2012; Sledge, J Clin Oncol 33:133–135, 2015) and requires much in-depth knowledge on different components of tumor microenvironment and how these stromal cells are interacting and communicating to each other. Therefore, understanding pericyte heterogeneity and their differential functional contribution will shed light on new potential approaches to treat breast cancer.

Keywords Pericyte · Breast cancer · Heterogeneity · Tumor microenvironment Blood vessels · Angiogenesis · Metastasis · Tumor immunity · Perivascular phenotypes · PDGFRβ · Vascular normalization

J. Kim (✉)
Department of Biological Sciences, North Dakota State University, Fargo, ND, USA
e-mail: jiha.kim.1@ndsu.edu

A. Birbrair (ed.), *Pericyte Biology in Disease*, Advances in Experimental Medicine and Biology 1147, https://doi.org/10.1007/978-3-030-16908-4_3

Pericytes

Endothelial cells and pericytes are the fundamental units of blood vessels. While endothelial cells make up the inner lining of the vessel wall, pericytes are responsible for enveloping the surface of the vessels and providing structural support. Recently, pericytes have gained much attention due to their diverse roles during vessels formation, vessels maturation, and endothelial support. Pericytes can be identified not only by their distinct morphological features but also by the sets of molecular markers, namely αSMA, desmin, PDGFRβ, NG2, and RGS-5 (Bergers and Song 2005). The expression pattern of these markers can be varying in different tissues or be dynamic during various developmental stages. RGS-5, desmin, and αSMA are intracellular proteins, of which desmin and αSMA are contractile filaments, and RGS-5 is a GTPases-activating protein. Neuron-glial 2 (NG2) and platelet-derived growth factor receptor beta (PDGFRβ) are cell-surface proteins. PDGFRβ is one of the most studied molecules expressed in pericyte due to its paracrine signaling through ligand PDGF-BB to control pericyte recruitment to the growing vessels (Hellstrom et al. 1999; Enge et al. 2002). The composition of these markers is various in different tissues, potentially linked to their diverse functions in the different microenvironment. Also, two distinct types of pericytes based on their marker expression (Type 1, Nestin-/NG2+ and Type 2, Nestin+/NG2+) were shown to exert different angiogenic capacity in vitro and in vivo (Birbrair et al. 2014a). Another in vitro study using tumor-derived PDGFR-β+ perivascular progenitors showed that these progenitors could differentiate into more mature phenotype (NG2+ or αSMA+) upon culture, whereas desmin expression was only induced when they were cultured with endothelial cells together (Song et al. 2005). Such phenotypic conversion indicates that specific pericyte phenotypes are largely influenced by their local environment and different cell states. Pericyte density also varies in different parts of the body based on the unique needs and pressure the blood vessels need to withstand (Sims 2000). In particular, the highest pericyte density is observed in the central nervous system, brain, and retina, to create blood–brain barrier (Ballabh et al. 2004). Pericytes have complex ontogeny including neuro crest, bone-marrow-derived mesenchymal stem cells (BM-MSCs), or onsite proliferation based on the tissues where they are residing in and their specific functions (Armulik et al. 2011; Hall 2006). However, recent studies have suggested an alternate source of pericytes in tumor microenvironment including epithelial-to-pericytes transition (EPT) in breast cancer adding another layer of complexity (Shenoy et al. 2016).

Just as normal pericytes perform a diverse function and express differential markers in different tissues, pericytes in tumor microenvironment exhibit great diversity in marker expression as well as functional contribution.

Pericyte Landscape (Investment) in Pathological Environment

Poorly invested angiogenesis in tumors results in the chaotic and disorganized vasculature that presents with tortuous, leaky, and permeable vessels often with functionally incompetent (Li et al. 2013; Eberhard et al. 2000). Under the constant influence of cancer cells and tumor microenvironment, both endothelial cells and pericytes appear morphologically and molecularly differ from normal counterparts which, in part, leads defective vasculature (Bergers and Song 2005). Different types of tumors present strikingly different vascular architecture, including vessels size, dilation, and pericyte coverage (Morikawa et al. 2002). Notably, tumor-associated pericytes are loosely associated with the endothelial cells, with cytoplasmic processes that penetrate deep in the tumor parenchyma. Although the exact mechanism for such aberrant phenotype is still not clear, it has been proposed that abnormal expression of growth factors or signaling molecules such as VEGF, TGF-β, PDGF-BB, and Ang-2 has a significant impact on pericyte morphology, quality, and investment (Kim et al. 2016; Keskin et al. 2015; Song et al. 2005; Raza et al. 2010). In particular, PDGF-BB/PDGFRβ is one of the most well studied and prominent signaling pathways that is involved in pericyte recruitment and survivor. These signaling molecules also have been implicated in other pathological condition related to vascular abnormalities, such as diabetic retinopathy (Enge et al. 2002; Benjamin 2001; Rangasamy et al. 2011), wound healing (Lin et al. 2008), or stroke, a cerebrovascular disorder associated with blood–brain barrier (BBB) disruption (Suzuki et al. 2016).

Perivascular Signature in Breast Cancer at a Glance

In breast cancer, despite enormously enlarged and thickened vessels appearance, relatively large percentage of vessels are covered by pericytes (Eberhard et al. 2000; Bergers and Song 2005). While a great deal of heterogeneity present on pericytes in breast cancer, all these markers have been detected in human cancer tissues while it is still not clear what such heterogeneity means. Several studies have reported the mean microvascular pericyte coverage index (MPI) in breast cancer to be from 32% up to 80% by quantifying αSMA expressing pericytes (Eberhard et al. 2000; Shrivastav et al. 2016). No pan-pericyte marker can identify all pericyte populations (Armulik et al. 2011; Hall 2006) and infect, and a recent study has shown the various flavors of pericytes existing within human breast tumor tissues (Kim et al. 2016). Several studies have attempted to measure MPI using immunohistochemistry for αSMA (Eberhard et al. 2000; Shrivastav et al. 2016), NG2 (Cooke et al. 2012), PDGFRβ (Shrivastav et al. 2016; Paulsson et al. 2009), desmin (Kim et al. 2016), and CD 248 (Viski et al. 2016) using breast cancer patient's tissue samples. It is worth noting that some of these analyses include both fibroblast and pericyte as both cell types tend to express these markers. αSMA is the marker mostly explored by many investigators due to its abundance. However, it is noteworthy that αSMA expression is lacking in quiescent pericytes in normal tissues (Gerhardt and Betsholtz 2003) and hence contributes to the pathological phenotype. TGFβ, involved in smooth muscle cell maturation, is known to be responsible for ectopic expression of αSMA in tumor pericytes (Song et al. 2005). Often pericyte composition/phenotype is used to indicate the functional status of the tumor vasculature, and thus many attempts have been made to correlate it with patient outcome. However, the results have not been consistent between analyses primarily due to different markers and methods used to quantify pericyte phenotypes. Other studies have indicated that pericyte phenotypes are differentially regulated between the primary tumor and metastases, suggesting the influence of tumor microenvironment (TME) in perivascular investment (Lyle et al. 2016). Such discordance between primary and secondary cancers also implies that perivascular phenotype in primary tumor site might be a prognostic factor for the metastasis (Jubb et al. 2011).

Pericytes Phenotype Conversion

PDGF-BB/PDGFRβ signaling is one of the most studied pathways known to be crucial for pericyte recruitment, survivor, and clinical implication (Ostman and Heldin 2007). PDGFRβ gene expression levels (high vs. low) in 3455 patients with breast cancer show a correlation between high PDGFRβ expression and the recurrence-free survival probability of patients (Keskin et al. 2015). Another study has used the immunohistological analysis of PDGFRβ expression in 512 breast cancer samples. Although in this particular study PDGFRβ expression was scored in entire stroma including fibroblast and pericytes, it has shown that high PDGFRβ expression in stroma was correlated with high pathological grade, estrogen receptor negativity, and high HER2 expression, as well as shorter recurrence-free and breast cancer-specific survival (Paulsson et al. 2009). In a separate study, 75 breast cancer samples were also analyzed for PDGFRβ expression using immunohistological assay to evaluate the pericytes as a prognostic factor for lymph node metastasis and molecular subtypes. However, this study failed to show any correlation between MPI and the known prognostic and predictive factors (Shrivastav et al. 2016). To establish the significance of PDGFRβ expressing pericytes, PDGFRβ-TK (thymidine kinase) mice in which PDGFRβ+ pericytes are specifically eliminated were explored in the context of murine mammary tumor model using 4T1 cells. While primary tumor progression was repressed upon PDGFRβ+ pericytes depletion due to anti-angiogenic effects, metastatic incidences were significantly increased via increased hypoxia, vascular leakiness, and epithelial–mesenchymal transition (EMT) (Keskin et al. 2015). It is noteworthy that TK system physically eliminates all pericytes expressing PDGFRβ; therefore, instead of changing pericyte phenotype, it exerts an anti-angiogenic effect at least on the primary tumor sites.

A study mentioned above using 75 breast cancer tissues samples revealed no correlation between αSMA expression and the known prognostic and predictive factors. NG2 chondroitin sulfate proteoglycan is expressed on the surface of pericytes during vasculogenic and angiogenic processes. NG2 is often considered to be a maker for mature pericytes, and its expression is observed in the large percentage of tumor pericytes despite the abnormal phenotype and function (Armulik et al. 2011). In breast cancer, low NG2+ pericyte coverage was significantly associated with the presence of metastasis, and low NG2+/high c-Met expression was correlated with poor survival of breast cancer patients (Cooke et al. 2012). Endosialin

(CD248) is a transmembrane glycoprotein, and its expression is known to be upregulated in tumor-associated pericytes and myofibroblasts in breast cancer (Viski et al. 2016). CD248 expression in microdissected breast tumor stroma was associated with decreased recurrence-free survival, and high CD248 expression was correlated with low distant metastasis-free survival. Such observation was confirmed by a 4T1 orthotopic mammary tumor in CD248 knockout mice background. Interestingly, CD248 knockout had a significant effect on decreasing metastasis but had no effect on primary tumor growth, revealing a specific function of CD248 on intravasation process.

Another example of abnormal/pathological pericyte phenotype was described in a mouse model for melanoma, breast cancer, and rhabdomyosarcoma (detail reviewed by others (Paiva et al. 2018)). In this study, increased expression of the pluripotency gene Klf4 in pericytes induced the phenotypic switch from mature/quiescent pericytes (NG2+) to a less differentiated state with increased proliferation, migration, and extracellular matrix (ECM) (e.g., fibronectin) production, which contributes to a prometastatic fibronectin-rich environment. Pericyte-specific knockout of Klf4 decreased premetastatic niche formation and metastasis (Murgai et al. 2017). In the lung, pericytes also exhibit heterogeneity, and two different subtypes of pericyte as previously mentioned (Type 1, Nestin−/NG2+ and Type 2, Nestin+/NG2+) were shown to be present on pulmonary blood vessels. Whether or not different subtypes contribute to the formation of premetastatic niche differently is not known (Paiva et al. 2018). However, only type-1 pericytes, but not type 2, were accumulated and producing collagen at the injury site of lung contributing to pulmonary fibrosis (Birbrair et al. 2014b). On the other hand, type-2 pericytes were shown to be actively engaged in the angiogenic process during orthotopic glioblastoma progression (Birbrair et al. 2014a). Thus, it would be attractive to explorer if type-1 pericytes are mainly contributing to the formation of the premetastatic niche by converting to Klf4+ phenotype (e.g., depositing ECM) at the beginning, and type-2 pericytes will be recruited to support secondary tumor formation once tumor started to grow at the niche.

In both cases for CD248 and Klf4, their expression was abnormally upregulated in response to the tumor microenvironment and had a significant influence on metastatic behavior rather than primary tumor growth. Suggesting genes that are differentially regulated compared to normal pericyte are of great interest to understand the fundamental impact of pericyte on distant metastasis and perhaps organ-specific tropism.

More pieces of evidence are emerging to indicate a large percentage of pericytes express multiple markers rather than a single marker. Therefore, it makes more sense to define pericyte phenotype using a combination of different markers. A recent study using multispectral images of multiplex stained tissue microarray of breast cancer provides a more comprehensive understanding of perivascular heterogeneity and phenotyping. In this study, tissue microarray (TMA) was co-stained for PDGFRβ, desmin, and CD31, and imaging analysis was performed to find a significance of ratio between PDGFRβ and desmin (Kim et al. 2016). Based on two separate cohort of breast cancer samples, this study has shown that the ratio of PDGFRβ and desmin is significantly different between subtypes of breast cancer, TNBC, and luminal, and that the high desmin to PDGFRβ ratio was considered to be a predictive factor for higher relapse-free survivor and higher breast cancer-specific survivor of patient who was treated with epirubicin but not with paclitaxel. Although the underlying mechanism remains to be determined, it provides new ways to understand the perivascular landscape and explains, in part, the discrepancy of previous different studies where a single marker was analyzed.

Origin of Pericyte in Breast Cancer

The origin of tumor pericytes has been investigated in different types of tumors, including fibrosarcoma, melanoma, and colorectal cancer. It has been shown that tumor pericytes can be recruited from local immature mesenchymal cells, bone marrow-derived cells, and onsite proliferation (Abramsson et al. 2002; Du et al. 2008; Rajantie et al. 2004). A recent study in breast cancer has proposed an alternative source of pericyte in TME through epithelial-to pericytes transition (EPT) adding another layer of complexity (Shenoy et al. 2016). In this study, MCF10DCIS cells were forced to undergo EMT, and its fate was followed in vivo and in vitro. EMT cells acquired mesenchymal phenotype (expressing PDGFR-β and N-cadherin) and physical contact with endothelium contributing tumor vasculature. Although in normal tumor context, cells undergo EMT is a small population and therefore attribute to a fraction of tumor-associated pericyte pool, it is an interesting observation to identify an alternative source of pericyte in breast cancer. A similar result was also shown in the case of glioblastoma (GBM) in which glioma stem cells (GSCs) give rise to pericytes to support vessel function and tumor growth (Cheng et al. 2013). In this case, human GBM specimens showed phenotypically switched pericyte populations containing the same mutational status with cancer cells. These studies suggest the alternative source of pericyte in the context of TME and thus new therapeutic targets. Considering what we have observed regarding pericyte heterogeneity and their functional contribution, it will be a great interest to investigate the phenotype of such converted pericytes and its correlation with the mutational status of cancer cells.

On the other hand, pericytes have been speculated for its stem cell capacities to differentiate into adipocytes and fibroblasts in different organs (Crisan et al. 2008). Cancer stem cells are often observed in perivascular niches (Calabrese et al. 2007; Pietras et al. 2008), and thus it is tempting to speculate that tumor vasculature-associated pericytes might hold mesenchymal stem cell properties although direct evidence for this proposition is still unclear. In case of renal cell carcinoma, PDGFR-β expressing pericytes were shown to transit its fate to fibroblasts (Pericyte-fibroblast transition) in response to tumor-derived PDGF-BB and contributed to tumor growth and metastasis (Hosaka et al. 2016), demonstrating phenotypic switching of pericytes in response to TME or malignancy. An interesting phenomenon has been observed in the study of breast cancer that stromal cells (adipocytes, fibroblasts, and myoepithelial cells) gained somatic mutation on GT198, a steroid hormone receptor coactivator, independent with mutational status of cancer cells. It was suggested that the progenitor cells with GT198 mutation (GT198+) is mostly capillary pericytes and differentiated into GT198+ stroma cells collectively contributing to malignant tumor microenvironment (Yang et al. 2016). Mutant GT198 expressing cells are shown to induce VEGF expression, which in turn influence cancer cells attributing reciprocal communication between cancer cells and TME.

Anti-Angiogenic (Anti-Vascular) Treatment

Angiogenesis and co-optive vascular remodeling are prerequisites of solid tumor growth. Breast cancer is one of the highly vascularized tumors with fairly high pericyte coverage (Eberhard et al. 2000) and largely dependent on vascular support for the survivor and growth. The level of neovascularization in aggressive breast cancer correlated with metastatic disease and may serve as an independent predictor for metastasis (Weidner et al. 1991). Therefore, it is a quite attractive approach to target tumor vasculature. However, anti-angiogenic treatment has been largely unsuccessful with marginal benefit (Sledge 2015; Aalders et al. 2017). Many of anti-angiogenic treatment involves targeting endothelial cells or pro-angiogenic factors, aiming to eliminate vessels and thus starving tumors. Tumor angiogenesis is accompanied by an increased level of pro-angiogenic factors such as HIf1α and VEGF (Aalders et al. 2017; Bos et al. 2001); thus, blocking VEGF pathway has been most extensively studied and considered as anti-angiogenic treatment including a monoclonal antibody against VEGF, bevacizumab (Sledge 2015). Other types of anti-angiogenic approaches include tyrosine kinase inhibitors such as sunitinib, sorafenib, imatinib, and axitinib. Anti-VEGF treatment in tumors led to a partial elimination of tumor blood vessels that are not covered by pericytes (Tolaney et al. 2015; Benjamin et al. 1999). To overcome such limitation, dual targeting of endothelial cells and pericytes has been proposed (Bergers et al. 2003). However, pericyte depletion did not provide an additive effect in some models (Nisancioglu et al. 2010; Sennino et al. 2007) or has proven to increase metastasis in breast cancer models (Meng et al. 2015; Cooke et al. 2012; Keskin et al. 2015).

In fact, low NG2+ pericyte coverage of tumor vasculature was significantly correlated with increased metastasis in clinical samples of breast cancer (Cooke et al. 2012). A separate study analyzing TNBC vs. luminal breast cancer has shown that TNBC tumor vasculature exhibits poor pericyte coverage compared to luminal tumor vasculature, suggesting lower pericyte coverage might be an indication of aggressive nature of tumor types (Kim et al. 2016). Such notion, perhaps, indicates that nonselective elimination of pericyte may not yield benefit but rather promote tumor aggressiveness and metastasis. Thus, a better understanding of pericyte heterogeneity in response to TME changes may provide insight to pericyte targeting strategy.

Vascular Normalization Using Pericyte Landscape

Despite the aberrant morphology, marker expression, and function, eliminating pericytes as a whole did not result in any beneficial effect and, in fact, did more harm by increasing metastasis. Instead, vascular normalization concept takes advantage of pericytes by only eliminating vessels that are not covered by pericytes (immature), leaving healthy pericyte covered functional vasculature (Goel et al. 2012). Several clinical trials of anti-angiogenic therapy (anti-VEGF) suggest vascular normalization phenomenon in many solid tumors. However, no significant survival benefit has been warranted so far (Mackey et al. 2012). A recent clinical trial of neoadjuvant bevacizumab and chemotherapy in breast cancer patients has shown limited efficacy despite the clear vascular normalization effect. Analysis of αSMA+ pericyte coverage in pretreatment vs. posttreatment showed significantly increased αSMA+ pericyte-associated vessels although not a clinically significant contribution to overall outcome (Tolaney et al. 2015). However, this study suggests that patient might benefit from bevacizumab treatment if sufficient numbers of vessels are initially present. Collectively, it is a plausible explanation that vascular normalization approach using bevacizumab might only benefit patients with specific vascular phenotype defined by pericyte investment.

Interesting results from anti-angiogenic treatment have been observed in melanoma case. Subpopulations of pericytes that were characterized by distinct marker expression (high αSMA and PDGFRβ) and loose attachment to endothelial cells showed a more significant effect on combinatorial treatment using the VEGFR inhibitor PTK787 and the PDGFR inhibitor STI571 in PDGF-BB overexpressing tumor (Hasumi et al. 2007). However, desmin + pericytes that are usually more mature and tightly bound to endothelium remained intact. This study indicates that different subpopulations of pericyte responded differently not only to anti-angiogenic drug treatment but also to the intrinsic nature of cancer cell themselves, in this case, the expression level of PDGF-BB. Therefore, pericyte landscape might be a predictable marker for patients who are more likely respond to anti-angiogenic treatment. However, pericytes are well recognized for its controversial function on metastasis, and further analysis of metastatic behavior upon treatment should be followed. Such finding is in accordance with the study mentioned above in breast cancer, where the ratio between PDGFRβ+ and desmin + pericyte on treatment naïve biopsy has predictive power for patient outcome upon treatment with the specific drug (Kim et al. 2016). Despite the promising preclinical results and rational to justify anti-angiogenic therapy, the overall benefit is marginal, and the toxicity and cost are not outweighed. As pointed out by others, anti-angiogenic therapy such as bevacizumab should only be considered when we have a better idea on the predictive biomarker for sufficient benefit. Considering the emerging data on perivascular phenotype can have a profound effect on vascular functionality, perhaps pericyte landscape should be explored as a valid predictive marker for the success of anti-angiogenesis or other drugs.

Their Contribution to Breast Cancer: Friend or Foe?

Originally pericytes were proposed to be a gatekeeper ***(friend)*** of metastasis based on several studies where low pericyte coverage or depletion of pericyte leads to increased hypoxia, pro-metastatic factors, vascular leakiness, and metastasis (Gerhardt & Semb 2008; Xian et al. 2006). Therefore, simply targeting tumor pericytes will not produce many beneficial outcomes. However, increasing pieces of evidence indicate that subpopulations of tumor pericytes undergo phenotype switching by altered gene expression, leading to a pathological characteristic ***(foe)*** as we have discussed previously and summarized in Table 3.1. Thus, a better understanding of pathological phenotypes of pericyte will open up the opportunity for us to target these abnormal pericytes, which potentially leads to more efficient vascular normalization and anti-angiogenic approaches. Also, pericyte contribution to the vascular function might be context dependent. For instance, brain vasculature holds a unique structure called blood–brain barrier (BBB) of which pericyte is one of the major components. Truth hold in part in brain metastasis or brain tumor and it acts as an obstacle of drug delivery efficacy. Therefore, a specific function of the different subset of pericytes should be considered in a context-dependent manner.

We have already discussed the potential contribution of pericytes in metastasis in breast cancer. However, the newly emerging role of pericytes in the pathological/inflammatory environment gained much attention recently in the field of breast cancer. Several recent studies have shown reciprocal communication between tumor pericytes and immune components of the stroma. Regulator of G-protein signaling 5 (Rgs5) is one of the pericyte markers that are known to be expressed in PDGFRβ+ progenitor perivascular cells and overexpressed in the aberrant tumor vasculature. In RIP1-Tag5 mouse model, a large population of tumor pericytes expresses Rgs5 and PDGFRβ, representing immature progenitor status and small populations that express αSMA/NG2/desmin representing mature pericytes. In genetic deletion of Rgs-5 tumor context, pericytes phenotype was shifted toward the more mature state, αSMA/NG2/desmin leading to vascular normalization. Such phenotypic switching resulted in increased tumor infiltration by CD4+ and CD8+ T cells and immune control (Hamzah et al. 2008). This finding proposes the connection between subpopulation of pericytes (mature pericytes) and immune cells infiltration in a mouse model for pancreatic cancer, and it should be explored in breast cancer. One of the critical mechanisms by which tumors can escape from immune surveillance is the recruitment of myeloid-derived suppressor cells (MDSCs) (Gabrilovich and Nagaraj 2009). In mice defective for PDGFB retention ($PDGF\beta^{ret/ret}$), the loss of PDGFRβ+ pericytes hence decreased pericyte recruitment and enhanced intratumoral trafficking of MDSCs in IL-6-dependent manner (Hong et al. 2015). Gene expression analysis from patients with breast cancer showed that increased expression of human MDSC markers such as CD33 and S100A9 was correlated with decreased expression of pericyte marker genes. Moreover, the group of patients with low pericyte poor/MDSC rich was associated with poor long-term breast cancer-specific survival. Most recent finding in breast cancer emphasizes the importance of mutual

Table 3.1 Functional contribution of various pericytes in breast cancer

Friend				Foe			
Pericyte (marker)	Functional contribution			Pericyte (marker)	Functional contribution		
	IM	MET	VF		IM	MET	VF
NG2	Increase tumor immunity	Inhibit metastasis	Less hypoxia Better perfusion	CD248	NS	Increase intravasation	No effect
PDGFRβ	Suppress MDSC recruitment	Inhibit metastasis	Less hypoxia Better perfusion	Kfl4	NS	Metastatic niche formation	No effect
Desmin	Increase tumor immunity	Inhibit metastasis	Less hypoxia Better perfusion	Rgs-5	Immunosuppressive	Increase metastasis	Induce hypoxia Leaky vessels
αSMA[a]	?	?	?	GT198[b]	NS	NS	Leaky vessels
				EPT[c]	NS	Promote tumor growth	Less permeable vessels

IM tumor immunity, *MET* metastasis, *VF* vascular function, *NS* not studied
Pericyte phenotype is color-coded based on Fig. 3.1
[a]αSMA expression in pericytes of capillary beds is abnormal and a hallmark of pathological pericytes and its contribution to tumor microenvironment has not been explored in breast cancer
[b]Mutant form GT198 expression is the marker for malignant pericyte that can manifest the tumor microenvironment by affecting other stromal cells
[c]Tumor-specific type of pericyte arose by "Epithelial to pericyte phenotypic transition"

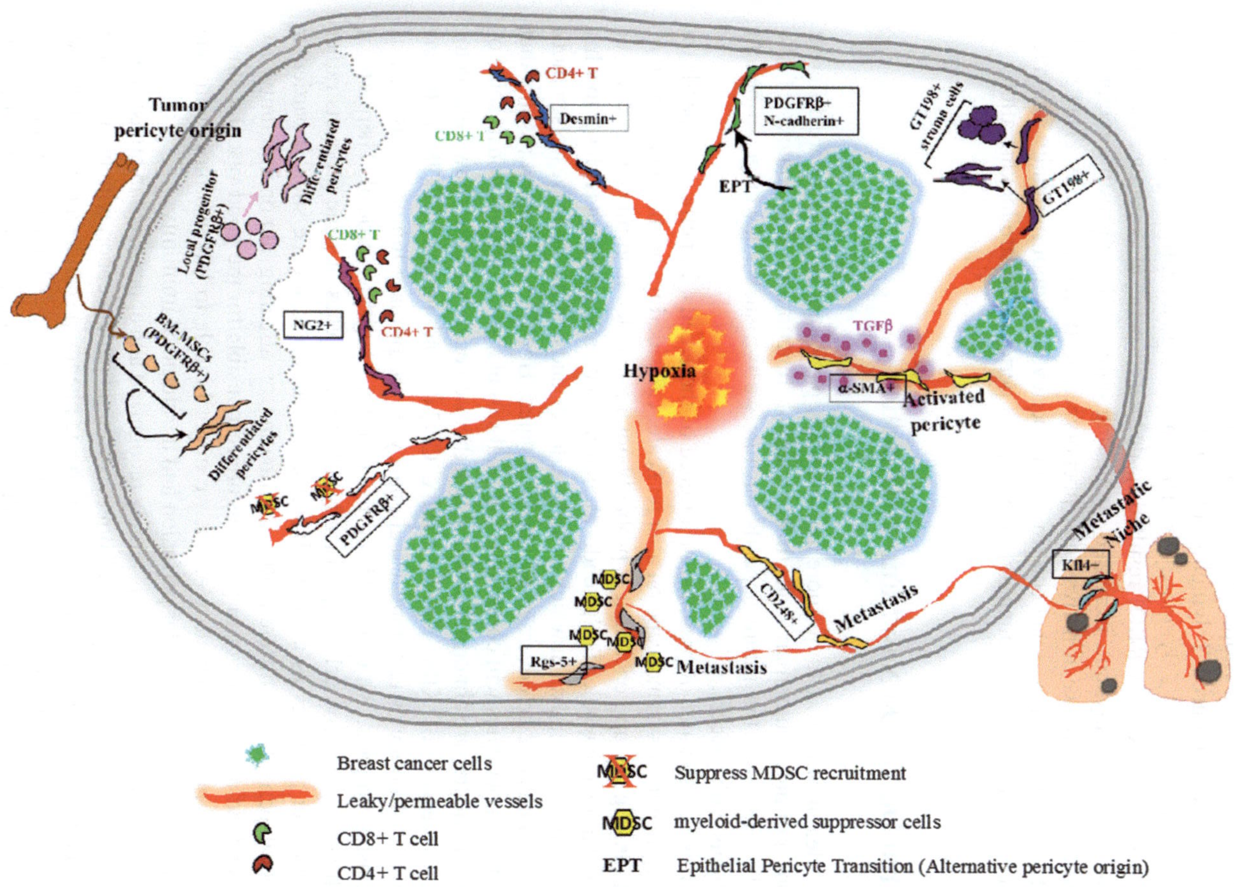

Fig. 3.1 Various pericyte phenotypes and their functional contribution in the context of breast cancer. Each phenotype is defined based on studied marker expression. In general, pericytes with ectopic expression of markers (e.g., Klf4, Rgs-5, GT198) confer adverse effects on tumor progression, metastasis, intra-tumoral immunity, and vascular function (Table 3.1). Details and references are described in the main text

regulation of tumor vascular normalization and tumor immunity. In this study, bioinformatic analysis data indicated that gene expression related to vascular normalization correlate with immune-stimulatory pathways and such hypothesis was further validated in a various model system. Loss of mature pericyte (NG2+ pericyte) leads to reduced T lymphocytes infiltration into orthotopic breast tumor, E0771. In reverse, T lymphocytes deficiency in genetically engineered mice (CD4KO, CD8KO, TCRKO) resulted in decreased pericyte coverage, increased vessels permeability, and increased circulating tumor cells, suggesting reciprocal regulatory loop between perivascular phenotype and tumor immunity (Tian et al. 2017). A different functional contribution of pericyte in tumor immunity reiterates that we should pay more attention to the type of pericytes we are looking after.

What Do We Do Now? Future Direction

It is clear now that we know more than ever how pathological pericytes are different from normal counterpart and their potential function in the context of the TME. However, the future depends on how we use such knowledge to benefit patients with cancers.

It is not a matter of presence or absence of these cells. It, perhaps, depends on their phenotype or landscape on a larger scale. Considering the marginal benefit of anti-angiogenic approaches, it is probably not a good idea to eliminate pericytes or vasculature as a whole. To this end, vascular normalization concept is closer to what we want to accomplish in which immature, leaky, and nonfunctional vessels without appropriate pericyte coverage will be eliminated. However, we cannot assume that all the left pericytes will contribute to normal vascular structure/function as some of these pericytes are abnormal or malignant themselves. It has been shown that ectopic expression of Klf4, Rgs-5, αSMA, CD248, and mutant GT198 in tumor-associated pericytes can spread malignancy to the primary tumor site as well as metastatic organs. We cannot afford to keep these pericytes around. Therefore, it will be a safer approach to specifically target these genes or gene products rather than target pericytes as a whole. Increasing pieces of evidence show a particular type of pericytes is differentially contributing or affecting tumor immunity. By understanding what flavor of pericytes are responsible for the good vascular structure and immune-stimulatory effect, we might be able to kill two birds with one stone by improving vascular perfusion and intratumoral immunity.

Perivascular heterogeneity is largely recognized and appreciated in breast cancer filed. A recent study using TMA from the different patient cohort with breast cancer has shown promising results in which pericyte phenotype can be a potential predictor for successful response to the specific type of drug. Although the underlying mechanism remains elusive, such results add a promising approach to map out personalized treatment. An additional approach might include reverting malignant pericyte phenotype to beneficial phenotype by molecular conversion. In this case, we do not have to kill anything. We just need to correct the problem.

The fact that different subtypes of breast cancer, namely TNBC and luminal, displayed significantly different perivascular phenotype might implicate that such perivascular landscape is either an intrinsic property of the particular type of cancer or heavily influenced by the distinct tumor microenvironment. Either way, we should consider identifying a connection between the properties of cancer cells and pericyte phenotypes.

We have accumulated enough pieces of evidence to be finally convinced that tumor-associated pericytes are heterogeneous and should not be considered as a single-cell population, as it can be a Jekyll or a Hyde at any moment depends on their phenotype, environment, and perhaps influence by cancer cells.

References

Aalders, K. C., et al. (2017). Anti-angiogenic treatment in breast cancer: Facts, successes, failures and future perspectives. *Cancer Treatment Reviews, 53*, 98–110.

Abramsson, A., et al. (2002). Analysis of mural cell recruitment to tumor vessels. *Circulation, 105*(1), 112–117.

Armulik, A., Genove, G., & Betsholtz, C. (2011). Pericytes: Developmental, physiological, and pathological perspectives, problems, and promises. *Developmental Cell, 21*(2), 193–215.

Ballabh, P., Braun, A., & Nedergaard, M. (2004). The blood-brain barrier: An overview: Structure, regulation, and clinical implications. *Neurobiology of Disease, 16*(1), 1–13.

Benjamin, L. E. (2001). Glucose, VEGF-A, and diabetic complications. *The American Journal of Pathology, 158*(4), 1181–1184.

Benjamin, L. E., et al. (1999). Selective ablation of immature blood vessels in established human tumors follows vascular endothelial growth factor withdrawal. *The Journal of Clinical Investigation, 103*(2), 159–165.

Bergers, G., & Song, S. (2005). The role of pericytes in blood-vessel formation and maintenance. *Neuro-Oncology, 7*(4), 452–464.

Bergers, G., et al. (2003). Benefits of targeting both pericytes and endothelial cells in the tumor vasculature with kinase inhibitors. *The Journal of Clinical Investigation, 111*(9), 1287–1295.

Birbrair, A., et al. (2014a). Type-2 pericytes participate in normal and tumoral angiogenesis. *American Journal of Physiology: Cell Physiology, 307*(1), C25–C38.

Birbrair, A., et al. (2014b). Type-1 pericytes accumulate after tissue injury and produce collagen in an organ-dependent manner. *Stem Cell Research & Therapy, 5*(6), 122.

Bos, R., et al. (2001). Levels of hypoxia-inducible factor-1 alpha during breast carcinogenesis. *Journal of the National Cancer Institute, 93*(4), 309–314.

Calabrese, C., et al. (2007). A perivascular niche for brain tumor stem cells. *Cancer Cell, 11*(1), 69–82.

Cheng, L., et al. (2013). Glioblastoma stem cells generate vascular pericytes to support vessel function and tumor growth. *Cell, 153*(1), 139–152.

Cooke, V. G., et al. (2012). Pericyte depletion results in hypoxia-associated epithelial-to-mesenchymal transition and metastasis mediated by met signaling pathway. *Cancer Cell, 21*(1), 66–81.

Crisan, M., et al. (2008). A perivascular origin for mesenchymal stem cells in multiple human organs. *Cell Stem Cell, 3*(3), 301–313.

Du, R., et al. (2008). HIF1alpha induces the recruitment of bone marrow-derived vascular modulatory cells to regulate tumor angiogenesis and invasion. *Cancer Cell, 13*(3), 206–220.

Eberhard, A., et al. (2000). Heterogeneity of angiogenesis and blood vessel maturation in human tumors: Implications for antiangiogenic tumor therapies. *Cancer Research, 60*(5), 1388–1393.

Enge, M., et al. (2002). Endothelium-specific platelet-derived growth factor-B ablation mimics diabetic retinopathy. *The EMBO Journal, 21*(16), 4307–4316.
Gabrilovich, D. I., & Nagaraj, S. (2009). Myeloid-derived suppressor cells as regulators of the immune system. *Nature Reviews. Immunology, 9*(3), 162–174.
Gerhardt, H., & Betsholtz, C. (2003). Endothelial-pericyte interactions in angiogenesis. *Cell and Tissue Research, 314*(1), 15–23.
Gerhardt, H., & Semb, H. (2008). Pericytes: Gatekeepers in tumour cell metastasis? *Journal of Molecular Medicine (Berlin, Germany), 86*(2), 135–144.
Goel, S., Wong, A. H., & Jain, R. K. (2012). Vascular normalization as a therapeutic strategy for malignant and nonmalignant disease. *Cold Spring Harbor Perspectives in Medicine, 2*(3), a006486.
Hall, A. P. (2006). Review of the pericyte during angiogenesis and its role in cancer and diabetic retinopathy. *Toxicologic Pathology, 34*(6), 763–775.
Hamzah, J., et al. (2008). Vascular normalization in Rgs5-deficient tumours promotes immune destruction. *Nature, 453*(7193), 410–414.
Hasumi, Y., et al. (2007). Identification of a subset of pericytes that respond to combination therapy targeting PDGF and VEGF signaling. *International Journal of Cancer, 121*(12), 2606–2614.
Hellstrom, M., et al. (1999). Role of PDGF-B and PDGFR-beta in recruitment of vascular smooth muscle cells and pericytes during embryonic blood vessel formation in the mouse. *Development, 126*(14), 3047–3055.
Hong, J., et al. (2015). Role of tumor pericytes in the recruitment of myeloid-derived suppressor cells. *Journal of the National Cancer Institute, 107*(10). https://doi.org/10.1093/jnci/djv209.
Hosaka, K., et al. (2016). Pericyte-fibroblast transition promotes tumor growth and metastasis. *Proceedings of the National Academy of Sciences of the United States of America, 113*(38), E5618–E5627.
Jubb, A. M., et al. (2011). Vascular phenotypes in primary non-small cell lung carcinomas and matched brain metastases. *British Journal of Cancer, 104*(12), 1877–1881.
Keskin, D., et al. (2015). Targeting vascular pericytes in hypoxic tumors increases lung metastasis via angiopoietin-2. *Cell Reports, 10*(7), 1066–1081.
Kim, J., et al. (2016). Heterogeneous perivascular cell coverage affects breast cancer metastasis and response to chemotherapy. *JCI Insight, 1*(21), e90733.
Li, Y. J., et al. (2013). Perfusion heterogeneity in breast tumors for assessment of angiogenesis. *Journal of Ultrasound in Medicine, 32*(7), 1145–1155.
Lin, S. L., et al. (2008). Pericytes and perivascular fibroblasts are the primary source of collagen-producing cells in obstructive fibrosis of the kidney. *The American Journal of Pathology, 173*(6), 1617–1627.
Lyle, L. T., et al. (2016). Alterations in pericyte subpopulations are associated with elevated blood-tumor barrier permeability in experimental brain metastasis of breast cancer. *Clinical Cancer Research, 22*(21), 5287–5299.
Mackey, J. R., et al. (2012). Controlling angiogenesis in breast cancer: A systematic review of anti-angiogenic trials. *Cancer Treatment Reviews, 38*(6), 673–688.
Meng, M. B., et al. (2015). Pericytes: A double-edged sword in cancer therapy. *Future Oncology, 11*(1), 169–179.
Morikawa, S., et al. (2002). Abnormalities in pericytes on blood vessels and endothelial sprouts in tumors. *The American Journal of Pathology, 160*(3), 985–1000.
Murgai, M., et al. (2017). KLF4-dependent perivascular cell plasticity mediates pre-metastatic niche formation and metastasis. *Nature Medicine, 23*(10), 1176–1190.
Nisancioglu, M. H., Betsholtz, C., & Genove, G. (2010). The absence of pericytes does not increase the sensitivity of tumor vasculature to vascular endothelial growth factor-A blockade. *Cancer Research, 70*(12), 5109–5115.
Ostman, A., & Heldin, C. H. (2007). PDGF receptors as targets in tumor treatment. *Advances in Cancer Research, 97*, 247–274.

Paiva, A. E., et al. (2018). Pericytes in the premetastatic niche. *Cancer Research, 78*(11), 2779–2786.

Paulsson, J., et al. (2009). Prognostic significance of stromal platelet-derived growth factor beta-receptor expression in human breast cancer. *The American Journal of Pathology, 175*(1), 334–341.

Pietras, A., et al. (2008). High levels of HIF-2alpha highlight an immature neural crest-like neuroblastoma cell cohort located in a perivascular niche. *The Journal of Pathology, 214*(4), 482–488.

Rajantie, I., et al. (2004). Adult bone marrow-derived cells recruited during angiogenesis comprise precursors for periendothelial vascular mural cells. *Blood, 104*(7), 2084–2086.

Rangasamy, S., et al. (2011). A potential role for angiopoietin 2 in the regulation of the blood-retinal barrier in diabetic retinopathy. *Investigative Ophthalmology & Visual Science, 52*(6), 3784–3791.

Raza, A., Franklin, M. J., & Dudek, A. Z. (2010). Pericytes and vessel maturation during tumor angiogenesis and metastasis. *American Journal of Hematology, 85*(8), 593–598.

Sennino, B., et al. (2007). Sequential loss of tumor vessel pericytes and endothelial cells after inhibition of platelet-derived growth factor B by selective aptamer AX102. *Cancer Research, 67*(15), 7358–7367.

Shenoy, A. K., et al. (2016). Epithelial-to-mesenchymal transition confers pericyte properties on cancer cells. *The Journal of Clinical Investigation, 126*(11), 4174–4186.

Shrivastav, S., et al. (2016). Tumor angiogenesis in breast cancer: Pericytes and maturation does not correlate with lymph node metastasis and molecular subtypes. *Clinical Breast Cancer, 16*(2), 131–138.

Sims, D. E. (2000). Diversity within pericytes. *Clinical and Experimental Pharmacology & Physiology, 27*(10), 842–846.

Sledge, G. W. (2015). Anti-vascular endothelial growth factor therapy in breast cancer: Game over? *Journal of Clinical Oncology, 33*(2), 133–135.

Song, S., et al. (2005). PDGFRbeta+ perivascular progenitor cells in tumours regulate pericyte differentiation and vascular survival. *Nature Cell Biology, 7*(9), 870–879.

Suzuki, Y., Nagai, N., & Umemura, K. (2016). A review of the mechanisms of blood-brain barrier permeability by tissue-type plasminogen activator treatment for cerebral ischemia. *Frontiers in Cellular Neuroscience, 10*, 2.

Tian, L., et al. (2017). Mutual regulation of tumour vessel normalization and immunostimulatory reprogramming. *Nature, 544*(7649), 250–254.

Tolaney, S. M., et al. (2015). Role of vascular density and normalization in response to neoadjuvant bevacizumab and chemotherapy in breast cancer patients. *Proceedings of the National Academy of Sciences of the United States of America, 112*(46), 14325–14330.

Viski, C., et al. (2016). Endosialin-expressing pericytes promote metastatic dissemination. *Cancer Research, 76*(18), 5313–5325.

Weidner, N., et al. (1991). Tumor angiogenesis and metastasis—correlation in invasive breast carcinoma. *The New England Journal of Medicine, 324*(1), 1–8.

Xian, X., et al. (2006). Pericytes limit tumor cell metastasis. *The Journal of Clinical Investigation, 116*(3), 642–651.

Yang, Z., et al. (2016). GT198 expression defines mutant tumor stroma in human breast cancer. *The American Journal of Pathology, 186*(5), 1340–1350.

Chapter 4
Pericytes in Sarcomas and Other Mesenchymal Tumors

Leslie Chang, Michelle A. Scott, Carolyn A. Meyers, and Aaron W. James

Abstract Tumors of mesenchymal origin are a diverse group, with >130 distinct entities currently recognized by the World Health Organization. A subset of mesenchymal tumors grow or invade in a perivascular fashion, and their potential relationship to pericytes is a matter of ongoing interest. In fact, multiple intersections exist between pericytes and tumors of mesenchymal origin. First, pericytes are the likely cell of origin for a group of mesenchymal tumors with a common perivascular growth pattern. These primarily benign tumors grow in a perivascular fashion and diffusely express canonical pericyte markers such as CD146, smooth muscle actin (SMA), platelet-derived growth factor receptor beta (PDGFR-β), and RGS5. These benign tumors include glomus tumor, myopericytoma, angioleiomyoma, and myofibroma. Second and as suggested by animal models, pericytes may give rise to malignant sarcomas. This is not a suggestion that all sarcomas within a certain subtype arise from pericytes, but that genetic modifications within a pericyte cell type may give rise to sarcomas. Third, mesenchymal tumors that are likely not a pericyte derivative co-opt pericyte markers in certain contexts. These include the PEComa family of tumors and liposarcoma. Fourth and finally, as "guardians" that enwrap the microvasculature, nonneoplastic pericytes may be important in sarcoma disease progression.

Keywords Pericyte · Mural cell · Glomus tumor · Myopericytoma Angioleiomyoma · Myofibroma · Sarcoma · Liposarcoma · Angiomyolipoma PEComa · Perivascular epithelioid cell tumor · Pericyte mimicry · Extravascular migratory metastasis

---

L. Chang · C. A. Meyers · A. W. James (✉)
Division of Surgical Pathology, Department of Pathology, Johns Hopkins University, Baltimore, MD, USA
e-mail: awjames@jhmi.edu

M. A. Scott
Division of Plastic Surgery, Department of Surgery, Children's Hospital of Pennsylvania, Philadelphia, PA, USA

A. Birbrair (ed.), *Pericyte Biology in Disease*, Advances in Experimental Medicine and Biology 1147, https://doi.org/10.1007/978-3-030-16908-4_4

Introduction

Tumors of mesenchymal origin are a diverse group, with >130 recognized by the World Health Organization (Fletcher et al. 2013). Broadly, they can be divided into those tumors that arise from the skeleton (tumors of bone) and those that do not (tumors of soft tissue). Malignant tumors of mesenchymal origin are termed sarcomas and are further classified based on distinguishing cytomorphology, production of matrix, immunohistochemical profile, or characteristic genetic changes. Although immunohistochemistry and molecular pathology allow for distinguishing the many sarcoma subtypes, little is known regarding the cell type of origin for most sarcomas.

Pericytes are mural cells within the microvasculature that interact with endothelium, with important physiologic and pathologic roles in vascular growth and homeostasis. Multiple intersections exist between pericytes and tumors of mesenchymal origin. Each of these intersections will be described in more detail in the present chapter. First, pericytes are the likely cell of origin for a group of mesenchymal tumors with a common perivascular growth pattern. These primarily benign tumors grow in a perivascular fashion and diffusely express canonical pericyte markers such as CD146, smooth muscle actin (SMA), and platelet-derived growth factor receptor beta (PDGFR-β) (Corselli et al. 2012). Second, pericytes may give rise to malignant sarcomas. This is not a suggestion that all sarcomas within a certain subtype arise from pericytes, but that genetic modifications within a pericyte cell type may give rise to sarcomas. Third, mesenchymal tumors that are likely not a pericyte derivative co-opt pericyte markers in certain contexts. Here, the PEComa family of tumors and liposarcoma will be discussed, with analogies drawn to other nonmesenchymal tumors. Fourth and finally, nonneoplastic pericytes may be important in sarcoma disease progression.

Pericytes as the Likely Cell of Origin for Mesenchymal Neoplasms/Tumorigenesis

First, pericytes are the likely cell of origin for a group of mesenchymal tumors with a common perivascular growth pattern. These primarily benign tumors diffusely express canonical pericyte markers. These features strongly suggest a pericyte or modified pericyte cell of origin.

The pericytic/perivascular family of soft-tissue tumors are generally composed of modified vascular smooth muscle cells. These consist of glomus tumor and its variants, myopericytoma, angioleiomyoma, and sinonasal hemangiopericytoma. Myofibroma also has overlapping histologic and immunohistochemical features, but will not be discussed further. We will describe the clinical and histologic appearance of these tumors as well as their potential perivascular origin. Later discussed is the PEComa (perivascular epithelioid cell tumor) family of tumors. Perivascular markers are more focal in the PEComa family tumors, and their pericyte antigen expression is more likely a manifestation of pericyte mimicry, a co-option of pericyte markers by a neoplastic cell of different origin.

Classic (Sporadic) Glomus Tumor

Classic glomus tumor (GT) are benign, less than 1 cm, red-blue nodules that are typically located on the distal upper and lower extremities (Weiss and Goldblum 2008). They are most often found in a subcutaneous, subungual location and clinically present with pain and temperature sensitivity (Nuovo et al. 1990; Van Geertruyden et al. 1996). Histologically, GTs resemble the thermoregulatory glomus body and demonstrate small uniform glomus cells in a variably prominent perivascular growth pattern (Fig. 4.1). This architectural feature, combined with ultrastructural and immunohistochemical demonstration of smooth muscle differentiation, shows some overlap with microvascular pericytes. For example, glomus cells demonstrate immunoreactivity for α-smooth muscle actin, vimentin, and up to 20% have focal CD34 expression (Folpe et al. 2001). Electron microscopy is also suggestive of modified pericytic or smooth muscle differentiation and reveals abundant cytoplasmic actin filaments and numerous micropinocytotic vesicles (Ghadially 1980; Erlandson 1994a). Like other perivascular tumor members, GTs demonstrate diffuse immunoreactivity for pericyte markers, including SMA, CD146, PDGFR-β (Shen et al. 2015a), and the more novel pericyte marker RGS5 (Shen et al. 2016) (Table 4.1). These findings support the notion of a pericyte cell of origin for GTs as well as GT variants, discussed below. Recently, *NOTCH1*, *NOTCH2*, and *NOTCH3* gene rearrangements were described in glomus tumors (Mosquera et al. 2013). Although speculative, it is possible that these characteristic gene rearrangements in a pericyte cell type give rise to classic GT.

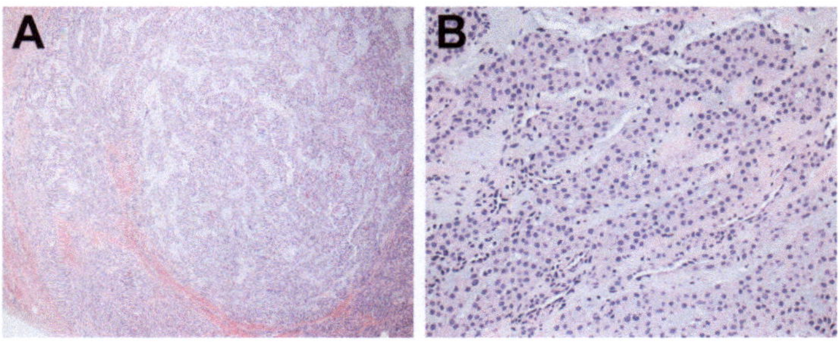

Fig. 4.1 Representative histologic appearance of classic glomus tumor (GT). Representative images of classic glomus tumor, demonstrating a solid growth pattern, monomorphic rounded tumor cells, and a myxoid background. (**a**) 4×. (**b**) 20×

Glomus Tumor Variants

Although a number of GT variants diverge from the typical appearance and presenting features of GT, they still express similar perivascular tumor growth patterns and a pericytic/myoid immunophenotype. One such variant, glomangiomyoma, displays focal smooth muscle differentiation and lies on a spectrum with similarity to angiomyoma. Another variant, known as glomuvenous malformation or glomangioma, is made up of typical glomus cells, but has architecture resembling a cavernous hemangioma (Fig. 4.2). Lesions generally arise during childhood and are less painful than classic GT (Weiss and Goldblum 2008). Glomangiomatosis is a rare type of invasive and diffuse lesions that are typically located in adipose tissue (Weiss and Goldblum 2008). Also exceedingly rare, atypical, and malignant glomus tumors are deep-seated, proliferative, larger tumors and have a combination of infiltrative growth, atypia, and increased mitotic activity. Folpe et al. suggested the following scheme to classify glomus tumor malignancy: deep-seated location >2 cm in size, or atypical mitotic figures, or moderate to high nuclear grade and ≥ 5 mitotic figures/50 hpf (Folpe et al. 2001). Malignant GT demonstrates a similar pattern of immunoreactivity for pericyte antigens, with at least some staining for SMA, CD146, PDGFR-β, and RGS5 in most tumors (Shen et al. 2015a, 2016). Interestingly, malignant GT shows a marked reduction in the intensity and distribution of pericyte markers in comparison to classic GTs (including SMA, CD146, PDGFR-β) (Shen et al. 2015a) (Table 4.1). These findings may reflect de-differentiation in malignant GT. Like classic GT, malignant GT has been observed to harbor characteristic *NOTCH* gene rearrangements (Mosquera et al. 2013).

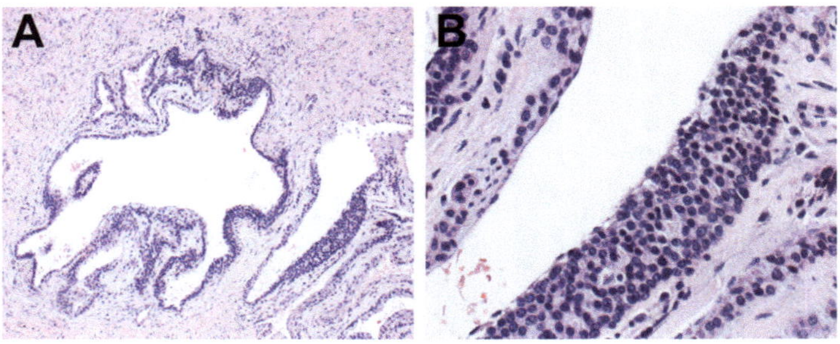

Fig. 4.2 Representative images of glomuvenous malformation, which demonstrate the same rounded tumor cells as glomus tumor. A cuff of round, monomorphic tumor cells is seen bordering the dilated vascular spaces. (**a**) 10×. (**b**) 40×

Myopericytoma

Myopericytomas are dermal or subcutaneous tumors that are benign, painless, and well circumscribed. They are typically found on the lower extremities and are known to have pericytic/smooth muscle differentiation (Weiss and Goldblum 2008). Similar to the glomus tumor, histological analysis of myopericytoma shows myoid cell proliferation around the vasculature (Fig. 4.3). However, myopericytoma has a characteristic whorled pattern of perivascular, ovoid to spindled cell with eosinophilic cytoplasm. Immunohistochemical analysis of myopericytoma includes αSMA and h-caldesmon positivity in a diffuse or perivascular pattern (Mentzel et al. 2006; Granter et al. 1998). Similar to the classic GT, focal CD34 expression has been found (Granter et al. 1998). Cytologic analysis illustrates smaller nuclei and more eosinophilic cytoplasm in myopericytoma cells as compared to glomus cells. Evidence of pericytic/myoid differentiation can be observed via electron microscopy, which reveals thin filaments, subplasmalemmal densities, and pinocytotic vessels (Erlandson 1994a). Like glomus tumor, myopericytoma demonstrates diffuse expression of pericyte antigens, including SMA, CD146, PDGFR-β, and RGS5 in most tumors (Shen et al. 2015a, 2016) (Table 4.1).

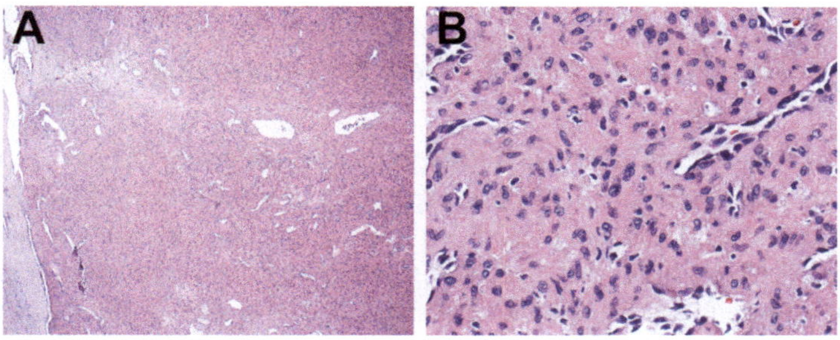

Fig. 4.3 Representative histologic appearance of myopericytoma, which demonstrate the monomorphic round to ovoid tumor cells with eosinophilic cytoplasm and indistinct borders. As in glomus tumors, vascular spaces are numerous. Perivascular cell growth outside the main tumor mass is not uncommon. (**a**) 4×. (**b**) 40×

Table 4.1 Summary of immunohistochemical staining results for perivascular markers in perivascular tumors

	SMA	CD146	PDGFRB	h-Caldesmon	RGS5
Classic GT	100% (9/9)[a]	88% (8/9)	67% (6/9)	44% (4/9)	67% (4/6)
Malignant GT	75% (3/4)	100% (4/4)	25% (1/4)	25% (1/4)	50% (2/4)
Myopericytoma	100% (3/3)	100% (3/3)	67% (2/3)	100% (3/3)	100% (3/3)
Angioleiomyoma	100% (9/9)	78% (7/9)	11% (1/9)	78% (7/9)	67% (6/9)
SFT	0% (0/10)	0% (0/10)	0% (0/10)	10% (1/10)	10% (1/10)
PEComa family	58% (11/19)	58% (11/19)	37% (7/19)	NR	0% (0/19)

GT glomus tumor, *NR* not recorded, *PDGFRB* platelet-derived growth factor receptor beta, *SFT* solitary fibrous tumor, *SMA* smooth muscle actin. Data derived from Shen et al. (2015a, b, 2016)
[a]Positive staining defined as 50% distribution or greater of intermediate (2+) staining intensity or greater

Angioleiomyoma

Angioleiomyoma generally presents as a painful subcutaneous nodule, with a histological appearance of more differentiated smooth muscle cells in comparison to either myopericytoma or glomus tumor. The tumor cells of angioleiomyoma can be arranged in perivascular, fascicular, or cavernous growth patterns. Tumors are found in the dermis and superficial soft tissues, most commonly of the distal extremities. Tumors typically show characteristics of both solid and venous growth patterns, and they are composed of eosinophilic tumor cells with indistinct cell borders, consistent with smooth muscle differentiation (Fig. 4.4). Clinical immunohistochemical stains generally are diffusely positive for markers of smooth muscle differentiation, including MSA and desmin. All angioleiomyoma specimens demonstrated at least some focal immunoreactivity for each pericyte marker examined, including αSMA, CD146, and PDGFRβ (Shen et al. 2015a, 2016). In comparison to other perivascular tumors, angioleiomyoma showed more weak and focal immunoreactivity for PDGFRβ (Shen et al. 2015a). Interestingly, pericyte markers were more consistently expressed in the prominent perivascular/venous areas, rather than in areas of fascicular tumor growth (Shen et al. 2015a).

Solitary Fibrous Tumor (SFT)/Hemangiopericytoma (HPC)

Solitary fibrous tumor (SFT), previously termed hemangiopericytoma, is a translocation sarcoma resulting in STAT6 overexpression (Robinson et al. 2013; Chmielecki et al. 2013). Several features of SFT led investigators to previously hypothesize a pericyte cell of origin for SFT, including the prominent vasculature of SFT with branching (or staghorn-shaped) blood vessels, as well as previously described ultrastructural features of SFT suggesting a pericyte-like cell type (Ghadially 1980; Erlandson 1994b). In general, SFT is currently categorized as a fibrous rather than

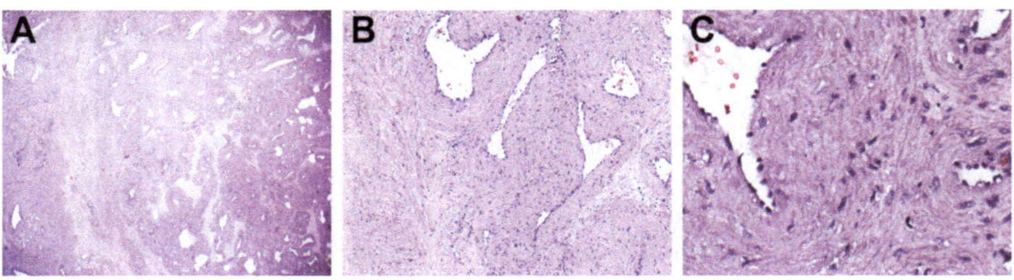

Fig. 4.4 Representative histologic appearance of angioleiomyoma, which demonstrates a smooth muscle neoplasm with fascicular (leiomyoma-like) and venous growth patterns. Dilated venous spaces show concentric layers of smooth muscle, and it is in these areas that pericyte markers are most expressed. (**a**) 4×. (**b**) 10×. (**c**) 40×

pericytic tumor, characterized by CD34 expression. The exception to this is potentially in the sinonasal region (see below).

Sinonasal Hemangiopericytoma

Sinonasal hemangiopericytoma, also termed hemangiopericytoma-like tumor of nasal passages (HTNPs), is a distinct entity which arises in the nasal passages, often in female patients (Weiss and Goldblum 2008). Histologically, sinonasal hemangiopericytoma is similar in appearance to typical SFT/HPC; however, most HTNPs lack CD34 immunoreactivity (Yasui et al. 2001). Nuclear beta-catenin expression is a common feature in sinonasal hemangiopericytoma, although this is a nonspecific feature (Jo and Fletcher 2017). These tumors are similar to glomangiopericytoma of soft tissue in both morphology and expression of αSMA and vimentin, leading to the adoption of the term glomangiopericytoma for a potentially more accurate classification (Thompson 2004). Although HTNP has been hypothesized to display true pericytic/myoid differentiation (Kuo et al. 2005), this idea remains largely uninvestigated.

Pericytes as a Potential Cell of Origin for Sarcomas

Although the cell of origin for mesenchymal tumors is unclear, we and others have postulated that pericytes may give rise to malignant mesenchymal tumors (sarcomas). This is not a suggestion that all sarcomas within a certain subtype arise from pericytes, but that genetic modifications within a pericyte cell type may give rise to sarcomas. To examine the hypothesis experimentally, Sato et al. used a tracing approach using the pericyte marker *Ng2* to examine potential pericyte contribution to sarcoma (Sato et al. 2016). Pericyte reporter expression was observed among *Trp53* deficiency-induced osteosarcoma and *Trp53* deficiency-induced undifferentiated pleomorphic sarcoma (Sato et al. 2016). In this study, stabilization of β-catenin resulted in tumors with phenotypic similarity to fibromatosis (desmoid tumor), which were also diffusely labeled with the pericyte marker Ng2 (Sato et al. 2016). Microarray analysis demonstrated that these spontaneous tumors also demonstrated similar gene expression to human sarcomas (Sato et al. 2016). This direct evidence of pericyte contribution *Trp53*-driven sarcomas has parallels in observations among human sarcoma. For example, NG2 expression is observed in soft-tissue sarcomas (Benassi et al. 2009), while the canonical pericyte marker CD146 is expressed in human osteosarcoma and osteosarcoma cell lines (Schiano et al. 2012). Importantly, no known pericyte marker in either mouse or human is entirely specific for a pericyte cell. For example, NG2 expression is found in neural progenitor cells, chondrocytes, chondroblasts, and cardiomyocytes (Levine and Nishiyama 1996). Thus, the direct contribution of pericytes to sarcomas, while an intriguing hypothesis supported by some indirect evidence, is as yet unproven in either mouse or man.

Aberrant Adoption of Pericyte Markers

In addition, mesenchymal tumors that are likely not a pericyte derivative co-opt pericyte markers in certain contexts. Here, the PEComa family of tumors and liposarcoma will be discussed, with analogies drawn to other nonmesenchymal tumors.

PEComa Family of Tumors

The PEComa family tumors are a unique and diverse group of tumors with dual smooth muscle and melanocytic differentiation and unknown histogenesis (Weiss and Goldblum 2008; Armah and Parwani 2009). The most common tumors in the PEComa family are renal and extrarenal angiomyolipomas. Angiomyolipomas have a characteristic triphasic appearance including thick-walled blood vessels, myoid-appearing perivascular cells, and lipid-distended cells resembling adipocytes

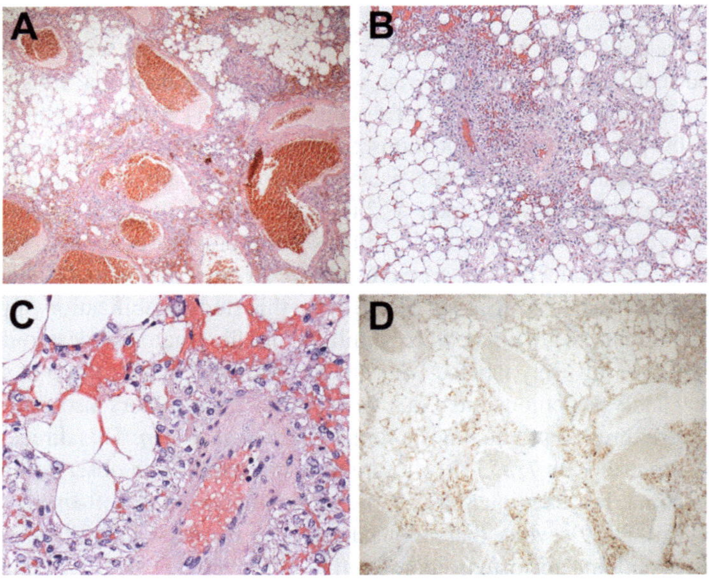

Fig. 4.5 Representative histologic appearance of angiomyolipoma (AML). Among the PEComa family tumors, AML is the most common. AML has a triphasic appearance, with prominent vasculature, perivascular epithelioid cell proliferation, and adipose-like tissue. (**a**) 4×. (**b**). 10×. (**c**) 40×. (**d**) MART1 immunohistochemical staining, 4×

(Fig. 4.5) (Weiss and Goldblum 2008). Examples of other variants include lymphangiomyomatosis and clear cell sugar tumor of the lung (Frack et al. 1968; Liebow and Castleman 1971). PEComa tumor cells are all characterized by their perivascular growth pattern and dual myomelanocytic differentiation and can be either spindled or epithelioid in cytomorphology. Immunohistochemical features of PEComa tumors are distinct and demonstrate co-expression of pericyte markers including CD146, αSMA, and PDGFR-β (Shen et al. 2015b) and melanocytic markers including HMB-45, Melan-A, MiTF, and S100 (Pea et al. 1991; Folpe et al. 2005). Studies examining pericyte antigens on the PEComa family of tumors suggest that differential expression of markers may reflect changes in pericyte differentiation and tumor behavior (Shen et al. 2015b). Interesting similarities exist between angiotropic melanoma, discussed below, and PEComa. However, while angiotropism in melanoma is associated with a poor prognosis (Barnhill et al. 2002), PEComa family tumors are defined by their vasculocentric growth, and a similar prognostic significance has not been established.

Pericyte Mimicry/Extravascular Migratory Metastasis

In the past 15 years, a novel metastatic paradigm has been described in extravascular tumors. In contrast to the classical mechanism of lymphovascular invasion, extravascular migratory metastasis (EVMM) involves metastasis outside of the bloodstream. Also known as "angiotropism" or "pericytic mimicry," tumor cells adopt characteristic pericyte cell surface markers and perivascular migration patterns along the external or abluminal surfaces of vascular channels, without intravasation (Lugassy and Barnhill 2007; Lugassy et al. 2013a). By this mechanism, tumor cells may spread to nearby or more distant sites. Perivascular invasion through extravascular migratory metastasis is an underrecognized route of tumor spread.

Pericyte mimicry has been identified in common malignancies of the skin, brain, pancreas, and prostate and most thoroughly studied in melanoma (Lugassy and Barnhill 2007; Lugassy et al. 2002, 2005, 2013a, b; Bald et al. 2014). Lugassy et al. first described melanoma cells invading the blood vessel basal lamina (Lugassy et al. 1998, 1999, 2000). Pericyte mimicry in melanoma has since been confirmed at the immunophenotypic level by the expression of pericyte antigens: PDGFR-β, NG2, and CD146 (Lugassy et al. 2013b). Microarray analysis of migrating of melanoma cells along the abluminal endothelial tubules found upregulated expression of malignancy-associated genes linked to metastasis (including CCL2, ICAM1, and IL6) and disease progression (including CCL2, ICAM1, SELE, TRAF1, IL6, SERPINB2, and CXCL6) (Lugassy et al. 2013b). Additionally, angiotropism is clinically significant and a predictor of metastasis in malignant melanoma (Barnhill and Lugassy 2004).

Parallels may be drawn between angiotropism of melanoma cells and that of other tumors with perivascular invasion. For example, malignant glial cells in glioblastoma multiforme (GBM) adopt an angiotropic distribution without entering the luminal space (Lugassy et al. 2002; Giese and Westphal 1996). Using in vivo cell tracking, Cheng et al. showed that glioblastoma stem cells (GSC) can differentiate into an immunophenotype similar to pericytes (Cheng et al. 2013). Recent studies have found that GBM cell angiotropism is associated with dramatic changes in the structure of preexisting blood vessels, termed "vessel co-option" (Caspani et al. 2014). Additionally, pancreatic and prostate adenocarcinomas have also displayed pericyte mimicry with regional invasion and EVVM (Lugassy et al. 2005; Levy et al. 2009). Well-differentiated liposarcoma often shows evidence of perivascular condensation/myoid proliferation, and we observed that a similar phenotype of pericytic mimicry is also seen here (Shen et al. 2016). Like melanoma, well-differentiated liposarcoma tumor cells adjacent to blood vessels result in aberrant adoption of pericyte markers, including αSMA, CD146, and RGS5 (Shen et al. 2016). In sum, these studies suggest that common malignancies of the skin, brain, pancreas, prostate, and fat tissue all show evidence of pericyte mimicry/EVMM. Other tumor types, although not yet described, likely have a similar ability to invade in a similar perivascular fashion.

Pericytes and Disease Progression

Lastly, tumor-associated pericytes may play an important role in sarcoma formation and metastasis. Below is a discussion of the role of nonneoplastic pericytes in both mesenchymal malignancies (sarcomas) and epithelial malignancies (carcinomas).

Multiple lines of evidence suggest that pericytes play a naturally protective role against vascular invasion by tumor cells. This "barrier" function is relatively intuitive, given intimate association of the pericytes with the endothelium. First, established evidence has shown that pericytes around tumor vessels demonstrate abnormalities including disordered arrangement, cell shape, and loosened attachment (Lugassy et al. 2005; Abramsson et al. 2002; Allt and Lawrenson 2001). By electron and confocal microscopy, tumor-associated pericytes have been observed to have increased distance from associating blood vessels as well as extended cytoplasmic processes and altered endothelial interactions (Morikawa et al. 2002; Barlow et al. 2013). Morphological and phenotypic changes of pericytes in tumors are thought to result in increased vascular permeability and reduced barrier to vascular invasion. In fact, reduced pericyte coverage along tumor-associated vessels has been associated with poor prognosis in breast adenocarcinoma (Cooke et al. 2012). Conversely, higher levels of markers of pericyte stability including CD34+ and PDGF-β, a cell-surface tyrosine kinase receptor critical to pericyte formation and migration, have been found as a favorable prognostic factor (Wang et al. 2015). Tumors may also recruit perivascular precursors from the bone marrow for vascular remodeling (Song et al. 2005). These structural changes are associated with abnormal protein expression in tumor-associated pericytes. For example, aberrant desmin immunoreactivity was observed in pericytes in pancreatic adenocarcinoma, whereas normal pancreatic pericytes are not immunoreactive (Morikawa et al. 2002).

Multiple independent investigators have studied the role of pericytes or the pericyte–endothelial cell complex in tumors and found increased vascular permeability, tumor cell vascular invasion, and tumor metastasis. Xian et al. examined the role of pericytes in limiting vascular invasion in two murine pancreatic β-cell tumor models (Xian et al. 2006). First, NCAM (neural cell adhesion molecule)-deficient mice displayed perturbed pericyte–endothelial cell interaction, resulting in increased vascular permeability and tumor metastasis (Xian et al. 2006). Similarly, pericyte-deficient *Pdgfbret/ret* mouse resulted in widespread metastasis of pancreatic β cell in lymph nodes and distant sites (Xian et al. 2006). Conversely, NCAM gain of function limits tumor metastasis and vascular invasion through pericyte–endothelial interaction (Xian et al. 2006). Cooke et al. examined the effects of pericyte deficiency in breast adenocarcinoma metastasis, using a genetically engineered mouse model with ablation of NG2+ PDGFR-β+ pericytes using ganciclovir as well as Imatinib and Sunitinib (Cooke et al. 2012). Similar to Xian et al. reduced pericyte coverage resulted in decreased tumor volumes; however, it also increased the rate of vascular invasion and metastatic burden.

Disruption or dissociation of pericytes from their endothelial cells also results in a hypoxic tumor microenvironment, which may induce tumor cell epithelial-to-mesenchymal transition (EMT) and promotion of metastasis (Harris 2002). Hypoxic conditions inhibit degradation of hypoxia-inducible factor-1alpha (HIF-1α), which then binds to the proximal promoter of *TWIST,* an EMT master regulator (Yang et al. 2008). HIF-1α also binds to the *Met* promoter and amplifies signaling of hepatocyte growth factor (HGF) receptor protein, promoting tumor invasion (Fujiuchi et al. 2003; Ren et al. 2005) and EMT (Ren et al. 2005) in multiple cancer cell lines in vitro. One of the mechanisms leading to tumor hypoxia is reduced pericyte coverage. Cooke et al. found that pericyte depletion through ganciclovir treatment led to an EMT shift with increased expression mesenchymal transcription factors, such as *HIF-1α* and *Met,* rather than an epithelial phenotype (Cooke et al. 2012). Thus, pericyte loss likely has multiple effects enhancing tumor metastasis, including hypoxia-induced EMT cascade leading to tumor invasiveness as well as increased vascular permeability and concomitant vascular invasion. Collectively, these carcinoma models illustrate that the disruption of pericyte–endothelial cell interaction leads to hyperpermeable vessels, increased vascular invasion, and tumor metastasis.

Pathways important in pericyte biology have been studied in the setting of sarcomagenesis, and the Notch signaling pathway plays a vital role in angiogenesis and vasculogenesis of neoplasms. Notch signaling is crucial for pericyte recruitment, and pericytes express Notch1-3. Ewing's sarcoma upregulates Notch signaling through RE1-silencing transcription factor (REST) and activation of ligand Delta-like4 (DLL4). Investigators that have targeted inhibition of REST and DLL4 discovered altered vascular morphology and increased hypoxia (Zhou et al. 2014; Stewart et al. 2011; Schadler et al. 2010). Notch signaling is also a known oncogene for osteosarcoma and plays a role in tumor invasion and metastasis (Engin et al. 2009; Zhang et al. 2010; Hughes 2009). The study by Tanaka et al. found that upregulation of Notch1 and Notch2 in vitro has been correlated with increase in metastasis

(Tanaka et al. 2009). Conversely, inhibition of Notch through gamma secretase inhibitor reduced osteosarcoma invasion and migration (Tanaka et al. 2009). Specific targeting of pericyte signaling cascades demonstrates the complex role pericytes play in cancer progression and metastasis.

Additionally, it has been shown that pericytes play a role in the resistance of antiangiogenic therapy. Overall, it appears that immature blood vessels without pericyte coverage are more susceptible to VEGF-targeted agents (Benjamin et al. 1999). Conversely, anti-VEGF agents tend to be ineffective in mature vessels with pericyte coverage (Inai et al. 2004; Bergers et al. 2003; Erber et al. 2004). Therefore, the concept of a combination therapy has been developed, using both VEGF and PDGFR inhibitors (Bergers et al. 2003; Erber et al. 2004). It appears that in many clinical scenarios approaches using both types of inhibitors are more effective against tumor vessels than either approach alone (Bergers et al. 2003; Erber et al. 2004). However, this synergistic effect has not been observed by all investigators (Hainsworth et al. 2007; Kuhnert et al. 2008).

Recently, investigators have attempted to subcategorize pericytes by cell surface marker expression. For example, Birbair et al. have described type 1 and 2 pericytes based on differential expression of Nestin and NG2 (Birbrair et al. 2014). Interestingly, only type-2 pericytes (Nestin+/NG2+) seem to participate in tumoral angiogenesis in this model, although selective targeting of type-2 pericytes has not yet been employed (Birbrair et al. 2014). Given that pericytes are multipotent, they may be a heterogeneous population and may vary in function and require more specific cellular targeting during tumor angiogenesis.

Summary

Far from being inert bystanders in tumorigenesis and tumor spread, pericytes and other perivascular stem cells have significant function in tumor biology. As their biologic relevance continues to expand, more and more investigators have turned to pericytes as potential regulators of local and distant tumor spread. Pericytic regulation of tumor spread has been studied across diverse malignancies, including melanoma, liposarcoma, various adenocarcinomas, and glioblastoma, to name a few. In particular to regulation of tumor spread, it seems that pericytes play at least a dual role: including (1) prevention/regulation of vascular invasion, but also (2) probable regulation of angiotropism/pericyte mimicry. From these standpoints, methods to reinforce the pericyte–endothelial cell interaction may be an important future adjunct to traditional chemotherapeutic agents. As investigators have already shown, pericytes may be subcategorized based on cell surface marker expression. Selected targeting of pericyte subpopulations may be a future promising avenue for tumor-specific effects. Finally, in terms of tumorigenesis, several lines of evidence suggest pericytic differentiation in a number of soft-tissue tumors. While interesting to posit that the pericyte is the cell of origin for select perivascular/mesenchymal tumors, further investigation in this area is required.

Acknowledgments AWJ was supported by the NIH/NIAMS (grants R01 AR070773, K08 AR068316), the Musculoskeletal Transplant Foundation, the Maryland Stem Cell Research Fund, and Orthopaedic Research and Education Foundation with funding provided by the Musculoskeletal Transplant Foundation.

References

Abramsson, A., et al. (2002). Analysis of mural cell recruitment to tumor vessels. *Circulation, 105*(1), 112–117.

Allt, G., & Lawrenson, J. G. (2001). Pericytes: Cell biology and pathology. *Cells, Tissues, Organs, 169*(1), 1–11.

Armah, H. B., & Parwani, A. V. (2009). Perivascular epithelioid cell tumor. *Archives of Pathology & Laboratory Medicine, 133*(4), 648–654.

Bald, T., et al. (2014). Ultraviolet-radiation-induced inflammation promotes angiotropism and metastasis in melanoma. *Nature, 507*(7490), 109–113.

Barlow, K. D., et al. (2013). Pericytes on the tumor vasculature: Jekyll or hyde? *Cancer Microenvironment, 6*(1), 1–17.

Barnhill, R. L., & Lugassy, C. (2004). Angiotropic malignant melanoma and extravascular migratory metastasis: Description of 36 cases with emphasis on a new mechanism of tumour spread. *Pathology, 36*(5), 485–490.

Barnhill, R., Dy, K., & Lugassy, C. (2002). Angiotropism in cutaneous melanoma: A prognostic factor strongly predicting risk for metastasis. *The Journal of Investigative Dermatology, 119*(3), 705–706.

Benassi, M. S., et al. (2009). NG2 expression predicts the metastasis formation in soft-tissue sarcoma patients. *Journal of Orthopaedic Research, 27*(1), 135–140.

Benjamin, L. E., et al. (1999). Selective ablation of immature blood vessels in established human tumors follows vascular endothelial growth factor withdrawal. *The Journal of Clinical Investigation, 103*(2), 159–165.

Bergers, G., et al. (2003). Benefits of targeting both pericytes and endothelial cells in the tumor vasculature with kinase inhibitors. *The Journal of Clinical Investigation, 111*(9), 1287–1295.

Birbrair, A., et al. (2014). Type-2 pericytes participate in normal and tumoral angiogenesis. *American Journal of Physiology. Cell Physiology, 307*(1), C25–C38.

Caspani, E. M., et al. (2014). Glioblastoma: A pathogenic crosstalk between tumor cells and pericytes. *PLoS One, 9*(7), e101402.

Cheng, L., et al. (2013). Glioblastoma stem cells generate vascular pericytes to support vessel function and tumor growth. *Cell, 153*(1), 139–152.

Chmielecki, J., et al. (2013). Whole-exome sequencing identifies a recurrent NAB2-STAT6 fusion in solitary fibrous tumors. *Nature Genetics, 45*(2), 131–132.

Cooke, V. G., et al. (2012). Pericyte depletion results in hypoxia-associated epithelial-to-mesenchymal transition and metastasis mediated by met signaling pathway. *Cancer Cell, 21*(1), 66–81.

Corselli, M., et al. (2012). The tunica adventitia of human arteries and veins as a source of mesenchymal stem cells. *Stem Cells and Development, 21*(8), 1299–1308.

Engin, F., et al. (2009). Notch signaling contributes to the pathogenesis of human osteosarcomas. *Human Molecular Genetics, 18*(8), 1464–1470.

Erber, R., et al. (2004). Combined inhibition of VEGF and PDGF signaling enforces tumor vessel regression by interfering with pericyte-mediated endothelial cell survival mechanisms. *The FASEB Journal, 18*(2), 338–340.

Erlandson, R. (1994a). *Diagnostic transmission electron microscopy of tumors*. New York: Raven Press.

Erlandson, R. A. (1994b). *Diagnostic transmission electron microscopy of tumors: With clinicopathological, immunohistochemical, and cytogenetic correlations* (xvii, 857 p). New York: Raven Press.

Fletcher, C. D. M., World Health Organization, & International Agency for Research on Cancer. (2013). *WHO classification of tumours of soft tissue and bone* (4th ed.). Lyon: IARC Press. 468 p.

Folpe, A. L., et al. (2001). Atypical and malignant glomus tumors: Analysis of 52 cases, with a proposal for the reclassification of glomus tumors. *The American Journal of Surgical Pathology, 25*(1), 1–12.

Folpe, A. L., et al. (2005). Perivascular epithelioid cell neoplasms of soft tissue and gynecologic origin: A clinicopathologic study of 26 cases and review of the literature. *The American Journal of Surgical Pathology, 29*(12), 1558–1575.

Frack, M. D., Simon, L., & Dawson, B. H. (1968). The lymphangiomyomatosis syndrome. *Cancer, 22*(2), 428–437.

Fujiuchi, Y., et al. (2003). Effect of hepatocyte growth factor on invasion of prostate cancer cell lines. *Oncology Reports, 10*(4), 1001–1006.

Ghadially, F. N. (1980). In A. Boyde (Ed.), *Diagnostic electron microscopy of tumours*. London: Butterworth.

Giese, A., & Westphal, M. (1996). Glioma invasion in the central nervous system. *Neurosurgery, 39*(2), 235–250. discussion 250–252.

Granter, S. R., Badizadegan, K., & Fletcher, C. D. (1998). Myofibromatosis in adults, glomangiopericytoma, and myopericytoma: A spectrum of tumors showing perivascular myoid differentiation. *The American Journal of Surgical Pathology, 22*(5), 513–525.

Hainsworth, J. D., et al. (2007). Treatment of advanced renal cell carcinoma with the combination bevacizumab/erlotinib/imatinib: A phase I/II trial. *Clinical Genitourinary Cancer, 5*(7), 427–432.

Harris, A. L. (2002). Hypoxia—a key regulatory factor in tumour growth. *Nature Reviews Cancer, 2*(1), 38–47.

Hughes, D. P. (2009). How the NOTCH pathway contributes to the ability of osteosarcoma cells to metastasize. *Cancer Treatment and Research, 152*, 479–496.

Inai, T., et al. (2004). Inhibition of vascular endothelial growth factor (VEGF) signaling in cancer causes loss of endothelial fenestrations, regression of tumor vessels, and appearance of basement membrane ghosts. *The American Journal of Pathology, 165*(1), 35–52.

Jo, V. Y., & Fletcher, C. D. M. (2017). Nuclear beta-catenin expression is frequent in sinonasal hemangiopericytoma and its mimics. *Head and Neck Pathology, 11*(2), 119–123.

Kuhnert, F., et al. (2008). Soluble receptor-mediated selective inhibition of VEGFR and PDGFRbeta signaling during physiologic and tumor angiogenesis. *Proceedings of the National Academy of Sciences of the United States of America, 105*(29), 10185–10190.

Kuo, F. Y., et al. (2005). Sinonasal hemangiopericytoma-like tumor with true pericytic myoid differentiation: A clinicopathologic and immunohistochemical study of five cases. *Head & Neck, 27*(2), 124–129.

Levine, J. M., & Nishiyama, A. (1996). The NG2 chondroitin sulfate proteoglycan: A multifunctional proteoglycan associated with immature cells. *Perspectives on Developmental Neurobiology, 3*(4), 245–259.

Levy, M. J., Gleeson, F. C., & Zhang, L. (2009). Endoscopic ultrasound fine-needle aspiration detection of extravascular migratory metastasis from a remotely located pancreatic cancer. *Clinical Gastroenterology and Hepatology, 7*(2), 246–248.

Liebow, A. A., & Castleman, B. (1971). Benign clear cell ("sugar") tumors of the lung. *The Yale Journal of Biology and Medicine, 43*(4–5), 213–222.

Lugassy, C., & Barnhill, R. L. (2007). Angiotropic melanoma and extravascular migratory metastasis: A review. *Advances in Anatomic Pathology, 14*(3), 195–201.

Lugassy, C., et al. (1998). Angio-tumoral laminin in murine tumors derived from human melanoma cell lines. Immunohistochemical and ultrastructural observations. *Journal of Submicroscopic Cytology and Pathology, 30*(2), 231–237.
Lugassy, C., et al. (1999). Tumor microvessels in melanoma express the beta-2 chain of laminin. Implications for melanoma metastasis. *Journal of Cutaneous Pathology, 26*(5), 222–226.
Lugassy, C., Barnhill, R. L., & Christensen, L. (2000). Melanoma and extravascular migratory metastasis. *Journal of Cutaneous Pathology, 27*(9), 481.
Lugassy, C., et al. (2002). Pericytic-like angiotropism of glioma and melanoma cells. *The American Journal of Dermatopathology, 24*(6), 473–478.
Lugassy, C., et al. (2005). Angiotropism of human prostate cancer cells: Implications for extravascular migratory metastasis. *BJU International, 95*(7), 1099–1103.
Lugassy, C., et al. (2013a). Could pericytic mimicry represent another type of melanoma cell plasticity with embryonic properties? *Pigment Cell & Melanoma Research, 26*(5), 746–754.
Lugassy, C., et al. (2013b). Pilot study on "pericytic mimicry" and potential embryonic/stem cell properties of angiotropic melanoma cells interacting with the abluminal vascular surface. *Cancer Microenvironment, 6*(1), 19–29.
Mentzel, T., et al. (2006). Myopericytoma of skin and soft tissues: Clinicopathologic and immunohistochemical study of 54 cases. *The American Journal of Surgical Pathology, 30*(1), 104–113.
Morikawa, S., et al. (2002). Abnormalities in pericytes on blood vessels and endothelial sprouts in tumors. *The American Journal of Pathology, 160*(3), 985–1000.
Mosquera, J. M., et al. (2013). Novel MIR143-NOTCH fusions in benign and malignant glomus tumors. *Genes, Chromosomes & Cancer, 52*(11), 1075–1087.
Nuovo, M., Grimes, M., & Knowles, D. (1990). Glomus tumors: A clinicopathologic and immunohistochemical analysis of forty cases. *Surgical Pathology, 3*, 31–45.
Pea, M., et al. (1991). Melanocyte-marker-HMB-45 is regularly expressed in angiomyolipoma of the kidney. *Pathology, 23*(3), 185–188.
Ren, Y., et al. (2005). Hepatocyte growth factor promotes cancer cell migration and angiogenic factors expression: A prognostic marker of human esophageal squamous cell carcinomas. *Clinical Cancer Research, 11*(17), 6190–6197.
Robinson, D. R., et al. (2013). Identification of recurrent NAB2-STAT6 gene fusions in solitary fibrous tumor by integrative sequencing. *Nature Genetics, 45*(2), 180–185.
Sato, S., et al. (2016). Mesenchymal tumors can derive from Ng2/Cspg4-expressing pericytes with beta-catenin modulating the neoplastic phenotype. *Cell Reports, 16*(4), 917–927.
Schadler, K. L., et al. (2010). Delta-like ligand 4 plays a critical role in pericyte/vascular smooth muscle cell formation during vasculogenesis and tumor vessel expansion in Ewing's sarcoma. *Clinical Cancer Research, 16*(3), 848–856.
Schiano, C., et al. (2012). Different expression of CD146 in human normal and osteosarcoma cell lines. *Medical Oncology, 29*(4), 2998–3002.
Shen, J., et al. (2015a). Pericyte antigens in perivascular soft tissue tumors. *International Journal of Surgical Pathology, 23*(8), 638–648.
Shen, J., et al. (2015b). Pericyte antigens in angiomyolipoma and PEComa family tumors. *Medical Oncology, 32*(8), 210.
Shen, J., et al. (2016). The pericyte antigen RGS5 in perivascular soft tissue tumors. *Human Pathology, 47*(1), 121–131.
Song, S., et al. (2005). PDGFRbeta+ perivascular progenitor cells in tumours regulate pericyte differentiation and vascular survival. *Nature Cell Biology, 7*(9), 870–879.
Stewart, K. S., et al. (2011). Delta-like ligand 4-Notch signaling regulates bone marrow-derived pericyte/vascular smooth muscle cell formation. *Blood, 117*(2), 719–726.
Tanaka, M., et al. (2009). Inhibition of Notch pathway prevents osteosarcoma growth by cell cycle regulation. *British Journal of Cancer, 100*(12), 1957–1965.
Thompson, L. D. (2004). Sinonasal tract glomangiopericytoma (hemangiopericytoma). *Ear, Nose, & Throat Journal, 83*(12), 807.

Van Geertruyden, J., et al. (1996). Glomus tumours of the hand. A retrospective study of 51 cases. *Journal of Hand Surgery (British Volume), 21*(2), 257–260.

Wang, W., et al. (2015). Effect of platelet-derived growth factor-B on renal cell carcinoma growth and progression. *Urologic Oncology, 33*(4), 168.e17–168.e27.

Weiss, S. W., & Goldblum, J. (2008). *Enzinger and Weiss's soft tissue tumors*. Philadelphia: Mosby Elsevier.

Xian, X., et al. (2006). Pericytes limit tumor cell metastasis. *The Journal of Clinical Investigation, 116*(3), 642–651.

Yang, M. H., et al. (2008). Direct regulation of TWIST by HIF-1alpha promotes metastasis. *Nature Cell Biology, 10*(3), 295–305.

Yasui, R., et al. (2001). Nasal septal hemangiopericytoma-like tumor: A case report with an immunohistochemical study. *American Journal of Rhinology, 15*(4), 267–270.

Zhang, P., et al. (2010). Regulation of NOTCH signaling by reciprocal inhibition of HES1 and Deltex 1 and its role in osteosarcoma invasiveness. *Oncogene, 29*(20), 2916–2926.

Zhou, Z., Yu, L., & Kleinerman, E. S. (2014). EWS-FLI-1 regulates the neuronal repressor gene REST, which controls Ewing sarcoma growth and vascular morphology. *Cancer, 120*(4), 579–588.

Chapter 5
Pericytes in Metastasis

Zalitha Pieterse, Devbarna Sinha, and Pritinder Kaur

Abstract Pericytes have long been known to contribute indirectly to tumour growth by regulating angiogenesis. Thus, remodelling tumour blood vessels to maintain blood supply is critical for continued tumour growth. A role for pericytes in restricting leakage of tumour cells through blood vessels has also become evident given that adequate pericyte coverage of these blood vessels is critical for maintaining vascular permeability. Interestingly, the relocation of pericytes from blood vessels to the tumour microenvironment results in the emergence of different properties in these cells that actively promote tumour growth and metastasis—functions not associated with their well-studied role in vascular stability and permeability. These form the focus of this review.

Keywords Cancer · Cancer cell invasion · Cancer stem cells · Epithelial-mesenchymal interactions · Mesenchymal stem cells · Metastasis · Ovarian cancer Pericytes · Tumour microenvironment · Vascular permeability · Tumour vasculature

Introduction

The role of the tumour microenvironment (TME) in promoting tumour growth and metastasis is widely recognised and consists of a variety of cells including cancer-associated fibroblasts (CAFs), bone marrow-derived mesenchymal stem cells (BM-MSC), endothelial cells, pericytes and immune cells and the growth factors and proteins they produce. Studies of various types of cancer including ovarian, colorectal, pancreatic and breast demonstrate that stromal signatures predict relapse and recurrence in patients lending strong support to the notion that the TME is a

Z. Pieterse · P. Kaur (✉)
Curtin Health Innovation Research Institute, Curtin University, Perth, WA, Australia
e-mail: zalitha.pieterse@postgrad.curtin.edu.au; pritinder.kaur@curtin.edu.au

D. Sinha
Peter MacCallum Cancer Centre, Melbourne, VIC, Australia

A. Birbrair (ed.), *Pericyte Biology in Disease*, Advances in Experimental Medicine and Biology 1147, https://doi.org/10.1007/978-3-030-16908-4_5

strong contributor to malignant progression (Tothill et al. 2008; Finak et al. 2008; Tsujino et al. 2007; Fujita et al. 2010; Calon et al. 2015). It is possible that not all elements of the TME are pro-tumourigenic—indeed, many tumours are encapsulated by fibroblasts perhaps limiting metastatic spread. It has been suggested that the TME has a role in contributing to resistance against anti-cancer therapeutics (Frame and Serrels 2015). It is therefore important that we gain a better understanding of the contributions of specific subsets of cells found in the TME to cancer progression. This discussion focuses on the role of pericytes given the relatively recent discovery that they have angiogenesis-independent roles in promoting malignant cancer.

Difficulties in Distinguishing Pericytes from CAFs in the TME

CAFs are defined as fibroblasts that universally promote tumour growth, invasiveness and metastasis compared to normal "fibroblasts" (Olumi et al. 1999; Cunha et al. 2003; Kalluri and Zeisberg 2006; Pietras and Ostman 2010). CAFs can originate from diverse sources including tissue-resident myofibroblasts, activated adipocytes and distal bone marrow-derived MSCs/BM-MSCs (Kalluri and Zeisberg 2006; Cirri and Chiarugi 2011), and are mostly identified retrospectively using equivocal "CAF markers", such as α-SMA (smooth muscle actin) which is activated in all mesenchymal cell types in conditions of stress. Thus, α-SMA is expressed in cultured or activated fibroblasts, myofibroblasts, pericytes and most BM-MSCs (Kalluri and Zeisberg 2006; Cirri and Chiarugi 2011). Lineage-marking studies in animals show that GFP-tagged BM-MSCs home to developing tumours inducing increased metastases (Karnoub et al. 2007; Hung et al. 2005; Mishra et al. 2008; Studeny et al. 2002; Quante et al. 2011), making up about ***~20–50%*** of CAFs. Thus, 50–80% of CAFs are not BM derived and may originate from local fibroblasts or other MSC-like populations such as pericytes. Indeed, our lab has shown that pericytes accelerate tumour growth rates and promote metastatic spread in a xenograft model of ovarian cancer (Sinha et al. 2016). Moreover, we showed that pericytes recruited BM-MSCs to developing tumours, suggesting that they may act upstream of BM-MSCs.

Classic Functions of Pericytes in Cancer: Stabilising Tumour Blood Supply and Limiting Hypoxia

In cancer, pericytes have been widely studied in the context of their well-known capacity to stabilise blood vessel structure and permeability. Dual targeting of vascular endothelial cells and pericytes using kinase inhibitors or anti-VEGF and anti-PDGFβ antibodies, has a synergistic anti-angiogenic/anti-tumour effect, resulting in increased tumour cell killing in animal models attributed to destabilisation of the tumour microvasculature (Bergers et al. 2003; Druker 2002; Erber et al. 2004; Kuhnert et al. 2008; Maciag et al. 2008). Other studies claiming that tumour growth was unaffected after pericyte removal are equivocal given that a maximal 50% pericyte knockdown (KD) was achieved using AX102, an inhibitor of PDGFB signalling (Sennino et al. 2007), or in PDGF$^{ret/ret}$ mice that harbour a mutation in the PDGFB retention motif (Nisancioglu et al. 2010; Lindblom et al. 2003). On the other hand, complete KD of pericytes with NG2 promoter driven thymidine kinase, caused tumour hypoxia which led to epithelial-mesenchymal transition and metastases to the lungs in mouse models of breast/renal cell carcinoma and melanoma (Cooke et al. 2012). These data support the idea that pericytes limit metastatic spread through otherwise leaky tumour blood vessels (Xian et al. 2006).

Tumour Blood Vessel Remodelling Leads to Displacement of Pericytes from Their Vascular Niche

Blood vessel remodelling during tissue repair is a dynamic process requiring the initial detachment of pericytes from endothelial cells from pre-existing vessels resulting in the removal of paracrine signalling between the two cell types that keep vessels in a homeostatic state. Pericyte detachment permits endothelial cell sprouting and proliferation, and subsequent re-association of the two cell types in the newly extended blood vessels – processes driven by angiopoietin-1/2 and Tie2 (Ang/Tie2), transforming growth factor-β (TGF-β), and platelet-derived growth factor-B (PDGFB) and its receptor PDGFR-β (Lindblom et al. 2003; Stapor et al. 2014). Similar mechanisms underlie tumour vessel remodelling although tumour vasculature is typically disorganised with torturous vessels, excessive branching and altered gene expression resulting in impaired vascular structure and increased vessel leakiness (Ruoslahti 2002). Notably, pericytes are more loosely attached to endothelial cells with cytoplasmic projections invading the tumour stroma (Morikawa et al. 2002). It has been shown that detachment of perivascular cells (and subsequent endothelial cell sprouting) is mediated by angiopoietin-2 secreted by activated endothelial cells (Scharpfenecker et al. 2005). Similarly, it has been shown that tumour cells overexpressing PDGFBB xenografted onto mice, induced dissociation of pericytes from tumour blood vessels in a dose-dependent manner increasing vascular permeability leading to vascular impairment (Hosaka et al. 2013). Notably, continued exposure of pericytes to PDGFBB led to down-regulation of PDGFβR that in turn decreased the expression of the $\alpha1\beta1$ integrin receptor from the cell surface of pericytes, abrogating their adhesion to extracellular matrix proteins in the blood vessel walls resulting in their detachment from them (Hosaka et al. 2013). Thus, paracrine signalling between pericytes and cancer cells provides an important mechanism by which pericytes can be persuaded to leave their normal microenvironment within microvessels and associate more closely with tumour cells. Given the reports that the precise location of pericytes can alter their functionality, it is clear that pericytes may be able to act directly on tumour cells as part of their mesenchymal microenvironment.

Mesenchymal Stem Cell Properties of Pericytes: Similarities and Distinctions

Given that many tissues are well vascularised, it has been speculated that perivascular cells throughout the body serve as a reservoir of multipotent mesenchymal stem cells that can be recruited upon tissue injury. Certainly, the phenotypic and functional similarities between pericytes and mesenchymal stem cells have been widely reported, serving to underpin the idea that the two cell types are in fact one and the same. Thus, both pericytes and BM-MSCs are $CD45^{-}CD31^{-}\alpha SMA^{+}CD146^{+}NG2^{+}PDGFRB^{+}CD73^{+}CD90^{+}$, located in a perivascular niche and can differentiate into fat, bone, cartilage, muscle, and neuronal cells (Crisan et al. 2008; Caplan 2008; Paquet-Fifield et al. 2009). However, important distinctions exist between pericytes and MSCs in that an immunosuppressive role has been described for MSCs (Shi et al. 2018) whereas the indications are that in animal studies pericytes or perivascular cells are pro-inflammatory (Mills et al. 2015; Dulauroy et al. 2012) and likely to contribute to delayed healing and fibrosis.

Pericytes and Fibrosis

Pericytes are involved in various fibrosis-related pathologies in the kidneys, liver, and skin, acting as progenitors to myofibroblasts, which are the main mediators for extracellular matrix deposition, leading to fibrogenesis and ultimately fibrosis during the healing process (Greenhalgh et al. 2015; Kramann and Humphreys 2014). In a transgenic reporter mouse model, coll1α1-GFP-expressing pericytes were shown to be the main source of myofibroblasts leading to kidney fibrosis (Lin et al. 2008). Similar studies in the liver, where pericytes are known as hepatic stellate cells, also showed that they were the main source of myofibroblasts, and a major player in liver fibrosis (Greenhalgh et al. 2015). Consistent with this, studies with a Cre-transgenic mouse model which labelled hepatic stellate cells in various models of liver injury, demonstrated that they accounted for 82–96% of the myofibroblast pool, which contributes to liver fibrosis (Mederacke et al. 2013). Pericytes have a similar role in skin fibrosis and scar formation as illustrated by genetic fate mapping, revealing that the majority of collagen producing myofibroblasts originate from ADAM12 expressing cells, derived from $PDGFRB^{+}$ $NG2^{+}$ perivascular cells or pericytes (Dulauroy et al. 2012). Notably, ablation or knockdown of $ADAM12^{+}$ cells, was sufficient to limit collagen production in the healing site of injury and reduce fibrosis. These studies demonstrate the ability for pericytes to differentiate into myofibroblasts thereby contributing to fibrosis. They also illustrate the ability of pericytes to contribute to the remodelling of tissue stroma to achieve wound repair thus pointing to ways in which these cells can affect biological processes in an angiogenesis-independent manner.

Pericytes in Cancer and Metastasis

It has been variously postulated that pericytes affect tumour growth and metastasis both positively and negatively. Many of the tumour growth promoting effects are related to establishing a stable vascular network thus ensuring delivery of nutrients to rapidly growing tumour cells and preventing tumour cell dissemination through blood vessels by maintaining vascularity permeability. Both these aspects require adequate pericyte investment on the abluminal surface of tumour blood vessels—experimental depletion of pericytes does indeed result in tumour regression (Bergers et al. 2003) but also leads to hypoxia-induced epithelial-mesenchymal transition increasing metastasis (Cooke et al. 2012). Consistent with this, normalising tumour vasculature by abrogating RGS5 expression (a cell surface protein that is abnormally expressed in tumour vessels), makes tumours more susceptible to chemotherapeutic agents (Hamzah et al. 2008). Moreover, the context in which pericyte dissociation from tumour vessels occurs also affects the response to chemotherapeutic agents as shown for variable levels of PDGFBB expression by various tumours (Hosaka et al. 2013).

Angiogenesis-Independent Mechanisms by Which Pericytes Promote Metastasis

A more direct role for pericytes in promoting cancer growth and metastasis without impact on angiogenesis has recently emerged from several laboratories including our own. It has become increasingly evident that pericytes are potent mesenchymal stem cell-like cells with an ability to promote organ repair and regeneration and multiple mesenchymal lineage differentiation capacity (Crisan et al. 2008; Sa da Bandeira et al. 2017). In the haemopoietic system, pericytes are an integral part of the haemopoietic stem cell niche regulating their maintenance and quiescence through paracrine effects (Sacchetti et al. 2007) supporting haemopoiesis both in vitro and in vivo (Birbrair and Frenette 2016; Morrison and Scadden 2014), as reviewed in Sa da Bandeira et al. (2017). In the process of studying the cellular microenvironment of epithelial renewal in human skin, we discovered a novel, paracrine role for pericytes in influencing skin tissue regeneration in 3D organotypic cultures completely lacking any blood vessels (Paquet-Fifield et al. 2009).

In view of the fact that pericytes are MSC-like and that MSCs had been reported to promote breast cancer metastasis (Karnoub et al. 2007) and ovarian cancer growth (McLean et al. 2011), we sought to establish whether pericytes may be a critical element of the TME with a more direct role in cancer progression. Recognising that the stromal signature of serous ovarian cancer patients reported by the Australian Ovarian Cancer Study Group (AOCS) (Tothill et al. 2008) had markers of both fibroblasts and pericytes, we used the molecular signature previously generated by us for both individual cell types (Paquet-Fifield et al. 2009) to interrogate the AOCS patient dataset annotated for patient outcomes (Sinha et al. 2016). Remarkably, the pericyte signature outperformed the ovarian cancer stromal signature at predicting early relapse revealing that those serous ovarian cancer patients carrying a high pericyte score (evidenced by a set of 146 genes co-expressed by both pericytes and ovarian cancer stromal cells), relapsed significantly earlier with a mean progression-free survival/PFS time of 9 months (vs. 29 months in those with a low pericyte score; $n = 215$), despite similar treatment ($p = 0.00067$ vs. $p = 0.0011$ from Tothill et al. 2008). Notably, the fibroblast signature was relatively poorer at predicting relapse ($p = 0.01$). Subsequently, we used a xenograft model to demonstrate that pericytes could act as CAFs when co-injected with ovarian cancer cell lines and that critically the tumour vasculature derived entirely from host murine cells remained unaffected with respect to the number of blood vessels or pericyte investment. Thus, co-injection of human pericytes with OVCAR-5 or OVCAR-8 cells, accelerated tumour growth rates and caused rapid dissemination to local tissues increasing metastases in a dose-dependent manner, typical of ovarian cancer spread clinically (Sinha et al. 2016). Notably, the human pericytes remained in the tumour stroma not associating with the tumour vasculature presumably due to species-specific incompatibility of signals that might otherwise result in their incorporation into tumour microvessels. This study provided the first clear evidence uncoupling the pro-angiogenic versus pro-metastatic function of pericytes in cancer. This suggests to us that when pericytes are *dissociated* from blood vessels they promote metastasis—a novel site of pericyte action, whereas their normal location in blood vessels restricts metastasis (Fig. 5.1). Consistent with a paracrine role for pericytes in promoting tumour cell metastasis, transwell co-culture experiments showed that pericytes increased ovarian cancer cell migration and invasion through matrigel; in other experiments we were able to demonstrate increased ovarian cancer cell proliferation with pericyte co-culture (Sinha et al. 2016). Interestingly, recent work with a variety of epithelial cancers has shown that pericytes contribute to cancer progression by giving rise to CAFs when dissociated from tumour blood vessels (Hosaka et al. 2013).

It is also likely that pericytes promote aggressive tumour growth by affecting the cancer stem cell compartment within tumours preferentially by secreting paracrine regulators. In support of this, it has been reported that human cancer-associated MSCs found in the TME of ovarian cancer increased the incidence of $ALDH^{+}CD133^{+}$ cancer stem cells via BMP-2 (McLean et al. 2011). In recent work from our laboratory, we have shown that pericytes can dictate the orientation of cell divisions within the skin's proliferative compartment, i.e. the basal layer, increasing planar presumed

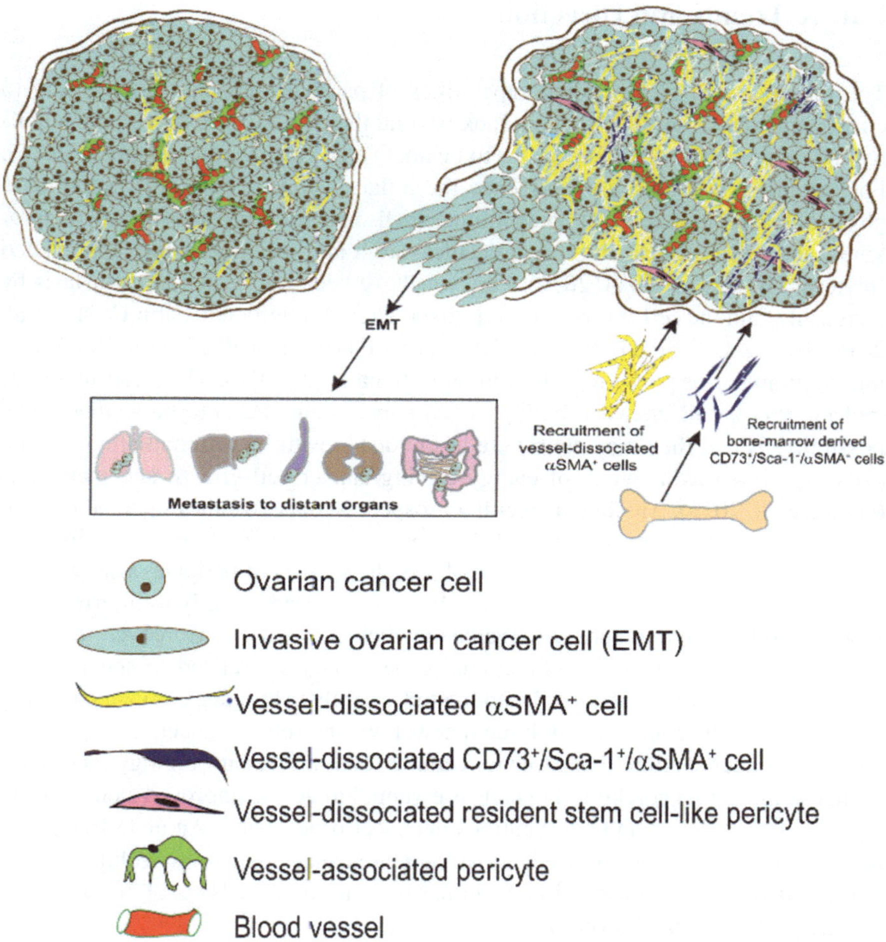

Fig. 5.1 Schematic model of the role of pericytes in promoting malignant cancer progression. Localisation of pericytes in intimate association with blood vessels results in a tightly encapsulated ovarian cancer tumour (left) whilst placing pericytes directly within the tumour microenvironment leads to increased tumour cell proliferation, increased recruitment of αSMA^+ stromal cells including cells expressing markers of BM-MSCs, induction of EMT and invasion, and metastatic spread to distant organs. Based on Sinha et al. (2016)

symmetric cell divisions delaying differentiation thereby maintaining epidermal cells in a more primitive state via BMP-2 (Zhuang et al. 2018). These data and work from lower organisms such as *Drosophila* demonstrating a role for BMP signalling in maintaining "stem-ness" of neighbouring cells (Kawase et al. 2004; Song et al. 2004) points to a conserved mechanism for the stem cell niche in regulating the fate of adjacent cells.

Future Trends and Directions

The full spectrum of functional capabilities of pericytes are only just starting to emerge and the closer investigators look beyond their classic role in vascular stability and permeability, the more seems to be uncovered (Ribeiro and Okamoto 2015). For instance, their ability to influence the inflammatory response by acting as a physical barrier to extravasation of immune cells and secreting a vast array of cytokines and extracellular matrix molecules are only recently being appreciated (reviewed in Navarro et al. 2016). Pericytes have been implicated in metastasis by increasing tumour cell intravasation at distal sites through endosialin (Viski et al. 2016), by regulating the metastatic niche via KLF4 (Paiva et al. 2018) and suppressing the immune response to brain tumours (Sena et al. 2018). The exact nature of molecular crosstalk between pericytes and cancer cells needs to be studied—one possibility is that they may contribute to tumour growth by differentiating into fat cells, which act as a source of energy driving cancer cell growth and metastasis (Huang et al. 2018). Another an exciting prospect is that exosomes secreted by cancer cells signal pericytes to become CAFs (Ning et al. 2018). Thus, a clear driver of future work has to be the identification of specific subsets of pericytes with cutting-edge technologies such as single cell RNA seq as reported recently for murine brain vascular cells including pericytes (Vanlandewijck et al. 2018). Moreover, the commonalities and distinctions in functional gene expression related to specific anatomical sites and organs needs to be addressed urgently to broaden our understanding of how these cells contribute to tissue renewal, wound repair, cancer, and ageing. A thorough understanding of pericyte cellular and molecular biology and their immense impact on neighbouring cells is essential to devise improved stem cell and regenerative medicine and interventions in cancer progression. An underlying concept is that just as we acknowledge that there are cancer stem cells within tumours that can drive cancer progression, a similar recognition of subsets of "cancer stromal stem cells" is much overdue.

References

Bergers, G., et al. (2003). Benefits of targeting both pericytes and endothelial cells in the tumor vasculature with kinase inhibitors. *The Journal of Clinical Investigation, 111*(9), 1287–1295.

Birbrair, A., & Frenette, P. S. (2016). Niche heterogeneity in the bone marrow. *Annals of the New York Academy of Sciences, 1370*(1), 82–96.

Calon, A., et al. (2015). Stromal gene expression defines poor-prognosis subtypes in colorectal cancer. *Nature Genetics, 47*(4), 320–329.

Caplan, A. I. (2008). All MSCs are pericytes? *Cell Stem Cell, 3*(3), 229–230.

Cirri, P., & Chiarugi, P. (2011). Cancer associated fibroblasts: the dark side of the coin. *American Journal of Cancer Research, 1*(4), 482–497.

Cooke, V. G., et al. (2012). Pericyte depletion results in hypoxia-associated epithelial-to-mesenchymal transition and metastasis mediated by met signaling pathway. *Cancer Cell, 21*(1), 66–81.

Crisan, M., et al. (2008). A perivascular origin for mesenchymal stem cells in multiple human organs. *Cell Stem Cell, 3*(3), 301–313.
Cunha, G. R., et al. (2003). Role of the stromal microenvironment in carcinogenesis of the prostate. *International Journal of Cancer, 107*(1), 1–10.
Druker, B. J. (2002). STI571 (Gleevec) as a paradigm for cancer therapy. *Trends in Molecular Medicine, 8*(4 Suppl), S14–S18.
Dulauroy, S., et al. (2012). Lineage tracing and genetic ablation of ADAM12(+) perivascular cells identify a major source of profibrotic cells during acute tissue injury. *Nature Medicine, 18*(8), 1262–1270.
Erber, R., et al. (2004). Combined inhibition of VEGF and PDGF signaling enforces tumor vessel regression by interfering with pericyte-mediated endothelial cell survival mechanisms. *The FASEB Journal, 18*(2), 338–340.
Finak, G., et al. (2008). Stromal gene expression predicts clinical outcome in breast cancer. *Nature Medicine, 14*(5), 518–527.
Frame, M. C., & Serrels, A. (2015). FAK to the rescue: activated stroma promotes a "safe haven" for BRAF-mutant melanoma cells by inducing FAK signaling. *Cancer Cell, 27*(4), 429–431.
Fujita, H., et al. (2010). alpha-Smooth muscle actin expressing stroma promotes an aggressive tumor biology in pancreatic ductal adenocarcinoma. *Pancreas, 39*, 1254–1262.
Greenhalgh, S. N., Conroy, K. P., & Henderson, N. C. (2015). Healing scars: targeting pericytes to treat fibrosis. *QJM: An International Journal of Medicine, 108*(1), 3–7.
Hamzah, J., et al. (2008). Vascular normalization in Rgs5-deficient tumours promotes immune destruction. *Nature, 453*(7193), 410–414.
Hosaka, K., et al. (2013). Tumour PDGF-BB expression levels determine dual effects of anti-PDGF drugs on vascular remodelling and metastasis. *Nature Communications, 4*, 2129.
Huang, J., et al. (2018). Adipocyte p62/SQSTM1 suppresses tumorigenesis through opposite regulations of metabolism in adipose tissue and tumor. *Cancer Cell, 33*(4), 770–784 e6.
Hung, S. C., et al. (2005). Mesenchymal stem cell targeting of microscopic tumors and tumor stroma development monitored by noninvasive in vivo positron emission tomography imaging. *Clinical Cancer Research, 11*(21), 7749–7756.
Kalluri, R., & Zeisberg, M. (2006). Fibroblasts in cancer. *Nature Reviews. Cancer, 6*(5), 392–401.
Karnoub, A. E., et al. (2007). Mesenchymal stem cells within tumour stroma promote breast cancer metastasis. *Nature, 449*(7162), 557–563.
Kawase, E., et al. (2004). Gbb/Bmp signaling is essential for maintaining germline stem cells and for repressing bam transcription in the Drosophila testis. *Development, 131*(6), 1365–1375.
Kramann, R., & Humphreys, B. D. (2014). Kidney pericytes: Roles in regeneration and fibrosis. *Seminars in Nephrology, 34*(4), 374–383.
Kuhnert, F., et al. (2008). Soluble receptor-mediated selective inhibition of VEGFR and PDGFRbeta signaling during physiologic and tumor angiogenesis. *Proceedings of the National Academy of Sciences of the United States of America, 105*(29), 10185–10190.
Lin, S.-L., et al. (2008). Pericytes and perivascular fibroblasts are the primary source of collagen-producing cells in obstructive fibrosis of the kidney. *The American Journal of Pathology, 173*(6), 1617–1627.
Lindblom, P., et al. (2003). Endothelial PDGF-B retention is required for proper investment of pericytes in the microvessel wall. *Genes & Development, 17*(15), 1835–1840.
Maciag, P. C., et al. (2008). Cancer immunotherapy targeting the high molecular weight melanoma-associated antigen protein results in a broad antitumor response and reduction of pericytes in the tumor vasculature. *Cancer Research, 68*(19), 8066–8075.
McLean, K., et al. (2011). Human ovarian carcinoma-associated mesenchymal stem cells regulate cancer stem cells and tumorigenesis via altered BMP production. *The Journal of Clinical Investigation, 121*(8), 3206–3219.
Mederacke, I., et al. (2013). Fate-tracing reveals hepatic stellate cells as dominant contributors to liver fibrosis independent of its etiology. *Nature Communications, 4*, 2823–2823.

Mills, S. J., et al. (2015). Effects of human pericytes in a murine excision model of wound healing. *Experimental Dermatology, 24*(11), 881–882.
Mishra, P. J., et al. (2008). Carcinoma-associated fibroblast-like differentiation of human mesenchymal stem cells. *Cancer Research, 68*(11), 4331–4339.
Morikawa, S., et al. (2002). Abnormalities in pericytes on blood vessels and endothelial sprouts in tumors. *The American Journal of Pathology, 160*(3), 985–1000.
Morrison, S. J., & Scadden, D. T. (2014). The bone marrow niche for haematopoietic stem cells. *Nature, 505*(7483), 327–334.
Navarro, R., et al. (2016). Immune regulation by pericytes: Modulating innate and adaptive immunity. *Frontiers in Immunology, 7*, 480.
Ning, X., et al. (2018). Exosomes released by gastric cancer cells induce transition of pericytes into cancer-associated fibroblasts. *Medical Science Monitor, 24*, 2350–2359.
Nisancioglu, M. H., Betsholtz, C., & Genove, G. (2010). The absence of pericytes does not increase the sensitivity of tumor vasculature to vascular endothelial growth factor-A blockade. *Cancer Research, 70*(12), 5109–5115.
Olumi, A. F., et al. (1999). Carcinoma-associated fibroblasts direct tumor progression of initiated human prostatic epithelium. *Cancer Research, 59*(19), 5002–5011.
Paiva, A. E., et al. (2018). Pericytes in the premetastatic niche. *Cancer Research, 78*(11), 2779–2786.
Paquet-Fifield, S., et al. (2009). A role for pericytes as microenvironmental regulators of human skin tissue regeneration. *The Journal of Clinical Investigation, 119*(9), 2795–2806.
Pietras, K., & Ostman, A. (2010). Hallmarks of cancer: interactions with the tumor stroma. *Experimental Cell Research, 316*(8), 1324–1331.
Quante, M., et al. (2011). Bone marrow-derived myofibroblasts contribute to the mesenchymal stem cell niche and promote tumor growth. *Cancer Cell, 19*(2), 257–272.
Ribeiro, A. L., & Okamoto, O. K. (2015). Combined effects of pericytes in the tumor microenvironment. *Stem Cells International, 2015*, 868475.
Ruoslahti, E. (2002). Specialization of tumour vasculature. *Nature Reviews. Cancer, 2*(2), 83–90.
Sa da Bandeira, D., Casamitjana, J., & Crisan, M. (2017). Pericytes, integral components of adult hematopoietic stem cell niches. *Pharmacology & Therapeutics, 171*, 104–113.
Sacchetti, B., et al. (2007). Self-renewing osteoprogenitors in bone marrow sinusoids can organize a hematopoietic microenvironment. *Cell, 131*(2), 324–336.
Scharpfenecker, M., et al. (2005). The Tie-2 ligand angiopoietin-2 destabilizes quiescent endothelium through an internal autocrine loop mechanism. *Journal of Cell Science, 118*(Pt 4), 771–780.
Sena, I. F. G., et al. (2018). Glioblastoma-activated pericytes support tumor growth via immunosuppression. *Cancer Medicine, 7*(4), 1232–1239.
Sennino, B., et al. (2007). Sequential loss of tumor vessel pericytes and endothelial cells after inhibition of platelet-derived growth factor B by selective aptamer AX102. *Cancer Research, 67*(15), 7358–7367.
Shi, Y., et al. (2018). Immunoregulatory mechanisms of mesenchymal stem and stromal cells in inflammatory diseases. *Nature Reviews Nephrology, 14*, 493–507.
Sinha, D., et al. (2016). Pericytes promote malignant ovarian cancer progression in mice and predict poor prognosis in serous ovarian cancer patients. *Clinical Cancer Research, 22*(7), 1813–1824.
Song, X., et al. (2004). Bmp signals from niche cells directly repress transcription of a differentiation-promoting gene, bag of marbles, in germline stem cells in the Drosophila ovary. *Development, 131*(6), 1353–1364.
Stapor, P. C., et al. (2014). Pericyte dynamics during angiogenesis: new insights from new identities. *Journal of Vascular Research, 51*(3), 163–174.
Studeny, M., et al. (2002). Bone marrow-derived mesenchymal stem cells as vehicles for interferon-beta delivery into tumors. *Cancer Research, 62*(13), 3603–3608.

Tothill, R. W., et al. (2008). Novel molecular subtypes of serous and endometrioid ovarian cancer linked to clinical outcome. *Clinical Cancer Research, 14*(16), 5198–5208.

Tsujino, T., et al. (2007). Stromal myofibroblasts predict disease recurrence for colorectal cancer. *Clinical Cancer Research, 13*(7), 2082–2090.

Vanlandewijck, M., et al. (2018). A molecular atlas of cell types and zonation in the brain vasculature. *Nature, 554*(7693), 475–480.

Viski, C., et al. (2016). Endosialin-expressing pericytes promote metastatic dissemination. *Cancer Research, 76*(18), 5313–5325.

Xian, X., et al. (2006). Pericytes limit tumor cell metastasis. *The Journal of Clinical Investigation, 116*(3), 642–651.

Zhuang, L., Lawlor, K. T., Schlueter, H., Pieterse, Z., Yu, Y., & Kaur, P. (2018). Pericytes promote skin regeneration by inducing epidermal cell polarity and planar cell divisions. *Life Science Alliance, 1*(4), e201700009.

Chapter 6
The Role of Pericytes in Amyotrophic Lateral Sclerosis

Giuliana Castello Coatti, Natale Cavaçana, and Mayana Zatz

Abstract In amyotrophic lateral sclerosis (ALS), motor neurons die selectively. Therefore, initial symptoms that include fasciculation, spasticity, muscle atrophy, and weakness emerge following axons retraction and consequent muscles' denervation. Patients lose the ability to talk and swallow and rely on parenteral nutrition and assisted ventilation to survive. The degeneration caused by ALS is progressive and irreversible. In addition to the autonomous mechanism of neuronal cell death, non-autonomous mechanisms have been proved to be toxic for motor neurons, such as the activation of astrocytes and microglia. Among the cells being studied to unveil these toxic mechanisms are pericytes, cells that help keep the integrity of the blood–brain barrier and blood–spinal cord barrier. In this chapter, we aim to discuss the role of pericytes in ALS.

Keywords Amyotrophic lateral sclerosis · Lou Gehrig's disease · Motor neuron · Pericyte · Perivascular cells · Blood–brain barrier · Blood–spinal barrier · Microhemorrhage · SOD1 mice · Pericyte loss

What is Amyotrophic Lateral Sclerosis (ALS)?

Fundamentally, motor neurons are cells responsible for sending signals from the brain to muscles to initiate movement. In amyotrophic lateral sclerosis (ALS), also known as Lou Gehrig's disease, these cells die selectively. Therefore, initial symptoms that include fasciculation, spasticity, muscle atrophy, and weakness emerge following axons retraction and consequent muscles' denervation. This retraction is initially compensated by the sprouting of more resistant neurons. However, this mechanism eventually fails, and the neuronal cell bodies become abnormal and die (Robberecht and Philips 2013). ALS is usually a late-onset disease

G. C. Coatti
Department of Genetics and Genome Sciences, Case Western Reserve University, Cleveland, Ohio, USA

N. Cavaçana · M. Zatz (✉)
Human Genome and Stem Cell Research Center, Biosciences Institute, University of São Paulo, São Paulo, Brazil
e-mail: mayazatz@usp.br

A. Birbrair (ed.), *Pericyte Biology in Disease*, Advances in Experimental Medicine and Biology 1147, https://doi.org/10.1007/978-3-030-16908-4_6

in which both upper and lower motor neurons are affected (Goodall and Morrison 2006). Despite many years of investigation around the world, relatively little is known about the cause of this progressively degenerative neurological disease.

Annually, the incidence of ALS is 1–2.6 cases per 100,000 persons, with death occurring within 3–4 years after the onset of symptoms (Talbott et al. 2016). Sporadic cases of ALS (SALS) comprise about 90% of all cases. Regarding the familial cases (FALS), mutations in C9ORF chromosomal region, causing a hexanucleotide repeat, comprise the most common cause of FALS discovered to date (DeJesus-Hernandez et al. 2011). Described more than 20 years ago, *SOD1* gene was the first to be linked to the disease (Rosen et al. 1993). *SOD1* gene encodes the superoxide dismutase-1, a major cytoplasmic antioxidant enzyme that, when mutated, forms intracellular aggregates in a trimeric form that interfere in motor neuron survival (Proctor et al. 2016). An important ALS animal model is SOD1 mice, which has overexpression of the human transgene carrying G93A mutation, an alteration already observed in familial cases (Doble and Kennel 2000). In addition to progressive muscular symptoms, these mice exhibit selective death of motor neurons and their astrocytes and microglia become reactive (Kassa et al. 2009).

In the great majority of sporadic cases (SALS), it is not possible to identify a single causative agent. Recent studies using high-throughput technologies such as exome and genome sequencing have identified not only novel mutations in these known ALS-causing genes, but also additional genetic causes of ALS. The Project MinE is a new large-scale effort to better understand the genetic basis of ALS, aiming to analyze the DNA of at least 15,000 ALS patients and 7500 control subjects (https://www.projectmine.com).

Including *SOD1*, about 30 genes have been linked to ALS, providing important clues on the pathophysiological mechanisms of the disease (ALSoD—http://alsod.iop.kcl.ac.uk). In recent studies, the combination of exome sequencing and bioinformatic filtering is being used to investigate new genes linked to ALS, such as the newly discovered *NEK1* variants (Kenna et al. 2016). *NEK1* encodes the serine/threonine kinase, NIMA (never in mitosis gene-A)-related kinase that acts in several cellular functions, including cilia formation, DNA-damage response, microtubule stability, neuronal morphology, and axonal polarity. The loss of function of the *NEK1* gene can lead to ALS (Kenna et al. 2016).

Recently, the ALS ice bucket challenge, aiming to raise funding for ALS research, received great attention in the media (https://www.als.net/icebucketchallenge/, https://www.mndassociation.org/). This illustrates the current need in uncovering many aspects of the disease pathophysiology in order to discover an effective cure, since the available drugs are only effective for slowing slightly the disease progression.

Riluzole is the main FDA-approved drug used for ALS treatment. This compound acts in the reuptake of glutamate in the synaptic cleft reducing thus the excitotoxic effect on motor neurons (Miller et al. 2012). However, the overall effect of riluzole is limited, resulting in a survival increase of only 2–4 months (Bensimon et al. 1994; Miller et al. 2012). Recently, FDA approved Radicava (edaravone), an antioxidant compound, to treat ALS patients. This approval is based on data obtained from a

Phase 3 clinical trial (NCT01492686), which demonstrated Radicava's ability to slow the decline in the daily functions of ALS patients.

In addition to the autonomous mechanism of neuronal cell death, non-autonomous mechanisms have been proved to be toxic for motor neurons. Reactive astrocytes and microglia can trigger neuroinflammation, a key contributor to motor neuron damage in ALS (Henkel et al. 2004; Sargsyan et al. 2005; Wang et al. 2003). In addition, these cells can secrete inflammatory cytokines that ultimately may facilitate glutamate excitotoxicity (Pickering et al. 2005; Tilleux and Hermans 2007). Among the cells being studied to unveil these toxic mechanisms are pericytes, cells that help keep the integrity of the blood–brain barrier and blood–spinal cord barrier. In this chapter, we aim to discuss the role of pericytes in ALS.

What are Pericytes?

Pericytes are cells found within the perivascular region throughout the body. These cells participate in diverse biological functions such as vascular stability and angioarchitecture (Winkler et al. 2011). Pericytes can be characterized by the presence of molecular markers such as CD146, alkaline phosphatase (ALP) and NG2 proteoglycan, with absence of endothelial markers like CD45, CD31, and CD34 (Corselli et al. 2011; Dellavalle et al. 2007). Given this profile, pericytes can be obtained from a mixed population of cells by cell sorting.

The exact origin of pericytes differs between tissues (Asahina et al. 2011; Bergwerff et al. 1998; Simon et al. 2012) and also among tissues (Dias Moura Prazeres et al. 2017). Evidences indicate that pericytes can originate mesenchymal stem cells (MSCs), after being liberated from their basement membrane surrounding blood vessels upon injury or inflammation (Caplan and Hariri 2015). Functional heterogeneity was also observed among different subpopulation of pericytes: type 1 pericytes can originate fat and fibrous cells while type 2 pericytes can differentiate into skeletal muscle, endothelial, and neural cells (Birbrair et al. 2015).

The Blood–Brain Barrier (BBB) and the Blood–Spinal Cord Barrier (BSCB)

Differently from the systemic capillaries, capillaries within the brain and spinal cord are not leaky (Mann 1985). Instead, central nervous system (CNS) endothelial cells are connected by different types of tight and adherens junctions (Zlokovic 2008), forming a semipermeable monolayer barrier, the blood–brain barrier (BBB) and the blood–spinal cord barrier (BSCB), that allows the passage of water, gases, and small lipid-soluble molecules, which can cross via lipid-mediated diffusion, and limits the entry of plasma components, red blood cells, and leukocytes into the brain and the spinal cord (Zlokovic 2008). This barrier also blocks the entrance of toxic molecules into the CNS, but this limits the entry of small molecules drugs and all large molecule that could be used for therapy, e.g., recombinant peptides, proteins, antisense agents, and genetic vectors (Zlokovic 2011).

In addition to the endothelial cells, BBB is composed of the extracellular matrix, astrocytes, neurons, microglia, and pericytes. The interaction between these cells is known as neurovascular unity. The endothelial cells are surrounded by basal lamina and astrocytic perivascular endfeet. Astrocytes provide the cellular link to neurons. (Abbott et al. 2006; Armulik et al. 2010). Pericytes extend long cytoplasmic processes on the surface of endothelial cells, making interdigitating contacts. Endothelial cells and pericytes are separated by the basement membrane, and at points of contact, the pericyte communicate directly with endothelial cells through 'peg and socket' contacts (Allt and Lawrenson 2001).

Pericytes act in the physical stabilization of vessels, regulation of microcirculation, and capillary blood flow. They affect blood clotting and immune function, and may regulate the activation of lymphocytes, participate in angiogenesis and vasculogenesis (Birbrair et al. 2015). The pericytes interact with astrocytes by regulating proliferation, migration and differentiation of endothelial cells (Armulik et al. 2010). Pericytes regulate endothelial cell tight junction formation and permeability through tight junction and transendothelial cell transport (Kamouchi et al. 2011).

BSCB represent functional and morphological extension of the BBB in the spinal cord, but with some differences. For example, the BSCB has glycogen deposits not seen in the BBB, and they seem to differ in their permeability with BSCB being more permeable than the BBB (Bartanusz et al. 2011).

Pericytes and ALS

In some neurodegenerative diseases, the BBB/BSCB breakdown allows blood components to infiltrate the neuronal environment, aggravating the existing inflammation and accelerating the progression of symptoms (Bell et al. 2010; Winkler et al. 2011) (Fig. 6.1).

Analysis of the cerebrospinal fluid (CSF) of ALS patients revealed the presence of albumin and serum-derived proteins (Annunziata and Volpi 1985; Apostolski et al. 1991; Brettschneider et al. 2006). Evidences of the BBB/BSCB breakdown can also be found in *postmortem* tissue of spinal cord or motor cortex of these patients (Donnenfeld et al. 1984; Engelhardt and Appel 1990; Engelhardt et al. 1993; Sasaki 2015).

In addition to the well-known degeneration, the analysis of *postmortem* spinal cord of SALS patients also revealed edematous changes in the cytoplasm of endothelial cells and pericytes and also diminished areas of capillary lumens. Higher accumulation of collagen fibers in SALS patients may indicate a compensatory mechanism for maintaining vascular integrity (Sasaki 2015). Yamaneda and coworkers have also shown that microvascular density was increased, pericyte coverage was decreased and that there was abnormal angiogenesis in *postmortem* spinal cord

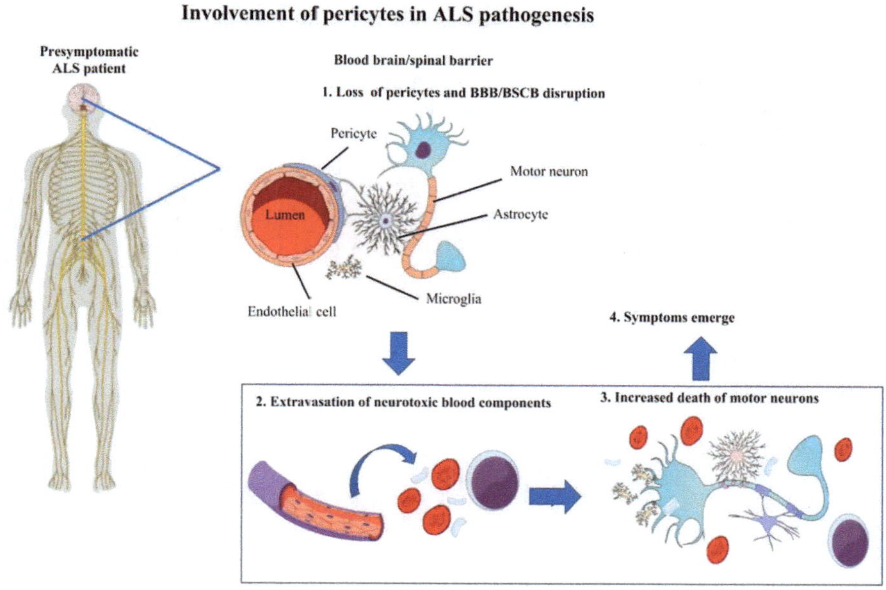

Fig. 6.1 How pericytes may be involved in ALS pathology. The loss of pericytes leads to BBB/BSCB disruption resulting in micro hemorrhages with the release of neurotoxic hemoglobin-derived products. Figure created in the Mind the Graph platform (www.mindthegraph.com)

tissue of SALS patients. Pericyte loss may have influenced these abnormalities (Yamadera et al. 2015).

Henkel and coworkers demonstrated decreased expression of tight junction proteins in lumbar spinal cords of ALS patients, including zonula occludens-1 (ZO-1) and occludin (Henkel et al. 2009) Another alteration observed in ALS patients serum (Beuche et al. 2000; Demestre et al. 2005; Niebroj-Dobosz et al. 2010) and in post-mortem tissue (Lim et al. 1996) is a higher amount of matrix metalloproteinase-9 (MMP-9), an enzyme responsible for degrading tight junctions of endothelial cells (Bell et al. 2012). It has been demonstrated that pericytes under stress conditions can secrete MMP-9, leading to vascular fragility (Bell et al. 2012). Importantly, analysis of postmortem spinal cord tissue from ALS patients indicated a 54% reduction in pericyte number when compared to healthy subjects. The level of the pericyte reduction correlates with the magnitude of BSCB damage (Winkler et al. 2013).

Regarding the animal model, SOD1 mice present a spontaneous infiltration of erythrocytes due to a spontaneous breakdown of the BBB (Garbuzova-Davis et al. 2007a, b, 2011; Miyazaki et al. 2011) that anticipates motor symptoms and neuronal loss (Miyazaki et al. 2011; Zhong et al. 2008, 2009) Another study showed that BSCB breakdown plays an important role in the early-stage disease pathogenesis in SOD1 mice (Winkler et al. 2014). Microvascular lesions contain hemoglobin that releases free iron, which can catalyze the formation of free radical species, molecules toxic for motor neurons (Regan and Guo 1998; Winkler et al. 2014; Zhong et al. 2009) (Fig. 6.2). Based on these results, two main therapeutical strategies were

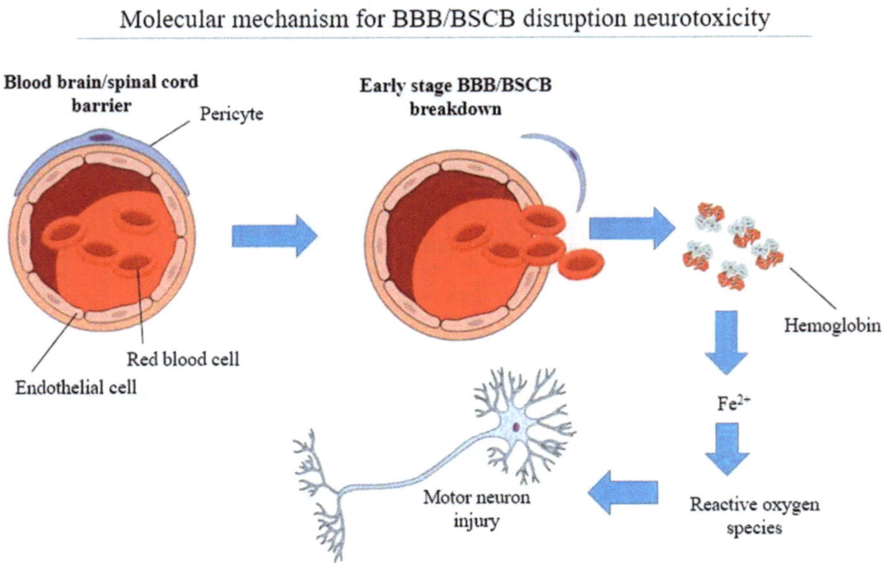

Fig. 6.2 Molecular mechanism of neuronal toxicity after BBB/BSCB disruption. Following pericyte loss and BBB/BSCB breakdown, red blood cells leak into the neuronal microenvironment and lyse, liberating hemoglobin. Hemoglobin releases free iron, which can catalyze the formation of free-radical species, molecules toxic for motor neurons. Figure created in the Mind the Graph platform (www.mindthegraph.com)

proven to be efficient to alleviate the motor symptoms of SOD1 mice: (1) early treatment with an activated protein C analog, aiming to restore the BSCB integrity and (2) early chelation of blood-derived iron and antioxidant treatment (Winkler et al. 2014).

Concluding Remarks

Recently, a growing number of studies have shown structural and functional changes in BBB and BSCB in ALS patients. In SOD1 mice, this endothelial damage is observed before motor neuron degeneration, indicating that this event is a central contributor to disease initiation (Miyazaki et al. 2011; Zhong et al. 2008). Even though it is not what actually causes ALS, it represents an important mechanism related to its pathogenesis.

Following the endothelial damage, higher accumulation of collagen fibers in ALS patients may indicate a compensatory mechanism for maintaining vascular integrity (Sasaki 2015). This feature can hinder the design of new drugs for ALS since the accumulation of collagen, especially type IV, occurring over a long period of time may represent an obstacle to the diffusion of various substances into the CNS (Garbuzova-Davis et al. 2016).

The etiology of ALS is complex, and the development of therapeutic strategies is even more complicated. Ideally, an effective therapeutic protocol for ALS should provide a cocktail of drugs, targeting not only the intrinsic mechanism of neuronal death but also these other non-autonomous changes. Preclinical experiments aiming to recover the disrupted BBB/BSCB are ongoing with exciting results. Winkler and coworkers observed that restoration of BSCB integrity or the chelation of blood-derived iron and antioxidant treatment reduced early injury on motor neurons (Winkler et al. 2014).

Cell therapy has also been evaluated to treat SOD1 mice. We observed that males that received weekly injections of adipose-derived pericytes lived longer than untreated males. Interestingly, treated animals presented increased expression of antioxidant enzymes in their brains. A similar effect was observed in ALS-derived motor neurons after co-culture with pericytes (Coatti et al. 2017). This study shows that, besides their important function on BBB/BSCB integrity, pericytes may secrete soluble factors of importance for maintaining a healthy neuronal environment. More studies are needed to understand why pericytes are lost in ALS.

In this same line of thought, other strategies may be used to avoid pericyte loss in ALS patients. Prostacyclin for example, may be efficient for ALS since it has been shown to prevent pericyte loss and demyelination (Muramatsu et al. 2015). Even Riluzole, the conventional drug used to treat ALS, has been found to diminish pericytes loss in the retina of diabetic mice (Choi et al. 2017). It would be important to investigate whether this effect also occurs in ALS patients' CNS.

One important issue that may hinder the progress of drug discovery for ALS is the fact that the diagnosis for this disease may take up to 14 months (Brooks 2000). As mentioned before, endothelial damage, a mechanism that accelerates motor neuron death, occurs before the onset of the symptoms. Because of that, strategies to find an effective molecular marker for ALS are highly relevant. Early diagnosis would increase the chances of more successful clinical trials, since neuronal degeneration may be advanced when symptoms start.

References

Abbott, N. J., Rönnbäck, L., & Hansson, E. (2006). Astrocyte–endothelial interactions at the blood–brain barrier. *Nature Reviews. Neuroscience, 7*, 41–53.

Allt, G., & Lawrenson, J. G. (2001). Pericytes: Cell biology and pathology. *Cells, Tissues, Organs, 169*, 1–11.

Annunziata, P., & Volpi, N. (1985). High levels of C3c in the cerebrospinal fluid from amyotrophic lateral sclerosis patients. *Acta Neurologica Scandinavica, 72*, 61–64.

Apostolski, S., Nikolić, J., Bugarski-Prokopljević, C., Miletić, V., Pavlović, S., & Filipović, S. (1991). Serum and CSF immunological findings in ALS. *Acta Neurologica Scandinavica, 83*, 96–98.

Armulik, A., Genové, G., Mäe, M., Nisancioglu, M. H., Wallgard, E., Niaudet, C., He, L., Norlin, J., Lindblom, P., Strittmatter, K., et al. (2010). Pericytes regulate the blood–brain barrier. *Nature, 468*, 557–561.

Asahina, K., Zhou, B., Pu, W. T., & Tsukamoto, H. (2011). Septum transversum-derived mesothelium gives rise to hepatic stellate cells and perivascular mesenchymal cells in developing mouse liver. *Hepatology (Baltimore, Md.), 53*, 983–995.

Bartanusz, V., Jezova, D., Alajajian, B., & Digicaylioglu, M. (2011). The blood–spinal cord barrier: Morphology and clinical implications. *Annals of Neurology, 70*, 194–206.
Bell, R. D., Winkler, E. A., Sagare, A. P., Singh, I., LaRue, B., Deane, R., & Zlokovic, B. V. (2010). Pericytes control key neurovascular functions and neuronal phenotype in the adult brain and during brain aging. *Neuron, 68*, 409–427.
Bell, R. D., Winkler, E. A., Singh, I., Sagare, A. P., Deane, R., Wu, Z., Holtzman, D. M., Betsholtz, C., Armulik, A., Sallstrom, J., et al. (2012). Apolipoprotein E controls cerebrovascular integrity via cyclophilin A. *Nature, 485*, 512–516.
Bensimon, G., Lacomblez, L., & Meininger, V. (1994). A controlled trial of riluzole in amyotrophic lateral sclerosis. ALS/Riluzole Study Group. *The New England Journal of Medicine, 330*, 585–591.
Bergwerff, M., Verberne, M. E., DeRuiter, M. C., Poelmann, R. E., & Gittenberger-de-Groot, A. C. (1998). Neural crest cell contribution to the developing circulatory system: Implications for vascular morphology? *Circulation Research, 82*, 221–231.
Beuche, W., Yushchenko, M., Mäder, M., Maliszewska, M., Felgenhauer, K., & Weber, F. (2000). Matrix metalloproteinase-9 is elevated in serum of patients with amyotrophic lateral sclerosis. *Neuroreport, 11*, 3419–3422.
Birbrair, A., Zhang, T., Wang, Z.-M., Messi, M. L., Mintz, A., & Delbono, O. (2015). Pericytes at the intersection between tissue regeneration and pathology. *Clinical Science, 128*, 81–93.
Brettschneider, J., Petzold, A., Süssmuth, S. D., Ludolph, A. C., & Tumani, H. (2006). Axonal damage markers in cerebrospinal fluid are increased in ALS. *Neurology, 66*, 852–856.
Brooks, B. R. (2000). Risk factors in the early diagnosis of ALS: North American epidemiological studies. ALS CARE Study Group. *Amyotrophic Lateral Sclerosis Motor Neuron Disease: Official Publication of the World Federation of Neurology Research Group on Motor Neuron Diseases, 1*(Suppl 1), S19–S26.
Caplan, A. I., & Hariri, R. (2015). Body management: Mesenchymal stem cells control the internal regenerator. *Stem Cells Translational Medicine, 4*, 695–701.
Choi, J. A., Chung, Y.-R., Byun, H.-R., Park, H., Koh, J.-Y., & Yoon, Y. H. (2017). The anti-ALS drug riluzole attenuates pericyte loss in the diabetic retinopathy of streptozotocin-treated mice. *Toxicology and Applied Pharmacology, 315*, 80–89.
Coatti, G. C., Frangini, M., Valadares, M. C., Gomes, J. P., Lima, N. O., Cavaçana, N., Assoni, A. F., Pelatti, M. V., Birbrair, A., de Lima, A. C. P., et al. (2017). Pericytes extend survival of ALS SOD1 mice and induce the expression of antioxidant enzymes in the murine model and in IPSCs derived neuronal cells from an ALS patient. *Stem Cell Reviews, 13*, 686–698.
Corselli, M., Chen, C.-W., Sun, B., Yap, S., Rubin, J. P., & Péault, B. (2011). The tunica adventitia of human arteries and veins as a source of mesenchymal stem cells. *Stem Cells and Development, 21*, 1299–1308.
DeJesus-Hernandez, M., Mackenzie, I. R., Boeve, B. F., Boxer, A. L., Baker, M., Rutherford, N. J., Nicholson, A. M., Finch, N. A., Flynn, H., Adamson, J., et al. (2011). Expanded GGGGCC hexanucleotide repeat in noncoding region of C9ORF72 causes chromosome 9p-linked FTD and ALS. *Neuron, 72*, 245–256.
Dellavalle, A., Sampaolesi, M., Tonlorenzi, R., Tagliafico, E., Sacchetti, B., Perani, L., Innocenzi, A., Galvez, B. G., Messina, G., Morosetti, R., et al. (2007). Pericytes of human skeletal muscle are myogenic precursors distinct from satellite cells. *Nature Cell Biology, 9*, 255–267.
Demestre, M., Parkin-Smith, G., Petzold, A., & Pullen, A. H. (2005). The pro and the active form of matrix metalloproteinase-9 is increased in serum of patients with amyotrophic lateral sclerosis. *Journal of Neuroimmunology, 159*, 146–154.
Dias Moura Prazeres, P. H., Sena, I. F. G., Borges, I. d. T., de Azevedo, P. O., Andreotti, J. P., de Paiva, A. E., de Almeida, V. M., de Paula Guerra, D. A., Pinheiro Dos Santos, G. S., Mintz, A., et al. (2017). Pericytes are heterogeneous in their origin within the same tissue. *Developmental Biology, 427*, 6–11.
Doble, A., & Kennel, P. (2000). Animal models of amyotrophic lateral sclerosis. *Amyotrophic Lateral Sclerosis Motor Neuron Disease: Official Publication of the World Federation of Neurology Research Group on Motor Neuron Diseases, 1*, 301–312.

Donnenfeld, H., Kascsak, R. J., & Bartfeld, H. (1984). Deposits of IgG and C3 in the spinal cord and motor cortex of ALS patients. *Journal of Neuroimmunology, 6*, 51–57.
Engelhardt, J. I., & Appel, S. H. (1990). IgG reactivity in the spinal cord and motor cortex in amyotrophic lateral sclerosis. *Archives of Neurology, 47*, 1210–1216.
Engelhardt, A., Lörler, H., & Neundörfer, B. (1993). Immunohistochemical findings in vasculitic neuropathies. *Acta Neurologica Scandinavica, 87*, 318–321.
Garbuzova-Davis, S., Saporta, S., Haller, E., Kolomey, I., Bennett, S. P., Potter, H., & Sanberg, P. R. (2007a). Evidence of compromised blood-spinal cord barrier in early and late symptomatic SOD1 mice modeling ALS. *PLoS One, 2*, e1205.
Garbuzova-Davis, S., Haller, E., Saporta, S., Kolomey, I., Nicosia, S. V., & Sanberg, P. R. (2007b). Ultrastructure of blood–brain barrier and blood–spinal cord barrier in SOD1 mice modeling ALS. *Brain Research, 1157*, 126–137.
Garbuzova-Davis, S., Rodrigues, M. C. O., Hernandez-Ontiveros, D. G., Louis, M. K., Willing, A. E., Borlongan, C. V., & Sanberg, P. R. (2011). Amyotrophic lateral sclerosis: A neurovascular disease. *Brain Research, 1398*, 113–125.
Garbuzova-Davis, S., Thomson, A., Kurien, C., Shytle, R. D., & Sanberg, P. R. (2016). Potential new complication in drug therapy development for amyotrophic lateral sclerosis. *Expert Review of Neurotherapeutics, 16*, 1397–1405.
Goodall, E. F., & Morrison, K. E. (2006). Amyotrophic lateral sclerosis (motor neuron disease): Proposed mechanisms and pathways to treatment. *Expert Reviews in Molecular Medicine, 8*, 1–22.
Henkel, J. S., Engelhardt, J. I., Siklós, L., Simpson, E. P., Kim, S. H., Pan, T., Goodman, J. C., Siddique, T., Beers, D. R., & Appel, S. H. (2004). Presence of dendritic cells, MCP-1, and activated microglia/macrophages in amyotrophic lateral sclerosis spinal cord tissue. *Annals of Neurology, 55*, 221–235.
Henkel, J. S., Beers, D. R., Wen, S., Bowser, R., & Appel, S. H. (2009). Decreased mRNA expression of tight junction proteins in lumbar spinal cords of patients with ALS. *Neurology, 72*, 1614–1616.
Kamouchi, M., Ago, T., & Kitazono, T. (2011). Brain pericytes: Emerging concepts and functional roles in brain homeostasis. *Cellular and Molecular Neurobiology, 31*, 175–193.
Kassa, R. M., Mariotti, R., Bonaconsa, M., Bertini, G., & Bentivoglio, M. (2009). Gene, cell, and axon changes in the familial amyotrophic lateral sclerosis mouse sensorimotor cortex. *Journal of Neuropathology and Experimental Neurology, 68*, 59–72.
Kenna, K. P., van Doormaal, P. T. C., Dekker, A. M., Ticozzi, N., Kenna, B. J., Diekstra, F. P., van Rheenen, W., van Eijk, K. R., Jones, A. R., Keagle, P., et al. (2016). NEK1 variants confer susceptibility to amyotrophic lateral sclerosis. *Nature Genetics, 48*, 1037–1042.
Lim, G. P., Backstrom, J. R., Cullen, M. J., Miller, C. A., Atkinson, R. D., & Tökés, Z. A. (1996). Matrix metalloproteinases in the neocortex and spinal cord of amyotrophic lateral sclerosis patients. *Journal of Neurochemistry, 67*, 251–259.
Mann, D. M. (1985). The neuropathology of Alzheimer's disease: a review with pathogenetic, aetiological and therapeutic considerations. *Mechanisms of Ageing and Development, 31*, 213–255.
Miller, R. G., Mitchell, J. D., & Moore, D. H. (2012). Riluzole for amyotrophic lateral sclerosis (ALS)/motor neuron disease (MND). *Cochrane Database of Systematic Reviews, 3*, CD001447.
Miyazaki, K., Ohta, Y., Nagai, M., Morimoto, N., Kurata, T., Takehisa, Y., Ikeda, Y., Matsuura, T., & Abe, K. (2011). Disruption of neurovascular unit prior to motor neuron degeneration in amyotrophic lateral sclerosis. *Journal of Neuroscience Research, 89*, 718–728.
Muramatsu, R., Kuroda, M., Matoba, K., Lin, H., Takahashi, C., Koyama, Y., & Yamashita, T. (2015). Prostacyclin prevents pericyte loss and demyelination induced by lysophosphatidylcholine in the central nervous system. *The Journal of Biological Chemistry, 290*, 11515–11525.
Niebroj-Dobosz, I., Janik, P., Sokołowska, B., & Kwiecinski, H. (2010). Matrix metalloproteinases and their tissue inhibitors in serum and cerebrospinal fluid of patients with amyotrophic lateral sclerosis. *European Journal of Neurology, 17*, 226–231.

Pickering, M., Cumiskey, D., & O'Connor, J. J. (2005). Actions of TNF-α on glutamatergic synaptic transmission in the central nervous system. *Experimental Physiology, 90*, 663–670.

Proctor, E. A., Fee, L., Tao, Y., Redler, R. L., Fay, J. M., Zhang, Y., Lv, Z., Mercer, I. P., Deshmukh, M., Lyubchenko, Y. L., et al. (2016). Nonnative SOD1 trimer is toxic to motor neurons in a model of amyotrophic lateral sclerosis. *Proceedings of the National Academy of Sciences of the United States of America, 113*, 614–619.

Regan, R. F., & Guo, Y. (1998). Toxic effect of hemoglobin on spinal cord neurons in culture. *Journal of Neurotrauma, 15*, 645–653.

Robberecht, W., & Philips, T. (2013). The changing scene of amyotrophic lateral sclerosis. *Nature Reviews Neuroscience, 14*, 248–264.

Rosen, D. R., Siddique, T., Patterson, D., Figlewicz, D. A., Sapp, P., Hentati, A., Donaldson, D., Goto, J., O'Regan, J. P., & Deng, H. X. (1993). Mutations in Cu/Zn superoxide dismutase gene are associated with familial amyotrophic lateral sclerosis. *Nature, 362*, 59–62.

Sargsyan, S. A., Monk, P. N., & Shaw, P. J. (2005). Microglia as potential contributors to motor neuron injury in amyotrophic lateral sclerosis. *Glia, 51*, 241–253.

Sasaki, S. (2015). Alterations of the blood-spinal cord barrier in sporadic amyotrophic lateral sclerosis. *Neuropathology, 35*, 518–528.

Simon, C., Lickert, H., Götz, M., & Dimou, L. (2012). Sox10-iCreERT2: A mouse line to inducibly trace the neural crest and oligodendrocyte lineage. *Genesis, 50*, 506–515.

Talbott, E. O., Malek, A. M., & Lacomis, D. (2016). The epidemiology of amyotrophic lateral sclerosis. *Handbook of Clinical Neurology, 138*, 225–238.

Tilleux, S., & Hermans, E. (2007). Neuroinflammation and regulation of glial glutamate uptake in neurological disorders. *Journal of Neuroscience Research, 85*, 2059–2070.

Wang, Z., Pekarskaya, O., Bencheikh, M., Chao, W., Gelbard, H. A., Ghorpade, A., Rothstein, J. D., & Volsky, D. J. (2003). Reduced expression of glutamate transporter EAAT2 and impaired glutamate transport in human primary astrocytes exposed to HIV-1 or gp120. *Virology, 312*, 60–73.

Winkler, E. A., Bell, R. D., & Zlokovic, B. V. (2011). Central nervous system pericytes in health and disease. *Nature Neuroscience, 14*, 1398.

Winkler, E. A., Sengillo, J. D., Sullivan, J. S., Henkel, J. S., Appel, S. H., & Zlokovic, B. (2013). Blood–spinal cord barrier breakdown and pericyte reductions in amyotrophic lateral sclerosis. *Acta Neuropathologica (Berlin), 125*, 111–120.

Winkler, E. A., Sengillo, J. D., Sagare, A. P., Zhao, Z., Ma, Q., Zuniga, E., Wang, Y., Zhong, Z., Sullivan, J. S., Griffin, J. H., et al. (2014). Blood–spinal cord barrier disruption contributes to early motor-neuron degeneration in ALS-model mice. *Proceedings of the National Academy of Sciences, 111*, E1035–E1042.

Yamadera, M., Fujimura, H., Inoue, K., Toyooka, K., Mori, C., Hirano, H., & Sakoda, S. (2015). Microvascular disturbance with decreased pericyte coverage is prominent in the ventral horn of patients with amyotrophic lateral sclerosis. *Amyotrophic Lateral Sclerosis and Frontotemporal Degeneration, 16*, 393–401.

Zhong, Z., Deane, R., Ali, Z., Parisi, M., Shapovalov, Y., O'Banion, M. K., Stojanovic, K., Sagare, A., Boillee, S., Cleveland, D. W., et al. (2008). ALS-causing SOD1 mutants generate vascular changes prior to motor neuron degeneration. *Nature Neuroscience, 11*, 420–422.

Zhong, Z., Ilieva, H., Hallagan, L., Bell, R., Singh, I., Paquette, N., Thiyagarajan, M., Deane, R., Fernandez, J. A., Lane, S., et al. (2009). Activated protein C therapy slows ALS-like disease in mice by transcriptionally inhibiting SOD1 in motor neurons and microglia cells. *The Journal of Clinical Investigation, 119*, 3437–3449.

Zlokovic, B. V. (2008). The blood-brain barrier in health and chronic neurodegenerative disorders. *Neuron, 57*, 178–201.

Zlokovic, B. V. (2011). Neurovascular pathways to neurodegeneration in Alzheimer's disease and other disorders. *Nature Reviews Neuroscience, 12*, 723–738.

Chapter 7
Pericytes in Alzheimer's Disease: Novel Clues to Cerebral Amyloid Angiopathy Pathogenesis

Alla B. Salmina, Yulia K. Komleva, Olga L. Lopatina, and Alexander Birbrair

Abstract Pericytes in the central nervous system attract growing attention of neurobiologists because of obvious opportunities to use them as target cells in numerous brain diseases. Functional activity of pericytes includes control of integrity of the endothelial cell layer, regeneration of vascular cells, and regulation of microcirculation. Pericytes are well integrated in the so-called neurovascular unit (NVU) serving as a platform for effective communications of neurons, astrocytes, endothelial cells, and pericytes. Contribution of pericytes to the establishment and maintaining the structural and functional integrity of blood–brain barrier is confirmed in numerous experimental and clinical studies. The review covers current understandings on the role of pericytes in molecular pathogenesis of NVU/BBB dysfunction in Alzheimer's disease with the special focus on the development of cerebral amyloid angiopathy, deregulation of cerebral angiogenesis, and progression of BBB breakdown seen in Alzheimer's type neurodegeneration.

Keywords Pericyte · Alzheimer's disease · Amyloid angiopathy · Blood–brain barrier · Cerebral angiogenesis

A. B. Salmina (✉) · Y. K. Komleva · O. L. Lopatina
Department of Biochemistry, Medical, Pharmaceutical & Toxicological Chemistry, Krasnoyarsk State Medical University named after Prof. V.F. Voino-Yasenetsky, Krasnoyarsk, Russia

Research Institute of Molecular Medicine & Pathobiochemistry, Krasnoyarsk State Medical University named after Prof. V.F. Voino-Yasenetsky, Krasnoyarsk, Russia

A. Birbrair
Department of Radiology, Columbia University Medical Center, New York, NY, USA

Department of Pathology, Federal University of Minas Gerais, Belo Horizonte, MG, Brazil

A. Birbrair (ed.), *Pericyte Biology in Disease*, Advances in Experimental Medicine and Biology 1147, https://doi.org/10.1007/978-3-030-16908-4_7

Introduction

Pericytes represent one of the vascular cell populations located next to endothelial cells at their abluminal part. Their functioning covers a wide spectrum of activities, i.e., control of integrity of the endothelial cell layer, regeneration of vascular cells, and regulation of microcirculation (Trost et al. 2016). There is a growing evidence that intrinsic plasticity of pericytes is very important for their role on vascular remodeling since they are able to control endothelial cells proliferation, apoptosis, vascular sprouting, and corresponding regression (Simonavicius et al. 2012; Stapor et al. 2014). Pericytes express many surface antigens that are used for phenotyping of cells, i.e., nestin, angiopoietin, chondroitin sulfate proteoglycan 4 NG2, CD146, CD31 (PECAM-1), platelet-derived growth factor receptor-beta PDGFRβ, and desmin (Armulik et al. 2011). However, the majority of markers can also be detected on other cell types, i.e., CD31 on endothelial cells, NG2 on oligodendrocytes, PDGFRβ on fibroblasts. In the context of expression pattern, pericytes seem to be very close to mesenchymal stromal cells, and it was suggested that CD146(+) CD34(−) cells isolated from some (i.e., bone marrow, placenta) but not all the tissues may represent the population of pericytes (Blocki et al. 2013). Generation of pericytes occurs in embryonic and postnatal period throughout the life, but the origin of pericytes remains to be unresolved question in vascular biology, and current data suggest that even within one tissue pericytes seem to be heterogenous in their developmental story (Dias Moura Prazeres et al. 2017). Application of immunostaining for detection of surface antigens, using of transgenic markers (i.e., XlacZ4 and NG2 dsRED) and numerous functional assays allowed identifying several subtypes of pericytes, particularly, type-1 pericytes contribute to fibrogenesis and production of collagen (Birbrair et al. 2013, 2014a, b, c) whereas type-2 pericytes take part in angiogenesis (Birbrair et al. 2014a, b, c).

Pericytes in the central nervous system (CNS) attract the growing attention of neurobiologists because of obvious opportunities to use them as target cells in numerous brain diseases. Cerebral microvessels have higher pericytes/endothelial cells ratio (10–30-fold) than other tissues (Winkler et al. 2014), thereby the role of pericytes in controlling brain microvessel endothelial cell (BMECs) functional activity is rather significant. Brain pericytes are sparsely distributed and occupied the middle of the capillary bed, keep relative stability of their somata but demonstrate dynamic changes in their processes in the adult brain, and may effectively recover after damage (Berthiaume et al. 2018). Being in a close contact with endothelial cells, pericytes are surrounded by basement membrane and extend processes both along and around capillaries (Attwell et al. 2016), therefore, there are no doubts on the key role of pericytes in the control of endothelial cells. Thus, in the brain tissue, they are well integrated in the concept of neurovascular unit (NVU) as a platform for effective communications of neurons, astrocytes, endothelial cells, and pericytes which is required for the maintenance of metabolic coupling, gliovascular control as well as the integrity of the blood–brain barrier (BBB) in (patho) physiological conditions (Salmina et al. 2014). In addition, the activity of pericytes attributes to the regulation of cerebral angiogenesis (experience-induced and reparative), acquisition of specific phenotype of BMECs, establishment of neurovascular coupling providing adequate blood supply in active brain regions, promotion of neurogenesis within neurogenic niches or oligodendrogenesis within oligovascular niches (Hall et al. 2014; Trost et al. 2016). Moreover, pericytes are recognized as a key cellular component of BBB models in vitro suggested for BBB-on-chip or brain-on-chip microphysiological systems as well as a promising tool for nervous system regeneration (Yamamizu et al. 2017; Greenwood-Goodwin et al. 2016; Tian et al. 2017). Thus, pericyte dysfunction in neurodevelopmental and neurodegenerative diseases is gradually becoming a "hot topic" in neurosciences.

Pericyte Dysfunction in Cerebral Amyloid Angiopathy

Dysfunction of NVU is a well-recognized feature of Alzheimer's disease (AD) (Salmina et al. 2010, 2015a, b). In addition to other hypothesis of AD development (amyloid, calcium, or gliocentric), for more than two recent decades, pathogenesis of AD has been discussed in the context of prominent vascular alterations culminating in the establishment of the so-called cerebral amyloid angiopathy (CAA) caused by the accumulation of beta-amyloid (Aβ) in small-sized and medium-sized vessels, mostly arterial (Biffi and Greenberg 2011) predominantly in leptomeningeal and cortical vessels of cerebral lobes and cerebellum (Yamada 2015). In severe angiopathy, amyloid deposits replace degenerating vessel smooth muscle cells leading to microaneurysms formation and hemorrhages (Yamada 2000; Jellinger 2002). In general, vascular nature of AD has been confirmed in numerous epidemiological, neuroimaging, pathological, experimental, and clinical studies (de la Torre 2004) (Fig. 7.1). The vascular hypothesis of AD underlies the initial role of chronic cerebral hypoperfusion, abnormal microvascular remodeling, BBB breakdown, development of ischemic lesions and microhemorrhages associated with Aβ deposition, neuroinflammation, NVU disorganization, loss of neuroplasticity, and synaptic plasticity, thereby resulting in progressive cognitive and behavioral deficits (Salmina et al. 2015a, b) (Fig. 7.2).

It is clear that all types of cells within the NVU could be affected in CAA; therefore, it is easy to observe endothelial alterations and smooth muscle cells degeneration in medium-sized cerebral vessels, as well as endothelial dysfunction, pathology of perivascular astroglia and prominent pericytes loss in cerebral microvessels along the time-course of AD progression. These changes affect the viability of neuronal cells (Grammas et al. 1999), alter microcirculation (Pluta et al. 2013), and lead to chronic ischemia and neurodegeneration. In a contrast to other NVU cells, pericytes have not been recognized as dramatically affected cells in the pathogenesis of AD, but recent findings suggest their significant role in the progression of AD-associated microvascular alterations and impairment of plasticity in AD brain.

At this point, it should be clarified that morphology and physiology of pericytes might be greatly compromised in aged brain, therefore, one should take care while

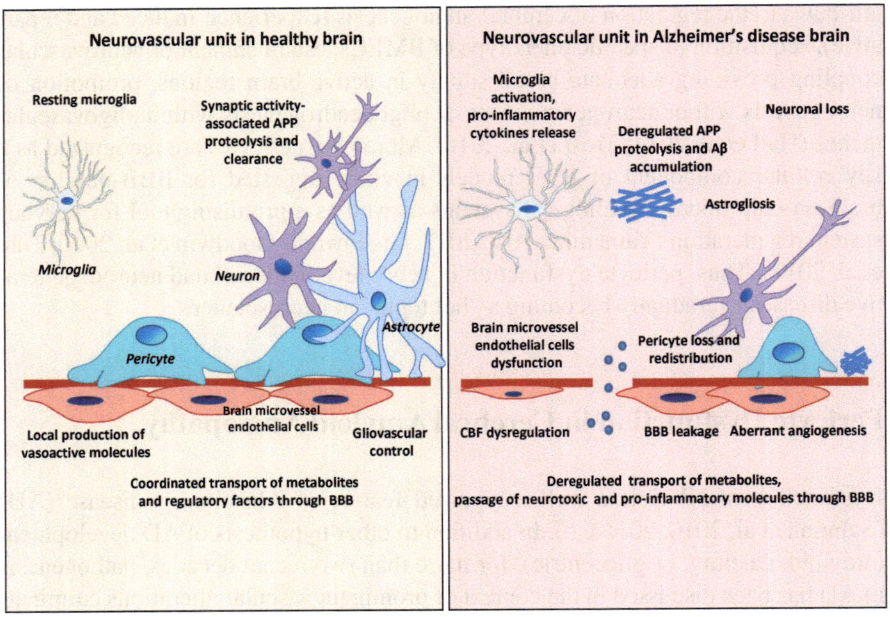

Fig. 7.1 Mechanisms of neurovascular unit dysfunction in Alzheimer's disease. NVU dysfunction is an obligatory component of Alzheimer's disease pathogenesis. Compared with the normal brain, Alzheimer's diseases brain is characterized by prominent neuronal loss, reactive gliosis, dysfunction and death of brain microvessel endothelial cells and pericytes caused by excessive accumulation of Aβ in brain parenchyma and perivascular region. As a result, structural and functional integrity of BBB is compromised, thereby supporting the establishment of circulus vitiosus

talking about pericytes dysfunction in a time-course of Alzheimer's type of dementia. As an example, aging is always associated with pericytes loss in various tissues and dramatic changes in their myogenic or angiogenic capacity, and decreased number of effective endothelial–pericyte interactions (Hughes et al. 2006; Birbrair et al. 2014a, b, c; Stefanska et al. 2015). Pericytes aging is also important in the brain tissue where reduction of pericytes number might be partially compensated by extension of their processes, presumably, due to the activity of PDGF/PDGFR signaling. Particularly, Berthiaume and colleagues demonstrated that pericytes can participate in vascular remodeling in the adult brain. The authors revealed pericytes' plasticity in the adult brain by using elegant state-of-the-art techniques, including two-photon microscopy in combination with sophisticated Cre/loxP in vivo tracing technologies. They imaged at high resolution over several weeks cerebral pericytes in NG2-CreER/TdTomato, Myh11-CreER/TdTomato, and PDGFRβ-Cre/YFP mice. These experiments unveiled that pericytes compose a quasi-continuous, not overlapping, network along the entire length of blood vessels. Interestingly, the pericytes' prolongations were not stable in length, extending or retracting during the period of analysis. Then, the authors explored the effect of pericyte's death on its neighbor pericytes. After pericyte' ablation, using targeted two-photon irradiation, Berthiaume and colleagues showed that adjacent pericytes extend their processes

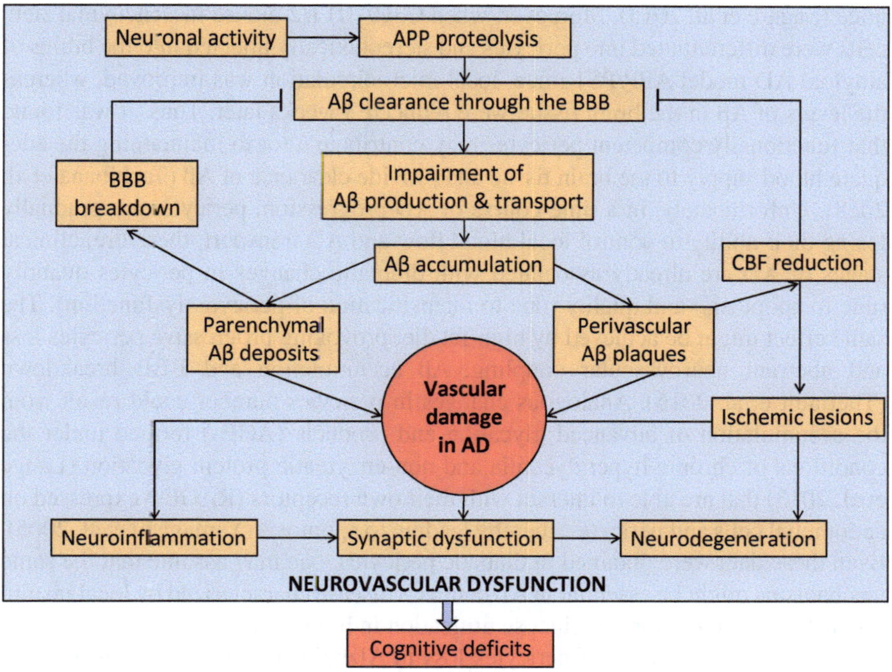

Fig. 7.2 Molecular pathogenesis of cerebral amyloid angiopathy (CAA) development in Alzheimer's disease. Various mechanisms contribute to initiation and progression of CAA in Alzheimer's disease, including excessive Aβ production, inadequate Aβ clearance, insufficient blood supply in affected brain regions, and neuroinflammation

into the uncovered area, covering the exposed blood vessel. Strikingly, neighboring pericytes are able to inhibit vascular dilatation that happens after pericyte depletion (Berthiaume et al. 2018). Thus, one may assume that aging- or Alzheimer's-associated loss of pericytes in cerebral microvessels could lead to compensatory extension of processes from neighboring pericytes in order to cover the most achievable area at the abluminal side of BMECs. However, since both aging and Alzheimer's type of neurodegeneration are often accompanied by elevation in PDGF levels or dysfunction in PDGF/PDGFR signaling (Vazquez-Padron et al. 2004; Liu et al. 2018), remodeling of remaining pericytes might be inefficient to compensate for pericytes loss, thereby leading to incomplete coverage of BMECs and aberrant BBB structural and functional integrity.

Indeed, Alzheimer's type of neurodegeneration is marked with significant and progressive pericytes loss (Giannoni et al. 2016). Several recent reviews were focused on the role of pericytes in the pathogenesis of AD (Winkler et al. 2014; Kisler et al. 2017a, b). Aβ is toxic for cerebral pericytes, thereby, pericyte loss further contributes to amyloid neurotoxicity as it was confirmed in mice overexpressing the Swedish mutation of human Aβ-precursor protein (APPsw/0) and crossed with pericyte-deficient platelet-derived growth factor receptor-β (PDGFRβ+/−)

mice (Sagare et al. 2013). Moreover, when C3H/10T1/2 mouse mesenchymal stem cells were differentiated into pericytes and stereotaxically injected into the brains of amyloid AD model APP/PS1 mice, local microcirculation was improved, whereas the levels of Aβ in the brain tissue were reduced 3 weeks later. Thus, it was found that functionally competent pericytes may contribute a lot to maintaining the adequate blood supply to the brain tissue and provide clearance of Aβ (Tachibana et al. 2018). Unfortunately, in a time-course of AD progression, pericytes are gradually losing their ability to control local blood flow and AD transport, therefore, clinical stages of AD are already associated with dramatic changes in pericytes quantity (due to apoptosis) and quality (due to intensification of pericyte dysfunction). The same effect might be achieved by high-fat diet provoking progressive pericytes loss and aberrant neurovascular coupling, Aβ accumulation and BBB breakdown (Thériault et al. 2016). Analogous changes in pericytes number could result from the accumulation of advanced glycation end products (AGEs) formed under the conditions of chronic hyperglycemia and non-enzymatic protein glycation (Lange et al. 2013) that are able to interact with their own receptors (RAGEs) expressed on endothelial cells and pericytes, thereby leading to apoptosis (Yamagishi et al. 2005). Even these data were obtained in diabetic pericytes, one may assume that the same mechanisms might be essential in Alzheimer's disease characterized by local insulin resistance and alterations in glucose utilization in brain cells.

Thus, the main causes of pericytes loss in Alzheimer's disease might be summarized as follows: (1) toxic action of supraphysiological concentrations of Aβ leading to pericytes injury and cell death; (2) toxic action of reactive oxygen species produced in the conditions of oxidative stress and excitotoxicity within the NVU; (3) pro-apoptotic action of AGEs; (4) aberrant signaling pathways that contribute to the control of pericytes' functional activity and viability (i.e., PDGF-, TGFβ1-, or angiopoietin/Tie2-mediated); (5) metabolic alterations in pericytes; (6) excessive pericytes remodeling due to stimulation of angiogenesis leading to hypervascularity.

Pericytes take an active part in the translocation of Aβ through the BBB. In normal conditions, low-density lipoprotein receptor-related protein-1 (LRP1) acting as Aβ translocator is abundantly expressed in BMECs and pericytes being involved in the transfer of Aβ which is constantly produced in the brain tissue to the peripheral blood (Winkler et al. 2014). AD-associated loss of pericytes results in impaired clearance of Aβ in the brain, thereby leading to amyloid deposition and development of CAA. Moreover, incubation of cerebrovascular cells in vitro with toxic concentrations of Aβ results in the increase of LRP1 expression, probably, for better internalization of amyloid by pericytes, but excess of Aβ leads to loss of pericytes (Wilhelmus et al. 2007).

In general, pericytes appear as a nice model to study Alzheimer's type of neurodegeneration in vitro. In very early studies, pericytes have been shown to produce and metabolize amyloid precursor protein (APP) as well as some other APP-associated molecules (ApoE, complement factor C1q (Verbeek et al. 1999)). At the same time, pericytes serve as a target for the cytotoxic action of Aβ but seem to be more resistant to its action comparing to other brain cells, i.e., in terms of Ca^{2+} signaling: basal levels of intracellular Ca^{2+} are greatly affected by the exposure of pericytes to Aβ in vitro, however, calcium machinery controlled by G-protein-coupled receptors remains unaffected (Piegsa et al. 2017).

In *APP*$^{sw/0}$; *Pdgfrβ*$^{+/-}$, pericyte deficiency leads to progression of signs of tau pathology and an early neuronal loss further resulting in accelerated cognitive decline due to complimentary harmful effects of Aβ accumulation and pericytes loss (Sagare et al. 2013).

Interesting properties of pericytes have been described in hypothalamus where these cells can specifically increase insulin sensitivity of hypothalamic neurons (Takahashi et al. 2015). Taking into consideration the current view on the pathogenesis of AD as a local insulin resistance and impairment of glucose metabolism in the brain tissue (An et al. 2018), one can suggest that pericytes loss in the defined brain regions could be responsible for reduced susceptibility of neurons to insulin action and glucose utilization. Pericytes serve as important regulators of insulin transport through the BBB (Banks et al. 2012), and, vice versa, insulin stabilizes BMECs-pericytes interactions and integrity of tight junctions within the BBB in a phosphoinositide-3 kinase/protein kinase B/glycogen synthase kinase-3β-dependent manner (Ito et al. 2017). Thus, dysfunction of pericytes in AD might lead to poor entry of peripherally produced insulin into the brain tissue and aggravation of local insulin resistance state. It is interesting that insulin may protect cerebral pericytes from Aβ cytotoxicity (Rensink et al. 2004a), whereas the treatment of pericytes with toxic concentrations of Aβ results in decreased expression of insulin-like growth factor-binding protein-2 mRNA (Rensink et al. 2004b).

Since functional insulin receptors are expressed in pericytes (Escudero et al. 2017), local insulin resistance in AD would have more pronounced effects on pericytes viability and functioning by limiting their glycolytic activity. It was shown that the inhibition of 6-phosphofructo-2-kinase/fructose-2,6-biphosphatase 3 (PFKFB3) activity (regulatory glycolytic enzyme) in pericytes resulted in dramatic changes such as suppression of motility and increasing adhesion to endothelial cells, thereby resulting in better coverage of endothelial layer with pericytes and, obviously, in the establishment of anti-angiogenic microenvironment (Cantelmo et al. 2016). However, no GluT4 expression has been detected in brain pericytes; therefore, insulin action on pericyte glycolytic metabolism needs in further careful evaluation.

Pericytes contribute a lot to the local control of cerebral blood flow, particularly, they are able to dilate in response to neuronal stimulation to provide adequate blood supply in active brain regions (Winkler et al. 2014). When functional activity of pericytes is compromised, i.e., in PDGFRβ+/− mice, such responses to neuronal stimulation is abolished within the NVU without obvious changes in the activity of other cells regulating functional hyperemia (BMEC, perivascular astroglia). As a result of hemodynamic alterations, NVU dysfunction develops and leads to neurodegeneration. PDGFRβ−/− also demonstrate delayed capillary but not arteriolar dilation to various stimuli corresponding to lack of pericyte coverage of BMECs (Kisler et al. 2017a, b). However, very recent data obtained with an optogenetic approach to pericytes stimulation suggest that pericyte-controlled diameter of small vessels in the brain tissue might not predominate over smooth-muscle actin (SMA)-regulated constriction of arterioles (Sweeney et al. 2018).

Other factors contributing to pericyte dysfunction and promotion of CAA are as follows: (1) oxidative stress due to Aβ-mediated cytotoxicity, overproduction of reactive oxygen species (ROS) and CAA-associated cerebrovascular deficits (Han et al. 2015), microglia-induced activation of pericytes leading to ROS production (Ding et al. 2017); (2) aberrant PDGFRβ expression and signaling induced by Aβ in neuronal cells (Liu et al. 2018) and in pericytes (Miners et al. 2018); (3) mitochondrial dysfunction caused by the deposition of Aβ and progression of CAA culminating in pericytes degeneration (Szpak et al. 2007); (4) hypoxia-induced HIF-1-mediated changes in pericytes motility and adhesion (Mayo and Bearden 2015) leading to excessive angiogenesis and hypervascularity of brain tissue.

Figure 7.3 summarizes current understandings of alterations of pericytes biology in CAA.

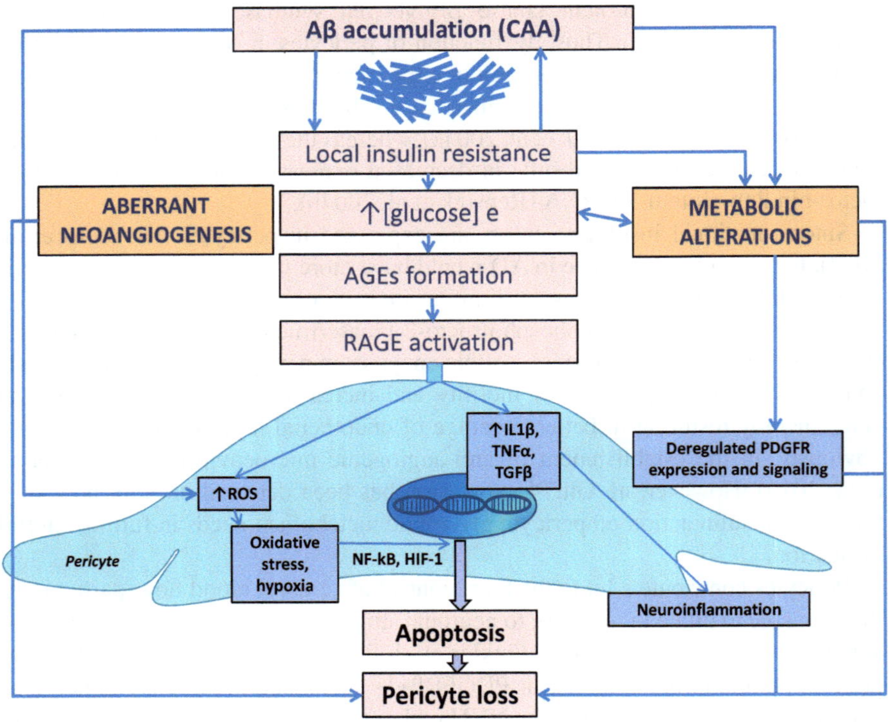

Fig. 7.3 Molecular mechanisms leading to pericytes loss in cerebral amyloid angiopathy (CAA). Neurotoxic action of accumulated Aβ in Alzheimer's disease leads to local insulin resistance which further promotes Aβ production and deposition. Chronic high levels of extracellular glucose result in non-enzymatic protein glycation, accumulation of advanced glycation end products (AGEs) and persistent activation of their receptors (RAGE) in pericytes. Then, pericytes respond to these changes by activation transcription of oxidative stress- and hypoxia-controlled transcription factors, thereby switching on the programmed cell death. Aberrant neoangiogenesis and neuroinflammation serve as mechanisms supporting non-reversible pericytes dysfunction

Pericyte Dysfunction in Cerebral Hypervascularity and BBB Breakdown Associated with Alzheimer's Disease

Role of pericytes in controlling angiogenesis is well established. Particularly, they may participate in different stages of angiogenic process: (1) establishment of local pro-angiogenic microenvironment stimulating endothelial cells to proliferate and to migrate along the newly forming vascular tube; (2) detachment from the abluminal part of endothelial cells to ensure effective contribution of tip-cells to the vessel elongation; (3) regulation of final stages of angiogenesis when maturation of newly established vessels is of key importance for the integration of vessel in the pre-existing vascular network and acquisition of selective permeability; (4) controlling processes of microvascular rarefaction that are essential for adequate remodeling of vessels; (5) support of endothelial cells survival and functional competence (Benjamin et al. 1998; Franco et al. 2011; Ribatti et al. 2011; Simonavicius et al. 2012; Eilken et al. 2017). To do this, pericytes utilize many signaling pathways, including PDGF, VEGF, Angiopoietin/Tie, MMP9, Notch, Endosialin/CD248, purinergic signaling, and gap junction machinery.

Recently, it became clear that AD is characterized by paradoxical hypervascularity occurring due to excessive neoangiogenesis and establishment of newly formed small vessels with leaky BBB (Biron et al. 2011). Cessation of neoangiogenesis by Aβ immunotherapy may prevent further brain tissue degeneration (Biron et al. 2013). Disruption of PDGFRβ-mediated signaling in brain pericytes results in early and progressive loss of pericytes, microvascular rarefaction and alterations in BB structural and functional integrity predominantly in the cortex, hippocampus and striatum (Nikolakopoulou et al. 2017), thus providing new insights in the pathogenic role of pericyte-controlled vascular factor in AD.

Unstimulated pericytes support vessel wall integrity, however, they convert into cells with evident pro-angiogenic potential being stimulated by various regulatory and damaging factors. Recent data suggest that pericytes may contribute to excessive angiogenesis in AD via several mechanisms. First of all, as it was mentioned above, detachment of pericytes from the endothelial cell layer associated with CypA-MMP9-mediated basal membrane destruction is required for tip-cell movement and vascular tube formation. Secondly, stimulated pericytes are able to produce numerous pro-angiogenic factors, i.e., ligands of chemokine receptor CXCR3 (probably, CXCL4/platelet factor 4, CXCL9/MIG, CXCL10/IP-10, or CXCL11/IP-9) expressed on endothelial cells (Bodnar et al. 2013), or angiopoietin involved in angiopoietin/Tie signaling between endothelial cells and pericytes (Teichert et al. 2017), thereby affecting cerebral angiogenesis being damaged by accumulating Aβ. Thirdly, pericytes may contribute to cerebral microvascular rarefaction seen in normal aging and AD and correlating with dementia progression and BBB impairment (Tucsek et al. 2014). The same phenomenon—microvascular rarefaction and pericytes deficiency—is also well recognized in hypertensive individuals with AD (Toth et al. 2013), thereby leading to microthrombosis, inadequate blood supply in active brain regions. Fourthly, metabolic disturbances caused by the impairment of glucose metabolism in AD brain could lead to suppressed glycolytic flux in BMECs and pericytes, thereby resulting in abnormal vessel sprouting and disorganization, reduced pericyte coverage, and breakdown of BBB (Cruys et al. 2016). Finally, insulin signaling which is absolutely required for developmental angiogenesis and is rather specific for pericytes (Warmke et al. 2017).

Basal and stimulated production of lactate in glycolysis and its transport between the NVU cells is an important mechanism of angiogenesis and barriergenesis regulation in the brain tissue (Salmina et al. 2015a, b). Astrocytes serve a major source of lactate which is utilized by neurons (for energy production) and by endothelial cells (for brain-to-blood transfer). Within the NVU, lactate acts at target cells via specific lactate receptor (HCAR1/GPR81) which is known as metabolic sensor regulating several processes in carbohydrate and lipid metabolism, partially in the insulin-dependent manner in some peripheral tissues and in the NVU as well (Ahmed et al. 2010; Lauritzen et al. 2014). In this context, lactate acts as autocrine or paracrine regulator. Recently, we have shown that long-lasting stimulation of GPR81 receptors in BMECs in vitro could activates mitochondrial biogenesis but suppresses expression of monocarboxylate transporter-1 (MCT-1) and CD147 (Khilazheva et al. 2017). Taking into consideration that cerebral pericytes are equipped with GPR81 whose stimulation results in the elevation of local VEGFA levels and promotion of angiogenesis associated with the activation of extracellular signal-regulated kinase (ERK1/2)- and Akt (Morland et al. 2017), one may suggest that local production of lactate in active brain regions (due to neuron-astroglia metabolic coupling) could activate both pericytes and BMEC to provide microenvironment favoring angiogenesis. Moreover, it may also have some additional meaning in the context of vascular tone regulation as it was shown in retinal pericytes subjected to extracellular lactate (Yamanishi et al. 2006). High concentrations of lactate is a hallmark of aging and AD progression (Kapogiannis and Reiter 2014; Liguori et al. 2015); therefore, it is tempting to speculate that permanent elevated levels of lactate could degenerating brain to excessive angiogenesis (presumably, via mechanism involving BMECs and pericytes activation via GPR81 receptors) resulting in the establishment of disorganized and defective microvessels with impaired barrier function.

The unresolved question remains whether such changes could have any relation to the mechanism of pro-angiogenic activity of hormones and growth factors (i.e., insulin or VEGF) which implies detachment of pericytes from endothelial cells as a prerequisite for effective tip-cell proliferation and migration along the growing vascular tube (Escudero et al. 2017). If so, Aβ-induced pericytes loss from the affected brain microvessels could serve as a mechanism of angiogenesis support. However, in a case of CAA, loss of pericytes would result in unbalanced pro- and anti-angiogenic activities in the perivascular space further leading to abnormal hypervascularity.

As expected, location of pericytes just next to the endothelial cell layer makes them ideal contributors to the control of BBB permeability. As an example, expression of CD146 which acts as a co-receptor for PDGFRβ is required for efficient pericyte–endothelial interactions and maturation of the BBB (Chen et al. 2017).

Cerebrovascular pericytes produce lipidated forms of ApoE acting together with perivascular astroglial cells, and ApoE suppresses motility of pericytes and their adhesion in LRP1- and RhoA-dependent manner (Casey et al. 2015). Experimental data obtained in mice with targeted replacement of murine ApoE with each human ApoE isoform or in ApoE−/− mice demonstrate that ApoE is required for the cerebrovascular integrity by regulating cyclophilin A (CypA)–NF-κB–matrix metalloproteinase (MMP9) pathway in pericytes, whereas insufficiency of ApoE expression leads to elevated production of CypA in pericytes and disruption of tight junctions in BMECs (Bell et al. 2012). Individuals with ApoE4 gene (genetic risk factor for AD) display severe alterations of BBB structural and functional integrity that are related to the degree of pericytes loss in brain microvessels due to ApoE4 leads to excessive activation of LRP1-dependent cyclophilin A (CypA)– matrix metalloproteinase 9 (MMP-9) signaling in pericytes (Halliday et al. 2016).

Degree of pericytes loss correlates with the impairment of BBB permeability in individuals with Alzheimer's type of neurodegeneration: high levels of sPDGFRβ originated from dysfunctional or destroyed pericytes have been detected in the cerebrovascular fluid of persons with mild cognitive impairment (MCI) compared to age-matched cognitively normal subjects, sPDGFRβ concentrations positively correlated with increased BBB permeability in the hippocampus of MCI patients (Montagne et al. 2015).

Elevated permeability of BBB caused (at least, partially) by functionally incompetent or damaged pericytes, and provides conditions for progression of neuroinflammation which is a key mechanism of AD pathogenesis. Contribution of glial cells and pericytes to the pathogenesis of neuroinflammation can be distinguished, i.e., in the 3D BBB-on-chip model in vitro (Herland et al. 2016). It should be noted that pericytes may act as macrophage-like cells to clean extracellular perivascular fluid in the brain tissue by means of phagocytosis and pinocytosis (Bergers and Song 2005). Besides, they have rather impressive secretome consisting of chemokines, interleukins (IL-9, -10, -12, -13), granulocyte-colony stimulating factor, granulocyte macrophage-colony stimulating factor, etc., particularly, being activated with pro-inflammatory stimuli (Kovac et al. 2011). On the other hand, neuroinflammation affects pericytes viability and functional activity, i.e., TGFβ1 whose dyscoordinated signaling has been detected in AD (von Bernhardi et al. 2015) may stimulate pericytes to release pro-inflammatory cytokines and extracellular matrix degrading enzymes. As a result, BBB integrity is compromised and microglia is attracted to the site of neuroinflammation (Rustenhoven et al. 2016). Some data suggest that due to multipotent stem cell activity of pericytes, they may serve as a source of microglial cells in brain ischemia (Sakuma et al. 2016), but whether this mechanism is active in chronic neurodegeneration is not clear yet.

In some clonogenic niches (i.e., in the bone marrow), CD146+ pericytes exist as multipotent cells (Mangialardi et al. 2016). In the brain, pericytes may differentiate to microglia in conditions with the demand for more immune cells, whereas neuronal cells could be also achieved from pericytes by means of genetic reprogramming procedure in vitro (Karow et al. 2012). Thus, the plasticity of pericytes allows achieving a phenotype which is mainly anticipated in the context of current conditions in the brain tissue. At the same time, pericyte-regulated BBB permeability may affect neuroplasticity as well. It is known that BBB serve as a platform for neurogenic and oligovascular niches providing optimal microenvironment for neurogenesis and oligodendrogenesis in (patho)physiological conditions. Partially compromised BBB may be rather important for the effective adjustment of niche microenvironment to the metabolic needs of neural stem cells and progenitors: leaky BBB could be an advantage for the population of actively proliferating cells with high intensity of metabolism. In a contrast to the structure of BBB in other brain regions or BBB in microvessels of hippocampal subgranular zone (SGZ), vascular scaffold in the subventricular zone (SVZ) is characterized by less expression of tight junction proteins and aquaporin-4 (AQP4), probably, for the direct contact of endothelium with stem or progenitors cells. Hence, BBB in this neurogenic niche might be functionally defective (Pozhilenkova et al. 2017), presumably, for maintaining the number of quiescent stem cells and preventing depletion of their pool (Ottone et al. 2014). Recent data reveal that pericytes play an important role in controlling stem cells proliferation within SVZ (Crouch et al. 2015). In SGZ, pericytes act in a coordination with astrocytes and endothelial cells being, probably, mainly involved in the regulation of stem cells adhesion (Ehret et al. 2015). However, how pericytes dysfunction may affect adult neurogenesis in Alzheimer's disease is not clear yet and requires scrupulous investigation.

In oligovascular niches, pericytes contact to oligodendrocyte progenitor cells (OPCs) and these two cell populations mutually regulate proliferation and support survival of each other in the perivascular region (Maki et al. 2015). Thus, pericytes may control the process of oligodendrogenesis and myelinization of axons in newly formed neurons. Since oligodendrocyte pathology is a very early sign of AD (Desai et al. 2010), one may assume that oligovascular niches might be compromised due to insufficient pericyte support of OPCs development.

In sum, pericyte dysfunction in AD results in the induction of aberrant angiogenesis, pathological hypervascularity, disorganized microvasculature, and leaky BBB. AD-affected pericytes that should control neurogenesis and oligodendrogenesis lose their ability to support stem cells population dynamics, thereby contributing to progressive cognitive deficits. Figure 7.4 shows mechanisms of pericyte-mediated control of BBB integrity, angiogenesis and neurogenesis in Alzheimer's disease.

Summary and Future Prospects

Deciphering a role of pericytes in the regulation of key mechanisms within the NVU, incl. gliovascular control, BBB integrity, immune defense etc., suggests novel approaches to the treatment of CNS disorders associated with NVU dysfunction (neurodegeneration, stroke, trauma, neuroinfection). As an example, contractile

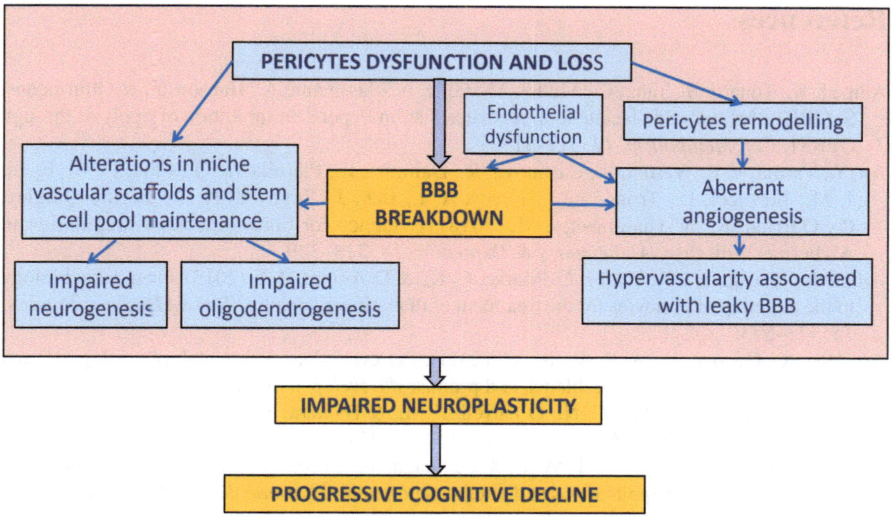

Fig. 7.4 Pericyte-mediated control of BBB integrity, angiogenesis and neurogenesis in Alzheimer's disease. Alterations of BBB structural and functional integrity caused by endothelial and pericyte dysfunction result in the impairment of neurogenic and oligovascular niches activity. In sum, aberrant neurogenesis, oligodendrogenesis and angiogenesis underlie neuroplasticity alterations and progressive cognitive decline seen in chronic neurodegeneration

pericytes in the brain vessels serve as cellular mediators of no-reflow phenomenon seen in ischemic brain tissue (O'Farrell et al. 2017); therefore, targeting pericytes in brain microvasculature could be helpful in restoring adequate blood supply in the ischemic regions. Such effects could be achieved by modulating pericytes response to neurotransmitters and gliotransmitters (serotonin, adenosine) (Li et al. 2017). Another intriguing possibility is to control PDGF/PDGFR2 and Angiopoietin-1/Tie2 signaling cascades in cerebral pericytes in a similar manner as was shown in retina (Arboleda-Velasquez et al. 2015) or in tumor tissue (Kang and Shin 2016) to stimulate or to reduce angiogenesis. This approach has not been tested in neurodegeneration, however, high degree of pericytes plasticity makes them very attractive tool for the promotion of CNS repair in neurodegenerative diseases (Lange et al. 2013). Finally, pericytes serve as a functional part of NVU/BBB models in vitro utilized for studying pharmacokinetics of novel drug candidates (Wang et al. 2016) or as a cell component for bioengineered constructs with a great potential in the regenerative medicine (Avolio et al. 2017).

Acknowledgment A.B.S., Y.K.K., and O.L.L. are supported by the grant given by the President of Russian Federation for the Leading Scientific Teams (N 6240.2018.7).

References

Ahmed, K., Tunaru, S., Tang, C., Muller, M., Gille, A., Sassmann, A., Hanson, J., & Offermanns, S. (2010). An autocrine lactate loop mediates insulin-dependent inhibition of lipolysis through GPR81. *Cell Metabolism, 11*, 311–319.

An, Y., Varma, V. R., Varma, S., Casanova, R., Dammer, E., Pletnikova, O., Chia, C. W., Egan, J. M., Ferrucci, L., Troncoso, J., Levey, A. I., Lah, J., Seyfried, N. T., Legido-Quigley, C., O'Brien, R., & Thambisetty, M. (2018). Evidence for brain glucose dysregulation in Alzheimer's disease. *Alzheimer's & Dementia, 14*, 318–329.

Arboleda-Velasquez, J. F., Valdez, C., Marko, C. K., & D'Amore, P. A. (2015). From pathobiology to the targeting of pericytes for the treatment of diabetic retinopathy. *Current Diabetes Reports, 15*, 573–573.

Armulik, A., Genové, G., & Betsholtz, C. (2011). Pericytes: Developmental, physiological, and pathological perspectives, problems, and promises. *Developmental Cell, 21*, 193–215.

Attwell, D., Mishra, A., Hall, C. N., O'Farrell, F. M., & Dalkara, T. (2016). What is a pericyte? *Journal of Cerebral Blood Flow & Metabolism, 36*, 451–455.

Avolio, E., Alvino, V. V., Ghorbel, M. T., & Campagnolo, P. (2017). Perivascular cells and tissue engineering: Current applications and untapped potential. *Pharmacology & Therapeutics, 171*, 83–92.

Banks, W. A., Owen, J. B., & Erickson, M. A. (2012). Insulin in the brain: There and back again. *Pharmacology & Therapeutics, 136*, 82–93.

Bell, R. D., Winkler, E. A., Singh, I., Sagare, A. P., Deane, R., Wu, Z., Holtzman, D. M., Betsholtz, C., Armulik, A., Sallstrom, J., Berk, B. C., & Zlokovic, B. V. (2012). Apolipoprotein E controls cerebrovascular integrity via cyclophilin A. *Nature, 485*, 512–516.

Benjamin, L. E., Hemo, I., & Keshet, E. (1998). A plasticity window for blood vessel remodelling is defined by pericyte coverage of the preformed endothelial network and is regulated by PDGF-B and VEGF. *Development, 125*, 1591–1598.

Bergers, G., & Song, S. (2005). The role of pericytes in blood-vessel formation and maintenance. *Neuro-Oncology, 7*, 452–464.

Berthiaume, A. A., Grant, R. I., McDowell, K. P., Underly, R. G., Hartmann, D. A., Levy, M., Bhat, N. R., & Shih, A. Y. (2018). Dynamic remodeling of pericytes in vivo maintains capillary coverage in the adult mouse brain. *Cell Reports, 22*, 8–16.

Biffi, A., & Greenberg, S. M. (2011). Cerebral amyloid angiopathy: A systematic review. *Journal of Clinical Neurology, 7*, 1–9.

Birbrair, A., Zhang, T., Wang, Z. M., Messi, M. L., Mintz, A., & Delbono, O. (2013). Type-1 pericytes participate in fibrous tissue deposition in aged skeletal muscle. *American Journal of Physiology. Cell Physiology, 305*, C1098–C1113.

Birbrair, A., Zhang, T., Files, D. C., Mannava, S., Smith, T., Wang, Z. M., Messi, M. L., Mintz, A., & Delbono, O. (2014a). Type-1 pericytes accumulate after tissue injury and produce collagen in an organ-dependent manner. *Stem Cell Research & Therapy, 5*, 122.

Birbrair, A., Zhang, T., Wang, Z.-M., Messi, M. L., Mintz, A., & Delbono, O. (2014b). Pericytes: Multitasking cells in the regeneration of injured, diseased, and aged skeletal muscle. *Frontiers in Aging Neuroscience, 6*, 245.

Birbrair, A., Zhang, T., Wang, Z. M., Messi, M. L., Olson, J. D., Mintz, A., & Delbono, O. (2014c). Type-2 pericytes participate in normal and tumoral angiogenesis. *American Journal of Physiology. Cell Physiology, 307*, C25–C38.

Biron, K. E., Dickstein, D. L., Gopaul, R., & Jefferies, W. A. (2011). Amyloid triggers extensive cerebral angiogenesis causing blood brain barrier permeability and hypervascularity in Alzheimer's disease. *PLoS One, 6*, e23789.

Biron, K. E., Dickstein, D. L., Gopaul, R., Fenninger, F., & Jefferies, W. A. (2013). Cessation of neoangiogenesis in Alzheimer's disease follows amyloid-beta immunization. *Scientific Reports, 3*, 1354.

Blocki, A., Wang, Y., Koch, M., Peh, P., Beyer, S., Law, P., Hui, J., & Raghunath, M. (2013). Not all MSCs can act as pericytes: Functional in vitro assays to distinguish pericytes from other mesenchymal stem cells in angiogenesis. *Stem Cells and Development, 22*, 2347–2355.

Bodnar, R. J., Rodgers, M. E., Chen, W. C. W., & Wells, A. (2013). Pericyte regulation of vascular remodeling through the CXC receptor 3. *Arteriosclerosis, Thrombosis, and Vascular Biology, 33*, 2818–2829.

Cantelmo, A. R., Conradi, L.-C., Brajic, A., Goveia, J., Kalucka, J., Pircher, A., Chaturvedi, P., Hol, J., Thienpont, B., Teuwen, L.-A., Schoors, S., Boeckx, B., Vriens, J., Kuchnio, A., Veys, K., Cruys, B., Finotto, L., Treps, L., Stav-Noraas, T. E., Bifari, F., Stapor, P., Decimo, I., Kampen, K., De Bock, K., Haraldsen, G., Schoonjans, L., Rabelink, T., Eelen, G., Ghesquière, B., Rehman, J., Lambrechts, D., Malik, A. B., Dewerchin, M., & Carmeliet, P. (2016). Inhibition of the glycolytic activator pfkfb3 in endothelial cells induces tumor vessel normalization, impairs metastasis and improves chemotherapy. *Cancer Cell, 30*, 968–985.

Casey, C. S., Atagi, Y., Yamazaki, Y., Shinohara, M., Tachibana, M., Fu, Y., Bu, G., & Kanekiyo, T. (2015). Apolipoprotein E inhibits cerebrovascular pericyte mobility through a RhoA protein-mediated pathway. *The Journal of Biological Chemistry, 290*, 14208–14217.

Chen, J., Luo, Y., Hui, H., Cai, T., Huang, H., Yang, F., Feng, J., Zhang, J., & Yan, X. (2017). CD146 coordinates brain endothelial cell–pericyte communication for blood–brain barrier development. *Proceedings of the National Academy of Sciences, 114*, E7622–E7631.

Crouch, E. E., Liu, C., Silva-Vargas, V., & Doetsch, F. (2015). Regional and stage-specific effects of prospectively purified vascular cells on the adult V-SVZ neural stem cell lineage. *The Journal of Neuroscience, 35*, 4528–4539.

Cruys, B., Wong, B. W., Kuchnio, A., Verdegem, D., Cantelmo, A. R., Conradi, L.-C., Vandekeere, S., Bouché, A., Cornelissen, I., Vinckier, S., Merks, R. M. H., Dejana, E., Gerhardt, H., Dewerchin, M., Bentley, K., & Carmeliet, P. (2016). Glycolytic regulation of cell rearrangement in angiogenesis. *Nature Communications, 7*, 12240.

de la Torre, J. C. (2004). Is Alzheimer's disease a neurodegenerative or a vascular disorder? Data, dogma, and dialectics. *Lancet Neurology, 3*, 184–190.

Desai, M. K., Mastrangelo, M. A., Ryan, D. A., Sudol, K. L., Narrow, W. C., & Bowers, W. J. (2010). Early oligodendrocyte/myelin pathology in Alzheimer's disease mice constitutes a novel therapeutic target. *The American Journal of Pathology, 177*, 1422–1435.

Dias Moura Prazeres, P. H., Sena, I. F. G., Borges, I. D. T., de Azevedo, P. O., Andreotti, J. P., de Paiva, A. E., de Almeida, V. M., de Paula Guerra, D. A., Pinheiro Dos Santos, G. S., Mintz, A., Delbono, C., & Birbrair, A. (2017). Pericytes are heterogeneous in their origin within the same tissue. *Developmental Biology, 427*, 6–11.

Ding, X., Zhang, M., Gu, R., Xu, G., & Wu, H. (2017). Activated microglia induce the production of reactive oxygen species and promote apoptosis of co-cultured retinal microvascular pericytes. *Graefe's Archive for Clinical and Experimental Ophthalmology, 255*, 777–788.

Ehret, F., Vogler, S., & Kempermann, G. (2015). A co-culture model of the hippocampal neurogenic niche reveals differential effects of astrocytes, endothelial cells and pericytes on proliferation and differentiation of adult murine precursor cells. *Stem Cell Research, 15*, 514–521.

Eilken, H. M., Dieguez-Hurtado, R., Schmidt, I., Nakayama, M., Jeong, H. W., Arf, H., Adams, S., Ferrara, N., & Adams, R. H. (2017). Pericytes regulate VEGF-induced endothelial sprouting through VEGFR1. *Nature Communications, 8*, 1574.

Escudero, C. A., Herlitz, K., Troncoso, F., Guevara, K., Acurio, J., Aguayo, C., Godoy, A. S., & Gonzalez, M. (2017). Pro-angiogenic role of insulin: From physiology to pathology. *Frontiers in Physiology, 8*, 204.

Franco, M., Roswall, P., Cortez, E., Hanahan, D., & Pietras, K. (2011). Pericytes promote endothelial cell survival through induction of autocrine VEGF-A signaling and Bcl-w expression. *Blood, 118*, 2906–2917.

Giannoni, P., Arango-Lievano, M., Neves, I. D., Rousset, M. C., Baranger, K., Rivera, S., Jeanneteau, F., Claeysen, S., & Marchi, N. (2016). Cerebrovascular pathology during the progression of experimental Alzheimer's disease. *Neurobiology of Disease, 88*, 107–117.

Grammas, P., Moore, P., & Weigel, P. H. (1999). Microvessels from Alzheimer's disease brains kill neurons in vitro. *The American Journal of Pathology, 154*, 337–342.

Greenwood-Goodwin, M., Yang, J., Hassanipour, M., & Larocca, D. (2016). A novel lineage restricted, pericyte-like cell line isolated from human embryonic stem cells. *Scientific Reports, 6*, 24403.

Hall, C. N., Reynell, C., Gesslein, B., Hamilton, N. B., Mishra, A., Sutherland, B. A., O'Farrell, F. M., Buchan, A. M., Lauritzen, M., & Attwell, D. (2014). Capillary pericytes regulate cerebral blood flow in health and disease. *Nature, 508*, 55–60.

Halliday, M. R., Rege, S. V., Ma, Q., Zhao, Z., Miller, C. A., Winkler, E. A., & Zlokovic, B. V. (2016). Accelerated pericyte degeneration and blood-brain barrier breakdown in apolipoprotein E4 carriers with Alzheimer's disease. *Journal of Cerebral Blood Flow and Metabolism, 36*, 216–227.

Han, B. H., Zhou, M. L., Johnson, A. W., Singh, I., Liao, F., Vellimana, A. K., Nelson, J. W., Milner, E., Cirrito, J. R., Basak, J., Yoo, M., Dietrich, H. H., Holtzman, D. M., & Zipfel, G. J. (2015). Contribution of reactive oxygen species to cerebral amyloid angiopathy, vasomotor dysfunction, and microhemorrhage in aged Tg2576 mice. *Proceedings of the National Academy of Sciences of the United States of America, 112*, E881–E890.

Herland, A., van der Meer, A. D., FitzGerald, E. A., Park, T.-E., Sleeboom, J. J. F., & Ingber, D. E. (2016). Distinct contributions of astrocytes and pericytes to neuroinflammation identified in a 3D human blood-brain barrier on a chip. *PLoS One, 11*, e0150360.

Hughes, S., Gardiner, T., Hu, P., Baxter, L., Rosinova, E., & Chan-Ling, T. (2006). Altered pericyte-endothelial relations in the rat retina during aging: Implications for vessel stability. *Neurobiology of Aging, 27*, 1838–1847.

Ito, S., Yanai, M., Yamaguchi, S., Couraud, P. O., & Ohtsuki, S. (2017). Regulation of tight-junction integrity by insulin in an in vitro model of human blood-brain barrier. *Journal of Pharmaceutical Sciences, 106*, 2599–2605.

Jellinger K. A. (2002). Alzheimer disease and cerebrovascular pathology: an update. *Journal of Neural Transmission (Vienna), 109*(5–6), 813–836.

Kang, E., & Shin, J. W. (2016). Pericyte-targeting drug delivery and tissue engineering. *International Journal of Nanomedicine, 11*, 2397–2406.

Kapogiannis, D., & Reiter, D. (2014). Low glucose utilization and high lactate production in the Alzheimer's disease brain. *Alzheimer's & Dementia: The Journal of the Alzheimer's Association, 10*, P62.

Karow, M., Sanchez, R., Schichor, C., Masserdotti, G., Ortega, F., Heinrich, C., Gascon, S., Khan, M. A., Lie, D. C., Dellavalle, A., Cossu, G., Goldbrunner, R., Gotz, M., & Berninger, B. (2012). Reprogramming of pericyte-derived cells of the adult human brain into induced neuronal cells. *Cell Stem Cell, 11*, 471–476.

Khilazheva, E. D., Pisareva, N. V., Morgun, A. V., Boytsova, E. B., Taranushenko, T. E., Frolova, O. V., & Salmina, A. B. (2017). Activation of GPR81 lactate receptors stimulates mitochondrial biogenesis in cerebral microvessel endothelial cells. *Annals of Clinical and Experimental Neurology, 11*, 34–39.

Kisler, K., Nelson, A. R., Montagne, A., & Zlokovic, B. V. (2017a). Cerebral blood flow regulation and neurovascular dysfunction in Alzheimer disease. *Nature Reviews Neuroscience, 18*, 419–434.

Kisler, K., Nelson, A. R., Rege, S. V., Ramanathan, A., Wang, Y., Ahuja, A., Lazic, D., Tsai, P. S., Zhao, Z., Zhou, Y., Boas, D. A., Sakadžić, S., & Zlokovic, B. V. (2017b). Pericyte degeneration leads to neurovascular uncoupling and limits oxygen supply to brain. *Nature Neuroscience, 20*, 406–416.

Kovac, A., Erickson, M. A., & Banks, W. A. (2011). Brain microvascular pericytes are immunoactive in culture: Cytokine, chemokine, nitric oxide, and LRP-1 expression in response to lipopolysaccharide. *Journal of Neuroinflammation, 8*, 139–139.

Lange, S., Trost, A., Tempfer, H., Bauer, H.-C., Bauer, H., Rohde, E., Reitsamer, H. A., Franklin, R. J. M., Aigner, L., & Rivera, F. J. (2013). Brain pericyte plasticity as a potential drug target in CNS repair. *Drug Discovery Today, 18*, 456–463.

Lauritzen, K. H., Morland, C., Puchades, M., Holm-Hansen, S., Hagelin, E. M., Lauritzen, F., Attramadal, H., Storm-Mathisen, J., Gjedde, A., & Bergersen, L. H. (2014). Lactate receptor sites link neurotransmission, neurovascular coupling, and brain energy metabolism. *Cerebral Cortex, 24*, 2784–2795.

Li, Y., Lucas-Osma, A. M., Black, S., Bandet, M. V., Stephens, M. J., Vavrek, R., Sanelli, L., Fenrich, K. K., Di Narzo, A. F., Dracheva, S., Winship, I. R., Fouad, K., & Bennett, D. J. (2017). Pericytes impair capillary blood flow and motor function after chronic spinal cord injury. *Nature Medicine, 23*, 733–741.

Liguori, C., Stefani, A., Sancesario, G., Sancesario, G. M., Marciani, M. G., & Pierantozzi, M. (2015). CSF lactate levels, tau proteins, cognitive decline: A dynamic relationship in Alzheimer's disease. *Journal of Neurology, Neurosurgery, and Psychiatry, 86*, 655–659.

Liu, H., Saffi, G. T., Vasefi, M. S., Choi, Y., Kruk, J. S., Ahmed, N., Gondora, N., Mielke, J., Leonenko, Z., & Beazely, M. A. (2018). Amyloid-beta inhibits PDGFbeta receptor activation and prevents PDGF-BB-induced neuroprotection. *Current Alzheimer Research, 15*, 618–627.

Maki, T., Maeda, M., Uemura, M., Lo, E. K., Terasaki, Y., Liang, A. C., Shindo, A., Choi, Y. K., Taguchi, A., Matsuyama, T., Takahashi, R., Ihara, M., & Arai, K. (2015). Potential interactions between pericytes and oligodendrocyte precursor cells in perivascular regions of cerebral white matter. *Neuroscience Letters, 597*, 164–169.

Mangialardi, G., Cordaro, A., & Madeddu, P. (2016). The bone marrow pericyte: An orchestrator of vascular niche. *Regenerative Medicine, 11*, 883–895.

Mayo, J. N., & Bearden, S. E. (2015). Driving the hypoxia inducible pathway in human pericytes promotes vascular density in an exosome dependent manner. *Microcirculation (New York, NY), 22*, 711–723.

Miners, J. S., Schulz, I., & Love, S. (2018). Differing associations between Abeta accumulation, hypoperfusion, blood-brain barrier dysfunction and loss of PDGFRB pericyte marker in the precuneus and parietal white matter in Alzheimer's disease. *Journal of Cerebral Blood Flow and Metabolism, 38*, 103–115.

Montagne, A., Barnes, S. R., Sweeney, M. D., Halliday, M. R., Sagare, A. P., Zhao, Z., Toga, A. W., Jacobs, R. E., Liu, C. Y., Amezcua, L., Harrington, M. G., Chui, H. C., Law, M., & Zlokovic, B. V. (2015). Blood-brain barrier breakdown in the aging human hippocampus. *Neuron, 85*, 296–302.

Morland, C., Andersson, K. A., Haugen, Ø. P., Hadzic, A., Kleppa, L., Gille, A., Rinholm, J. E., Palibrk, V., Diget, E. H., Kennedy, L. H., Stølen, T., Hennestad, E., Moldestad, O., Cai, Y., Puchades, M., Offermanns, S., Vervaeke, K., Bjørås, M., Wisløff, U., Storm-Mathisen, J., & Bergersen, L. H. (2017). Exercise induces cerebral VEGF and angiogenesis via the lactate receptor HCAR1. *Nature Communications, 8*, 15557.

Nikolakopoulou, A. M., Zhao, Z., Montagne, A., & Zlokovic, B. V. (2017). Regional early and progressive loss of brain pericytes but not vascular smooth muscle cells in adult mice with disrupted platelet-derived growth factor receptor-β signaling. *PLoS One, 12*, e0176225.

O'Farrell, F. M., Mastitskaya, S., Hammond-Haley, M., Freitas, F., Wah, W. R., & Attwell, D. (2017). Capillary pericytes mediate coronary no-reflow after myocardial ischaemia. *eLife, 6*, e29280.

Ottone, C., Krusche, B., Whitby, A., Clements, M., Quadrato, G., Pitulescu, M. E., Adams, R. H., & Parrinello, S. (2014). Direct cell-cell contact with the vascular niche maintains quiescent neural stem cells. *Nature Cell Biology, 16*, 1045–1056.

Piegsa, J., Miners, S., Love, S., Shore, A., Tamagnini, F., & Randall, A. (2017). Ca^{2+} signalling in human brain pericytes: Investigation of effects of amyloid beta treatment. *Alzheimer's & Dementia: The Journal of the Alzheimer's Association, 13*, P331.

Pluta, R., Jabłoński, M., Ułamek-Kozioł, M., Kocki, J., Brzozowska, J., Januszewski, S., Furmaga-Jabłońska, W., Bogucka-Kocka, A., Maciejewski, R., & Czuczwar, S. J. (2013). Sporadic Alzheimer's disease begins as episodes of brain ischemia and ischemically dysregulated Alzheimer's disease genes. *Molecular Neurobiology, 48*, 500–515.

Pozhilenkova, E. A., Lopatina, O. L., Komleva, Y. K., Salmin, V. V., & Salmina, A. B. (2017). Blood-brain barrier-supported neurogenesis in healthy and diseased brain. *Reviews in the Neurosciences, 28*, 397–415.

Rensink, A. A., Otte-Holler, I., de Boer, R., Bosch, R. R., ten Donkelaar, H. J., de Waal, R. M., Verbeek, M. M., & Kremer, B. (2004a). Insulin inhibits amyloid beta-induced cell death in cultured human brain pericytes. *Neurobiology of Aging, 25*, 93–103.

Rensink, A. A., Otte-Holler, I., ten Donkelaar, H. J., De Waal, R. M., Kremer, B., & Verbeek, M. M. (2004b). Differential gene expression in human brain pericytes induced by amyloid-beta protein. *Neuropathology and Applied Neurobiology, 30*, 279–291.

Ribatti, D., Nico, B., & Crivellato, E. (2011). The role of pericytes in angiogenesis. *The International Journal of Developmental Biology, 55*, 261–268.

Rustenhoven, J., Aalderink, M., Scotter, E. L., Oldfield, R. L., Bergin, P. S., Mee, E. W., Graham, E. S., Faull, R. L. M., Curtis, M. A., Park, T. I. H., & Dragunow, M. (2016). TGF-beta1 regulates human brain pericyte inflammatory processes involved in neurovasculature function. *Journal of Neuroinflammation, 13*, 37.

Sagare, A. P., Bell, R. D., Zhao, Z., Ma, Q., Winkler, E. A., Ramanathan, A., & Zlokovic, B. V. (2013). Pericyte loss influences Alzheimer-like neurodegeneration in mice. *Nature Communications, 4*, 2932.

Sakuma, R., Kawahara, M., Nakano-Doi, A., Takahashi, A., Tanaka, Y., Narita, A., Kuwahara-Otani, S., Hayakawa, T., Yagi, H., Matsuyama, T., & Nakagomi, T. (2016). Brain pericytes serve as microglia-generating multipotent vascular stem cells following ischemic stroke. *Journal of Neuroinflammation, 13*, 57.

Salmina, A. B., Inzhutova, A. I., Malinovskaya, N. A., & Petrova, M. M. (2010). Endothelial dysfunction and repair in Alzheimer-type neurodegeneration: Neuronal and glial control. *Journal of Alzheimer's Disease, 22*, 17–36.

Salmina, A. B., Morgun, A. V., Kuvacheva, N. V., Lopatina, O. L., Komleva, Y. K., Malinovskaya, N. A., & Pozhilenkova, E. A. (2014). Establishment of neurogenic microenvironment in the neurovascular unit: The connexin 43 story. *Reviews in the Neurosciences, 25*, 97–111.

Salmina, A. B., Komleva, Y. K., Szijártó, I. A., Gorina, Y. V., Lopatina, O. L., Gertsog, G. E., Filipovic, M. R., & Gollasch, M. (2015a). H(2)S- and NO-signaling pathways in Alzheimer's amyloid vasculopathy: Synergism or antagonism? *Frontiers in Physiology, 6*, 361.

Salmina, A. B., Kuvacheva, N. V., Morgun, A. V., Komleva, Y. K., Pozhilenkova, E. A., Lopatina, O. L., Gorina, Y. V., Taranushenko, T. E., & Petrova, L. L. (2015b). Glycolysis-mediated control of blood-brain barrier development and function. *The International Journal of Biochemistry & Cell Biology, 64*, 174–184.

Simonavicius, N., Ashenden, M., van Weverwijk, A., Lax, S., Huso, D. L., Buckley, C. D., Huijbers, I. J., Yarwood, H., & Isacke, C. M. (2012). Pericytes promote selective vessel regression to regulate vascular patterning. *Blood, 120*, 1516–1527.

Stapor, P. C., Sweat, R. S., Dashti, D. C., Betancourt, A. M., & Murfee, W. L. (2014). Pericyte dynamics during angiogenesis: New insights from new identities. *Journal of Vascular Research, 51*, 163–174.

Stefanska, A., Eng, D., Kaverina, N., Duffield, J. S., Pippin, J. W., Rabinovitch, P., & Shankland, S. J. (2015). Interstitial pericytes decrease in aged mouse kidneys. *Aging, 7*, 370–382.

Sweeney, P. W., Walker-Samuel, S., & Shipley, R. J. (2018). Insights into cerebral haemodynamics and oxygenation utilising in vivo mural cell imaging and mathematical modelling. *Scientific Reports, 8*, 15.

Szpak, G. M., Lewandowska, E., Wierzba-Bobrowicz, T., Bertrand, E., Pasennik, E., Mendel, T., Stepień, T., Leszczyńska, A., & Rafałowska, J. (2007). Small cerebral vessel disease in familial amyloid and non-amyloid angiopathies: FAD-PS-1 (P117L) mutation and CADASIL. Immunohistochemical and ultrastructural studies. *Folia Neuropathologica, 45*, 192–204.

Tachibana, M., Yamazaki, Y., Liu, C. C., Bu, G., & Kanekiyo, T. (2018). Pericyte implantation in the brain enhances cerebral blood flow and reduces amyloid-beta pathology in amyloid model mice. *Experimental Neurology, 300*, 13–21.

Takahashi, H., Takata, F., Matsumoto, J., Machida, T., Yamauchi, A., Dohgu, S., & Kataoka, Y. (2015). Brain pericyte-derived soluble factors enhance insulin sensitivity in GT1-7 hypothalamic neurons. *Biochemical and Biophysical Research Communications, 457*, 532–537.

Teichert, M., Milde, L., Holm, A., Stanicek, L., Gengenbacher, N., Savant, S., Ruckdeschel, T., Hasanov, Z., Srivastava, K., Hu, J., Hertel, S., Bartol, A., Schlereth, K., & Augustin, H. G. (2017). Pericyte-expressed Tie2 controls angiogenesis and vessel maturation. *Nature Communications, 8*, 16106.

Thériault, P., ElAli, A., & Rivest, S. (2016). High fat diet exacerbates Alzheimer's disease-related pathology in APPswe/PS1 mice. *Oncotarget, 7*, 67808–67827.

Tian, X., Brookes, O., & Battaglia, G. (2017). Pericytes from mesenchymal stem cells as a model for the blood-brain barrier. *Scientific Reports, 7*, 39676.

Toth, P., Tucsek, Z., Sosnowska, D., Gautam, T., Mitschelen, M., Tarantini, S., Deak, F., Koller, A., Sonntag, W. E., Csiszar, A., & Ungvari, Z. (2013). Age-related autoregulatory dysfunction and cerebromicrovascular injury in mice with angiotensin II-induced hypertension. *Journal of Cerebral Blood Flow and Metabolism, 33*, 1732–1742.

Trost, A., Lange, S., Schroedl, F., Bruckner, D., Motloch, K. A., Bogner, B., Kaser-Eichberger, A., Strohmaier, C., Runge, C., Aigner, L., Rivera, F. J., & Reitsamer, H. A. (2016). Brain and retinal pericytes: Origin, function and role. *Frontiers in Cellular Neuroscience, 10*, 20.

Tucsek, Z., Toth, P., Tarantini, S., Sosnowska, D., Gautam, T., Warrington, J. P., Giles, C. B., Wren, J. D., Koller, A., Ballabh, P., Sonntag, W. E., Ungvari, Z., & Csiszar, A. (2014). Aging exacerbates obesity-induced cerebromicrovascular rarefaction, neurovascular uncoupling, and cognitive decline in mice. *The Journals of Gerontology. Series A, Biological Sciences and Medical Sciences, 69*, 1339–1352.

von Bernhardi, R., Cornejo, F., Parada, G. E., & Eugenín, J. (2015). Role of TGFβ signaling in the pathogenesis of Alzheimer's disease. *Frontiers in Cellular Neuroscience, 9*, 426.

Vazquez-Padron, R. I., Lasko, D., Li, S., Louis, L., Pestana, I. A., Pang, M., Liotta, C., Fornoni, A., Aitouche, A., & Pham, S. M. (2004). Aging exacerbates neointimal formation, and increases proliferation and reduces susceptibility to apoptosis of vascular smooth muscle cells in mice. *Journal of Vascular Surgery, 40*, 1199–1207.

Verbeek, M. M., Otte-Holler, I., Ruiter, D. J., & de Waal, R. M. (1999). Human brain pericytes as a model system to study the pathogenesis of cerebrovascular amyloidosis in Alzheimer's disease. *Cellular and Molecular Biology (Noisy-le-Grand, France), 45*, 37–46.

Wang, J. D., Khafagy el, S., Khanafer, K., Takayama, S., & ElSayed, M. E. (2016). Organization of endothelial cells, pericytes, and astrocytes into a 3D microfluidic in vitro model of the -assod-brain barrier. *Molecular Pharmaceutics, 13*, 895–906.

Warmke, N., Griffin, K. J., Slater, C. I., Walker, A. M. N., Yuldasheva, N. Y., & Cubbon, R. M. (2017). Reduced pericyte insulin signalling causes abnormal developmental angiogenesis. *Heart, 103*, A127–A128.

Wilhelmus, M. M., Otte-Holler, I., van Triel, J. J., Veerhuis, R., Maat-Schieman, M. L., Bu, G., de Waal, R. M., & Verbeek, M. M. (2007). Lipoprotein receptor-related protein-1 mediates amyloid-beta-mediated cell death of cerebrovascular cells. *The American Journal of Pathology, 171*, 1989–1999.

Winkler, E. A., Sagare, A. P., & Zlokovic, B. V. (2014). The pericyte: A forgotten cell type with important implications for Alzheimer's disease? *Brain Pathology (Zurich, Switzerland), 24*, 371–386.

Yamada, M. (2000). Cerebral amyloid angiopathy: an overview. *Neuropathology, 20*(1), 8–22.

Yamada, M. (2015). Cerebral amyloid angiopathy: emerging concepts. *Journal of Stroke, 17*, 17–30.

Yamagishi, S., Takeuchi, M., Matsui, T., Nakamura, K., Imaizumi, T., & Inoue, H. (2005). Angiotensin II augments advanced glycation end product-induced pericyte apoptosis through RAGE overexpression. *FEBS Letters, 579*, 4265–4270.

Yamamizu, K., Iwasaki, M., Takakubo, H., Sakamoto, T., Ikuno, T., Miyoshi, M., Kondo, T., Nakao, Y., Nakagawa, M., Inoue, H., & Yamashita, J. K. (2017). In vitro modeling of blood-brain barrier with human iPSC-derived endothelial cells, pericytes, neurons, and astrocytes via notch signaling. *Stem Cell Reports, 8*, 634–647.

Yamanishi, S., Katsumura, K., Kobayashi, T., & Puro, D. G. (2006). Extracellular lactate as a dynamic vasoactive signal in the rat retinal microvasculature. *American Journal of Physiology. Heart and Circulatory Physiology, 290*, H925–H934.

Chapter 8
Pericytes in Multiple Sclerosis

Francisco J. Rivera, Bryan Hinrichsen, and Maria Elena Silva

Abstract Multiple sclerosis (MS) is an autoimmune inflammatory demyelinating disease that affects the central nervous system (CNS), particularly, in young adults. Current MS treatments aim to reduce demyelination; however, these have limited efficacy, display side effects and lack of regenerative activities. Oligodendrocyte progenitor cells (OPCs) represents the major source for new myelin. Upon demyelination, OPCs get activated, proliferate, migrate towards the lesion, and differentiate into remyelinating oligodendrocytes. Although myelin repair (remyelination) represents a robust response to myelin damage, during MS, this

F. J. Rivera (✉)
Laboratory of Stem Cells and Neuroregeneration, Institute of Anatomy, Histology and Pathology, Faculty of Medicine, Universidad Austral de Chile, Valdivia, Chile

Center for Interdisciplinary Studies on the Nervous System (CISNe), Universidad Austral de Chile, Valdivia, Chile

Institute of Molecular Regenerative Medicine, Paracelsus Medical University Salzburg, Salzburg, Austria

Spinal Cord Injury and Tissue Regeneration Center Salzburg (SCI-TReCS), Paracelsus Medical University Salzburg, Salzburg, Austria
e-mail: francisco.rivera@uach.cl

B. Hinrichsen
Laboratory of Stem Cells and Neuroregeneration, Institute of Anatomy, Histology and Pathology, Faculty of Medicine, Universidad Austral de Chile, Valdivia, Chile

Center for Interdisciplinary Studies on the Nervous System (CISNe), Universidad Austral de Chile, Valdivia, Chile

M. E. Silva
Laboratory of Stem Cells and Neuroregeneration, Institute of Anatomy, Histology and Pathology, Faculty of Medicine, Universidad Austral de Chile, Valdivia, Chile

Center for Interdisciplinary Studies on the Nervous System (CISNe), Universidad Austral de Chile, Valdivia, Chile

Institute of Pharmacy, Faculty of Sciences, Universidad Austral de Chile, Valdivia, Chile

A. Birbrair (ed.), *Pericyte Biology in Disease*, Advances in Experimental Medicine and Biology 1147, https://doi.org/10.1007/978-3-030-16908-4_8

regenerative phenomenon decays in efficiency or even fails. CNS-resident pericytes (CNS-PCs) are essential for vascular homeostasis regulating blood-brain barrier (BBB) permeability and stability as well as endothelial cells (ECs) function during angiogenesis and neovascularization. Recent studies indicate that CNS-PCs also play a crucial role regulating OPC function during remyelination, and very importantly, these cells are substantially affected in MS. This chapter summarizes important aspects of MS and CNS remyelination as well as it provides new insights supporting the contribution of CNS-PCs to myelin regeneration and to MS pathology. Currently, there is evidence arguing in favor of CNS-PCs as novel therapeutic targets for the development of future treatments for MS.

Keywords Multiple sclerosis · Remyelination · Oligodendrocyte progenitor cells · CNS-resident pericytes · Blood-brain barrier · Neurovascular unit · Extracellular matrix · Therapy

Abbreviation

BBB	Blood-brain barrier
CC	Corpus callosum
CNS	Central nervous system
CNS-PCs	Central nervous system-resident pericytes
EAE	Experimental autoimmune encephalomyelitis
ECM	Extracellular matrix
ECs	Endothelial cells
FGF	Fibroblast growth factor
LPC	Lysophosphatidylcholine
MS	Multiple sclerosis
MSCs	Mesenchymal stem cells
MS-PP	Multiple sclerosis primary progressive
MS-RR	Multiple sclerosis relapsing-remitting
MS-SP	Multiple sclerosis secondary progressive
NSCs	Neural stem cells
NVU	Neurovascular unit
OB	Olfactory bulb
OPCs	Oligodendroglial precursor/progenitor cells
PDGFRalpha	Platelet-derived growth factor receptor alpha
PDGFRbeta	Platelet-derived growth factor receptor beta
PLCs	Pericyte like cells
RMS	Rostral migratory stream
SVZ	Subventricular zone

Multiple Sclerosis

Multiple sclerosis (MS) is an inflammatory demyelinating disease of the central nervous system (CNS) (Hoftberger and Lassmann 2017). This disease is classified as an autoimmune disease affecting around 2.5 million people worldwide and being the most common cause of acquired neurological disability in young adults. As other diseases, MS prevalence differs among races, geographical locations, and gender. The Caucasian population from the northern part of the boreal hemisphere are the most affected (Ebers and Sadovnick 1993; Noseworthy et al. 2000). Within this population, this disease is predominant in women (Alonso and Hernan 2008; Orton et al. 2006; Ramagopalan et al. 2010). Since there are no evidences for the existence of MS-associated genes in the X chromosome, the higher MS incidence in females might be related to their specific physiology (i.e., hormones) (Whitacre 2001).

In 1996 the US national MS society classified the disease in three major clinical forms depending on its progression: Primary Progressive (PP), Secondary Progressive (SP) and Relapsing-Remitting (RR). RR-MS is the most frequent type and it is characterized by relapses (acute episodes of neurological dysfunctions) followed by remission (periods of variable recovery and clinical stability) (Compston and Coles 2008). More than half of the RR-MS patients develop progressive neurological symptoms, called SP variety of MS, and show sustained deterioration without an evident remission period. The PP clinical type, which is characterized by the absence of remission periods, affects the 10–15% of MS patients (Lublin and Reingold 1996). The clinical progression of MS patients is very different between the three clinical forms: patients with PP-MS worsen at similar speeds, while those with the RR-MS and SP-MS may have very different clinical courses. This clinical progression suggests that, while RR-MS and SP-MS are most likely distinct phases of the same disease, PP-MS may imply completely different processes. Up to now, the etiology of this immune mediated disease remains still unclear. The current discussion contemplates: (1) environmental and genetic risk factors, (2) dysregulation of the immune system leading to an induction of an autoimmune response, and (3) viral infections as the initial trigger (Compston and Coles 2008). However, an autoimmune response in which T and B cells react against myelin certainly involves MS pathology. In particular, in MS, an altered immune system produces lymphocytes that infiltrate into the CNS and recognize myelin proteins, targeting oligodendrocytes and causing myelin destruction. Demyelination leads to axonal injury, conduction block and progressive neuronal loss (Ferguson et al. 2004; Kornek and Lassmann 2003).

Current available treatments for MS are disease-modifying drugs, which have limited efficacy in RR phase of MS with considerable side effects. Among these medicines, we find immunosuppressive cytokines interferon β-1a and interferon β-1b, the immune-modulating drug glatiramer acetate and the immunosuppressant mitoxantrone. Besides these, monoclonal antibodies able to control immune cells function also represent an attractive alternative for the treatment of MS. For instance, natalizumab (a monoclonal anti α4 integrin antibody) reduces the ability of immune cells to cross the blood-brain barrier (BBB) or alemtuzumab (a monoclonal anti CD52 antibody) causes a pan-lymphocyte depletion (Compston and Coles 2008). Recently, clinical trials using ocrelizumab (a monoclonal anti-CD20 antibody), which causes mature B cells depletion, have shown efficacy even for the PP form of MS (Tintore et al. 2019). Overall, these treatments focus mainly in the inflammatory response having limited efficacy and displaying side effects with detrimental consequences, such as progressive multifocal leukoencephalopathy (PML) (Lassmann 2007b; Lassmann et al. 2012; Tintore et al. 2019). Moreover, current MS treatments have no repair-promoting activity. To increase the efficacy of a MS therapy and to avoid severe side effects new alternative therapeutic strategies that involve immunomodulation, neuroprotection, and promotion of structural and functional repair mechanisms, are currently under study and development. For example, autologous mesenchymal stem cells (MSCs) transplantation has been proposed for the treatment of MS (Connick et al. 2012; Jadasz et al. 2012; Rivera and Aigner 2012). However, to achieve this final therapeutic goal, it is imperative to understand the mechanisms that rule myelin repair.

Remyelination

At the end of the nineteenth century Joseph Babinski studying MS pathology described demyelinating axons that displayed short areas with thin myelin sheaths, suggesting for the first time that remyelination exists in the CNS. Today, it is clear that myelin sheaths are re-established along demyelinated axons in humans and in the various animal models (Franklin and Ffrench-Constant 2008; Lassmann et al. 1997; Smith et al. 1979; Woodruff and Franklin 1999) but it is still unknown the reason why remyelinated sheaths end up to be thinner than the myelin sheaths produced during development (Blakemore 1974; Ludwin and Maitland 1984). Currently, two hypotheses might explain this observation. The first one involves differences in axonal properties (Franklin and Hinks 1999) while the second considers that adult oligodendrocyte progenitors might show a weaker remyelination capability compared to the ones of developmental progenitors (Wolswijk and Noble 1989).

Oligodendrocyte Precursor/Progenitor Cells (OPCs)

In the early 1980s, Martin Raff and colleagues provided the first evidence on the existence of oligodendroglial precursor/progenitor cells (OPCs). They isolated these proliferating cells from the optic nerve and discovered their ability to differentiate into oligodendrocytes and type 2 astrocytes, fulfilling all criteria for OPCs in vitro (Raff et al. 1983, 1984). A few years later, Raff and colleagues described OPCs as the cells responsible for CNS remyelination (Ffrench-Constant and Raff 1986a, b). OPCs represent 5–8% of total glial cells and are widely spread throughout the CNS in the white and gray matter (Levine et al. 2001). OPCs are known to express specific markers such as ganglioside antigens recognized by the A2B5 antibody (Wolswijk and Noble 1989), chondroitin sulfate NG2 (Dawson et al. 2000; Keirstead et al. 1998), platelet-derived growth factor receptor alpha (PDGFRalpha) (Redwine and Armstrong 1998), and the transcription factor olig1 (Arnett et al. 2004). The statement that OPCs are the major source for new myelinating oligodendrocytes in adult CNS (Franklin and Kotter 2008) is endorsed by four substantial evidences: (1) lacZ-encoding retroviral tracing studies demonstrated that focal lysolecithin-induced demyelination in the white matter labeled proliferating cells that give rise to remyelinating oligodendrocytes (Gensert and Goldman 1997); (2) transplanted adult OPCs into a myelin-deficient (md) rat were shown to remyelinate bared axons (Zhang et al. 1999); (3) upon focal demyelination OPCs repopulation was observed before new mature oligodendrocytes appear (Levine and Reynolds 1999; Sim et al. 2002; Watanabe et al. 2002); (4) the existence of cells with a transitional expression of markers for OPCs and mature oligodendrocyte argues for OPCs as the source of newly generated myelin-producing cells in the adult CNS (Fancy et al. 2004; Zawadzka et al. 2010). Using a genetic fate mapping strategy and chemical-induced demyelination, demonstrated that CNS-resident PDGFRalpha/NG2-expressing cells (OPCs) give rise not only to remyelinating oligodendrocytes, but also to Schwann cells (Zawadzka et al. 2010).

In general, remyelination process via OPCs, involves three steps regulated by extrinsic and intrinsic factors that may act as either remyelination inhibitors or activators (Rivera et al. 2010): activation, recruitment and differentiation (Bruce et al. 2010; Franklin and Kotter 2008). Upon demyelination, OPCs become mitotically active and induce the expression of oligodendrogenic genes such as Olig2 and Nkx2.2 (Fancy et al. 2004; Levine and Reynolds 1999; Reynolds et al. 2002). Astrocytes and microglia get activated upon demyelination and release mitogens, probably, acting on OPCs as proliferation mediators (Redwine and Armstrong 1998; Schonrock et al. 1998; Wilson et al. 2006). OPCs' recruitment is intrinsically modulated by the cell cycle regulatory protein p27Kip1 (Crockett et al. 2005) and promoted by platelet-derived growth factor (PDGF) and fibroblast growth factor (FGF) (Murtie et al. 2005; Woodruff et al. 2004; Zhou et al. 2006). In addition to this, the coordinated interaction between cell surface molecules and extracellular matrix (ECM) is crucial for OPC recruitment (Larsen et al. 2003). Oligodendroglial differentiation and maturation is further subdivided into three steps: (1) OPCs establish contact with bare axons; (2) OPCs activate myelin genes expression and generate the myelin membrane; and (3) Myelin membrane compactly wraps around the axons forming the myelin sheath (Franklin and Kotter 2008).

It is well known that the generation of new oligodendrocytes declines with aging, and age is considered a limiting factor for spontaneous and endogenous CNS myelin repair (Sim et al. 2002; Franklin and Ffrench-Constant 2008; Shen et al. 2008; Shields et al. 1999). Nevertheless, it has been shown that remyelination efficiency in aged animals can be rescued after exposure to a youthful systemic milieu via heterochronic parabiosis (Ruckh et al. 2012 or after treatment with molecules such as 9-*cis* retinoic acid (Huang 2011, p. 1160). In summary, all these observations suggest that age-related myelin regeneration restrictions can be reverted by the contribution of a strong regenerative microenvironment or stimulus.

Subventricular Zone-Derived Oligodendrogenesis

Apparently, OPCs are not the only immature cells within the adult CNS, which can generate new oligodendrocytes. Altman and Das (Altman and Das 1964; Altman 1969; Altman and Das 1965) discovered newly generated neurons in the adult brain revealing the existence of an undifferentiated population of cells responsible for this. Neural stem cells (NSCs) reside mainly in a particular cellular and extracellular microenvironment termed stem cell niche in the subventricular zone (SVZ) of the wall of the lateral ventricles (Alvarez-Buylla and Garcia-Verdugo 2002; Doetsch and Scharff 2001; Gage 2000). In the SVZ, NSCs proliferate and differentiate into neuronal precursors migrating along the rostral migratory stream (RMS) to the olfactory bulb (OB), where they functionally integrate and differentiate into granule and periglomerular neurons (Carleton et al. 2003; Doetsch and Scharff 2001; Lois et al. 1996). Interestingly, the SVZ shows a very peculiar cellular organization that can be classified relying on three cell types (Alvarez-Buylla and colleagues). Based on the cell's location, expression profile, ultrastructural features, proliferation rate and their function in the SVZ they are classified in type A, B or C (Doetsch 2003; Doetsch and Alvarez-Buylla 1996; Doetsch et al. 1997, 1999). Type B cells are assumed to be the NSCs and are slowly proliferating GFAP-expressing cells in close proximity to ependymal cells. Type B cells give rise to transit-amplifying precursors type C cells that are fast proliferating cells with an elongated morphology. These type C cells generate neuroblasts, known as the type A cells. These type A cells migrate as homotypic chains along the RMS towards the OB.

In addition to neurons, SVZ-residing NSCs are able to generate oligodendrocytes. For example, type B cells give rise to a small subpopulation of Olig2-expressing transit-amplifying type C cells that in turn generate PSA-NCAM/PDGFRalpha-positive cells (Menn et al. 2006). This subpopulation migrates towards the corpus callosum (CC), the striatum and to the fimbria fornix where they differentiate into oligodendrocytes. NSCs from SVZ differentiate into oligodendrocytes also in response to a demyelinating lesion. For example, it has been shown that upon lysolecithin-induced demyelination of the CC, PSA-NCAM-expressing progenitors in the RMS, proliferate, migrate towards the lesioned CC and differentiate into oligodendrocytes and astrocytes (Nait-Oumesmar et al. 1999). In an EAE model, characterized by a chronic demyelination, SVZ-derived progenitors do respond to demyelination following the same pattern mentioned earlier. In conclusion, the SVZ stem cell niche constitutes a second source for new oligodendrocytes.

Animal Models for Multiple Sclerosis and Remyelination

Experimental Encephalomyelitis Models

Animal models aiming to reveal the cellular and molecular mechanisms of demyelinating diseases and/or to develop a novel therapy for MS treatment have substantially contributed to this research field. Experimental encephalomyelitis can be achieved by different means. For example, inflammatory demyelination can be provoked by the use of Theiler's murine encephalomyelitis virus (TMEV), which infects neurons and glia and induces lethal encephalomyelitis (Brahic 2002; Dal Canto and Rabinowitz 1982; Scheikl et al. 2010). Nevertheless, the most commonly used animal model to study MS pathophysiology is the experimental autoimmune encephalomyelitis (EAE) (Lassmann 2007a). This was first described in non-human primates (Rivers, 1933, p. 866), but now EAE is very frequently carried out in rodent species. EAE is induced either through the adoptive transfer of myelin-reactive T lymphocytes or by active immunization with myelin-derived antigens, such as myelin basic protein (MBP), myelin oligodendrocytes protein (MOG), myelin proteolipid protein (PLP) (Kabat et al. 1951; Kuchroo et al. 2002). Alternatively, immunization may be performed with immunodominant peptides derived from myelin antigens, such as MOG_{35-55}. The pathology as well as the clinical symptoms closely depend on the animal strain and the antigen used for immunization. However, typically around 2 weeks after immunization a susceptible rodent will debut with the first clinical symptoms developing a relapsing-remitting form of EAE. Thus, symptoms begin with tail atony leading in tail paralysis followed by reduced tonicity and progressive paralysis of hind- and finally forelimbs (quadriplegia) (Krishnamoorthy and Wekerle 2009). To a certain extent, EAE models resemble pathological features of MS, such as demyelination, inflammation, and neurodegeneration, which makes this model particularly attractive for the development of new MS therapies.

Chemically Induced Demyelination Models

Demyelination represents an extremely relevant aspect of MS pathology, especially when considering that remyelination is very limited or even it completely fails during this myelin disorder. Therefore, there are several animal models aiming to exclusively study the cellular and molecular mechanisms that may contribute to myelin repair as well as to develop therapeutic strategies able to enhance remyelination in MS. Overall, these models are based on the administration of chemicals that promote systemic or focal CNS demyelination. For example, cuprizone (bis-cyclohexanone-oxaldihydrazone) is one of the toxins widely used in preclinical MS research. It is easily administrated orally through food pellets, and once in the organism, it chelates copper, resulting in a systemic copper deficiency.

Copper deficiency affects in particular oligodendrocytes and induces a rapid and synchronous demyelination in various CNS regions such as, cortex, corpus callosum, hippocampus, superior cerebellar peduncles, optic chiasm, olfactory bulb, brainstem, etc. (Blakemore 1972, 1973; Kesterson and Carlton 1971; Komoly et al. 1987; Ludwin 1978; Matsushima and Morell 2001; Silvestroff et al. 2010; Skripuletz et al. 2008). This model is frequently used in mice as it induces a robust demyelination within weeks (Blakemore 1972; Ludwin 1978) and remyelination is quite evident 1–2 weeks after cuprizone removal and largely complete after 4 weeks (Matsushima and Morell 2001; Silvestroff et al. 2010). Also, the cuprizone model allows the analysis of de- and remyelination events without interference by inflammatory and immune-mediated mechanisms, since cuprizone treatment does not affect BBB integrity (Matsushima and Morell 2001).

Besides cuprizone, other toxic agents that are commonly used to investigate de- and remyelination are lysophosphatidylcholine (lysolecithin, LPC) and ethidium bromide (EtBr). In contrast to cuprizone, these chemicals are injected locally into the desired site of demyelination. 24 hours after stereotactic injection demyelination rapidly takes place and is reversible as remyelination is induced. LPC is a membrane-solubilizing agent that displays toxicity mainly on myelin-producing cells, while EtBr is a DNA intercalating agent that besides oligodendrocytes it also damages astrocytes (Woodruff and Franklin 1999). In general, these substances are injected into white matter CNS regions such as caudal cerebellar peduncle (CCP), spinal cord or corpus callosum (Jablonska et al. 2010; Woodruff and Franklin 1999; Zawadzka et al. 2010). Two relevant advantages of using these models when compared to cuprizone intoxication are that the demyelinated lesion is placed in a known location and OPC response to myelin damage as well as remyelination are easier to follow (Woodruff and Franklin 1999). Shortly after demyelination OPC proliferation and migration are induced, which peak at 5 days post-lesion-induction (dpl), followed by OPC differentiation, which is ongoing at 14 dpl. Remyelination is completed by 21 dpl. Although these models do not depict the full pathogenesis and pathology observed in MS, they allow to study an important CNS response to myelin damage, remyelination.

Pericytes in the Central Nervous System

Pericytes were firstly described by the end of the nineteenth century (Rouget 1874; Zimmermann 1923) as a cell population tightly associated with vasculature. Pericytes are flat cells with projections that wrap continuously around the capillaries (Fig. 8.1a) directly in contact with the endothelial cells (ECs) sharing a common basement membrane with microvascular endothelial cells. Under pathological condition, pericytes retract the projection wrapping the microvessels acquiring a migrating feature (Fig. 8.1b). (Diaz-Flores et al. 2009; Dore-Duffy and Cleary 2011; Kamouchi et al. 2011; Lange et al. 2013). The location of pericytes allows direct communication with ECs through gap junctions and peg-and-socket contacts

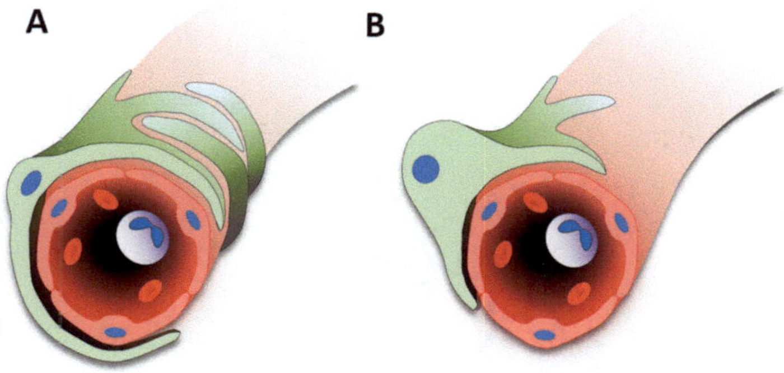

Fig. 8.1 Pericyte phenotype in microvessels. Pericytes are mural cells embedded in the basal membrane of capillaries and venules. (**a**) Normally they appear as flat cells with finger-like projections that wrap around microvessels. (**b**) When confronted with pathologic conditions, pericytes retract their projections and develop a surveillance-migration phenotype

stabilizing microvessels and regulating blood flow by their contractile and relaxant properties (Yemisci et al. 2009) (Peppiatt et al. 2006; Takata et al. 2009). In general, pericytes are known to express the immunologic markers such as a-smooth muscle actin, the chondroitin sulfate proteoglycan NG2, RGS-5, platelet-derived growth factor receptor beta (PDGFRbeta; CD140b), aminopeptidases-A and -N, Sca-1, CD34, 3G5, alkaline phosphatase, CD146 (MCAM), c-glutamyl transpeptidase, butyrylcholinesterase, FcR, CD4, CD11b, major histocompatibility complex (MHC) class I, II, and desmin (Dore-Duffy and Cleary 2011). However, not all pericytes display the same marker expression profile suggesting the existence of pericyte heterogeneity (Dore-Duffy 2008). Moreover, it seems that pericytes are not simple to distinguish as there are other cell types that have a similar marker expression profile and are located also at the vascularity, such as perivascular fibroblasts (Vanlandewijck et al. 2018). Therefore, a proper characterization is necessary to accurately identify bona fide pericytes.

In the CNS, pericytes (CNS-PCs) play an essential role in vascular homeostasis. Noteworthy, compared to other tissues and organs, the adult brain and spinal cord display a very high pericyte density with respect to EC population. Their distribution is around capillaries and rarely are found in bigger brain vessels. Depending on their location in the adult brain and spinal cord, CNS-PCs originate from different embryonic layers such as, neuroectoderm, neural crest and mesoderm (Winkler et al. 2011). CNS-PCs and other cells from the vasculature (ECs, and vascular smooth muscle cells) together with neural cells (neurons, astrocytes, oligodendrocytes, microglia) form a particular microenvironment, called neurovascular unit (NVU). The NVU facilitates the coupling between neural activity and vascular function through the BBB. This structure acts as a highly selective interface barrier between the systemic blood circulation and the interstitial fluid of the brain parenchyma, allowing tight regulation of brain homeostasis that maintains an optimal microenvironment for neuronal survival (ElAli et al. 2014; Tsai et al.

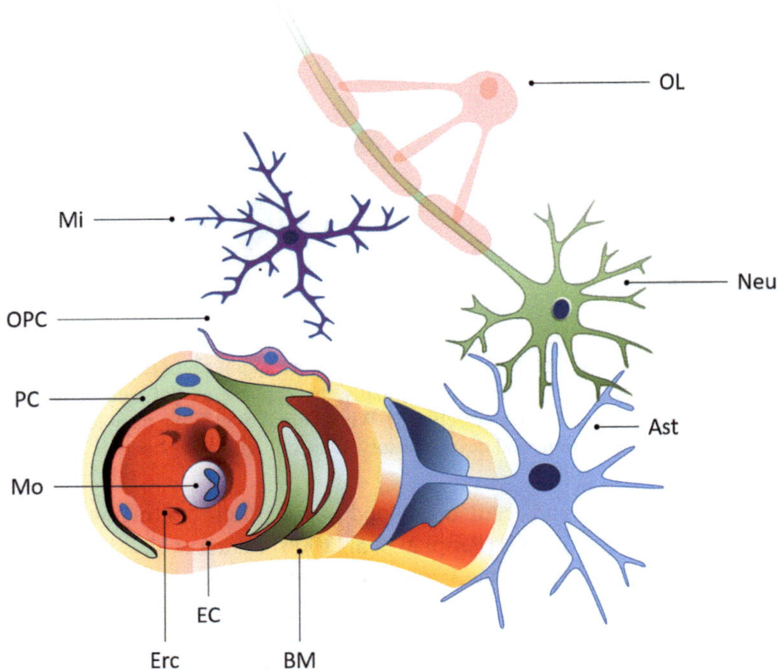

Fig. 8.2 The neurovascular unit. It consists of a collective of cells that populate the abluminal/parenchymal side of brain's microvasculature and contribute to the formation and support of the BBB. Endothelial Cells (EC) at the neurovascular unit establish a highly selective barrier of Claudin tight junctions facing the lumen of microvessels (apical side). Pericytes (PC) reside at the basal lamina of microvessels (abluminal side/basal side) and contribute to their maintenance and remodelling. These cells establish peg-socket connections with endothelial cells that allow the formation of gap junctions for cell communication. Basal Membrane (BM) is predominantly composed of laminins, collagen IV, nidogen, and heparan sulfate proteoglycans, produced by EC and PC, which provide a network for the presentation of trophic and immunological factors [4]. Astrocytes (Ast) end-foot partially encase the capillary network, producing a glial layer that metabolically couples with endothelial cells. Oligodendroglial Progenitor Cells (OPCs) are widely dispersed through gray and white matter and can be sited at the perivascular region of the NVU. Microglia (Mi) are resident immune cells of CNS that in response to demyelination produce TNF-α (Tumoral Necrosis Factor α) which in turn stimulate Pericyte proliferation and migration [5]. Neurons (Neu) often localize near to capillary networks for metabolic coupling with the NVU. Oligodendrocytes (OL) form the myelin ensheathing around neuronal axons in the CNS

2016) (Fig. 8.2). Overall, CNS-PCs express all pericyte markers but display few specific features that distinguish them from perivascular cells from other tissues and organs. For example, CNS-PCs express the potassium channel kir6.1 (Bondjers et al. 2006) conferring them specific physiological roles. Indeed, CNS-PCs display a number of specific functions that are essential for vascular homeostasis in the neural tissue. Animal models deficient in CNS-PCs show BBB leakage and breakdown due to a not well-established boundary between vessels (Lindahl et al. 1997), denoting their crucial role in BBB stability, permeability and in the regulation of capillary blood flow (Quaegebeur et al. 2010). In addition to this, prenatal CNS-PCs-deficient mice suffer of EC hyperplasia, evidencing the impact of CNS-PCs in the regulation of ECs proliferation, migration and stabilization during developmental angiogenesis and neovascularization (Ribatti et al. 2011; Armulik et al. 2005; Bell et al. 2010; Hellstrom et al. 2001; Krizbai et al. 2000). Besides vascular homeostasis, recent findings extend CNS-PCs' function towards neuroregeneration and, particularly, in myelin repair (De La Fuente et al. 2017) that will be further discussed in this chapter.

Pericytes Contribution to Myelin Repair in Multiple Sclerosis

Pericytes Support Myelin Development, Maintenance, and Regeneration

The NVU physiology is well coupled to OPC/oligodendrocyte function contributing to the regulation of myelin formation, maintenance, and repair. Indeed, the relation between brain vascularization and myelin formation rises early at embryonal stages (around gestational day 11.5 in mice and week 36 in humans), as the angiogenic sprouting of microvessels through the neuropil serves as scaffolds for OPC migration (Tsai et al. 2016) that, in turn, will give rise to myelinating oligodendrocytes. Particularly, CNS-PCs may regulate OPC function during myelin development. Histological examination in the white matter of adult OPC reporter mice as well as in human biopsies revealed that, within the perivascular region, OPCs localize in close proximity to microvessels containing bona fide CNS-PCs (Maki et al. 2015). This scenario may facilitate functional interactions between CNS-PCs and OPCs, especially when considering the capability of CNS-PCs to secrete trophic factors enabling them to exert a paracrine action on neighboring cells (Gaceb et al. 2017). Besides this, CNS-PCs can indirectly influence myelin sheath formation and stability. As CNS myelination is an energetically extremely demanding process, it depends on the metabolic coupling between cells from the oligodendrocyte lineage and the NVU (Fünfschilling et al. 2012; Rinholm et al. 2011), where blood supply results are essential. In this regard, CNS-PCs have emerged as essential regulators of the NVU function, since these cells substantially contribute to the formation and integrity of BBB as well as the control of CNS blood perfusion (Armulik et al. 2010a; Crawford et al. 2013). Hence, CNS-PCs malfunctions may alter myelin sheaths. In fact, a recent study has revealed in a pericyte-deficient animal model that alterations in the microvasculature are associated with an increase of BBB permeability through transcytosis and the accumulation of toxic deposits of fibrin(ogen) in the CNS parenchyma leading to white matter dysfunction (Montagne et al. 2018). This correlates with previous observations in which the deficit of pericytes causes changes in the genetic expression pattern of endothelial cells, leading to an increase in permeability and alterations in the cellular polarity of the astrocytes end-foots at the NVU (Armulik et al. 2010b; Montagne et al. 2018).

Therefore, CNS-PCs contribute to myelin development and maintenance by different means, either directly modulating OPCs/oligodendrocytes function or through the regulation of the BBB integrity and NVU biology.

As previously mentioned, remyelination in the CNS represents a regenerative response to demyelination by which myelin sheaths are restored around bared axons. This process mainly depends on the proliferative, migratory and differentiation capacities of OPCs. Contribution of CNS-PCs to myelin is not restricted to its formation and maintenance as these cells also impact on myelin regeneration. As during development, in the adult brain CNS-PCs can regulate OPC physiology through the remodelling of the extracellular matrix (ECM), and the secretion of trophic factors. CNS-PCs have been spotted as important source of pro-regenerative

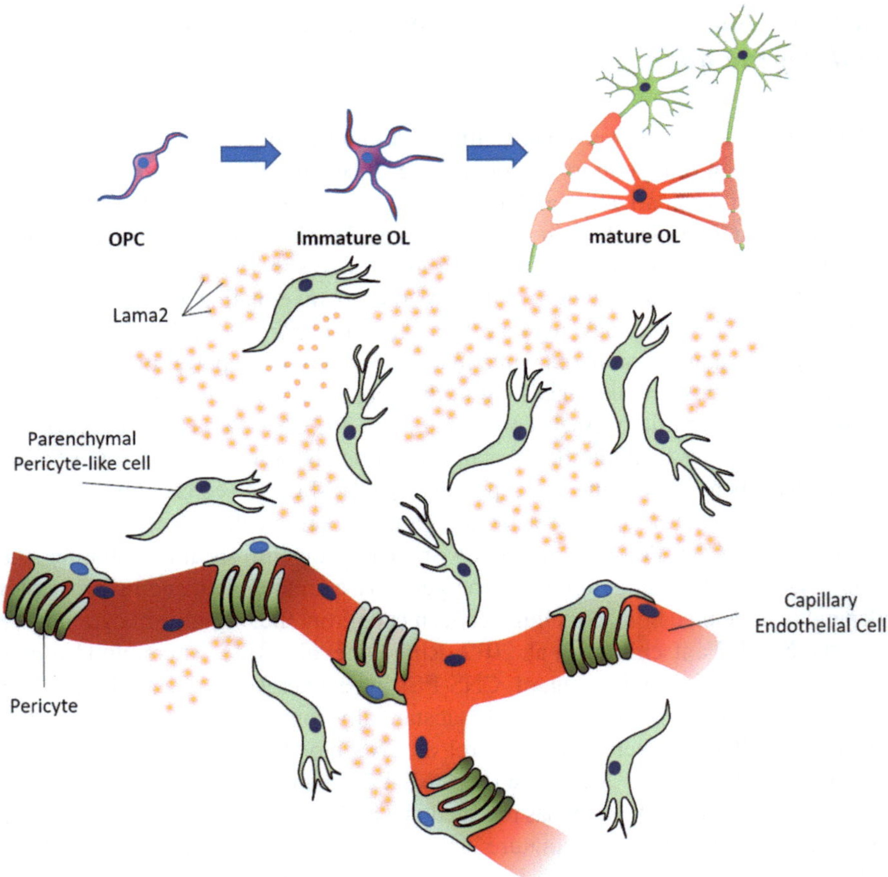

Fig. 8.3 Pericytes respond to demyelinated lesions. After a demyelinated insult, non-vessel associated pericyte-like cells (PDGFRbeta+ cells) develop at the core of the lesion and can be sited near to immature oligodendrocytes. In addition, pericytes can infuse paracrine stimulation to OPC differentiation by the secretion of LAMA2 subunit, a constituent of perivascular matrix of the basal lamina

molecules upon PDGF-BB/PDGFRbeta signalling (Gaceb et al. 2017). In response to focal demyelination in the CCP of adult rats, PDGFRbeta-expressing CNS-PCs proliferate and locate in close proximity to differentiating OPCs. In parallel, a new population of non-vessel associated PDGFRbeta+ cells (CNS-PCs-like cells, PLCs) develop within the lesion, however, is not known whether and how these PLCs may contribute to myelin repair. Notwithstanding this undisclosed issue, there is evidence showing that CNS-PCs secrete Laminin alpha2-chain (Lama2) modifying the ECM and favoring the generation of new oligodendrocytes (De La Fuente et al. 2017) (Fig. 8.3). Thus, recent studies have demonstrated that Lama2-derived from CNS-PCs promote OPC differentiation (De La Fuente et al. 2017) and oligodendrocyte fate choice in adult NSCs (Silva et al. 2019). Consistent with this, a CNS-PCs-deficient mouse model shows a delayed OPC differentiation during CNS remyelination (De La Fuente et al. 2017). A previous study has shown that the paracrine pro-oligodendrogenic effect exerted by CNS-PCs apparently depends on A-Kinase Associated Protein 12 (AKAP12). In the adult brain, AKAP12 is well expressed in CNS-PCs while only very few OPCs express this protein. AKAP12 is a PKA/CREB scaffold protein involved in the regulation of paracrine secretion and cell survival. Soluble factors derived from CNS-PCs that lackAKAP12 expression lost their capacity to promote OPC differentiation compared to conditioned media harvested from normal CNS-PCs (Takakuni Maki et al. 2018). These findings suggest that AKAP12 expression/activity is necessary for CNS-PCs to regulate OPC function. However, further studies are necessary to reveal the exact role and contribution of CNS-PCs (particularly, of PLCs) to remyelination.

Pericytes Abnormalities in MS Pathology

Pathological features of MS include the presence of infiltrated immune cells and blood-borne elements within the nervous parenchyma with a consistent disruption of the BBB that occurs at early stages and progressively manifests in areas of lesion (demyelinating plaques) (Zlokovic 2008). Imaging analysis in patients suffering from Clinically Isolated Syndrome (which often manifests before progressive forms of MS) revealed an early BBB breakdown suggesting that this event might precede the appearance of demyelinated plaques (Gündüz et al. 2018; Broman 1964). EAE, an animal model for MS, is characterized by significant angiodynamic changes during the course of the pathology. Upon exposure to chronic mild low oxygen apparently modulates these angiodynamic changes improving EAE clinical score (Esen et al. 2013). Due to their contribution to microvascular integrity, several studies have aimed at determining the role of CNS-PCs on microvascular remodeling during inflammation. Some of these studies have focused, particularly, in MS. PDGFRbeta-positive CNS-PCs specifically express the purinergic receptor P2X7R (ATP-gated P2X receptor cation channel) known to be relevant for inflammatory response (Grygorowicz et al. 2018). During the course of EAE, overexpression of P2X7R in CNS-PCs correlates with a downregulation of

PDGFRbeta as well as with lower levels of claudin-5 protein, a tight junction building element normally expressed by endothelial cells. The administration of a P2X7R antagonist increases the expression of PDGFRbeta and claudin-5, diminishing EAE clinical signs. Therefore, activation of P2X7R expressed by CNS-PCs contribute to microvascular abnormalities and BBB breakdown during EAE, suggesting that this particular receptor may participate in MS pathology (Grygorowicz et al. 2018). Furthermore, CNS-PCs well couple inflammation to vascular remodeling by different means. It seems that TNF-alpha, a pro-inflammatory cytokine, induces alpha2 integrin expression in CNS-PCs, apparently, allowing them to contribute to vascular remodeling in EAE (Tigges et al. 2013). Evidence supporting a role of CNS-PCs in MS is not restricted to animal models. Human biopsies obtained from patients suffering from both early progressive and late progressive MS (EPMS and LPMS, respectively) have revealed that perivascular cells subpopulations (including CNS-PCs) behave differently depending on the lesion type where they are located (Iacobaeus E et al. 2017). Overall, active lesions

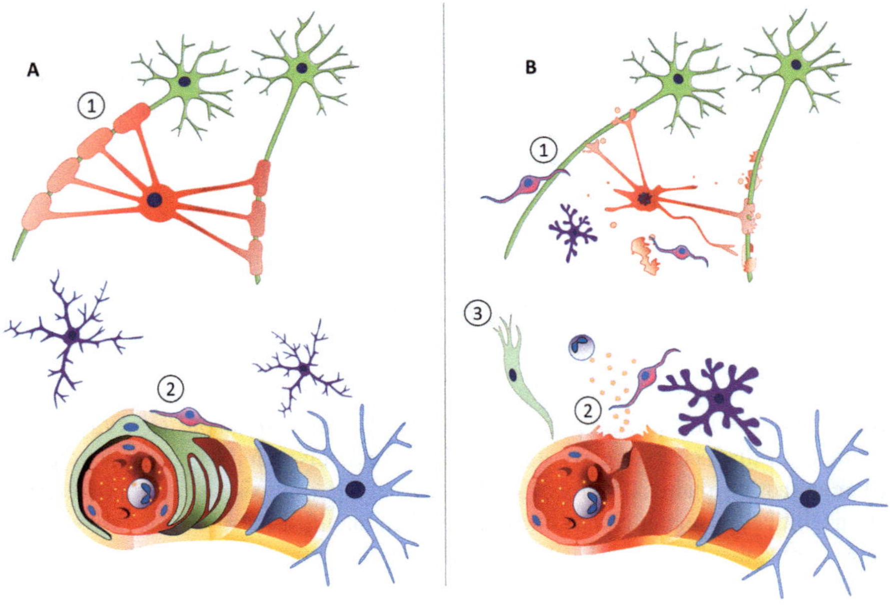

Fig. 8.4 BBB contribution to multiple sclerosis. (**a**) Healthy tissue: [1] OPCs represent the main source of myelinating oligodendrocytes of the CNS. Oligodendrocytes form a myelin sheathing that support the synaptic saltatory conduction of multiple neurons. By doing so, mature oligodendrocytes couple with the NVU. [2] Oligodendroglial progenitor cells (OPCs) reside near to the NVU with some of them attached to the basal membrane of capillaries, in which the pericytes are embedded. (**b**) Multiple sclerosis: [1] The autoimmune response elicits by microglia and leukocytes against myelin-derived constituents produce oligodendrocyte cell death and demyelination. [2] BBB breakdown of BBB manifests itself together with a lower coverage of pericytes in the capillaries. Higher permeability of disruption of BBB causes infiltration of toxic blood-derived elements and lymphocytes to CNS. [3] Different populations of non-vessel associated PDGFRbeta+ cells appear within the demyelinated plaque

contain high numbers of proliferative (CD146+/PDGFRbeta+/Ki67+) and quiescent (CD73+/CD271+/PDGFRbeta+/Ki67−) perivascular cells, compared to inactive lesions. However, chronic lesions display lower numbers of proliferative perivascular cells compared to normal-appearing white matter (healthy tissue). All these findings together indicate that during MS, CNS-PCs display phenotypic changes according to the pathological progression that involves inflammation, demyelination, eventual remyelination, axonal death, and scar formation (details in Fig. 8.4). Therefore, CNS-PCs have been suggested as an attractive therapeutic target for the treatment of MS (Azevedo et al. 2018).

Comments on Future Directions

As previously mentioned, recent studies indicate two main findings: (1) CNS-PCs react to demyelination and modulate OPCs function, probably, contributing to myelin repair and (2) CNS-PCs suffer of pathological changes that might alter their functionality during MS. In this scenario, it results in urgency to determine the mechanism(s) by which CNS-PCs might impact on remyelination and to find out how MS may alter CNS-PCs function impeding their contribution to myelin repair and/or affecting the NVU, eventually, favoring demyelination and neurodegeneration. Revealing these keys would allow to develop new therapies aiming to restore CNS-PCs function in MS.

Acknowledgments This work was supported by research funds from Chilean Comisión Nacional de Investigación Científica y Tecnológica (CONICYT) FONDECYT Program Regular Grant N° 1161787; Chilean CONICYT PCI Program Grant N° REDES170233 and Grant N° REDES180139; Chilean CONICYT FONDEF-IDeA Program Grant N° ID17AM0043; Chilean CONICYT Doctoral Scholarship 21171884.

References

Alonso, A., & Hernan, M. A. (2008). Temporal trends in the incidence of multiple sclerosis: A systematic review. *Neurology, 71*, 129–135.

Altman, J. (1969). Autoradiographic and histological studies of postnatal neurogenesis. IV. Cell proliferation and migration in the anterior forebrain, with special reference to persisting neurogenesis in the olfactory bulb. *The Journal of Comparative Neurology, 137*, 433–457.

Altman, J., & Das, G. (1965). Autoradiographic and histological evidence of postnatal hippocampal neurogenesis in rats. *The Journal of Comparative Neurology, 124*, 319–335.

Altman, J., & Das, G. D. (1964). Autoradiographic examination of the effects of enriched environment on the rate of glial multiplication in the adult rat brain. *Nature, 204*, 1161–1163.

Alvarez-Buylla, A., & Garcia-Verdugo, J. M. (2002). Neurogenesis in adult subventricular zone. *The Journal of Neuroscience, 22*, 629–634.

Armulik, A., Abramsson, A., & Betsholtz, C. (2005). Endothelial/pericyte interactions. *Circulation Research, 97*, 512–523.

Armulik, A., Genove, G., Mae, M., Nisancioglu, M. H., Wallgard, E., Niaudet, C., He, L., Norlin, J., Lindblom, P., Strittmatter, K., et al. (2010a). Pericytes regulate the blood-brain barrier. *Nature, 468*, 557–561.

Armulik, A., Genové, G., Mäe, M., Nisancioglu, M. H., Wallgard, E., Niaudet, C., He, L., Norlin, J., Lindblom, P., Strittmatter, K., et al. (2010b). Pericytes regulate the blood–brain barrier. *Nature, 468*, 557–561.

Arnett, H. A., Fancy, S. P., Alberta, J. A., Zhao, C., Plant, S. R., Kaing, S., Raine, C. S., Rowitch, D. H., Franklin, R. J., & Stiles, C. D. (2004). bHLH transcription factor Olig1 is required to repair demyelinated lesions in the CNS. *Science, 306*, 2111–2115.

Azevedo, P. O., Sena, I. F. G., Andreotti, J. P., Carvalho-Tavares, J., Alves, J. C., Cunha, T. M., Cunha, F. Q., Mintz, A., & Birbrair, A. (2018). Pericytes modulate myelination in the central nervous system. *Journal of Cellular Physiology, 233*, 5523–5529.

Bell, R. D., Winkler, E. A., Sagare, A. P., Singh, I., LaRue, B., Deane, R., & Zlokovic, B. V. (2010). Pericytes control key neurovascular functions and neuronal phenotype in the adult brain and during brain aging. *Neuron, 68*, 409–427.

Blakemore, W. F. (1972). Observations on oligodendrocyte degeneration, the resolution of status spongiosus and remyelination in cuprizone intoxication in mice. *Journal of Neurocytology, 1*, 413–426.

Blakemore, W. F. (1973). Demyelination of the superior cerebellar peduncle in the mouse induced by cuprizone. *Journal of the Neurological Sciences, 20*, 63–72.

Blakemore, W. F. (1974). Pattern of remyelination in the CNS. *Nature, 249*, 577–578.

Bondjers, C., He, L., Takemoto, M., Norlin, J., Asker, N., Hellstrom, M., Lindahl, P., & Betsholtz, C. (2006). Microarray analysis of blood microvessels from PDGF-B and PDGF-Rbeta mutant mice identifies novel markers for brain pericytes. *The FASEB Journal, 20*, 1703–1705.

Brahic, M. (2002). Theiler's virus infection of the mouse, or: Of the importance of studying animal models. *Virology, 301*, 1–5.

Broman, T. (1964). Blood-brain barrier damage in multiple sclerosis, supravital test observations. *Acta Neurologica Scandinavica, 40*, 21–24.

Bruce, C. C., Zhao, C., & Franklin, R. J. (2010). Remyelination – An effective means of neuroprotection. *Hormones and Behavior, 57*, 56–62.

Carleton, A., Petreanu, L. T., Lansford, R., Alvarez-Buylla, A., & Lledo, P.-M. (2003). Becoming a new neuron in the adult olfactory bulb. *Nature Neuroscience, 6*, 507–518.

Compston, A., & Coles, A. (2008). Multiple sclerosis. *Lancet, 372*, 1502–1517.

Connick, P., Kolappan, M., Crawley, C., Webber, D. J., Patani, R., Michell, A. W., Du, M. Q., Luan, S. L., Altmann, D. R., Thompson, A. J., et al. (2012). Autologous mesenchymal stem cells for the treatment of secondary progressive multiple sclerosis: An open-label phase 2a proof-of-concept study. *Lancet Neurology, 11*, 150–156.

Crawford, A. H., Chambers, C., & Franklin, R. J. M. (2013). Remyelination: The true regeneration of the central nervous system. *Journal of Comparative Pathology, 149*, 242–254.

Crockett, D. P., Burshteyn, M., Garcia, C., Muggironi, M., & Casaccia-Bonnefil, P. (2005). Number of oligodendrocyte progenitors recruited to the lesioned spinal cord is modulated by the levels of the cell cycle regulatory protein p27Kip-1. *Glia, 49*, 301–308.

Dal Canto, M. C., & Rabinowitz, S. G. (1982). Experimental models of virus-induced demyelination of the central nervous system. *Annals of Neurology, 11*, 109–127.

Dawson, M. R., Levine, J. M., & Reynolds, R. (2000). NG2-expressing cells in the central nervous system: Are they oligodendroglial progenitors? *Journal of Neuroscience Research, 61*, 471–479.

De La Fuente, A. G., Lange, S., Silva, M. E., Gonzalez, G. A., Tempfer, H., van Wijngaarden, P., Zhao, C., Di Canio, L., Trost, A., Bieler, L., et al. (2017). Pericytes stimulate oligodendrocyte progenitor cell differentiation during CNS remyelination. *Cell Reports, 20*, 1755–1764.

Diaz-Flores, L., Gutierrez, R., Madrid, J. F., Varela, H., Valladares, F., Acosta, E., Martin-Vasallo, P., & Diaz-Flores, L. (2009). Pericytes. Morphofunction, interactions and pathology in a quiescent and activated mesenchymal cell niche. *Histology and Histopathology, 24*, 909–969.

Doetsch, F. (2003). A niche for adult neural stem cells. *Current Opinion in Genetics & Development, 13*, 543–550.
Doetsch, F., & Alvarez-Buylla, A. (1996). Network of tangential pathways for neuronal migration in adult mammalian brain. *Proceedings of the National Academy of Sciences of the United States of America, 93*, 14895–14900.
Doetsch, F., Caille, I., Lim, D. A., Garcia-Verdugo, J. M., & Alvarez-Buylla, A. (1999). Subventricular zone astrocytes are neural stem cells in the adult mammalian brain. *Cell, 97*, 703–716.
Doetsch, F., Garcia-Verdugo, J. M., & Alvarez-Buylla, A. (1997). Cellular composition and three-dimensional organization of the subventricular germinal zone in the adult mammalian brain. *The Journal of Neuroscience, 17*, 5046–5061.
Doetsch, F., & Scharff, C. (2001). Challenges for brain repair: Insights from adult neurogenesis in birds and mammals. *Brain, Behavior and Evolution, 58*, 306–322.
Dore-Duffy, P. (2008). Pericytes: Pluripotent cells of the blood brain barrier. *Current Pharmaceutical Design, 14*, 1581–1593.
Dore-Duffy, P., & Cleary, K. (2011). Morphology and properties of pericytes. *Methods in Molecular Biology, 686*, 49–68.
Ebers, G. C., & Sadovnick, A. D. (1993). The geographic distribution of multiple sclerosis: A review. *Neuroepidemiology, 12*, 1–5.
ElAli, A., Thériault, P., & Rivest, S. (2014). The role of pericytes in neurovascular unit remodeling in brain disorders. *International Journal of Molecular Sciences, 15*, 6453.
Esen, N., Serkin, Z., & Dore-Duffy, P. (2013). Induction of vascular remodeling: A novel therapeutic approach in EAE. *Journal of the Neurological Sciences, 333*, 88–92.
Fancy, S. P., Zhao, C., & Franklin, R. J. (2004). Increased expression of Nkx2.2 and Olig2 identifies reactive oligodendrocyte progenitor cells responding to demyelination in the adult CNS. *Molecular and Cellular Neurosciences, 27*, 247–254.
Ferguson, A. R., Hook, M. A., Garcia, G., Bresnahan, J. C., Beattie, M. S., & Grau, J. W. (2004). A simple post hoc transformation that improves the metric properties of the BBB scale for rats with moderate to severe spinal cord injury. *Journal of Neurotrauma, 21*, 1601–1613.
Ffrench-Constant, C., & Raff, M. C. (1986a). The oligodendrocyte-type-2 astrocyte cell lineage is specialized for myelination. *Nature, 323*, 335–338.
Ffrench-Constant, C., & Raff, M. C. (1986b). Proliferating bipotential glial progenitor cells in adult rat optic nerve. *Nature, 319*, 499–502.
Franklin, R. J., & Ffrench-Constant, C. (2008). Remyelination in the CNS: From biology to therapy. *Nature Reviews. Neuroscience, 9*, 839–855.
Franklin, R. J., & Hinks, G. L. (1999). Understanding CNS remyelination: Clues from developmental and regeneration biology. *Journal of Neuroscience Research, 58*, 207–213.
Franklin, R. J., & Kotter, M. R. (2008). The biology of CNS remyelination: The key to therapeutic advances. *Journal of Neurology, 255*(Suppl 1), 19–25.
Fünfschilling, U., Supplie, L. M., Mahad, D., Boretius, S., Saab, A. S., Edgar, J., Brinkmann, B. G., Kassmann, C. M., Tzvetanova, I. D., Möbius, W., et al. (2012). Glycolytic oligodendrocytes maintain myelin and long-term axonal integrity. *Nature, 485*, 517.
Gaceb, A., Özen, I., Padel, T., Barbariga, M., & Paul, G. (2017). Pericytes secrete pro-regenerative molecules in response to platelet-derived growth factor-BB. *Journal of Cerebral Blood Flow & Metabolism, 38*, 45–57.
Gage, F. H. (2000). Mammalian neural stem cells. *Science, 287*, 1433–1438.
Gensert, J. M., & Goldman, J. E. (1997). Endogenous progenitors remyelinate demyelinated axons in the adult CNS. *Neuron, 19*, 197–203.
Grygorowicz, T., Dąbrowska-Bouta, B., & Strużyńska, L. (2018). Administration of an antagonist of P2X7 receptor to EAE rats prevents a decrease of expression of claudin-5 in cerebral capillaries. *Purinergic Signalling, 14*(4), 385–393.
Gündüz, T., Taryel Kiremitçi, T., Ulusoy, C., Kürtüncü, M., & Türkoğlu, R. (2018). Cerebrospinal fluid analysis of pericytic mediators in clinically isolated syndrome and multiple sclerosis: A preliminary study. *Journal of Experimental Medicine, 81*, 18–22.

Hellstrom, M., Gerhardt, H., Kalen, M., Li, X., Eriksson, U., Wolburg, H., & Betsholtz, C. (2001). Lack of pericytes leads to endothelial hyperplasia and abnormal vascular morphogenesis. *The Journal of Cell Biology, 153*, 543–553.

Hoftberger, R., & Lassmann, H. (2017). Inflammatory demyelinating diseases of the central nervous system. *Handbook of Clinical Neurology, 145*, 263–283.

Iacobaeus, E., Sugars, R. V., Andrén, A. T., Alm, J. J., Qian, H., Frantzen, J., Newcombe, J., Alkass, K., Druid, H., Bottai, M., Röyttä, M., Le Blanc, K. (2017). Dynamic Changes in Brain Mesenchymal Perivascular Cells Associate with Multiple Sclerosis Disease Duration, Active Inflammation, and Demyelination. *Stem Cells Translational Medicine, 6*(10),1840–1851.

Jablonska, B., Aguirre, A., Raymond, M., Szabo, G., Kitabatake, Y., Sailor, K. A., Ming, G. L., Song, H., & Gallo, V. (2010). Chordin-induced lineage plasticity of adult SVZ neuroblasts after demyelination. *Nature Neuroscience, 13*, 541–550.

Jadasz, J. J., Aigner, L., Rivera, F. J., & Kury, P. (2012). The remyelination Philosopher's Stone: Stem and progenitor cell therapies for multiple sclerosis. *Cell and Tissue Research, 349*(1), 331–347.

Kabat, E. A., Wolf, A., Bezer, A. E., & Murray, J. P. (1951). Studies on acute disseminated encephalomyelitis produced experimentally in rhesus monkeys. *The Journal of Experimental Medicine, 93*, 615–633.

Kamouchi, M., Ago, T., & Kitazono, T. (2011). Brain pericytes: Emerging concepts and functional roles in brain homeostasis. *Cellular and Molecular Neurobiology, 31*, 175–193.

Keirstead, H. S., Levine, J. M., & Blakemore, W. F. (1998). Response of the oligodendrocyte progenitor cell population (defined by NG2 labelling) to demyelination of the adult spinal cord. *Glia, 22*, 161–170.

Kesterson, J. W., & Carlton, W. W. (1971). Histopathologic and enzyme histochemical observations of the cuprizone-induced brain edema. *Experimental and Molecular Pathology, 15*, 82–96.

Komoly, S., Jeyasingham, M. D., Pratt, O. E., & Lantos, P. L. (1987). Decrease in oligodendrocyte carbonic anhydrase activity preceding myelin degeneration in cuprizone induced demyelination. *Journal of the Neurological Sciences, 79*, 141–148.

Kornek, B., & Lassmann, H. (2003). Neuropathology of multiple sclerosis-new concepts. *Brain Research Bulletin, 61*, 321–326.

Krishnamoorthy, G., & Wekerle, H. (2009). EAE: An immunologist's magic eye. *European Journal of Immunology, 39*, 2031–2035.

Krizbai, I. A., Bauer, H., Amberger, A., Hennig, B., Szabo, H., Fuchs, R., & Bauer, H. C. (2000). Growth factor-induced morphological, physiological and molecular characteristics in cerebral endothelial cells. *European Journal of Cell Biology, 79*, 594–600.

Kuchroo, V. K., Anderson, A. C., Waldner, H., Munder, M., Bettelli, E., & Nicholson, L. B. (2002). T cell response in experimental autoimmune encephalomyelitis (EAE): Role of self and cross-reactive antigens in shaping, tuning, and regulating the autopathogenic T cell repertoire. *Annual Review of Immunology, 20*, 101–123.

Lange, S., Trost, A., Tempfer, H., Bauer, H. C., Bauer, H., Rohde, E., Reitsamer, H. A., Franklin, R. J., Aigner, L., & Rivera, F. J. (2013). Brain pericyte plasticity as a potential drug target in CNS repair. *Drug Discovery Today, 18*, 456–463.

Larsen, P. H., Wells, J. E., Stallcup, W. B., Opdenakker, G., & Yong, V. W. (2003). Matrix metalloproteinase-9 facilitates remyelination in part by processing the inhibitory NG2 proteoglycan. *The Journal of Neuroscience, 23*, 11127–11135.

Lassmann, H. (2007a). Experimental models of multiple sclerosis. *Revue Neurologique (Paris), 163*, 651–655.

Lassmann, H. (2007b). Multiple sclerosis: Is there neurodegeneration independent from inflammation? *Journal of the Neurological Sciences, 259*, 3–6.

Lassmann, H., Bruck, W., Lucchinetti, C., & Rodriguez, M. (1997). Remyelination in multiple sclerosis. *Multiple Sclerosis, 3*, 133–136.

Lassmann, H., van Horssen, J., & Mahad, D. (2012). Progressive multiple sclerosis: Pathology and pathogenesis. *Nature Reviews. Neurology, 8*, 647–656.

Levine, J. M., & Reynolds, R. (1999). Activation and proliferation of endogenous oligodendrocyte precursor cells during ethidium bromide-induced demyelination. *Experimental Neurology, 160*, 333–347.
Levine, J. M., Reynolds, R., & Fawcett, J. W. (2001). The oligodendrocyte precursor cell in health and disease. *Trends in Neurosciences, 24*, 39–47.
Lindahl, P., Johansson, B. R., Leveen, P., & Betsholtz, C. (1997). Pericyte loss and microaneurysm formation in PDGF-B-deficient mice. *Science, 277*, 242–245.
Lois, C., Garcia-Verdugo, J. M., & Alvarez-Buylla, A. (1996). Chain migration of neuronal precursors. *Science, 271*, 978–981.
Lublin, F. D., & Reingold, S. C. (1996). Defining the clinical course of multiple sclerosis: Results of an international survey. National Multiple Sclerosis Society (USA) Advisory Committee on Clinical Trials of New Agents in Multiple Sclerosis. *Neurology, 46*, 907–911.
Ludwin, S. K. (1978). Central nervous system demyelination and remyelination in the mouse: An ultrastructural study of cuprizone toxicity. *Laboratory Investigation, 39*, 597–612.
Ludwin, S. K., & Maitland, M. (1984). Long-term remyelination fails to reconstitute normal thickness of central myelin sheaths. *Journal of the Neurological Sciences, 64*, 193–198.
Maki, T., Maeda, M., Uemura, M., Lo, E. K., Terasaki, Y., Liang, A. C., Shindo, A., Choi, Y. K., Taguchi, A., Matsuyama, T., et al. (2015). Potential interactions between pericytes and oligodendrocyte precursor cells in perivascular regions of cerebral white matter. *Neuroscience Letters, 597*, 164–169.
Matsushima, G. K., & Morell, P. (2001). The neurotoxicant, cuprizone, as a model to study demyelination and remyelination in the central nervous system. *Brain Pathology, 11*, 107–116.
Menn, B., Garcia-Verdugo, J. M., Yaschine, C., Gonzalez-Perez, O., Rowitch, D., & Alvarez-Buylla, A. (2006). Origin of oligodendrocytes in the subventricular zone of the adult brain. *The Journal of Neuroscience, 26*, 7907–7918.
Montagne, A., Nikolakopoulou, A. M., Zhao, Z., Sagare, A. P., Si, G., Lazic, D., Barnes, S. R., Daianu, M., Ramanathan, A., Go, A., et al. (2018). Pericyte degeneration causes white matter dysfunction in the mouse central nervous system. *Nature Medicine, 24*, 326.
Murtie, J. C., Zhou, Y. X., Le, T. Q., Vana, A. C., & Armstrong, R. C. (2005). PDGF and FGF2 pathways regulate distinct oligodendrocyte lineage responses in experimental demyelination with spontaneous remyelination. *Neurobiology of Disease, 19*, 171–182.
Nait-Oumesmar, B., Decker, L., Lachapelle, F., Avellana-Adalid, V., Bachelin, C., & Van Evercooren, A. B. (1999). Progenitor cells of the adult mouse subventricular zone proliferate, migrate and differentiate into oligodendrocytes after demyelination. *The European Journal of Neuroscience, 11*, 4357–4366.
Noseworthy, J. H., Lucchinetti, C., Rodriguez, M., & Weinshenker, B. G. (2000). Multiple sclerosis. *The New England Journal of Medicine, 343*, 938–952.
Orton, S. M., Herrera, B. M., Yee, I. M., Valdar, W., Ramagopalan, S. V., Sadovnick, A. D., & Ebers, G. C. (2006). Sex ratio of multiple sclerosis in Canada: A longitudinal study. *Lancet Neurology, 5*, 932–936.
Peppiatt, C. M., Howarth, C., Mobbs, P., & Attwell, D. (2006). Bidirectional control of CNS capillary diameter by pericytes. *Nature, 443*, 700–704.
Quaegebeur, A., Segura, I., & Carmeliet, P. (2010). Pericytes: Blood-brain barrier safeguards against neurodegeneration? *Neuron, 68*, 321–323.
Raff, M. C., Miller, R. H., & Noble, M. (1983). A glial progenitor cell that develops in vitro into an astrocyte or an oligodendrocyte depending on culture medium. *Nature, 303*, 390–396.
Raff, M. C., Williams, B. P. A., & Miller, R. H. (1984). The in vitro differentiation of a bipotential glial progenitor cell. *The EMBO Journal, 3*, 1857–1864.
Ramagopalan, S. V., Dobson, R., Meier, U. C., & Giovannoni, G. (2010). Multiple sclerosis: Risk factors, prodromes, and potential causal pathways. *Lancet Neurology, 9*, 727–739.
Redwine, J. M., & Armstrong, R. C. (1998). In vivo proliferation of oligodendrocyte progenitors expressing PDGFalphaR during early remyelination. *Journal of Neurobiology, 37*, 413–428.

Reynolds, R., Dawson, M., Papadopoulos, D., Polito, A., Di Bello, I. C., Pham-Dinh, D., & Levine, J. (2002). The response of NG2-expressing oligodendrocyte progenitors to demyelination in MOG-EAE and MS. *Journal of Neurocytology, 31*, 523–536.

Ribatti, D., Nico, B., & Crivellato, E. (2011). The role of pericytes in angiogenesis. *The International Journal of Developmental Biology, 55*, 261–268.

Rinholm, J. E., Hamilton, N. B., Kessaris, N., Richardson, W. D., Bergersen, L. H., & Attwell, D. (2011). Regulation of oligodendrocyte development and myelination by glucose and lactate. *The Journal of Neuroscience, 31*, 538–548.

Rivera, F. J., & Aigner, L. (2012). Adult mesenchymal stem cell therapy for myelin repair in multiple sclerosis. *Biological Research, 45*, 257–268.

Rivera, F. J., Steffenhagen, C., Kremer, D., Kandasamy, M., Sandner, B., Couillard-Despres, S., Weidner, N., Kury, P., & Aigner, L. (2010). Deciphering the oligodendrogenic program of neural progenitors: Cell intrinsic and extrinsic regulators. *Stem Cells and Development, 19*, 595–606.

Rouget, C.-M. B. (1874). Note sur le developpement de la tunique contractile des vaisseaux. *Comptes Rendus de l'Académie des Sciences, 59*, 4.

Ruckh, J. M., Zhao, J.-W., Shadrach, J. L., van Wijngaarden, P., Rao, T. N., Wagers, A. J., & Franklin, R. J. M. (2012). Rejuvenation of regeneration in the aging central nervous system. *Cell Stem Cell, 10*, 96–103.

Scheikl, T., Pignolet, B., Mars, L. T., & Liblau, R. S. (2010). Transgenic mouse models of multiple sclerosis. *Cellular and Molecular Life Sciences, 67*, 4011–4034.

Schonrock, L. M., Kuhlmann, T., Adler, S., Bitsch, A., & Bruck, W. (1998). Identification of glial cell proliferation in early multiple sclerosis lesions. *Neuropathology and Applied Neurobiology, 24*, 320–330.

Shen, S., Sandoval, J., Swiss, V. A., Li, J., Dupree, J., Franklin, R. J., & Casaccia-Bonnefil, P. (2008). Age-dependent epigenetic control of differentiation inhibitors is critical for remyelination efficiency. *Nature Neuroscience, 11*, 1024–1034.

Shields, S. A., Gilson, J. M., Blakemore, W. F., & Franklin, R. J. (1999). Remyelination occurs as extensively but more slowly in old rats compared to young rats following gliotoxin-induced CNS demyelination. *Glia, 28*, 77–83.

Silva, M. E., Lange, S., Hinrichsen, B., Philp, A. R., Reyes, C. R., Halabi, D., Mansilla, J. B., Rotheneichner, P., De la Fuente, A. G., Couillard-Despres, S., Bátiz, L. F., Franklin, R. J. M., Aigner, L., Francisco, J., & Rivera, F. J. (2019). Pericytes favor oligodendrocyte fate choice in adult neural stem cells. Frontiers in Cellular Neuroscience, https://doi.org/10.3389/fncel.2019.00085.

Silvestroff, L., Bartucci, S., Soto, E., Gallo, V., Pasquini, J., & Franco, P. (2010). Cuprizone-induced demyelination in CNP::GFP transgenic mice. *The Journal of Comparative Neurology, 518*, 2261–2283.

Sim, F. J., Zhao, C., Penderis, J., & Franklin, R. J. (2002). The age-related decrease in CNS remyelination efficiency is attributable to an impairment of both oligodendrocyte progenitor recruitment and differentiation. *The Journal of Neuroscience, 22*, 2451–2459.

Skripuletz, T., Lindner, M., Kotsiari, A., Garde, N., Fokuhl, J., Linsmeier, F., Trebst, C., & Stangel, M. (2008). Cortical demyelination is prominent in the murine cuprizone model and is strain-dependent. *The American Journal of Pathology, 172*, 1053–1061.

Smith, K. J., McDonald, W. I., & Blakemore, W. F. (1979). Restoration of secure conduction by central demyelination. *Transactions of the American Neurological Association, 104*, 25–29.

Takakuni Maki, Y. K. C., Miyamoto, N., Shindo, A., Liang, A. C., Ahn, B. J. U., Mandeville, E. T., Kaji, S., Itoh, K., Seo, J. I. H., Gelman, I. H., Lok, J., Takahashi, R., Kim, K.-W., Lo, E. H., & Arai, K. (2018). A-kinase anchor protein 12 is required for oligodendrocyte differentiation in adult white matter. *Stem Cells Translational Medicine, 36*, 751–760.

Takata, F., Dohgu, S., Nishioku, T., Takahashi, H., Harada, E., Makino, I., Nakashima, M., Yamauchi, A., & Kataoka, Y. (2009). Adrenomedullin-induced relaxation of rat brain pericytes is related to the reduced phosphorylation of myosin light chain through the cAMP/PKA signaling pathway. *Neuroscience Letters, 449*, 71–75.

Tigges, U., Boroujerdi, A., Welser-Alves, J. V., & Milner, R. (2013). TNF-α promotes cerebral pericyte remodeling in vitro, via a switch from α1 to α2 integrins. *Journal of Neuroinflammation, 10*, 33–33.

Tintore, M., Vidal-Jordana, A., & Sastre-Garriga, J. (2019). Treatment of multiple sclerosis – success from bench to bedside. *Nature Reviews Neurology, 15*(1), 53–58.

Tsai, H.-H., Niu, J., Munji, R., Davalos, D., Chang, J., Zhang, H., Tien, A.-C., Kuo, C. J., Chan, J. R., Daneman, R., et al. (2016). Oligodendrocyte precursors migrate along vasculature in the developing nervous system. *Science, 351*, 379–384.

Vanlandewijck, M., He, L., Mae, M. A., Andrae, J., Ando, K., Del Gaudio, F., Nahar, K., Lebouvier, T., Lavina, B., Gouveia, L., et al. (2018). Author Correction: A molecular atlas of cell types and zonation in the brain vasculature. *Nature, 560*, E3.

Watanabe, M., Toyama, Y., & Nishiyama, A. (2002). Differentiation of proliferated NG2-positive glial progenitor cells in a remyelinating lesion. *Journal of Neuroscience Research, 69*, 826–836.

Whitacre, C. C. (2001). Sex differences in autoimmune disease. *Nature Immunology, 2*, 777–780.

Wilson, H. C., Scolding, N. J., & Raine, C. S. (2006). Co-expression of PDGF alpha receptor and NG2 by oligodendrocyte precursors in human CNS and multiple sclerosis lesions. *Journal of Neuroimmunology, 176*, 162–173.

Winkler, E. A., Bell, R. D., & Zlokovic, B. V. (2011). Central nervous system pericytes in health and disease. *Nature Neuroscience, 14*, 1398–1405.

Wolswijk, G., & Noble, M. (1989). Identification of an adult-specific glial progenitor cell. *Development, 105*, 387–400.

Woodruff, R. H., & Franklin, R. J. (1999). Demyelination and remyelination of the caudal cerebellar peduncle of adult rats following stereotaxic injections of lysolecithin, ethidium bromide, and complement/anti-galactocerebroside: A comparative study. *Glia, 25*, 216–228.

Woodruff, R. H., Fruttiger, M., Richardson, W. D., & Franklin, R. J. M. (2004). Platelet-derived growth factor regulates oligodendrocyte progenitor numbers in adult CNS and their response following CNS demyelination. *Molecular and Cellular Neuroscience, 25*, 252–262.

Yemisci, M., Gursoy-Ozdemir, Y., Vural, A., Can, A., Topalkara, K., & Dalkara, T. (2009). Pericyte contraction induced by oxidative-nitrative stress impairs capillary reflow despite successful opening of an occluded cerebral artery. *Nature Medicine, 15*, 1031–1037.

Zawadzka, M., Rivers, L. E., Fancy, S. P., Zhao, C., Tripathi, R., Jamen, F., Young, K., Goncharevich, A., Pohl, H., Rizzi, M., et al. (2010). CNS-resident glial progenitor/stem cells produce Schwann cells as well as oligodendrocytes during repair of CNS demyelination. *Cell Stem Cell, 6*, 578–590.

Zhang, S. C., Ge, B., & Duncan, I. D. (1999). Adult brain retains the potential to generate oligodendroglial progenitors with extensive myelination capacity. *Proceedings of the National Academy of Sciences of the United States of America, 96*, 4089–4094.

Zhou, Y. X., Flint, N. C., Murtie, J. C., Le, T. Q., & Armstrong, R. C. (2006). Retroviral lineage analysis of fibroblast growth factor receptor signaling in FGF2 inhibition of oligodendrocyte progenitor differentiation. *Glia, 54*, 578–590.

Zimmermann, K. W. (1923). Der feinere Bau der Blutkapillaren. *Zeitschrift für Anatomie und Entwicklungsgeschichte, 68*, 5.

Zlokovic, B. V. (2008). The blood-brain barrier in health and chronic neurodegenerative disorders. *Neuron, 57*, 178–201.

Chapter 9
Pericytes in Ischemic Stroke

Turgay Dalkara, Luis Alarcon-Martinez, and Muge Yemisci

Abstract Recent stroke research has shifted the focus to the microvasculature from neuron-centric views. It is increasingly recognized that a successful neuroprotection is not feasible without microvascular protection. On the other hand, recent studies on pericytes, long-neglected cells on microvessels have provided insight into the regulation of microcirculation. Pericytes play an essential role in matching the metabolic demand of nervous tissue with the blood flow in addition to regulating the development and maintenance of the blood–brain barrier (BBB), leukocyte trafficking across the BBB and angiogenesis. Pericytes appears to be highly vulnerable to injury. Ischemic injury to pericytes on cerebral microvasculature unfavorably impacts the stroke-induced tissue damage and brain edema by disrupting microvascular blood flow and BBB integrity. Strongly supporting this, clinical imaging studies show that tissue reperfusion is not always obtained after recanalization. Therefore, prevention of pericyte dysfunction may improve the outcome of recanalization therapies by promoting microcirculatory reperfusion and preventing hemorrhage and edema. In the peri-infarct tissue, pericytes are detached from microvessels and promote angiogenesis and neurogenesis, and hence positively effect stroke outcome. Expectedly, we will learn more about the place of pericytes in CNS pathologies including stroke and devise approaches to treat them in the next decades.

T. Dalkara (✉)
Department of Neurology, Faculty of Medicine, Hacettepe University, Ankara, Turkey

Institute of Neurological Sciences and Psychiatry, Hacettepe University, Ankara, Turkey

Department of Radiology, Massachusetts General Hospital, Harvard University, Boston, MA, USA
e-mail: tdalkara@hacettepe.edu.tr

L. Alarcon-Martinez
Department of Neuroscience and Centre de Recherche du Centre Hospitalier de l'Université de Montréal, Université de Montréal, Montréal, QC, Canada

M. Yemisci
Department of Neurology, Faculty of Medicine, Hacettepe University, Ankara, Turkey

Institute of Neurological Sciences and Psychiatry, Hacettepe University, Ankara, Turkey

A. Birbrair (ed.), *Pericyte Biology in Disease*, Advances in Experimental Medicine and Biology 1147, https://doi.org/10.1007/978-3-030-16908-4_9

Keywords Pericyte · Microcirculation · Blood–brain barrier · Reperfusion · Recanalization · Thrombolysis · Angiogenesis · Neurogenesis · Cerebral ischemia · Post-stroke recovery · Retinal ischemia · CADASIL

Introduction: Pericytes and Stroke

Pericytes play a role in matching the metabolic demand of nervous tissue with the blood flow in addition to regulating the development and maintenance of the blood–brain barrier (BBB) (Abbott et al. 2006; Armulik et al. 2011; Attwell et al. 2010; Daneman et al. 2010; Iadecola 2004), leukocyte trafficking across the BBB (Proebstl et al. 2012; Dohgu and Banks 2013; Stark et al. 2013; Leick et al. 2014) and angiogenesis (Ozerdem and Stallcup 2003; Gerhardt and Betsholtz 2003; Díaz-Flores et al. 2009). Ischemic injury to pericytes on cerebral microvasculature unfavorably impacts the stroke-induced tissue damage and brain edema by disrupting microvascular blood flow and BBB integrity, whereas ischemia-triggered signaling in pericytes on the vasculature within the peri-infarct area positively effect stroke outcome by promoting post-stroke angiogenesis and neurogenesis (Ozerdem and Stallcup 2003; Gerhardt and Betsholtz 2003; Zhang et al. 2012).

CNS Pericytes

Pericytes are present on almost all microvessels in the body; however, their density is highest in the CNS and retina in accordance with their role in fine regulation of the microcirculatory blood flow and maintenance of the blood–brain/retina barrier (Frank et al. 1987; Shepro and Morel 1993; Winkler et al. 2011a; Armulik et al. 2011). Pericytes are located on pre-capillary arterioles, capillaries, and post-capillary venules (Sims 1986; Dore-Duffy and Cleary 2011a; Armulik et al. 2011) (Fig. 9.1a, b). Unlike smooth muscle cells (SMCs), pericytes are embedded within two layers of basement membrane (Shepro and Morel 1993). Adjoining membranes of the neighboring pericytes are interconnected with gap junctions, serving as a communication pathway along the microvascular wall (Peppiatt et al. 2006; Hamilton et al. 2010). Pericytes extend processes around microvessels, which are largely circumferential at the arteriole side of the microvascular bed and at branching points, more longitudinal in the middle of the capillary bed, and have a stellate morphology at the venular side (Fig. 9.1b) (Hartmann et al. 2015). Pericytes are structurally plastic cells (Berthiaume et al. 2018) and their morphology and protein expression vary along the course of microvasculature, presumably to accommodate differing functions (Nehls and Drenckhahn 1991; Armulik et al. 2011; Dore-Duffy and Cleary 2011a; Hill et al. 2015; Hartmann et al. 2015; Jung et al. 2018). Pericytes are heterogeneous in their origin (Dias Moura Prazeres et al. 2017). Several transitional forms are observed along the vascular bed at various developmental stages or after pathological stimuli (Sims 1986; Dore-Duffy and Cleary 2011a; Armulik

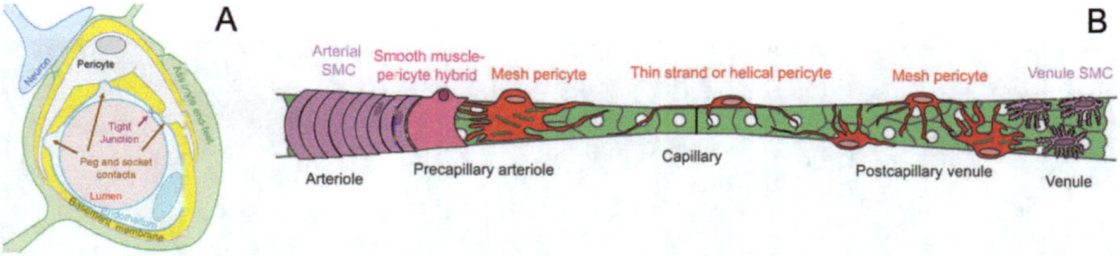

Fig. 9.1 Neurovascular unit and pericytes. (**a**) The neurovascular unit is composed of the endothelia and tight junctions between them, pericytes, the basal lamina encircling endothelia and pericytes, and astrocyte endfeet surrounding the microvessel. Note the peg and socket type contacts between endothelia and pericytes (Reproduced from Dalkara and Alarcon-Martinez 2015 with permission). (**b**) Pericyte processes are highly varied with shapes ranging from thin singular strands that run parallel to the microvasculature to more complex mesh processes that enwrap the entire vessel lumen. Pericytes located closer to the arteriolar end of the microcirculation exhibit more circular processes that may be essential to their contractile function (Reproduced from Hartmann et al. 2015 with permission)

et al. 2011; Sharma et al. 2012). The transition from smooth muscle cells to pericytes is not sharp. Smooth muscle-pericyte "hybrid" cells precede the prearteriolar pericytes having mesh-like circular processes (Sims 1986; Hartmann et al. 2015). Pericytes that give out more circumferential processes express more α-SMA, when assessed either with immunohistochemistry of brain sections ex vivo (Nehls and Drenckhahn 1991) or in mice cortex expressing reporter dyes under control of the α-SMA promoter in vivo (Hill et al. 2015; Hartmann et al. 2015). Detection of α-SMA in pericytes has been a controversial issue because of the some technical drawbacks that were missed in the methods used. First of all, It should be noted that reporter dyes expressed under the control of α-SMA promoter are membrane-bound, and therefore basically label the pericyte membrane, whereas immunohistochemistry directly detects the α-SMA protein (mainly in the cytoplasm of the soma and processes). However, the detection of the small pool of α-SMA in their relatively short processes by immunohistochemistry requires rapid fixation before α-SMA depolymerises during tissue processing (Alarcon-Martinez et al. 2018) (Fig. 9.2), whereas low level of α-SMA expression could be difficult to visualize due to dispersion of the limited amount of reporter fluorescent protein diffused over the large surface area of the pericyte membrane (Hill et al. 2015). Of note, α-SMA expression in pericytes is readily induced by tissue injury such as trauma, ischemia and injections (Dore-Duffy et al. 2000; Bai et al. 2018; Alarcon-Martinez et al. 2018).

Pericytes Regulate Microcirculatory Blood Flow in CNS and Retina

Functional hyperemia is an essential phenomenon in CNS by which oxygen and nutrients are supplied to tissue in accordance with metabolic demand generated by neuronal activity (Attwell et al. 2010). This tight pairing between the neural firing and blood flow, named neurovascular coupling, is provided by the neurovascular

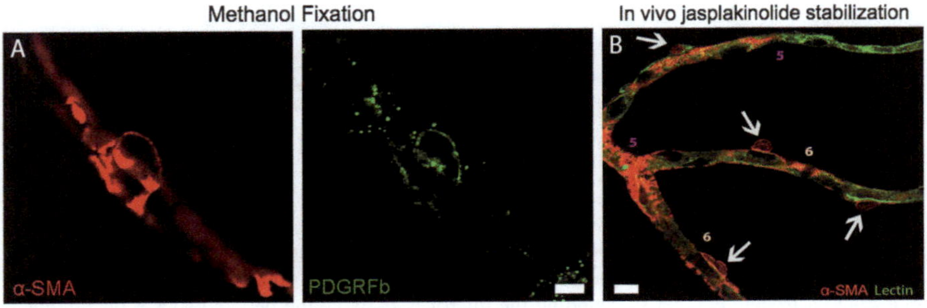

Fig. 9.2 Capillary pericytes express α-SMA. (**a**) Rapid fixation with methanol allowed visualization of α-SMA expression (red) in a deeper plexus capillary pericyte by preventing depolymerization of a small pool of α-SMA during tissue processing. This mural cell was also immunopositive for the pericyte marker PDGFRβ (green). Scale bar: 5 μm. (**b**) Stabilization of F-actin with intravitreal injection of Jasplakinolide before sacrificing the mouse disclosed α-SMA immunolabeling (red) on a sixth order retinal capillary visualized with lectin (green). Arrows point to pericyte somas and numbers indicate the branch order. Scale bar: 10 μm. (Reproduced from Alarcon-Martinez et al. 2018 with permission)

unit composed of neurons, astrocytes, vascular endothelia, smooth muscle cells, and pericytes (Iadecola 2004; Attwell et al. 2010; Dalkara and Alarcon-Martinez 2015). The neurovascular unit shows structural and functional differences between capillaries, which are covered by pericytes that respond to local activity in immediate vicinity and the intra-parenchymal arteries/arterioles, which are wrapped by smooth muscle cells and regulated by signals coming from a larger cohort of neurons as well as distal microvasculature (Hall et al. 2014; Dalkara and Alarcon-Martinez 2015; Biesecker et al. 2016; Mishra et al. 2016; Kisler et al. 2017). Both smooth muscle cells and pericytes contain α-SMA; a contractile protein that mediates the vascular diameter changes associated with neurovascular coupling (Herman and D'Amore 1985; Kelley et al. 1987; Hall et al. 2014; Alarcon-Martinez et al. 2018). The capability of pericytes to change vascular diameter in response to neural activation has been debated because blood flow is regulated at the level of arterioles in the peripheral circulation where a very focal flow regulation is not required unlike the CNS and retina. Failure of some laboratories to detect α-SMA in capillary pericytes strengthened these reservations. However, it has recently been proposed that this failure was caused by rapid depolymerization of small pool of α-SMA in capillary pericytes during slow tissue fixation with paraformaldehyde because rapid fixation methods disclosed that even small pericytes on high order capillaries expressed α-SMA (Alarcon-Martinez et al. 2018) as originally shown by Herman and D'Amore (Herman and D'Amore 1985). By using short interfering RNA, Alarcon-Martinez et al. readily suppressed α-SMA expression in distal capillary pericytes, but not in upstream larger microvessels where α-SMA is more abundant, supporting the view that the histological detection of the small pool of α-SMA in capillary pericytes is challenging compared to the α-SMA-rich pericytes on pre-capillary arterioles and vascular smooth muscle cells (Alarcon-Martinez et al. 2018). Indeed, as reviewed in detail by Díaz-Flores et al. (2009), the pericyte contractility is supported by several lines of evidence including their characteristic morphology with processes that envelop the microvessels as well as ultrastructural and immunohistochemical demonstration of contractile proteins (Wallow and Burnside 1980; Herman and D'Amore 1985; Joyce et al. 1985a, b; Fujimoto and Singer 1987; Kelley et al. 1987; Das et al. 1988; Nehls and Drenckhahn 1991; Shepro and Morel 1993; Allt and Lawrenson 2001; Bandopadhyay et al. 2001) in addition to the presence of receptors for vasoactive mediators on their surface (Peppiatt et al. 2006; Puro 2007; Hamilton et al. 2010). In vitro studies on cerebellar, cerebral, and retinal slices or on isolated microvessels or cultured pericytes and recent in vivo studies have disclosed that pericytes are indeed capable of contracting or dilating in response to vasoactive mediators and physiological stimuli (Herman and D'Amore 1985; Kelley et al. 1987; Peppiatt et al. 2006; Puro 2007; Fernandez-Klett et al. 2010; Hall et al. 2014; Biesecker et al. 2016; Mishra et al. 2016; Kisler et al. 2017). A recent in vivo study showed that cortical capillaries dilated before arterioles during sensory stimulation, supporting the view that microvascular blood flow in the CNS is regulated by pericytes in response to the very focal demand originating from a small group of nearby cells as a final step of flow regulation after the arterioles, which serve a larger cohort of cells (Hall et al. 2014). This flow regulation with fine spatial resolution may be essential for tissues with high functional specialization such as the brain and retina. However, it should be noted that all microvascular pericytes are not contractile and proportion of the contractile ones may vary with the tissue, species and developmental stage as well as along the arteriovenous axis (Krueger and Bechmann 2010; Fernández-Klett and Priller 2015; Hill et al. 2015).

When Rouget discovered the pericytes in 1873, he proposed that they might have contractile capability and regulate microcirculatory blood flow because of their shape and position on microvessels (Rouget 1873). This hypothesis has been supported and elaborated by many in vitro and in vivo studies over the years. The capillary diameter changes during metabolic demand were recently proposed to be mediated by astrocytic calcium increase through ATP-gated channels and pericyte relaxation on release of PGE2 from astrocyte endfeet (Mishra et al. 2016). In line with the findings from CNS, Biesecker et al. showed that calcium signaling in Müller cell endfeet was sufficient to evoke capillary but not arteriole dilation in the retina (Biesecker et al. 2016). Kisler et al. showed that transgenic mice with a decreased number of pericytes had a deficient neurovascular coupling, reinforcing the importance of pericytes in blood flow regulation (Kisler et al. 2017). Moreover, during ischemia, it was shown in situ that pericytes constricted capillaries by calcium-induced α-SMA contraction, impairing microcirculatory re-flow after recanalization (Yemisci et al. 2009; Hall et al. 2014) (Fig. 9.3). Hill et al. observed that most of the mural cells on the first 4 order capillaries expressed α-SMA and contracted in response to physiological stimuli or ischemia; however, they named these cells as smooth muscle cells because they expressed α-SMA and reserved the name pericyte for only strand-like mural cells lacking α-SMA (Hill et al. 2015). An opinion article entitled "What is a pericyte?" discusses this unconventional definition of pericyte and point to the fact that Hill et al.'s findings in fact confirm previous reports demonstrating pericyte contractility under physiological and ischemic conditions, once pericytes are defined as first described by Zimmerman in 1923 (Attwell et al. 2016).

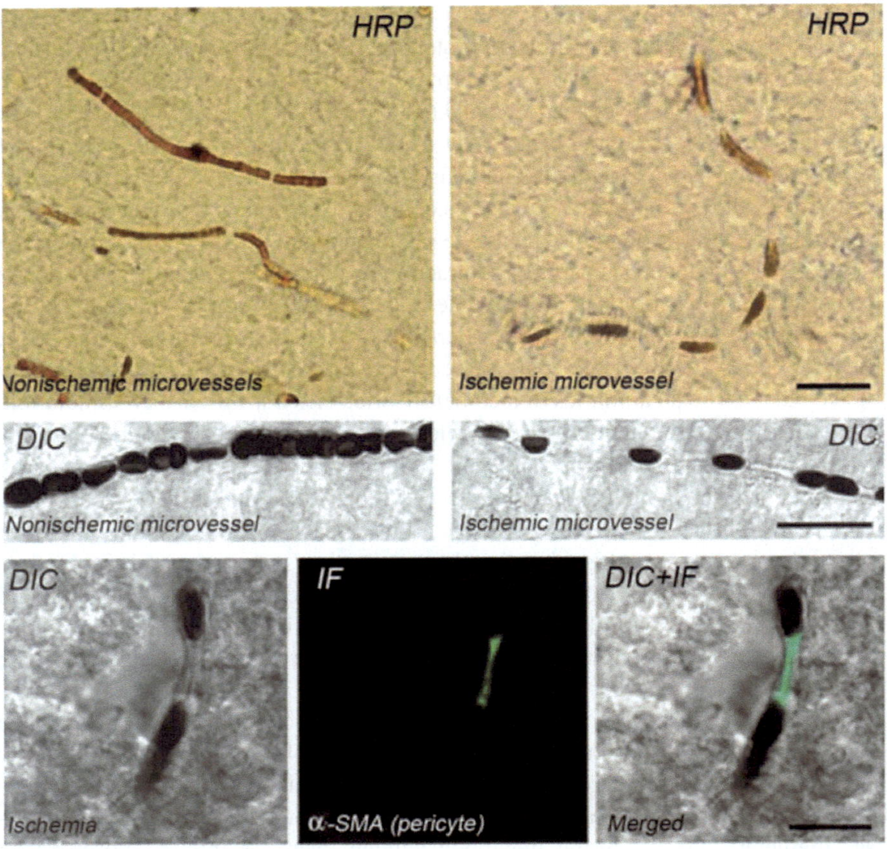

Fig. 9.3 Ischemia causes persistent pericyte contraction, which is not restored after complete recanalization of the occluded artery. Mice were subjected to 2 h of proximal MCA occlusion and intravenously injected with horseradish peroxidase (HRP) before decapitation 6 h after reopening of the MCA. HRP-filled microvessels exhibited sausage-like segmental constrictions in ischemic areas on brain sections (upper row). The differential interference contrast (DIC) microscopy images illustrate frequent interruptions in the erythrocyte column in an ischemic capillary contrary to a continuous row of erythrocytes flowing through an intact capillary (middle row). The constricted segments colocalized with α-smooth muscle actin (α-SMA) immunoreactive pericytes (bottom row). IF denotes immunofluorescence. Scale bar for upper and middle row, 20 μm; bottom row 10 μm (Reproduced from Yemisci et al. 2009 with permission)

Pericytes Are Vulnerable to Ischemic Injury

Pericyte contractility is regulated by intracellular Ca^{2+} concentrations (Kamouchi et al. 2004; Hamilton et al. 2010). The energy loss triggered by acute cerebral ischemia disrupts Ca^{2+} homeostasis and leads to an uncontrolled rise in intracellular Ca^{2+} in these metabolically dynamic cells (Hamilton et al. 2010). Calcium overload is likely to be potentiated by reactive oxygen species (ROS) (Kamouchi et al. 2007; Nakamura et al. 2009) coming from multiple sources during ischemia-reperfusion, including mitochondria in pericytes, astrocyte endfeet and endothelia (Gürsoy-Ozdemir et al. 2004, 2012) and, ROS generating enzymes on the microvascular wall. Pericytes express high quantities of a major superoxide-producing enzyme, nicotinamide adenine dinucleotide phosphate (NADPH) oxidase 4 (NOX4) (Manea et al. 2005; Kuroda et al. 2014; Nishimura et al. 2016). This enzyme was shown to be upregulated in microvascular pericytes of the peri-infarct region in a MCAo stroke model, which contributed activation of metalloproteinase-9 and BBB breakdown (Vallet et al. 2005; Nishimura et al. 2016). Reactive oxygen and nitrogen species, and their reaction product, the potent oxidant peroxynitrite are intensely generated on the microvascular wall during ischemia and reperfusion, and constrict microvessels by contracting pericytes (Chan 1996; Yemisci et al. 2009; Gursoy-Ozdemir et al. 2012). Importantly, pericytes on microvessels remain contracted after recanalization of the occluded artery; therefore, the microcirculatory flow cannot completely be restored. The impaired reperfusion despite recanalization, known as the 'no-reflow' phenomenon, negatively affects post-stroke tissue survival (Hallenbeck et al. 1986; del Zoppo et al. 1991a; Yemisci et al. 2009; del Zoppo et al. 2011; Dziennis et al. 2015) (Figs. 9.3 and 9.4). Therefore, the experimental evidence still warrants pursuit of this goal (Diener et al. 2008; Amaro and Chamorro 2011; Gursoy-Ozdemir et al. 2012; Taskiran-Sag et al. 2018) despite failure of an antioxidant agent in clinical trials (Diener et al. 2008).

Incomplete Microcirculatory Reflow After Recanalization

An impaired tissue reperfusion after recanalization of an occluded artery or restoration of blood flow following circulatory collapse was first noted more than half a century ago and named as no-reflow phenomenon (Ames et al. 1968; Crowell and Olsson 1972). Unfortunately, some later studies measuring capillary patency with serum flow claimed that all capillaries were reperfused after restoration of blood flow (Theilen et al. 1993; Li et al. 1998). Recent studies with modern imaging techniques recording from intact mice brain clearly illustrated that fluorescently labeled serum continued to flow at the periphery of clogged capillaries (though slowly), creating the illusion that capillaries remained patent when only serum was monitored (Yemisci et al. 2009; Hill et al. 2015). Fortunately, interest in no-reflow phenomenon was re-kindled with studies on post-ischemic microcirculatory failure caused by leukocytes, platelets, fibrin and, recently, by pericytes (Hallenbeck et al. 1986; del Zoppo et al. 1991a; 2011; Zhang et al. 1999; Yemisci et al. 2009; Hall et al. 2014).

The impaired reflow emerges as a function of the duration and severity of ischemia, which varies between brain regions. Ten to twenty minutes of global ischemia is sufficient to induce no-reflow. For focal ischemia, proximal MCA occlusion in the mouse induces nodal microvascular constrictions that generally do not recover after recanalization starting 1 h after ischemia and affecting half of the microvessels within 2 h (Yemisci et al. 2009; Hill et al. 2015) (Fig. 9.3). Capillary constrictions

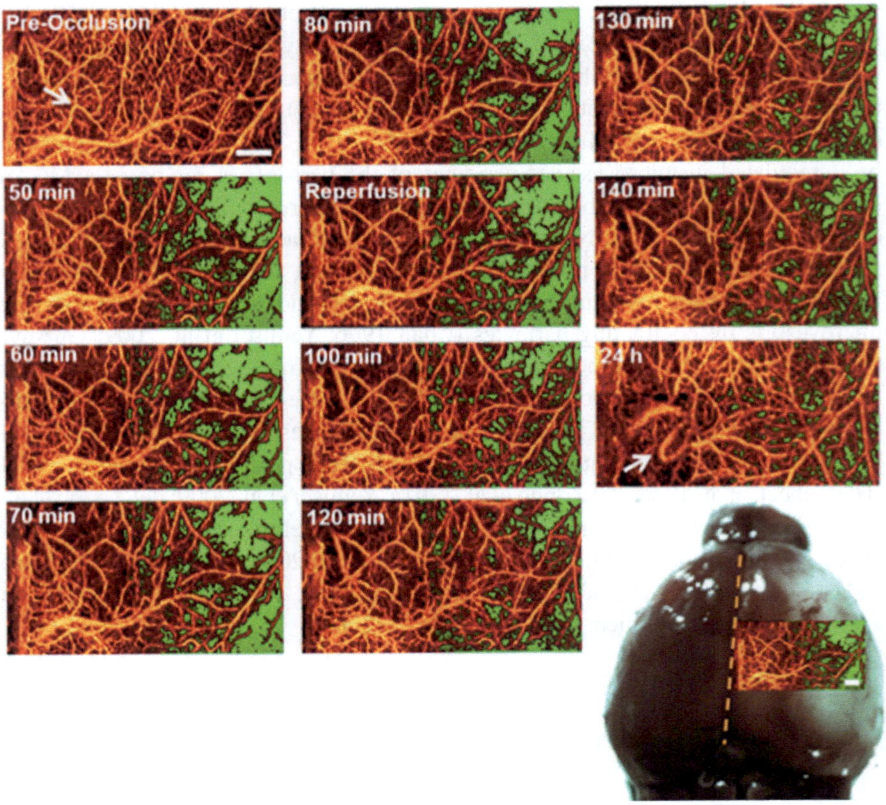

Fig. 9.4 Incomplete microcirculatory reflow after recanalization. Dynamic imaging of cortical blood flow using optical microangiography during 90-minute proximal MCA occlusion followed by recanalization illustrates the lack of microcirculatory blood flow in the MCA territory (the green area) during occlusion and its partial recovery after recanalization (incomplete microcirculatory reperfusion) in the mouse. Consecutive images are shown at 10-min intervals. Image size is $2.2 \times 4.4\ mm^2$. The image in the lower right is the optical microangiography image taken at 50 min overlaid on the 24 h infarct analysis by histological staining as the area of pallor. Scale bar = 500 μm (Reproduced from Dziennis et al. 2015 with permission)

emerge earlier when focal ischemia was induced by photothrombosis, perhaps due to an additional injury induced by this method (Underly et al. 2017). Microvessel lumina at the constricted segments are filled with entrapped erythrocytes (RBCs), leukocytes, and fibrin-platelet deposits (Little et al. 1976; Hallenbeck et al. 1986; Garcia et al. 1994; Zhang et al. 1999; Morris et al. 2000; Belayev et al. 2002). RBCs are the predominant cell type in aggregates possibly because they are the most prevalent cells in circulation. In addition to the constricted segments observed at the arteriolar end of microcirculation and capillaries, leukocytes adhered to postcapillary venules for entering to the parenchyma also induce luminal aggregates together with fibrin and platelets (Belayev et al. 2002; Zhang et al. 1999; del Zoppo et al. 1991b; Ritter et al. 2000).

Experimental data strongly suggest that incomplete restoration of the microcirculatory blood flow negatively impacts tissue recovery even if reopening of the occluded artery is achieved within the time window when there is still salvageable penumbral tissue (Yemisci et al. 2009; del Zoppo et al. 2011; McCabe et al. 2018). Recent clinical data from large prospective studies that examined recanalization as well as tissue reperfusion concluded that reperfusion was essential to achieve good functional recovery such that a satisfactory reperfusion was 4 times stronger in predicting the outcome than recanalization or collateral status (Eilaghi et al. 2013; Cho et al. 2015; Catanese et al. 2017). Interestingly, reperfusion provided through collaterals was also associated with good clinical outcome even when recanalization could not be attained (Makris et al. 2019). Anti-thrombotic agents and genetic manipulations reducing microvascular clogging by inhibiting leukocyte adherence, platelet activation, or fibrin-platelet interactions have been shown to restore microcirculation and improve stroke outcome in animal models (Hallenbeck et al. 1986; Mori et al. 1992; Choudhri et al. 1998; Belayev et al. 2002; Ishikawa et al. 2005). Current guidelines, however, do not recommend anti-thrombotic medication use in patients undergoing recanalization therapies because of increased risk of hemorrhage (Powers et al. 2018). Interestingly, adenosine-squalene nanoparticles have been shown to improve microcirculation by relaxing contracted pericytes during ischemia in mouse stroke models (Gaudin et al. 2014) (Fig. 9.5). Radiolabeling studies disclosed that adenosine nanoparticles did not enter the brain parenchyma but provided neuroprotection by improving microcirculation with slowly released adenosine in endothelia. Importantly, the neuroprotection was also obtained with other BBB-impermeable agents such as L-N^5-(1-iminoethyl)-ornithine (L-NIO) and 2-sulfo-phenyl-N-tert-butyl nitrone (S-PBN), strongly supporting the idea that restoring microvascular patency can alone improve stroke outcome independently of parenchymal mechanisms (Yemisci et al. 2009; Gaudin et al. 2014). A recent study by simultaneously imaging ROS formation in the parenchyma and vasculature, demonstrated that S-PBN, a BBB-impermeable analog of the ROS scavenger PBN provided neuroprotection by improving microcirculatory reperfusion and then secondarily reducing parenchymal ROS formation without entering parenchyma (Taskiran-Sag et al. 2018). Consequently, restitution of the microcirculatory reperfusion emerges as an exciting target to improve the success rate of recanalization (Dalkara and Arsava 2012) and neuroprotection therapies (Gursoy-Ozdemir et al. 2012).

In the past, ischemia-induced capillary constrictions were thought to be caused by swollen astrocyte endfeet around microvessels (Little et al. 1976; Garcia et al. 1994). However, this idea is hard to reconcile with the nodal character of constrictions because the endfeet homogenously encircle capillaries, hence, should lead to an even narrowing of the lumen. The pericyte contraction-induced segmental constrictions fit better with these observations as pericytes are intermittently spaced along the microvessels (Yemisci et al. 2009; Dore-Duffy and Cleary 2011b; Hall et al. 2014; Alarcon-Martinez et al. 2018) (Figs. 9.2b and 9.3). Nomenclature disagreements in naming capillary mural cells notwithstanding, the important point for the stroke pathophysiology is that contractile cells on brain microvessels impede

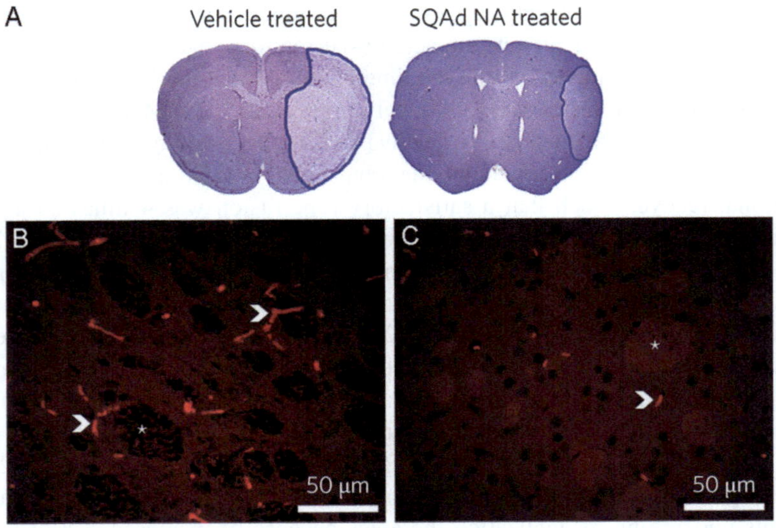

Fig. 9.5 Systemic administration of squalenoyl-adenosine (SQAd) nano-assemblies (NAs) provides significant neuroprotection in a mouse model of focal cerebral ischemia. (**a**) Infarct areas in control and treated mice subjected to transient (2 h MCAo and 22 h reperfusion) focal cerebral ischemia were identified by reduced Nissl staining under a light microscope (magnification ×10). (**b**) In untreated mice, capillaries in the ischemic brain were filled with trapped erythrocytes, whose hemoglobin was rendered fluorescent by treating brain sections with NaBH4 (**b**, red, arrowheads) 6 h after re-opening of the MCA following 2 h of occlusion, whereas the majority of capillaries were not clogged in SQAd nano-assemblies-treated mice (**c**). Unlike adenosine infusion, slow release from squalenoyladenosine nanoparticles did not cause cardiotoxicity or hypotension in the mouse model used. (Reproduced from Gaudin et al. 2014 with permission)

reperfusion after ischemia and unfavorably impact the outcome of recanalization. It should be noted that even small decreases in capillary radius caused by subtle pericyte contractions can lead to erythrocyte entrapments because capillary luminal size hardly allows passage of RBCs (Yemisci et al. 2009; Hamilton et al. 2010) (Fig. 9.3 middle row). Entrapped erythrocytes trigger platelet and fibrin aggregation by impeding passage of blood cells (Zhang et al. 1999; del Zoppo and Hamann 2011). The failure of erythrocyte circulation within some of the microvessels will lead to accelerated passage through the patent ones causing inefficient O_2 extraction (shunting). Modeling studies suggest that this increased heterogeneity of RBC transit times through patent capillaries (due to varying degrees of capillary resistances) can catastrophically reduce O_2 delivery to the tissue struggling to recover from ischemia-induced perturbations (Jespersen and Østergaard 2012). Since the plasma flow in constricted capillaries is relatively less restricted compared to RBC flux, glucose supply to some parts of the tissue may exceed O_2 supply and stimulate anaerobic glycolysis, hence, lactic acidosis (Yemisci et al. 2009; Dalkara et al. 2011) (please see supplementary movies mmc 5–7 in Hill et al. 2015). Current MR or CT techniques measure perfusion by detecting passage of contrast agent but not RBCs through the ischemic tissue, therefore, might be overestimating tissue oxygenation. However, since plasma flow is related to capillary resistance, perfusion parameters based on the transit time may still be used to assess the microcirculatory reperfusion and its disturbances (Engedal et al. 2017). Distal embolization and re-occlusion are not uncommonly encountered during thrombolysis or endovascular therapies (Alexandrov and Grotta 2002; Janjua et al. 2008); however, microcirculatory failure appears as an independent factor than flow reduction due to occlusion at proximal sites and predicts tissue to be infarcted in recanalized as well as non-recanalized patients (Engedal et al. 2017). In line with these findings, our group recently showed that the presence of microcirculatory failure distal to the thrombus prior to attempting recanalization is an unfavorable prognostic factor for a satisfactory reperfusion and clinical outcome in acute ischemic stroke patients treated with clot retrievers (Arsava et al. 2018).

Pericytes and Post-Stroke BBB Leakiness

The BBB is fundamental for normal functioning of the CNS. The sealed endothelial cells by tight junction proteins, astrocyte endfeet, and extracellular matrix form the main physical barrier between the blood and CNS parenchyma. A close communication between the pericytes and endothelia as well as astrocytes is required for development and functioning of the BBB (Armulik et al. 2010; Daneman et al. 2010). Pericytes regulate the expression of tight junction proteins and inhibit transendothelial vesicular transport and immune cell extravasation into CNS (Armulik et al. 2010; Daneman et al. 2010; Sweeney et al. 2016). Thereby, pericytes play a critical role in vascular stability at the microcirculatory level such that the number of pericytes per endothelial cell and the surface area of the vascular wall covered by pericytes determine the relative permeability of capillaries (Winkler et al. 2011b; Armulik et al. 2011). Accordingly, pericyte dysfunction as well as deficiency causes increased BBB permeability (Armulik et al. 2010, 2011; Daneman et al. 2010; Winkler et al. 2011a)

Injury to pericytes during acute ischemia contributes to BBB breakdown, hence brain edema in the ischemic territory in addition to impairing microcirculation (Simard et al. 2007; Underly et al. 2017). Death of the damaged pericytes may further aggravate BBB breakdown at later hours along with other factors such as MMP activation (Hall et al. 2014; Underly et al. 2017; Neuhaus et al. 2017). However, in the peri-infarct areas, pericytes were shown to migrate from microvessels within 1 h following ischemia. This migration may be protective by providing guidance for peri-infarct angiogenesis, but also be detrimental as it could increase microvascular permeability by disrupting the interaction of pericytes and tight junctions (Kamouchi et al. 2011; Liu et al. 2012). In the long run, however, post-stroke angiogenesis and neurogenesis in peri-infarct area plays an important role in stroke outcome (Wang et al. 2004; Ergul et al. 2012; Zhang et al. 2012; Cai et al. 2017).

Diabetic patients are prone to cerebral hemorrhage. This could be due to dysfunctional microvascular pericytes, as suggested for diabetic retinopathy manifested by retinal edema and hemorrhage (Wardlaw et al. 2009; Willard and Herman 2012; Desilles et al. 2013). Increased BBB permeability predisposes to intraparenchymal hemorrhage in about 5–6% of patients receiving recanalization treatments (Donnan et al. 2011). Diabetes is considered a risk factor for thrombolysis-related hemorrhage and adversely influences post-stroke recovery. These negative effects of diabetes are considered to be a reflection of microvascular dysfunction, and pericytes might play a central role in this. Most of the knowledge about diabetes and pericyte dysfunction comes from the observations in diabetic retinopathy (Ergul et al. 2014). The loss of pericyte coverage around retinal endothelial cells in diabetes has been shown to trigger pathological angiogenesis, endothelial cell apoptosis, and plasma leakage (Prakash et al. 2013; Ergul et al. 2014, 2015). Although the effects of diabetes on brain pericytes are not fully known, a decreased pericyte density has been reported within the cerebral microcirculation as well (Prakash et al. 2012). Experimental stroke in diabetic animals has led to an increase in hemorrhagic transformation after ischemia in diabetic mice (Ergul et al. 2007) and impairment in vascular repair mechanisms critical for neovascularization and angiogenesis (Prakash et al. 2013).

Interestingly, pericyte loss is increasingly reported for conditions that are risk factors for stroke, such as aging, hypertension as well as diabetes, the impact of which on stroke outcome needs to be clarified with future research (Østergaard et al. 2016). Cerebral pericytes in hypertensive animals show irregular profiles, associated with fragmentation of their processes and thickening in their basement membranes (Suzuki et al. 2003). These changes are reportedly led to decreased endothelial coverage by pericytes, capillary thrombotic occlusion, and luminal collapse (Suzuki et al. 2003). Capillary dysfunction induced by the above cerebrovascular disease risk factors has also been proposed to contribute to the risk of subsequent stroke and cognitive decline (Østergaard et al. 2016).

Post-Stroke Angioneurogenesis and Pericytes

Pericytes are essential, especially, for the early phase of neovascularization (angiogenic sprouting) (Ozerdem and Stallcup 2003; Gerhardt and Betsholtz 2003). Pericytes and endothelial cells communicate with each other for regulation of angiogenesis (Ozerdem and Stallcup 2003; Gerhardt and Betsholtz 2003; Díaz-Flores et al. 2009) (Fig. 9.6). Platelet-derived growth factor-β (PDGFβ), transforming growth factor-β (TGFβ), notch, angiopoietin and sphingosine-1-phosphate signaling, and the vascular endothelial grow factor and its receptor-2 (VEGF/VEGFR2) mediate this crosstalk (Gaengel et al. 2009; Armulik et al. 2011). Through those signaling pathways, pericyte may drive angiogenesis after stroke (Kokovay et al. 2006; Dore-Duffy et al. 2007; Beck and Plate 2009; Ergul et al. 2012; Zechariah et al. 2013a, b; Cai et al. 2017). First, endothelial cells start to proliferate and give off vessel sprouts 12–24 h after brain ischemia, leading to formation of new vessels in the peri-infarct region 3 days after ischemic injury

(Hayashi et al. 2003; Chopp et al. 2007; Beck and Plate 2009). Following a similar time course, the PDGFRβ expression is upregulated in pericytes, which increase in number and start migrating from the microvessel wall to the newly formed vessel sprouts to foster their maturation after ischemic injury (Takahashi et al. 1997; Dore-Duffy et al. 2000; Marti et al. 2000; Renner et al. 2003; Arimura et al. 2012; Dulmovits and Herman 2012). Renner et al. found that PDGFRβ increased in pericytes 48 hours after permanent ischemia (Renner et al. 2003). This upregulation of PDGFRβ in pericytes is proposed to be promoted by ischemia-induced increase in the basic fibroblast growth factor (bFGF) (Nakamura et al. 2016). Similarly, NG2+ or PDGFRβ+ pericytes are reportedly increased in peri-infarct areas 1–3 weeks after transient MCA occlusion (Fernández-Klett et al. 2013; Yang et al. 2013). A proportion of locally proliferating pericytes give rise to microglial cells (Özen et al. 2014). Corroborating these studies, conditional knockout of PDGFβ/PDGFRβ signaling in adult mice that have normally developed brain vasculature led to larger infarcts than controls when subjected to focal cerebral ischemia (Shen et al. 2012). Similarly, Zechariah et al. showed that pericytes did not appropriately cover the brain

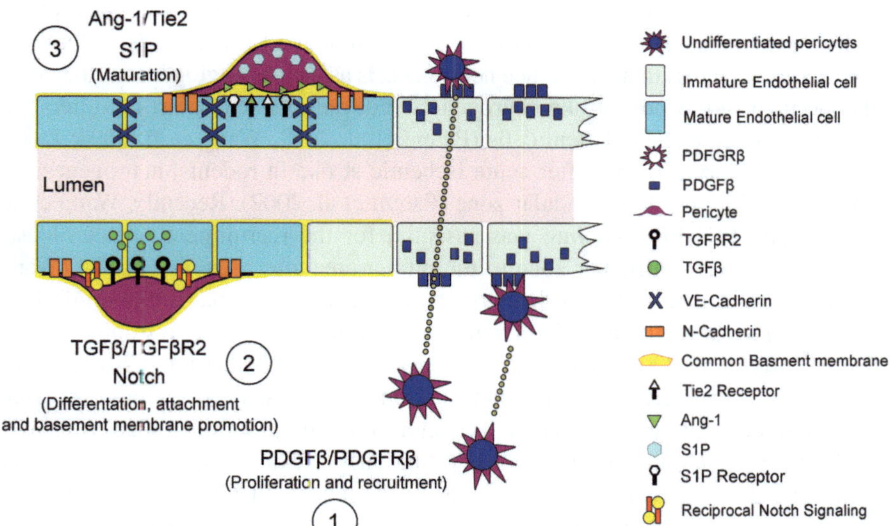

Fig. 9.6 Role of pericytes in angiogenesis. The interaction between PDGFβ secreted by the endothelium and its receptor localized on pericytes (PDGFRβ) is essential for recruitment of undifferentiated mesenchymal cells/pericytes to newly formed vessels. Once pericytes are at the vascular wall, reciprocal Notch signaling between the endothelia and pericytes as well as interactions between TGFβ secreted by endothelial cells and its receptor TGFβR2 located at pericytes differentiate mural cells and attach them to the newly formed vessels. The TGFβ/TGFβR2 interaction also promotes formation of the common basement membrane and stabilizes newly formed vessels by inhibiting endothelial proliferation. Ang-1, which is secreted by pericytes, activates its endothelial receptor Tie2 and promotes blood–brain barrier formation. Finally, S1P, whose receptor is abundantly expressed on pericytes down regulates genes related to vascular permeability and promotes both endothelial-endothelial (VE-cadherin) and pericyte-endothelial cell (N-cadherin) interconnections (Reproduced from Dalkara and Alarcon-Martinez 2015 with permission)

capillaries in hyperlipidemic mice exposed to ischemia, and this was associated with attenuation of post-stroke angiogenesis (Zechariah et al. 2013b). Moreover, after ischemic stroke, brain pericytes may also start to express other angiogenic signaling mediators. For instance, during hypoxia, pericytes begin to secrete VEGF (Dore-Duffy et al. 2007), which activates VEGFR2 in endothelial cells (Greenberg and Jin 2005; Beck and Plate 2009) and promote neovascularization in the peri-infarct region (Marti et al. 2000). Additionally, TGFβ, which is expressed in endothelial cells and pericytes (Gaengel et al. 2009), increased significantly in capillaries of the ischemic areas (Haqqani et al. 2005). Finally, following an ischemic injury, an upregulation of the angiopoietin and sphingosine-1-phosphate signaling pathway in the peri-infarct capillaries has been reported (Lin et al. 2000; Zhang and Chopp 2002).

Kokovay et al. showed that, following brain ischemia, bone marrow-derived cells with a pericytic phenotype and expressing angiogenic factors as VEGF and TGF-b were recruited to cerebral capillaries (Kokovay et al. 2006). Angiogenesis is also essential to promote neurogenesis after stroke (Palmer et al. 2000; Kamouchi et al. 2012; Nih et al. 2012). In fact, newly formed neurons are located near to the remodeled vessels (Okano et al. 2007), probably because vascular cells recruit and form a niche for neural stem cells (Palmer et al. 2000; Licht and Keshet 2015). Since pericytes express factors that can induce neurogenesis as well as angiogenesis, they may also be involved in post-stroke neurogenesis not only as neuroblast recruiters but also as a source of neural stem cells (Parent et al. 2002; Wang et al. 2004; Dore-Duffy and Cleary 2011b). After acute ischemic stroke in rodents, neurogenesis is activated within the subventricular zone (Parent et al. 2002). Recently, Wang et al. showed that PDGFR signaling was essential for the recruitment of neuroblasts formed at the subventricular zone to the infarct area after ischemic stroke (Wang et al. 2017). In line with this, in vitro studies have shown that the brain-derived pericytes have potential to differentiate into neurons in response to trophic factors (Dore-Duffy et al. 2006; Paul et al. 2012; Karow et al. 2012; Karow 2013). It has been reported that pericytes extracted from ischemic mouse brain regions expressed various stem cell markers or essential factors for reprogramming such as c-myc, Klf4, and Sox2 (Nakagomi et al. 2015). Similarly, Nakata et al. found that, after transient brain ischemia/reperfusion injury in the mouse, PDGFRβ+ pericytes were located within injured areas and commenced to expressed neural stem cell markers as nestin and immature neuronal markers as doublecortin (Nakata et al. 2017). In accordance with this, culture experiments showed that human brain pericytes under oxygen/glucose deprivation expressed not only pericyte markers as PDGFRb, NG2, or α-SMA but also Sox2 or Klf4 (Nakagomi et al. 2015). After examining post-stroke human brain tissue, Tatebayashi et al. also found the presence of nestin+ cells localized near blood vessels and co-expressing the pericytic markers α-SMA and NG2 (Tatebayashi et al. 2017). Finally, pericytes obtained from ischemic MCA tissue of adult animals or pericytes cultured under ischemic conditions also showed capability to differentiate to cells of neural as well as vascular lineage (Nakagomi et al. 2015).

Role of Pericytes in CADASIL

Cerebral autosomal-dominant arteriopathy with subcortical infarcts and leukoencephalopathy (CADASIL) is caused by mutations in the *NOTCH3* gene (Joutel et al. 1996). The protein encoded by the *NOTCH3* gene is expressed in pericytes and vascular SMCs. Studies in Notch3 transgenic mice expressing one of the human mutations showed that Notch3 aggregated around microvascular pericytes, leading to pericyte loss or reduced coverage of capillaries (Gu et al. 2012; Ghosh et al. 2015). These changes were associated with decrease in BBB impermeability and neurovascular dysfunction. In line with these findings, pericyte loss was also detected in skin and muscle biopsies of CADASIL patients (Dziewulska and Lewandowska 2012).

Future Trends and Directions

Recent research has clearly documented the important role of pericytes on microvascular physiology, especially in the brain and retina. Significant clues to the roles played by pericytes under several pathological conditions such as stroke, diabetic retinopathy, Alzheimer's disease, CADASIL have also been identified, creating novel targets for neuroprotection and restoring microvascular health. Expectedly, we will learn more about the place of pericytes in CNS pathologies and devise approaches to treat them in the next decades. It seems that it will be an exciting time for researchers interested in pericytes and microvasculature in health and disease.

Acknowledgments Dr. Turgay Dalkara's research is supported by The Turkish Academy of Sciences. Dr. Luis Alarcon-Martinez prepared Figs. 9.1a and 9.6.

References

Abbott, N. J., Ronnback, L., & Hansson, E. (2006). Astrocyte-endothelial interactions at the blood-brain barrier. *Nature Reviews. Neuroscience, 7*, 41–53.

Alarcon-Martinez, L., Yilmaz-Ozcan, S., Yemisci, M., Schallek, J., Kılıç, K., Can, A., Di Polo, A., & Dalkara, T. (2018). Capillary pericytes express α-smooth muscle actin, which requires prevention of filamentous-actin depolymerization for detection. *eLife, 7*, e34861. https://doi.org/10.7554/eLife.34861.

Alexandrov, A. V., & Grotta, J. C. (2002). Arterial reocclusion in stroke patients treated with intravenous tissue plasminogen activator. *Neurology, 59*, 862–867.

Allt, G., & Lawrenson, J. G. (2001). Pericytes: Cell biology and pathology. *Cells Tissues Organs, 169*, 1–11. https://doi.org/10.1159/000047855.

Amaro, S., & Chamorro, Á. (2011). Translational stroke research of the combination of thrombolysis and antioxidant therapy. *Stroke, 42*, 1495–1499. https://doi.org/10.1161/STROKEAHA.111.615039.

Ames, A., 3rd, Wright, R. L., Kowada, M., Thurston, J. M., & Majno, G. (1968). Cerebral ischemia. II. The no-reflow phenomenon. *The American Journal of Pathology, 52*, 437–453.

Arimura, K., Ago, T., Kamouchi, M., Nakamura, K., Ishitsuka, K., Kuroda, J., Sugimori, H., Ooboshi, H., Sasaki, T., & Kitazono, T. (2012). PDGF receptor β signaling in pericytes following ischemic brain injury. *Current Neurovascular Research, 9*, 1–9.

Armulik, A., Genové, G., & Betsholtz, C. (2011). Pericytes: Developmental, physiological, and pathological perspectives, problems, and promises. *Developmental Cell, 21*, 193–215. https://doi.org/10.1016/j.devcel.2011.07.001.

Armulik, A., Genove, G., Mae, M., Nisancioglu, M. H., Wallgard, E., Niaudet, C., He, L., Norlin, J., Lindblom, P., Strittmatter, K., Johansson, B. R., & Betsholtz, C. (2010). Pericytes regulate the blood-brain barrier. *Nature, 468*, 557–561.. nature09522 [pii]. https://doi.org/10.1038/nature09522.

Arsava, E. M., Arat, A., Topcuoglu, M. A., Peker, A., Yemisci, M., & Dalkara, T. (2018). Angiographic microcirculatory obstructions distal to occlusion signify poor outcome after endovascular treatment for acute ischemic stroke. *Translational Stroke Research, 9*, 44–50. https://doi.org/10.1007/s12975-017-0562-2.

Attwell, D., Buchan, A. M., Charpak, S., Lauritzen, M., Macvicar, B. A., & Newman, E. A. (2010). Glial and neuronal control of brain blood flow. *Nature, 468*, 232–243. https://doi.org/10.1038/nature09613.

Attwell, D., Mishra, A., Hall, C. N., O'Farrell, F. M., & Dalkara, T. (2016). What is a pericyte? *Journal of Cerebral Blood Flow and Metabolism, 36*, 451–455. https://doi.org/10.1177/0271678X15610340.

Bai, Y., Zhang, Y., Han, B., Yang, L., Chen, X., Huang, R., Wu, F., Chao, J., Liu, P., Hu, G., Zhang, J. H., & Yao, H. (2018). Circular RNA DLGAP4 ameliorates ischemic stroke outcomes by targeting miR-143 to regulate endothelial-mesenchymal transition associated with blood-brain barrier integrity. *The Journal of Neuroscience, 38*, 32–50. https://doi.org/10.1523/JNEUROSCI.1348-17.2017.

Bandopadhyay, R., Orte, C., Lawrenson, J. G., Reid, A. R., De Silva, S., & Allt, G. (2001). Contractile proteins in pericytes at the blood-brain and blood-retinal barriers. *Journal of Neurocytology, 30*, 35–44.

Beck, H., & Plate, K. H. (2009). Angiogenesis after cerebral ischemia. *Acta Neuropathologica, 117*, 481–496. https://doi.org/10.1007/s00401-009-0483-6.

Belayev, L., Pinard, E., Nallet, H., Seylaz, J., Liu, Y., Riyamongkol, P., Zhao, W., Busto, R., & Ginsberg, M. D. (2002). Albumin therapy of transient focal cerebral ischemia: In vivo analysis of dynamic microvascular responses. *Stroke, 33*, 1077–1084.

Berthiaume, A.-A., Grant, R. I., McDowell, K. P., Underly, R. G., Hartmann, D. A., Levy, M., Bhat, N. R., & Shih, A. Y. (2018). Dynamic remodeling of pericytes in vivo maintains capillary coverage in the adult mouse brain. *Cell Reports, 22*, 8–16. https://doi.org/10.1016/j.celrep.2017.12.016.

Biesecker, K. R., Srienc, A. I., Shimoda, A. M., Agarwal, A., Bergles, D. E., Kofuji, P., & Newman, E. A. (2016). Glial cell calcium signaling mediates capillary regulation of blood flow in the retina. *The Journal of Neuroscience, 36*, 9435–9445. https://doi.org/10.1523/JNEUROSCI.1782-16.2016.

Cai, W., Liu, H., Zhao, J., Chen, L. Y., Chen, J., Lu, Z., & Hu, X. (2017). Pericytes in brain injury and repair after ischemic stroke. *Translational Stroke Research, 8*, 107–121. https://doi.org/10.1007/s12975-016-0504-4.

Catanese, L., Tarsia, J., & Fisher, M. (2017). Acute ischemic stroke therapy overview. *Circulation Research, 120*, 541–558. https://doi.org/10.1161/CIRCRESAHA.116.309278.

Chan, P. H. (1996). Role of oxidants in ischemic brain damage. *Stroke, 27*, 1124–1129.

Cho, T.-H., Nighoghossian, N., Mikkelsen, I. K., Derex, L., Hermier, M., Pedraza, S., Fiehler, J., Østergaard, L., Berthezène, Y., & Baron, J.-C. (2015). Reperfusion within 6 hours outperforms recanalization in predicting penumbra salvage, lesion growth, final infarct, and clinical outcome. *Stroke, 46*, 1582–1589. https://doi.org/10.1161/STROKEAHA.114.007964.

Chopp, M., Zhang, Z. G., & Jiang, Q. (2007). Neurogenesis, angiogenesis, and MRI indices of functional recovery from stroke. *Stroke, 38*, 827–831. https://doi.org/10.1161/01.STR.0000250235.80253.e9.

Choudhri, T. F., Hoh, B. L., Zerwes, H. G., Prestigiacomo, C. J., Kim, S. C., Connolly, J. E., Kottirsch, G., & Pinsky, D. J. (1998). Reduced microvascular thrombosis and improved outcome in acute murine stroke by inhibiting GP IIb/IIIa receptor-mediated platelet aggregation. *The Journal of Clinical Investigation, 102*, 1301–1310. https://doi.org/10.1172/JCI3338.

Crowell, R. M., & Olsson, Y. (1972). Impaired microvascular filling after focal cerebral ischemia in the monkey. Modification by treatment. *Neurology, 22*, 500–504.

Dalkara, T., & Alarcon-Martinez, L. (2015). Cerebral microvascular pericytes and neurogliovascular signaling in health and disease. *Brain Research, 1623*, 3–17. https://doi.org/10.1016/j.brainres.2015.03.047.

Dalkara, T., & Arsava, E. M. (2012). Can restoring incomplete microcirculatory reperfusion improve stroke outcome after thrombolysis? *Journal of Cerebral Blood Flow and Metabolism, 32*, 2091–2099. https://doi.org/10.1038/jcbfm.2012.139.

Dalkara, T., Gursoy-Ozdemir, Y., & Yemisci, M. (2011). Brain microvascular pericytes in health and disease. *Acta Neuropathologica, 122*, 1–9. https://doi.org/10.1007/s00401-011-0847-6.

Daneman, R., Zhou, L., Kebede, A. A., & Barres, B. A. (2010). Pericytes are required for blood-brain barrier integrity during embryogenesis. *Nature, 468*, 562–566. https://doi.org/10.1038/nature09513.

Das, A., Frank, R. N., Weber, M. L., Kennedy, A., Reidy, C. A., & Mancini, M. A. (1988). ATP causes retinal pericytes to contract in vitro. *Experimental Eye Research, 46*, 349–362.

del Zoppo, G. J., & Hamann, G. F. (2011). The cerebral microvasculature and responses to ischemia. In J. P. Mohr, P. A. Wolf, J. C. Grotta, M. A. Moskowitz, M. R. Mayberg, & R. Von Kummer (Eds.), *Stroke: Pathophysiology, diagnosis and management* (5th ed., pp. 16–28). Philadelphia: Elsevier.

del Zoppo, G. J., Schmid-Schönbein, G. W., Mori, E., Copeland, B. R., & Chang, C. M. (1991a). Polymorphonuclear leukocytes occlude capillaries following middle cerebral artery occlusion and reperfusion in baboons. *Stroke, 22*, 1276–1283.

del Zoppo, G. J., Schmid-Schonbein, G. W., Mori, E., Copeland, B. R., & Chang, C. M. (1991b). Polymorphonuclear leukocytes occlude capillaries following middle cerebral artery occlusion and reperfusion in baboons. *Stroke, 22*, 1276–1283.

del Zoppo, G. J., Sharp, F. R., Heiss, W.-D., & Albers, G. W. (2011). Heterogeneity in the penumbra. *Journal of Cerebral Blood Flow and Metabolism, 31*, 1836–1851. https://doi.org/10.1038/jcbfm.2011.93.

Desilles, J.-P., Meseguer, E., Labreuche, J., Lapergue, B., Sirimarco, G., Gonzalez-Valcarcel, J., Lavallée, P., Cabrejo, L., Guidoux, C., Klein, I., Amarenco, P., & Mazighi, M. (2013). Diabetes mellitus, admission glucose, and outcomes after stroke thrombolysis: A registry and systematic review. *Stroke, 44*, 1915–1923. https://doi.org/10.1161/STROKEAHA.111.000813.

Dias Moura Prazeres, P. H., Sena, I. F. G., Borges, I. D. T., de Azevedo, P. O., Andreotti, J. P., de Paiva, A. E., de Almeida, V. M., de Paula Guerra, D. A., Pinheiro Dos Santos, G. S., Mintz, A., Delbono, O., & Birbrair, A. (2017). Pericytes are heterogeneous in their origin within the same tissue. *Developmental Biology, 427*, 6–11. https://doi.org/10.1016/j.ydbio.2017.05.001.

Díaz-Flores, L., Gutiérrez, R., Madrid, J. F., Varela, H., Valladares, F., Acosta, E., Martín-Vasallo, P., & Díaz-Flores, L. (2009). Pericytes. Morphofunction, interactions and pathology in a quiescent and activated mesenchymal cell niche. *Histology and Histopathology, 24*, 909–969.

Diener, H.-C., Lees, K. R., Lyden, P., Grotta, J., Davalos, A., Davis, S. M., Shuaib, A., Ashwood, T., Wasiewski, W., Alderfer, V., Hårdemark, H.-G., Rodichok, L., & SAINT I and II Investigators. (2008). NXY-059 for the treatment of acute stroke: Pooled analysis of the SAINT I and II trials. *Stroke, 39*, 1751–1758. https://doi.org/10.1161/STROKEAHA.107.503334.

Dohgu, S., & Banks, W. A. (2013). Brain pericytes increase the lipopolysaccharide-enhanced transcytosis of HIV-1 free virus across the in vitro blood-brain barrier: Evidence for cytokine-mediated pericyte-endothelial cell crosstalk. *Fluids Barriers CNS, 10*, 23. https://doi.org/10.1186/2045-8118-10-23.

Donnan, G. A., Davis, S. M., Parsons, M. W., Ma, H., Dewey, H. M., & Howells, D. W. (2011). How to make better use of thrombolytic therapy in acute ischemic stroke. *Nature Reviews. Neurology, 7*, 400–409. https://doi.org/10.1038/nrneurol.2011.89.

Dore-Duffy, P., & Cleary, K. (2011a). Morphology and properties of Pericytes. In S. Nag (Ed.), *The blood-brain and other neural barriers* (pp. 49–68). Totowa, NJ: Humana Press.

Dore-Duffy, P., & Cleary, K. (2011b). Morphology and properties of pericytes. *Methods in Molecular Biology, 686*, 49–68. https://doi.org/10.1007/978-1-60761-938-3_2.

Dore-Duffy, P., Katychev, A., Wang, X., & Van Buren, E. (2006). CNS microvascular pericytes exhibit multipotential stem cell activity. *Journal of Cerebral Blood Flow and Metabolism, 26*, 613–624.. 9600272 [pii]. https://doi.org/10.1038/sj.jcbfm.9600272.

Dore-Duffy, P., Owen, C., Balabanov, R., Murphy, S., Beaumont, T., & Rafols, J. A. (2000). Pericyte migration from the vascular wall in response to traumatic brain injury. *Microvascular Research, 60*, 55–69. https://doi.org/10.1006/mvre.2000.2244.

Dore-Duffy, P., Wang, X., Mehedi, A., Kreipke, C. W., & Rafols, J. A. (2007). Differential expression of capillary VEGF isoforms following traumatic brain injury. *Neurological Research, 29*, 395–403. https://doi.org/10.1179/016164107X204729.

Dulmovits, B. M., & Herman, I. M. (2012). Microvascular remodeling and wound healing: A role for pericytes. *The International Journal of Biochemistry & Cell Biology, 44*(11), 1800–1812. https://doi.org/10.1016/j.biocel.2012.06.031.

Dziennis, S., Qin, J., Shi, L., & Wang, R. K. (2015). Macro-to-micro cortical vascular imaging underlies regional differences in ischemic brain. *Scientific Reports, 5*, 10051. https://doi.org/10.1038/srep10051.

Dziewulska, D., & Lewandowska, E. (2012). Pericytes as a new target for pathological processes in CADASIL. *Neuropathology, 32*, 515–521. https://doi.org/10.1111/j.1440-1789.2011.01290.x.

Eilaghi, A., Brooks, J., d'Esterre, C., Zhang, L., Swartz, R. H., Lee, T.-Y., & Aviv, R. I. (2013). Reperfusion is a stronger predictor of good clinical outcome than recanalization in ischemic stroke. *Radiology, 269*, 240–248. https://doi.org/10.1148/radiol.13122327.

Engedal, T. S., Hjort, N., Hougaard, K. D., Simonsen, C. Z., Andersen, G., Mikkelsen, I. K., Boldsen, J. K., Eskildsen, S. F., Hansen, M. B., Angleys, H., Jespersen, S. N., Pedraza, S., Cho, T. H., Serena, J., Siemonsen, S., Thomalla, G., Nighoghossian, N., Fiehler, J., Mouridsen, K., & Østergaard, L. (2017). Transit time homogenization in ischemic stroke – A novel biomarker of penumbral microvascular failure? *Journal of Cerebral Blood Flow & Metabolism, 38*(11), 2006–2020. https://doi.org/10.1177/0271678X17721666.

Ergul, A., Abdelsaid, M., Fouda, A. Y., & Fagan, S. C. (2014). Cerebral neovascularization in diabetes: Implications for stroke recovery and beyond. *Journal of Cerebral Blood Flow and Metabolism, 34*, 553–563. https://doi.org/10.1038/jcbfm.2014.18.

Ergul, A., Alhusban, A., & Fagan, S. C. (2012). Angiogenesis: A harmonized target for recovery after stroke. *Stroke, 43*, 2270–2274. https://doi.org/10.1161/STROKEAHA.111.642710.

Ergul, A., Elgebaly, M. M., Middlemore, M.-L., Li, W., Elewa, H., Switzer, J. A., Hall, C., Kozak, A., & Fagan, S. C. (2007). Increased hemorrhagic transformation and altered infarct size and localization after experimental stroke in a rat model type 2 diabetes. *BMC Neurology, 7*, 33. https://doi.org/10.1186/1471-2377-7-33.

Ergul, A., Valenzuela, J. P., Fouda, A. Y., & Fagan, S. C. (2015). Cellular connections, microenvironment and brain angiogenesis in diabetes: Lost communication signals in the post-stroke period. *Brain Research, 1623*, 81–96. https://doi.org/10.1016/j.brainres.2015.02.045.

Fernandez-Klett, F., Offenhauser, N., Dirnagl, U., Priller, J., & Lindauer, U. (2010). Pericytes in capillaries are contractile in vivo, but arterioles mediate functional hyperemia in the mouse brain. *Proceedings of the National Academy of Sciences of the United States of America, 107*, 22290–22295.. 1011321108 [pii]. https://doi.org/10.1073/pnas.1011321108.

Fernández-Klett, F., Potas, J. R., Hilpert, D., Blazej, K., Radke, J., Huck, J., Engel, O., Stenzel, W., Genové, G., & Priller, J. (2013). Early loss of pericytes and perivascular stromal cell-induced scar formation after stroke. *Journal of Cerebral Blood Flow and Metabolism, 33*, 428–439. https://doi.org/10.1038/jcbfm.2012.187.

Fernández-Klett, F., & Priller, J. (2015). Diverse functions of pericytes in cerebral blood flow regulation and ischemia. *Journal of Cerebral Blood Flow and Metabolism, 35*, 883–887. https://doi.org/10.1038/jcbfm.2015.60.

Frank, R. N., Dutta, S., & Mancini, M. A. (1987). Pericyte coverage is greater in the retinal than in the cerebral capillaries of the rat. *Investigative Ophthalmology & Visual Science, 28*, 1086–1091.

Fujimoto, T., & Singer, S. J. (1987). Immunocytochemical studies of desmin and vimentin in pericapillary cells of chicken. *Journal of Histochemistry and Cytochemistry, 35*, 1105–1115.

Gaengel, K., Genové, G., Armulik, A., & Betsholtz, C. (2009). Endothelial-mural cell signaling in vascular development and angiogenesis. *Arteriosclerosis, Thrombosis, and Vascular Biology, 29*, 630–638. https://doi.org/10.1161/ATVBAHA.107.161521.

Garcia, J. H., Liu, K. F., Yoshida, Y., Chen, S., & Lian, J. (1994). Brain microvessels: Factors altering their patency after the occlusion of a middle cerebral artery (Wistar rat). *The American Journal of Pathology, 145*, 728–740.

Gaudin, A., Yemisci, M., Eroglu, H., Lepetre-Mouelhi, S., Turkoglu, O. F., Dönmez-Demir, B., Caban, S., Sargon, M. F., Garcia-Argote, S., Pieters, G., Loreau, O., Rousseau, B., Tagit, O., Hildebrandt, N., Le Dantec, Y., Mougin, J., Valetti, S., Chacun, H., Nicolas, V., Desmaële, D., Andrieux, K., Capan, Y., Dalkara, T., & Couvreur, P. (2014). Squalenoyl adenosine nanoparticles provide neuroprotection after stroke and spinal cord injury. *Nature Nanotechnology, 9*, 1054–1062. https://doi.org/10.1038/nnano.2014.274.

Gerhardt, H., & Betsholtz, C. (2003). Endothelial-pericyte interactions in angiogenesis. *Cell and Tissue Research, 314*, 15–23. https://doi.org/10.1007/s00441-003-0745-x.

Ghosh, M., Balbi, M., Hellal, F., Dichgans, M., Lindauer, U., & Plesnila, N. (2015). Pericytes are involved in the pathogenesis of CADASIL. *Annals of Neurology, 78*, 887. https://doi.org/10.1002/ana.24512.

Greenberg, D. A., & Jin, K. (2005). From angiogenesis to neuropathology. *Nature, 438*, 954–959. https://doi.org/10.1038/nature04481.

Gu, X., Liu, X.-Y., Fagan, A., Gonzalez-Toledo, M. E., & Zhao, L.-R. (2012). Ultrastructural changes in cerebral capillary pericytes in aged Notch3 mutant transgenic mice. *Ultrastructural Pathology, 36*, 48–55. https://doi.org/10.3109/01913123.2011.620220.

Gürsoy-Ozdemir, Y., Can, A., & Dalkara, T. (2004). Reperfusion-induced oxidative/nitrative injury to neurovascular unit after focal cerebral ischemia. *Stroke, 35*, 1449–1453. https://doi.org/10.1161/01.STR.0000126044.83777.f4.

Gursoy-Ozdemir, Y., Yemisci, M., & Dalkara, T. (2012). Microvascular protection is essential for successful neuroprotection in stroke. *Journal of Neurochemistry, 123*(Suppl 2), 2–11. https://doi.org/10.1111/j.1471-4159.2012.07938.x.

Hall, C. N., Reynell, C., Gesslein, B., Hamilton, N. B., Mishra, A., Sutherland, B. A., O'Farrell, F. M., Buchan, A. M., Lauritzen, M., & Attwell, D. (2014). Capillary pericytes regulate cerebral blood flow in health and disease. *Nature, 508*, 55–60. https://doi.org/10.1038/nature13165.

Hallenbeck, J. M., Dutka, A. J., Tanishima, T., Kochanek, P. M., Kumaroo, K. K., Thompson, C. B., Obrenovitch, T. P., & Contreras, T. J. (1986). Polymorphonuclear leukocyte accumulation in brain regions with low blood flow during the early postischemic period. *Stroke, 17*, 246–253.

Hamilton, N. B., Attwell, D., & Hall, C. N. (2010). Pericyte-mediated regulation of capillary diameter: a component of neurovascular coupling in health and disease. *Frontiers in Neuroenergetics, 2*, 5. https://doi.org/10.3389/fnene.2010.00005.

Haqqani, A. S., Nesic, M., Preston, E., Baumann, E., Kelly, J., & Stanimirovic, D. (2005). Characterization of vascular protein expression patterns in cerebral ischemia/reperfusion using laser capture microdissection and ICAT-nanoLC-MS/MS. *The FASEB Journal, 19*, 1809–1821. https://doi.org/10.1096/fj.05-3793com.

Hartmann, D. A., Underly, R. G., Grant, R. I., Watson, A. N., Lindner, V., & Shih, A. Y. (2015). Pericyte structure and distribution in the cerebral cortex revealed by high-resolution imaging of transgenic mice. *Neurophotonics, 2*, 41402. https://doi.org/10.1117/1.NPh.2.4.041402.

Hayashi, T., Noshita, N., Sugawara, T., & Chan, P. H. (2003). Temporal profile of angiogenesis and expression of related genes in the brain after ischemia. *Journal of Cerebral Blood Flow and Metabolism, 23*, 166–180.

Herman, I. M., & D'Amore, P. A. (1985). Microvascular pericytes contain muscle and nonmuscle actins. *The Journal of Cell Biology, 101*, 43–52.

Hill, R. A., Tong, L., Yuan, P., Murikinati, S., Gupta, S., & Grutzendler, J. (2015). Regional blood flow in the normal and ischemic brain is controlled by arteriolar smooth muscle cell contractility and not by capillary pericytes. *Neuron, 87*, 95–110. https://doi.org/10.1016/j.neuron.2015.06.001.

Iadecola, C. (2004). Neurovascular regulation in the normal brain and in Alzheimer's disease. *Nature Reviews. Neuroscience, 5*, 347–360. https://doi.org/10.1038/nrn1387. nrn1387 [pii].

Ishikawa, M., Vowinkel, T., Stokes, K. Y., Arumugam, T. V., Yilmaz, G., Nanda, A., & Granger, D. N. (2005). CD40/CD40 ligand signaling in mouse cerebral microvasculature after focal ischemia/reperfusion. *Circulation, 111*, 1690–1696. https://doi.org/10.1161/01.CIR.0000160349.42665.0C.

Janjua, N., Alkawi, A., Suri, M. F. K., & Qureshi, A. I. (2008). Impact of arterial reocclusion and distal fragmentation during thrombolysis among patients with acute ischemic stroke. *AJNR. American Journal of Neuroradiology, 29*, 253–258. https://doi.org/10.3174/ajnr.A0825.

Jespersen, S. N., & Østergaard, L. (2012). The roles of cerebral blood flow, capillary transit time heterogeneity, and oxygen tension in brain oxygenation and metabolism. *Journal of Cerebral Blood Flow and Metabolism, 32*, 264–277. https://doi.org/10.1038/jcbfm.2011.153.

Joutel, A., Corpechot, C., Ducros, A., Vahedi, K., Chabriat, H., Mouton, P., Alamowitch, S., Domenga, V., Cécillion, M., Marechal, E., Maciazek, J., Vayssiere, C., Cruaud, C., Cabanis, E. A., Ruchoux, M. M., Weissenbach, J., Bach, J. F., Bousser, M. G., & Tournier-Lasserve, E. (1996). Notch3 mutations in CADASIL, a hereditary adult-onset condition causing stroke and dementia. *Nature, 383*, 707–710. https://doi.org/10.1038/383707a0.

Joyce, N. C., Haire, M. F., & Palade, G. E. (1985a). Contractile proteins in pericytes. I. Immunoperoxidase localization of tropomyosin. *The Journal of Cell Biology, 100*, 1379–1386.

Joyce, N. C., Haire, M. F., & Palade, G. E. (1985b). Contractile proteins in pericytes. II. Immunocytochemical evidence for the presence of two isomyosins in graded concentrations. *The Journal of Cell Biology, 100*, 1387–1395.

Jung, B., Arnold, T. D., Raschperger, E., Gaengel, K., & Betsholtz, C. (2018). Visualization of vascular mural cells in developing brain using genetically labeled transgenic reporter mice. *Journal of Cerebral Blood Flow and Metabolism, 38*, 456–468. https://doi.org/10.1177/0271678X17697720.

Kamouchi, M., Ago, T., & Kitazono, T. (2011). Brain pericytes: Emerging concepts and functional roles in brain homeostasis. *Cellular and Molecular Neurobiology, 31*, 175–193. https://doi.org/10.1007/s10571-010-9605-x.

Kamouchi, M., Ago, T., Kuroda, J., & Kitazono, T. (2012). The possible roles of brain pericytes in brain ischemia and stroke. *Cellular and Molecular Neurobiology, 32*, 159–165. https://doi.org/10.1007/s10571-011-9747-5.

Kamouchi, M., Kitazono, T., Ago, T., Wakisaka, M., Kuroda, J., Nakamura, K., Hagiwara, N., Ooboshi, H., Ibayashi, S., & Iida, M. (2007). Hydrogen peroxide-induced Ca2+ responses in CNS pericytes. *Neuroscience Letters, 416*, 12–16. https://doi.org/10.1016/j.neulet.2007.01.039.

Kamouchi, M., Kitazono, T., Ago, T., Wakisaka, M., Ooboshi, H., Ibayashi, S., & Iida, M. (2004). Calcium influx pathways in rat CNS pericytes. *Brain Research. Molecular Brain Research, 126*, 114–120. https://doi.org/10.1016/j.molbrainres.2004.03.008.

Karow, M. (2013). Mountaineering pericytes--a universal key to tissue repair? *BioEssays: News and Reviews in Molecular, Cellular and Developmental Biology, 35*, 771–774. https://doi.org/10.1002/bies.201300055.

Karow, M., Sánchez, R., Schichor, C., Masserdotti, G., Ortega, F., Heinrich, C., Gascón, S., Khan, M. A., Lie, D. C., Dellavalle, A., Cossu, G., Goldbrunner, R., Götz, M., & Berninger, B. (2012). Reprogramming of pericyte-derived cells of the adult human brain into induced neuronal cells. *Cell Stem Cell, 11*, 471–476. https://doi.org/10.1016/j.stem.2012.07.007.

Kelley, C., D'Amore, P., Hechtman, H. B., & Shepro, D. (1987). Microvascular pericyte contractility in vitro: Comparison with other cells of the vascular wall. *The Journal of Cell Biology, 104*, 483–490.

Kisler, K., Nelson, A. R., Rege, S. V., Ramanathan, A., Wang, Y., Ahuja, A., Lazic, D., Tsai, P. S., Zhao, Z., Zhou, Y., Boas, D. A., Sakadžić, S., & Zlokovic, B. V. (2017). Pericyte degeneration leads to neurovascular uncoupling and limits oxygen supply to brain. *Nature Neuroscience, 20*, 406–416. https://doi.org/10.1038/nn.4489.

Kokovay, E., Li, L., & Cunningham, L. A. (2006). Angiogenic recruitment of pericytes from bone marrow after stroke. *Journal of Cerebral Blood Flow and Metabolism, 26*, 545–555. https://doi.org/10.1038/sj.jcbfm.9600214.

Krueger, M., & Bechmann, I. (2010). CNS pericytes: Concepts, misconceptions, and a way out. *Glia, 58*, 1–10. https://doi.org/10.1002/glia.20898.

Kuroda, J., Ago, T., Nishimura, A., Nakamura, K., Matsuo, R., Wakisaka, Y., Kamouchi, M., & Kitazono, T. (2014). Nox4 is a major source of superoxide production in human brain pericytes. *Journal of Vascular Research, 51*, 429–438. https://doi.org/10.1159/000369930.

Leick, M., Azcutia, V., Newton, G., & Luscinskas, F. W. (2014). Leukocyte recruitment in inflammation: Basic concepts and new mechanistic insights based on new models and microscopic imaging technologies. *Cell and Tissue Research, 355*, 647–656. https://doi.org/10.1007/s00441-014-1809-9.

Li, P. A., Vogel, J., Smith, M., He, Q. P., Kuschinsky, W., & Siesjö, B. K. (1998). Capillary patency after transient middle cerebral artery occlusion of 2 h duration. *Neuroscience Letters, 253*, 191–194.

Licht, T., & Keshet, E. (2015). The vascular niche in adult neurogenesis. *Mechanisms of Development, 138*, 56. https://doi.org/10.1016/j.mod.2015.06.001.

Lin, T. N., Wang, C. K., Cheung, W. M., & Hsu, C. Y. (2000). Induction of angiopoietin and tie receptor mRNA expression after cerebral ischemia-reperfusion. *Journal of Cerebral Blood Flow and Metabolism, 20*, 387–395. https://doi.org/10.1097/00004647-200002000-00021.

Little, J. R., Kerr, F. W. L., & Sundt, T. M., Jr. (1976). Microcirculatory obstruction in focal cerebral ischemia: An electron microscopic investigation in monkeys. *Stroke, 7*, 25–30.

Liu, S., Agalliu, D., Yu, C., & Fisher, M. (2012). The role of pericytes in blood-brain barrier function and stroke. *Current Pharmaceutical Design, 18*, 3653–3662.

Makris, N., Chamard, L., Mikkelsen, I. K., Hermier, M., Derex, L., Pedraza, S., Thomalla, G., Østergaard, L., Baron, J.-C., Nighoghossian, N., Berthezène, Y., & Cho, T.-H. (2019). Acute reperfusion without recanalization: Serial assessment of collaterals within 6 h of using perfusion-weighted magnetic resonance imaging. *Journal of Cerebral Blood Flow & Metabolism, 39*(2), 251–259. https://doi.org/10.1177/0271678X17744716.

Manea, A., Raicu, M., & Simionescu, M. (2005). Expression of functionally phagocyte-type NAD(P)H oxidase in pericytes: Effect of angiotensin II and high glucose. *Biology of the Cell, 97*, 723–734. https://doi.org/10.1042/BC20040107.

Marti, H. J., Bernaudin, M., Bellail, A., Schoch, H., Euler, M., Petit, E., & Risau, W. (2000). Hypoxia-induced vascular endothelial growth factor expression precedes neovascularization after cerebral ischemia. *The American Journal of Pathology, 156*, 965–976. https://doi.org/10.1016/S0002-9440(10)64964-4.

McCabe, C., Arroja, M. M., Reid, E., & Macrae, I. M. (2018). Animal models of ischaemic stroke and characterisation of the ischaemic penumbra. *Neuropharmacology, 134*, 169–177. https://doi.org/10.1016/j.neuropharm.2017.09.022.

Mishra, A., Reynolds, J. P., Chen, Y., Gourine, A. V., Rusakov, D. A., & Attwell, D. (2016). Astrocytes mediate neurovascular signaling to capillary pericytes but not to arterioles. *Nature Neuroscience, 19*, 1619–1627. https://doi.org/10.1038/nn.4428.

Mori, E., del Zoppo, G. J., Chambers, J. D., Copeland, B. R., & Arfors, K. E. (1992). Inhibition of polymorphonuclear leukocyte adherence suppresses no-reflow after focal cerebral ischemia in baboons. *Stroke, 23*, 712–718.

Morris, D. C., Davies, K., Zhang, Z., & Chopp, M. (2000). Measurement of cerebral microvessel diameters after embolic stroke in rat using quantitative laser scanning confocal microscopy. *Brain Research, 876*, 31–36. doi: S0006-8993(00)02543-9 [pii].

Nakagomi, T., Kubo, S., Nakano-Doi, A., Sakuma, R., Lu, S., Narita, A., Kawahara, M., Taguchi, A., & Matsuyama, T. (2015). Brain vascular pericytes following ischemia have multipotential stem cell activity to differentiate into neural and vascular lineage cells. *Stem Cells, 33*, 1962–1974. https://doi.org/10.1002/stem.1977.

Nakamura, K., Arimura, K., Nishimura, A., Tachibana, M., Yoshikawa, Y., Makihara, N., Wakisaka, Y., Kuroda, J., Kamouchi, M., Ooboshi, H., Kitazono, T., & Ago, T. (2016). Possible involvement of basic FGF in the upregulation of PDGFRβ in pericytes after ischemic stroke. *Brain Research, 1630*, 98–108. https://doi.org/10.1016/j.brainres.2015.11.003.

Nakamura, K., Kamouchi, M., Kitazono, T., Kuroda, J., Shono, Y., Hagiwara, N., Ago, T., Ooboshi, H., Ibayashi, S., & Iida, M. (2009). Amiloride inhibits hydrogen peroxide-induced Ca2+ responses in human CNS pericytes. *Microvascular Research, 77*, 327–334. https://doi.org/10.1016/j.mvr.2008.12.001.

Nakata, M., Nakagomi, T., Maeda, M., Nakano-Doi, A., Momota, Y., & Matsuyama, T. (2017). Induction of perivascular neural stem cells and possible contribution to neurogenesis following transient brain ischemia/reperfusion injury. *Translational Stroke Research, 8*, 131–143. https://doi.org/10.1007/s12975-016-0479-1.

Nehls, V., & Drenckhahn, D. (1991). Heterogeneity of microvascular pericytes for smooth muscle type alpha-actin. *The Journal of Cell Biology, 113*, 147–154.

Neuhaus, A. A., Couch, Y., Sutherland, B. A., & Buchan, A. M. (2017). Novel method to study pericyte contractility and responses to ischaemia in vitro using electrical impedance. *Journal of Cerebral Blood Flow and Metabolism, 37*, 2013–2024. https://doi.org/10.1177/0271678X16659495.

Nih, L. R., Deroide, N., Leré-Déan, C., Lerouet, D., Soustrat, M., Levy, B. I., Silvestre, J.-S., Merkulova-Rainon, T., Pocard, M., Margaill, I., & Kubis, N. (2012). Neuroblast survival depends on mature vascular network formation after mouse stroke: Role of endothelial and smooth muscle progenitor cell co-administration. *The European Journal of Neuroscience, 35*, 1208–1217. https://doi.org/10.1111/j.1460-9568.2012.08041.x.

Nishimura, A., Ago, T., Kuroda, J., Arimura, K., Tachibana, M., Nakamura, K., Wakisaka, Y., Sadoshima, J., Iihara, K., & Kitazono, T. (2016). Detrimental role of pericyte Nox4 in the acute phase of brain ischemia. *Journal of Cerebral Blood Flow and Metabolism, 36*, 1143–1154. https://doi.org/10.1177/0271678X15606456.

Okano, H., Sakaguchi, M., Ohki, K., Suzuki, N., & Sawamoto, K. (2007). Regeneration of the central nervous system using endogenous repair mechanisms. *Journal of Neurochemistry, 102*, 1459–1465. https://doi.org/10.1111/j.1471-4159.2007.04674.x.

Østergaard, L., Engedal, T. S., Moreton, F., Hansen, M. B., Wardlaw, J. M., Dalkara, T., Markus, H. S., & Muir, K. W. (2016). Cerebral small vessel disease: Capillary pathways to stroke and cognitive decline. *Journal of Cerebral Blood Flow and Metabolism, 36*, 302–325. https://doi.org/10.1177/0271678X15606723.

Özen, I., Deierborg, T., Miharada, K., Padel, T., Englund, E., Genové, G., & Paul, G. (2014). Brain pericytes acquire a microglial phenotype after stroke. *Acta Neuropathologica, 128*, 381–396. https://doi.org/10.1007/s00401-014-1295-x.

Ozerdem, U., & Stallcup, W. B. (2003). Early contribution of pericytes to angiogenic sprouting and tube formation. *Angiogenesis, 6*, 241–249. https://doi.org/10.1023/B:AGEN.0000021401.58039.a9.

Palmer, T. D., Willhoite, A. R., & Gage, F. H. (2000). Vascular niche for adult hippocampal neurogenesis. *The Journal of Comparative Neurology, 425*, 479–494.

Parent, J. M., Vexler, Z. S., Gong, C., Derugin, N., & Ferriero, D. M. (2002). Rat forebrain neurogenesis and striatal neuron replacement after focal stroke. *Annals of Neurology, 52*, 802–813. https://doi.org/10.1002/ana.10393.

Paul, G., Özen, I., Christophersen, N. S., Reinbothe, T., Bengzon, J., Visse, E., Jansson, K., Dannaeus, K., Henriques-Oliveira, C., Roybon, L., Anisimov, S. V., Renström, E., Svensson, M., Haegerstrand, A., & Brundin, P. (2012). The adult human brain harbors multipotent perivascular mesenchymal stem cells. *PLoS One, 7*, e35577. https://doi.org/10.1371/journal.pone.0035577.

Peppiatt, C. M., Howarth, C., Mobbs, P., & Attwell, D. (2006). Bidirectional control of CNS capillary diameter by pericytes. *Nature, 443*, 700–704. https://doi.org/10.1038/nature05193.

Powers, W. J., Rabinstein, A. A., Ackerson, T., Adeoye, O. M., Bambakidis, N. C., Becker, K., Biller, J., Brown, M., Demaerschalk, B. M., Hoh, B., Jauch, E. C., Kidwell, C. S., Leslie-Mazwi, T. M., Ovbiagele, B., Scott, P. A., Sheth, K. N., Southerland, A. M., Summers, D. V., Tirschwell, D. L., & American Heart Association Stroke Council. (2018). 2018 guidelines for the early Management of Patients with Acute Ischemic Stroke: A guideline for healthcare professionals from the American Heart Association/American Stroke Association. *Stroke, 49*, e46–e110. https://doi.org/10.1161/STR.0000000000000158.

Prakash, R., Li, W., Qu, Z., Johnson, M. A., Fagan, S. C., & Ergul, A. (2013). Vascularization pattern after ischemic stroke is different in control versus diabetic rats: Relevance to stroke recovery. *Stroke, 44*, 2875–2882. https://doi.org/10.1161/STROKEAHA.113.001660.

Prakash, R., Somanath, P. R., El-Remessy, A. B., Kelly-Cobbs, A., Stern, J. E., Dore-Duffy, P., Johnson, M., Fagan, S. C., & Ergul, A. (2012). Enhanced cerebral but not peripheral angiogenesis in the Goto-Kakizaki model of type 2 diabetes involves VEGF and peroxynitrite signaling. *Diabetes, 61*, 1533–1542. https://doi.org/10.2337/db11-1528.

Proebstl, D., Voisin, M.-B., Woodfin, A., Whiteford, J., D'Acquisto, F., Jones, G. E., Rowe, D., & Nourshargh, S. (2012). Pericytes support neutrophil subendothelial cell crawling and breaching of venular walls in vivo. *The Journal of Experimental Medicine, 209*, 1219–1234. https://doi.org/10.1084/jem.20111622.

Puro, D. G. (2007). Physiology and pathobiology of the pericyte-containing retinal microvasculature: New developments. *Microcirculation, 1994*(14), 1–10. https://doi.org/10.1080/10739680601072099.

Renner, O., Tsimpas, A., Kostin, S., Valable, S., Petit, E., Schaper, W., & Marti, H. H. (2003). Time- and cell type-specific induction of platelet-derived growth factor receptor-beta during cerebral ischemia. *Brain Research. Molecular Brain Research, 113*, 44–51.

Ritter, L. S., Orozco, J. A., Coull, B. M., McDonagh, P. F., & Rosenblum, W. I. (2000). Leukocyte accumulation and hemodynamic changes in the cerebral microcirculation during early reperfusion after stroke. *Stroke, 31*, 1153–1161.

Rouget, C. (1873). Memoire sur le development, la structure et les proprietes physiologiques des capillaries sanguins et lymphatiques. *Archives of Physiology Norm et Path, 5*, 603–663.

Sharma, V., Ling, T. W., Rewell, S. S., Hare, D. L., Howells, D. W., Kourakis, A., & Wookey, P. J. (2012). A novel population of α-smooth muscle actin-positive cells activated in a rat model of stroke: An analysis of the spatio-temporal distribution in response to ischemia. *Journal of Cerebral Blood Flow and Metabolism, 32*, 2055–2065. https://doi.org/10.1038/jcbfm.2012.107.

Shen, J., Ishii, Y., Xu, G., Dang, T. C., Hamashima, T., Matsushima, T., Yamamoto, S., Hattori, Y., Takatsuru, Y., Nabekura, J., & Sasahara, M. (2012). PDGFR-β as a positive regulator of tissue repair in a mouse model of focal cerebral ischemia. *Journal of Cerebral Blood Flow and Metabolism, 32*, 353–367. https://doi.org/10.1038/jcbfm.2011.136.

Shepro, D., & Morel, N. M. (1993). Pericyte physiology. *The FASEB Journal, 7*, 1031–1038.

Simard, J. M., Kent, T. A., Chen, M., Tarasov, K. V., & Gerzanich, V. (2007). Brain oedema in focal ischaemia: Molecular pathophysiology and theoretical implications. *Lancet Neurology, 6*, 258–268. https://doi.org/10.1016/S1474-4422(07)70055-8.

Sims, D. E. (1986). The pericyte--a review. *Tissue & Cell, 18*, 153–174.

Stark, K., Eckart, A., Haidari, S., Tirniceriu, A., Lorenz, M., von Brühl, M.-L., Gärtner, F., Khandoga, A. G., Legate, K. R., Pless, R., Hepper, I., Lauber, K., Walzog, B., & Massberg, S. (2013). Capillary and arteriolar pericytes attract innate leukocytes exiting through venules and "instruct" them with pattern-recognition and motility programs. *Nature Immunology, 14*, 41–51. https://doi.org/10.1038/ni.2477.

Suzuki, K., Masawa, N., Sakata, N., & Takatama, M. (2003). Pathologic evidence of microvascular rarefaction in the brain of renal hypertensive rats. *Journal of Stroke and Cerebrovascular Diseases, 12*, 8–16. https://doi.org/10.1053/jscd.2003.1.

Sweeney, M. D., Ayyadurai, S., & Zlokovic, B. V. (2016). Pericytes of the neurovascular unit: Key functions and signaling pathways. *Nature Neuroscience, 19*, 771–783. https://doi.org/10.1038/nn.4288.

Takahashi, A., Park, H. K., Melgar, M. A., Alcocer, L., Pinto, J., Lenzi, T., Diaz, F. G., & Rafols, J. A. (1997). Cerebral cortex blood flow and vascular smooth muscle contractility in a rat model of ischemia: A correlative laser Doppler flowmetric and scanning electron microscopic study. *Acta Neuropathologica, 93*, 354–368.

Taskiran-Sag, A., Yemisci, M., Gursoy-Ozdemir, Y., Erdener, S. E., Karatas, H., Yuce, D., & Dalkara, T. (2018). Improving microcirculatory reperfusion reduces parenchymal oxygen radical formation and provides neuroprotection. *Stroke, 49*, 1267–1275. https://doi.org/10.1161/STROKEAHA.118.020711.

Tatebayashi, K., Tanaka, Y., Nakano-Doi, A., Sakuma, R., Kamachi, S., Shirakawa, M., Uchida, K., Kageyama, H., Takagi, T., Yoshimura, S., Matsuyama, T., & Nakagomi, T. (2017). Identification of multipotent stem cells in human brain tissue following stroke. *Stem Cells and Development, 26*, 787–797. https://doi.org/10.1089/scd.2016.0334.

Theilen, H., Schröck, H., & Kuschinsky, W. (1993). Capillary perfusion during incomplete forebrain ischemia and reperfusion in rat brain. *The American Journal of Physiology, 265*, H642–H648. https://doi.org/10.1152/ajpheart.1993.265.2.H642.

Underly, R. G., Levy, M., Hartmann, D. A., Grant, R. I., Watson, A. N., & Shih, A. Y. (2017). Pericytes as inducers of rapid, matrix Metalloproteinase-9-dependent capillary damage during ischemia. *The Journal of Neuroscience, 37*, 129–140. https://doi.org/10.1523/JNEUROSCI.2891-16.2016.

Vallet, P., Charnay, Y., Steger, K., Ogier-Denis, E., Kovari, E., Herrmann, F., Michel, J.-P., & Szanto, I. (2005). Neuronal expression of the NADPH oxidase NOX4, and its regulation in mouse experimental brain ischemia. *Neuroscience, 132*, 233–238. https://doi.org/10.1016/j.neuroscience.2004.12.038.

Wallow, I. H., & Burnside, B. (1980). Actin filaments in retinal pericytes and endothelial cells. *Investigative Ophthalmology & Visual Science, 19*, 1433–1441.

Wang, J., Fu, X., Zhang, D., Yu, L., Lu, Z., Gao, Y., Liu, X., Man, J., Li, S., Li, N., Wang, M., Liu, X., Chen, X., Zang, W., Yang, Q., & Wang, J. (2017). Effects of crenolanib, a nonselective inhibitor of PDGFR, in a mouse model of transient middle cerebral artery occlusion. *Neuroscience, 364*, 202–211. https://doi.org/10.1016/j.neuroscience.2017.09.025.

Wang, L., Zhang, Z., Wang, Y., Zhang, R., & Chopp, M. (2004). Treatment of stroke with erythropoietin enhances neurogenesis and angiogenesis and improves neurological function in rats. *Stroke, 35*, 1732–1737. https://doi.org/10.1161/01.STR.0000132196.49028.a4.

Wardlaw, J. M., Murray, V., Berge, E., & Del Zoppo, G. J. (2009). Thrombolysis for acute ischaemic stroke. *Cochrane Database of Systematic Reviews, 7*, CD000213. https://doi.org/10.1002/14651858.CD000213.pub2.

Willard, A. L., & Herman, I. M. (2012). Vascular complications and diabetes: Current therapies and future challenges. *Journal of Ophthalmology, 2012*, 209538. https://doi.org/10.1155/2012/209538.

Winkler, E. A., Bell, R. D., & Zlokovic, B. V. (2011a). Central nervous system pericytes in health and disease. *Nature Neuroscience, 14*, 1398–1405.. doi: nn.2946 [pii]. https://doi.org/10.1038/nn.2946.

Winkler, E. A., Bell, R. D., & Zlokovic, B. V. (2011b). Lack of Smad or notch leads to a fatal game of brain pericyte hopscotch. *Developmental Cell, 20*, 279–280. https://doi.org/10.1016/j.devcel.2011.03.002.

Yang, Y., Thompson, J. F., Taheri, S., Salayandia, V. M., McAvoy, T. A., Hill, J. W., Yang, Y., Estrada, E. Y., & Rosenberg, G. A. (2013). Early inhibition of MMP activity in ischemic rat brain promotes expression of tight junction proteins and angiogenesis during recovery. *Journal of Cerebral Blood Flow and Metabolism, 33*, 1104–1114. https://doi.org/10.1038/jcbfm.2013.56.

Yemisci, M., Gursoy-Ozdemir, Y., Vural, A., Can, A., Topalkara, K., & Dalkara, T. (2009). Pericyte contraction induced by oxidative-nitrative stress impairs capillary reflow despite successful opening of an occluded cerebral artery. *Nature Medicine, 15*, 1031–1037. https://doi.org/10.1038/nm.2022.

Zechariah, A., ElAli, A., Doeppner, T. R., Jin, F., Hasan, M. R., Helfrich, I., Mies, G., & Hermann, D. M. (2013a). Vascular endothelial growth factor promotes pericyte coverage of brain capillaries, improves cerebral blood flow during subsequent focal cerebral ischemia, and preserves the metabolic penumbra. *Stroke, 44*, 1690–1697. https://doi.org/10.1161/STROKEAHA.111.000240.

Zechariah, A., ElAli, A., Hagemann, N., Jin, F., Doeppner, T. R., Helfrich, I., Mies, G., & Hermann, D. M. (2013b). Hyperlipidemia attenuates vascular endothelial growth factor-induced angiogenesis, impairs cerebral blood flow, and disturbs stroke recovery via decreased pericyte coverage of brain endothelial cells. *Arteriosclerosis, Thrombosis, and Vascular Biology, 33*, 1561–1567. https://doi.org/10.1161/ATVBAHA.112.300749.

Zhang, L., Zhang, Z. G., & Chopp, M. (2012). The neurovascular unit and combination treatment strategies for stroke. *Trends in Pharmacological Sciences, 33*, 415–422. https://doi.org/10.1016/j.tips.2012.04.006.

Zhang, Z., & Chopp, M. (2002). Vascular endothelial growth factor and angiopoietins in focal cerebral ischemia. *Trends in Cardiovascular Medicine, 12*, 62–66.

Zhang, Z. G., Chopp, M., Goussev, A., Lu, D., Morris, D., Tsang, W., Powers, C., & Ho, K. L. (1999). Cerebral microvascular obstruction by fibrin is associated with upregulation of PAI-1 acutely after onset of focal embolic ischemia in rats. *The Journal of Neuroscience, 19*, 10898–10907.

Chapter 10
Pericytes in Hereditary Hemorrhagic Telangiectasia

Georgios Galaris, Jérémy H. Thalgott, and Franck P. G. Lebrin

Abstract Hereditary hemorrhagic telangiectasia (HHT) is a genetic disorder characterized by multi-systemic vascular dysplasia affecting 1 in 5000 people worldwide. Individuals with HHT suffer from many complications including nose and gastrointestinal bleeding, anemia, iron deficiency, stroke, abscess, and high-output heart failure. Identification of the causative gene mutations and the generation of animal models have revealed that decreased transforming growth factor-β (TGF-β)/bone morphogenetic protein (BMP) signaling and increased vascular endothelial growth factor (VEGF) signaling activity in endothelial cells are responsible for the development of the vascular malformations in HHT. Perturbations in these key pathways are thought to lead to endothelial cell activation resulting in mural cell disengagement from the endothelium. This initial instability state causes the blood vessels to response inadequately when they are exposed to angiogenic triggers resulting in excessive blood vessel growth and the formation of vascular abnormalities that are prone to bleeding. Drugs promoting blood vessel stability have been reported as effective in preclinical models and in clinical trials indicating possible interventional targets based on a normalization approach for treating HHT. Here, we will review how disturbed TGF-β and VEGF signaling relates to blood vessel destabilization and HHT development and will discuss therapeutic opportunities based on the concept of vessel normalization to treat HHT.

G. Galaris · J. H. Thalgott
Department of Internal Medicine (Nephrology), Einthoven Laboratory for Experimental Vascular Medicine, Leiden University Medical Center, Leiden, The Netherlands

F. P. G. Lebrin (✉)
Department of Internal Medicine (Nephrology), Einthoven Laboratory for Experimental Vascular Medicine, Leiden University Medical Center, Leiden, The Netherlands

Physics for Medicine, ESPCI, INSERM U1273, CNRS, Paris, France

MEMOLIFE Laboratory of Excellence and PSL Research University, Paris, France
e-mail: f.lebrin@lumc.nl

A. Birbrair (ed.), *Pericyte Biology in Disease*, Advances in Experimental Medicine and Biology 1147, https://doi.org/10.1007/978-3-030-16908-4_10

Keywords Hereditary hemorrhagic telangiectasia · Vascular rare disease · Blood vessel stabilization · Transforming growth factor-β · Vascular endothelial growth factor

Introduction

Blood vessel instability and malfunction in combination with inflammatory triggers are the underlying cause of some of the most serious diseases affecting Western society. Conditions range from cardiovascular disease and heart failure to cognitive decline and dementia. Hereditary hemorrhagic telangiectasia (HHT) is one such condition. Albeit rare, vessels of HHT patients exhibit the salient features of other vascular disorders and as such provide tractable preclinical models for research and defined patient groups for clinical trials. This chapter introduces the current concepts that are proposed to explain the development of vascular malformations in HHT. Specifically, we highlight the roles of pericytes in the pathogenesis of HHT and examine how defective transforming growth factor-β (TGF-β)/bone morphogenetic protein (BMP) signaling pathways contribute to blood capillary destabilization. Finally, we will discuss recent blood capillary normalization-based therapies used in HHT patients to treat bleeding from vascular malformations.

Hereditary Hemorrhagic Telangiectasia (HHT)

Clinical Overview

HHT (also known as Osler-Weber-Rendu) is an autosomal dominant inherited disease characterized by multiple vascular malformations. The prevalence is estimated to be 1 in 5000–8000 affecting theoretically between 950,000 and 1,500,000 persons worldwide (Shovlin 2010; McDonald et al. 2011; Kroon et al. 2018). HHT is therefore relatively common. However, it remains a largely undiagnosed disease because people and doctors are not familiar with the wide variety of symptoms and only 500,000 people are indeed diagnosed with HHT around the world.

Individuals with HHT are affected by large Arteriovenous Malformations (AVMs) in the lungs, liver, and brain that consist of direct connections between arteries and veins without an intervening capillary bed. These AVMs are a potential source of serious morbidity and mortality as they can lead to ischemic stroke, abscess due to shunting through pulmonary AVMs, or to heart failure related to shunting effects of hepatic AVMs (Shovlin 2010). Many patients also suffer spontaneous and recurrent nose and gastrointestinal bleeding associated with severe anemia that significantly affect their quality of life. Bleeding occurs because of small and fragile dilated capillaries called telangiectasias that are near the surface of skin and mucous membranes (Shovlin 2010). To limit blood loss, several therapeutic options have been introduced in the last two decades including hormonal manipulation in the form of estrogen–progesterone or tamoxifen (Minami and Haji 2016; Yaniv et al. 2011), anti-fibrinolytic therapies (Zaffar et al. 2015), immunomodulatory therapy with tacrolimus (Sommer et al. 2018) and the use of inhibitors of angiogenesis such as bevacizumab (Halderman et al. 2018) or thalidomide (Lebrin et al. 2010). Unfortunately, there is as far, no demonstration of the long-term safety and benefit using these drugs in prevention of HHT hemorrhages (Shovlin 2010).

Genetic Basis of the Disease

Identification of the causative gene mutations has revealed that disruption of TGF-β/BMP signaling in endothelial cells underlines HHT. Mutations in the *ENG* gene (Endoglin) (McAllister et al. 1994) or in the *ACVRL1* gene (activin receptor-like kinase 1 or ALK1) (Johnson et al. 1996) are responsible for HHT1 or HHT2, respectively and account for more than 80% of cases of HHT. Both *ENG* and *ACVRL1* encode for receptors of TGF-β/BMP that are expressed in endothelial cells and share functions in signaling (Lebrin et al. 2005). All classical features of HHT can be seen in both HHT1 and HHT2, but the prevalence of specific vascular anomalies varies according to the genotype. Pulmonary and cerebral AVMs are more common in HHT1 than HHT2 while HHT2 patients have a higher incidence of hepatic AVMs and gastrointestinal hemorrhages (van Gent et al. 2010; Letteboer et al. 2006; Lesca et al. 2007). A rare form of HHT disease in which vascular lesions are combined with Juvenile Polyposis is associated with mutations in the gene *MADH4* (*Mothers Against Decapentaplegic Homolog 4*). *MADH4* encodes for SMAD4, a downstream effector of TGF-β/BMP family ligands (Gallione et al. 2004). While HHT3 (Cole et al. 2005) and HHT4 (Bayrak-Toydemir et al. 2006) have only been linked to a particular locus and no specific genes have been identified yet, HHT5 is due to mutations in the *Growth Differentiation Factor 2* (*GDF2*) gene (Wooderchak-Donahue et al. 2013). *GDF2* gene encodes for BMP9, a high-affinity ligand for ALK1 that controls endothelial cell quiescence (David et al. 2007).

Known gene mutations include deletion, insertion, and missense mutations as well as splice site changes and represent null allele indicating that haploinsufficiency is the underlying mechanism of HHT. As consequence, the remaining wild-type allele is unable to contribute sufficient protein for normal TGF-β/BMP signaling in endothelial cells leading to blood vessel dysfunctions (Abdalla and Letarte 2006).

Current Concepts for the Generation of AVM and Telangiectasia

Animal models have confirmed that *Gdf2*, *Eng*, *Acvrl1*, or *Smad4* mutations causes HHT and have brought important insights into the mechanisms by which HHT mutations lead to the development of vascular malformations. These models have employed classical null and heterozygous mice for either *Eng* or *Acvrl1* (Bourdeau et al. 1999; Li et al. 1999; Arthur et al. 2000; Srinivasan et al. 2003; Torsney et al. 2003), mice bearing conditional loxP knockout alleles for *Eng*, *Acvrl1*, or *Smad4* crossed with endothelial or mural tissue specific Cre-recombinase mouse lines (Park et al. 2009; Mahmoud et al. 2010; Walker et al. 2011; Choi et al. 2012, 2014; Chen et al. 2013; Garrido-Martin et al. 2014; Ola et al. 2016, 2018; Crist et al. 2018), mice injected with blocking antibodies targeting BMP9 and BMP10 (Ruiz et al. 2016) or *Zebrafish* embryos harboring a mutation in *alk1* or *eng* gene (Roman et al. 2002; Corti et al. 2011; Rochon et al. 2016; Sugden et al. 2017). From these studies, it appears that HHT mutations are deleterious predominantly during some forms of angiogenesis and that additional triggers to the gene mutations are required for AVMs to form (Fig. 10.1). The heterozygous for *Eng* or *Acvrl1* mice that are the closest genetic models of HHT patients in terms of genotype exhibit a very mild phenotype with HHT-like features appearing only at low frequency and in an unpredictable manner. Blood vessels develop and function normally in these mice, although they have a widespread abnormality of the vascular walls due to defective mural cell recruitment and attachment (Torsney et al. 2003; Lebrin et al. 2010; Li et al. 2011; Thalgott et al. 2015, 2018). Defective blood vessel stability represents the baseline situation in HHT and is caused by decreased TGF-β bioavailability (Carvalho et al. 2004) and increased VEGF signaling in endothelial cells (Thalgott et al. 2018) (Fig. 10.1). This is thought to favor inadequate responses of the endothelial cells to angiogenic triggers leading to excessive angiogenesis and the development of vascular malformations. Others and we have reported that a second event such as inflammation, infection, wound healing and/or angiogenesis is indeed required to initiate the formation of AV shunts (Xu et al. 2004; Peter et al. 2014; Gkatzis et al. 2016; Thalgott et al. 2018) (Fig. 10.1). Finally, advanced real-time imaging technologies using skinfold window chamber systems have revealed that the initial AV shunts are able to remodel due to hemodynamic changes with veins and arteries that dilate and adjacent blood capillaries that regress resulting in the maturation of the AV shunts (Park et al. 2009; Garrido-Martin et al. 2014; Han et al. 2014). Interestingly, recent studies point in the direction of a specific synergy between blood flow and Endoglin-Alk1 signaling pathway for the regulation of vessel caliber supporting the key role of shear stress in AVM formation and maturation (Baeyens et al. 2016; Gkatzis et al. 2016; Sugden et al. 2017; Jin et al. 2017). Additional events to the heterozygosity of HHT gene mutation and pro-angiogenic stimuli have been proposed to trigger the development of AVMs in HHT. These include the concept of local loss of heterozygosity where a somatic mutation in the *Eng* or *Acvrl1* gene would result in a group of endothelial cells that has lost the remaining wild-type allele precipitating the development of AV shunts (Fig. 10.1) (Tual-Chalot et al. 2015). This concept is supported by the recent generation of conditional knockout mice for *Eng, Acvrl1*, or *Smad4,* which develop robust AVMs resembling those seen in HHT individuals (Mahmoud et al. 2010; Tual-Chalot et al. 2014; Crist et al. 2018). However, the local loss of heterozygosity concept may be somehow simplistic to explain HHT pathogenesis. It seems really unlikely that many endothelial cells over the entire body could acquire somatic mutations in the

Eng or *Acvrl1* gene to explain the multiplicity of vascular malformations found in patients with HHT. Moreover, AVMs have been reported to express the same level of endoglin (approximately one half normal) than the unaffected blood vessels in HHT1 patients (Bourdeau et al. 2000; Matsubara et al. 2000). Alternatively, inflammatory cytokines such as tumor necrosis factor-α (TNF-α) that regulate receptor release from the cell surface have been proposed to result in a transient and local null-endoglin phenotype during inflammation, although this hypothesis needs to be experimentally confirmed (Mahmoud et al. 2010). Finally, several genetic modifiers

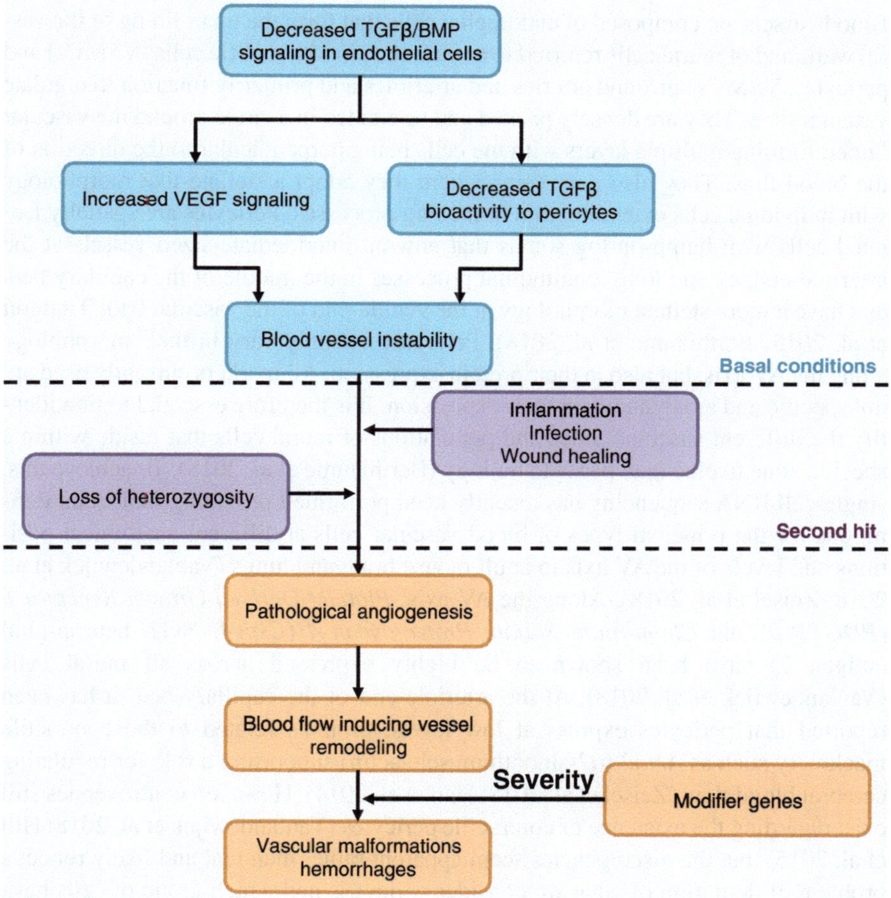

Fig. 10.1 How gene mutations in HHT leads to the development of AVMs. Decreased TGFβ/BMP signaling in endothelial cells leads is combined with increased VEGF signaling leading to blood capillary destabilization and poor mural cell attachment. A second hit such as inflammation, infection, wound healing or loss of heterozygosity is required to induce pathological angiogenesis and the formation of vessel abnormalities that subsequently remodel due to blood flow changes to form stable and mature AVMs. In addition, modifier genes are contributing to the severity of the disease

have also been described to play a role in susceptibility to HHT disease (Bourdeau et al. 2001; Letteboer et al. 2015; Gkatzis et al. 2016; Pawlikowska et al. 2018; Benzinou et al. 2012) (Fig. 10.1).

Blood Vessel Instability Is a Major Hallmark of HHT Disease

Mural Cell Organization and Function

Blood vessels are composed of endothelial cells that form the inner lining of the vessel walls and of mural cells referred to as vascular smooth muscle cells (VSMCs) and pericytes. VSMCs surround arteries and arterioles and primarily function to regulate vascular tone. They are densely packed and orientated in a circle around the vascular lumen forming multiple layers with the cells being perpendicular to the direction of the blood flow. They also coat veins where they adopt a stellate-like morphology with individual cells extending thin branching processes. Pericytes are spatially isolated cells with bump-on-log somas that enwrap intermediate-sized vessels at the arteriole-end, extend long longitudinal processes in the middle of the capillary bed, and have a more stellate morphology at the venule-end of the vascular bed (Thalgott et al. 2015; Berthiaume et al. 2018). Pericytes not only vary in their morphology along the AV axis, but also in their protein expression. All markers currently used are not specific and are dynamic in their expression. It is therefore essential to first identify the different vascular zones and populations of mural cells that reside within a specific zone to elucidate pericyte biology (Berthiaume et al. 2018). To achieve this, single-cell RNA sequencing has recently been performed providing molecular definitions for the principal types of blood vascular cells at different anatomical positions and levels of the AV axis in adult mouse brain and lungs (Vanlandewijck et al. 2018; Zeisel et al. 2018). Along the AV axis, *Platelet-Derived Growth Receptor-β (PDGFR-β)* and *Chondroitin Sulfate Proteoglycan 4 (Cspg4,* NG2, neuron-glial antigen 2*)* have been shown to be highly expressed across all mural cells (Vanlandewijck et al. 2018). At the arteriole-end of the capillary bed, it has been reported that pericytes express at low levels, proteins related to the contractile machinery such as *Acta2* (α2-smooth muscle actin) supporting a role for regulating cerebral blood flow (Zeisel et al. 2018; Hall et al. 2014). However, controversies still exist regarding the existence of contractile pericytes (Vanlandewijck et al. 2018; Hill et al. 2015), but the discrepancies seem apparent rather than real and likely reflect a problem of definition of what we consider pericytes, and which group of cells have been included as pericytes in the studies when analyzing their molecular signatures (Attwell et al. 2016). Pericytes from the middle capillary bed to the venule site have been found to highly express membrane transporters such as *abcc9* (*ATP-binding cassette sub-family C member 9*, SUR2, Sulfonylurea receptor 2) supporting their role for maintaining blood–brain barrier (BBB) (Vanlandewijck et al. 2018; Daneman et al. 2010). Recent findings suggest that pericytes are a dynamic cell population with high plasticity that do not only vary in their morphology, molecular signatures and function along the AV axis but also among vascular beds of different organs to acquire specializations and adaptations that match the local physiological needs (Vanlandewijck et al. 2018). This heterogeneity may also be expected in disease and should be systemically addressed adding a further dimension of complexity to pericyte biology.

Pericytes influence blood vessel formation, homeostasis and disease. They regulate angiogenesis by promoting blood vessel stabilization, modulate local blood flow and control vessel permeability to molecules and immune cells. In addition, pericytes may be pluripotent stem cells, remove debris through phagocytosis, regulate wound healing and provide trophic signals (Hall et al. 2014; Armulik et al. 2010; Daneman et al. 2010; Cupovic et al. 2016; Stapor et al. 2014; Dias et al. 2018; Gautam and Yao 2018; Guimaraes-Camboa et al. 2017). Recent studies have revealed that pericytes have important roles in numerous disorders including diabetic retinopathy and diabetic nephropathy, cancers or central nervous system (CNS) disorders such as Alzheimer's disease, Huntington's disease, radiation necrosis, epilepsy, or ischemic stroke (Cheng et al. 2018; Armulik et al. 2011). Pericyte dysfunctions include early pericyte constriction of blood capillaries and pericyte death or migration away from blood vessels. These lead to hypo-perfusion of the tissue, increase vascular permeability, excessive immune cell infiltration and inflammation combined with blood capillary death and/or abnormal angiogenesis.

Mural Cells Fail to Show Normal Blood Vessel Coverage in HHT

Eng or *Acvrl1* null mice exhibit similar phenotypes with embryos that die at midgestation due to severe cardiovascular defects. Vasculogenesis—the de novo production of endothelial cells and the initial formation of the primitive vascular plexus—appears to progress normally in these embryos but angiogenesis—the process through which new blood vessels form from existing vasculature—is impaired and major blood vessels of the $Eng^{-/-}$ or $Acvrl1^{-/-}$ mice at embryonic day E9.5–10.5 show reduced VSMCs coverage (Bourdeau et al. 1999; Li et al. 1999; Arthur et al. 2000; Oh et al. 2000). Moreover, the failure in blood vessel remodeling is not restricted to embryonic tissue. The vasculature of the yolk sac also failed to organize and has shown poor VSMC recruitment and differentiation (Li et al. 1999; Carvalho et al. 2004). Heterozygous *Eng* or *Acvrl1* mice develop to adulthood with no effect on survival (Tual-Chalot et al. 2015). However, these mice exhibit a general defect in the organization of the vessel walls. Irregular layers of VSMCs combined with variations in the number of collagen fibers and loss of integrity of elastin fibers have been reported in the skin vasculature of adult $Eng^{+/-}$ mice resulting in fragile vessels that are prone to bleeding (Fig. 10.2a). An increase in the number of perivascular degranulated mast cells in the reticular and papillary dermis was also observed in these mice supporting the idea that inflammation may exacerbate the development of HHT lesions (Torsney et al. 2003). Confocal images of the ear skin of

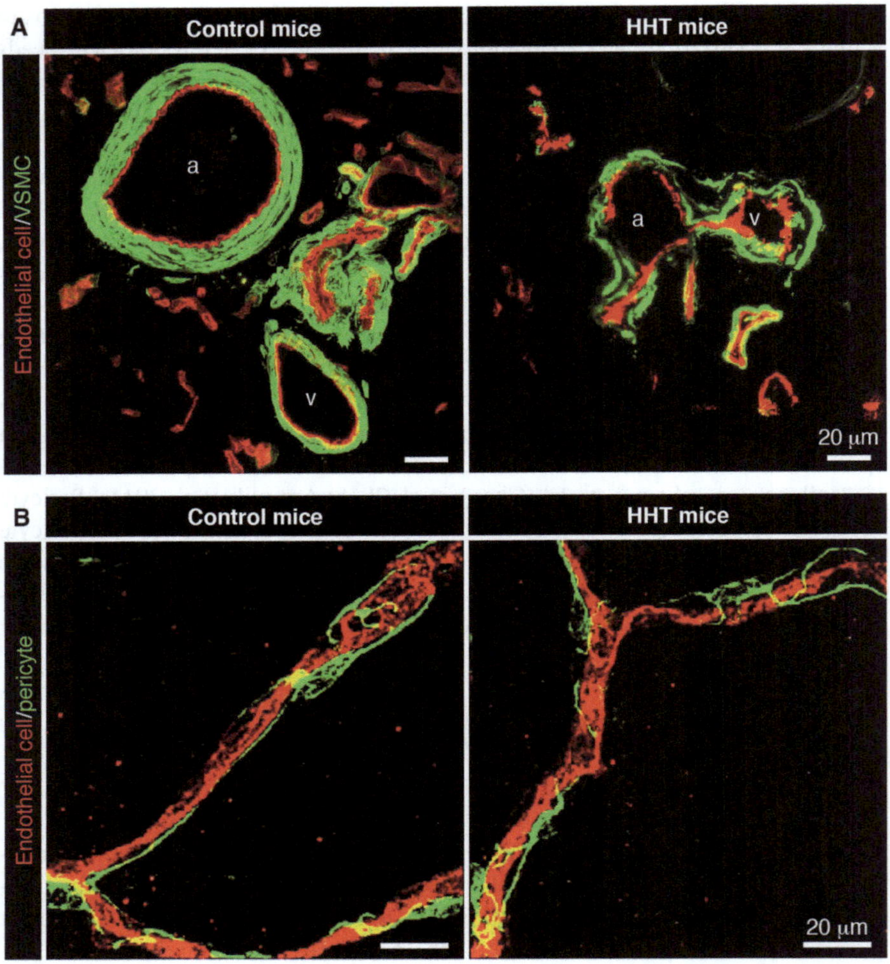

Fig. 10.2 Defective mural cell coverage in HHT mouse models. (**a**) Confocal image of skin sections of *Eng*$^{+/-}$ stained for PECAM-1 (Platelet Endothelial Cell Adhesion Molecule) (endothelial cells, red) and for α-SMA (VSMCs, green) showing defective coverage. (**b**) Confocal image of the tracheal vasculature of *Acvrl1*$^{+/-}$ mice stained for PECAM-1 (Platelet Endothelial Cell Adhesion Molecule) (endothelial cells, red) and Desmin (pericytes, green) revealing defective pericyte attachment

Eng$^{+/-}$ mice during the first postnatal month also confirmed defective VSMC coverage of the arteries (Lebrin et al. 2010). Defective maturation of the blood vasculature has been reported in *Acvrl1*$^{+/-}$ mice too. Histological examination of the *Acvrl1*$^{+/-}$ adult mice has revealed thin-walled dilated vessels in the skin and in multiple internal organs that resemble to those seen in HHT patients (Srinivasan et al. 2003). More recently, we have characterized the vasculature of the *Acvrl1*$^{+/-}$ tracheas. The mouse trachea has a segmented two-dimensional network of blood vessels aligned with a cartilage ring allowing a clear observation of the vasculature not only in healthy conditions but also in pathological situations such as inflammation when blood vessel changes occur (Thalgott et al. 2018). The overall vascular architecture of the *Acvrl1*$^{+/-}$ tracheas did not differ from control mice and *Acvrl1*$^{+/-}$ tracheas showed normal VSMC coverage. However, blood capillaries that are arranged in a ladder-like pattern in the mucosa overlying cartilage rings were significantly enlarged in *Acvrl1*$^{+/-}$ tracheas. Pericytes had shorter protrusions and were partially detached from the endothelium in the *Acvrl1*$^{+/-}$ tracheas, although the number of pericytes was unchanged compared to control mice (Thalgott et al. 2018) (Fig. 10.2b). The recent development of conditional knockout animals has provided important insights into the pathogenesis of HHT allowing us to study the effects of gene inactivation in different cell types and at different times. In early postnatal development, global or endothelial specific deletion of *Eng*, *Acvrl1*, or *Smad4* gene led to rapid death due to either pulmonary hemorrhage for *iKO-Acvrl1* mice or to gastrointestinal bleeding for *iKO-Eng* or *iKO-Smad4*. From these studies, it appeared that specific gene deletion in endothelial cells was a prerequisite for the development of HHT. In contrast, conditional deletion in mural cells did not trigger AVM formation in any organ studied (Garrido-Martin et al. 2014; Chen et al. 2014a). All conditional knockout mouse models reproducibly developed AVM combined with enlarged veins and hyper-branching of the capillary plexus in the neonatal retina. The radial expansion of the retinal vasculature was decreased in these mutants indicating migration defects (Crist et al. 2018; Kim et al. 2018; Tual-Chalot et al. 2014; Garrido-Martin et al. 2014; Mahmoud et al. 2010). The retinal vascular plexus is a well-established model amenable to studying sprouting angiogenesis. The vasculature develops in the first week of postnatal life to form a simple two-dimensional vascular network. It is highly organized showing an alternating pattern of arteries covered by α-smooth muscle actin (α-SMA) positive VSMCs and veins with an intervening capillary bed that includes endothelial tip cells at the sprouting front of the plexus (Fruttiger 2007). In contrast, the smooth muscle organization was found to not be longer arterial specific in mutant mice. α-SMA expression was shown to reproducibly follow the pattern of predicted blood flow across the AVM and into veins on the downstream side of the shunt. Elevated α-SMA expression is therefore likely a secondary response to increased blood flow (Crist et al. 2018; Kim et al. 2018; Tual-Chalot et al. 2014; Garrido-Martin et al. 2014; Mahmoud et al. 2010). Blood capillaries in the *iKO-Acvrl1* or *iKO-Smad4* also resulted in loss of pericyte coverage (Tual-Chalot et al. 2014; Crist et al. 2018). Consistent with this phenotype, focal *Acvrl1* deletion in the brain led to the formation of blood vessel abnormalities following VEGF stimulation that are similar to that of brain AVM in HHT patients. Vascular integrity was found impaired in the affected brain regions with a marked reduced number of pericytes. Consistent with this phenotype, BBB dysfunctions were observed as evidenced by increased fibrin and iron deposition, by the presence of small pockets of red blood cells (RBCs) outside the vessels and by infiltration of immune cells into the brain parenchyma (Chen et al. 2014b). Destabilization of the endothelial barrier has also been reported in *Eng* heterozygous mice. Although not elucidated in vivo, these results suggested a critical role for endoglin in integrin-mediated adhesion of mural cells (Rossi et al. 2016).

Despite the histological observations indicating that mural cell disengagement from the vessel wall is a hallmark of HHT disease, mural cell biology in HHT remains poorly understood and important questions are still unresolved. For example, the populations of mural cells that are affected along the AV axis and in different organs in the various HHT models are still not identified. How defective mural cell coverage affects blood vessel function is unclear. Is the loss of mural cell coverage implicated in the development of AVM or is it a secondary effect due to blood vessel remodeling? These questions are certainly important subjects of future research.

Mechanisms Underlying Blood Capillary Instability in HHT

Disturbed TGF-β/BMP Signaling in Endothelial Cells Impair Blood Vessel Stabilization

Identification of the causative gene mutations and the development of preclinical models have revealed that the primary cause of HHT is considered dysfunction of TGF-β/BMP signaling in endothelial cells with specific effects on blood vessel stability (Lebrin et al. 2005; Lebrin and Mummery 2008; Thalgott et al. 2015). The TGF-β/BMP superfamily of ligands comprises a large number of evolutionary conserved pleiotropic secreted cytokines that include TGF-βs, BMPs, activins, inhibins, growth differentiation factors (GDFs), glial derived neurotrophic factors (GDNFs), nodal, lefty and anti-Müllerian (AMH) hormone. They act on virtually of cell types of the body regulating multiple processes both during development and in adult life and have important roles in disease progression (Miyazono et al. 2010; Shi and Massague 2003). The cardiovascular system is no exception here, and both TGF-β and BMP receptors are expressed on endothelial cells and on mural cells where they play critical roles in the development and maintenance of blood vessels (Pardali and ten Dijke 2012; ten Dijke and Arthur 2007; Goumans and Mummery 2000). The action of TGF-β/BMP ligands is mediated by at least two trans-membrane serine/threonine kinases, type I and type II receptors. After ligand binding, type II receptors phosphorylate type I receptors leading to their activation. Activated type I receptors are then able to recruit and phosphorylate downstream transcription factors called R-Smads. On activation, R-Smads form heteromeric complexes with a related partner molecule Smad4 and accumulate in the nucleus where they participate in the transcriptional control of target genes (Miyazono et al. 2010; Shi and Massague 2003). Despite the large number and distinct functions of the TGF-β/BMP superfamily ligands, there is an enormous convergence in signaling to only five type II receptors (TβRII, BMPRII, ActRIIa, ActRIIb, and AMHRII) and seven type I receptors also described as activin receptor-like kinase (ALK) and two main SMAD intracellular pathways (ten Dijke and Arthur 2007). TGF-β/BMP ligands can also associate with type III receptors including Betaglycan and Endoglin (predominantly expressed in endothelial cells) that in turn are capable of fine-tuning the availability of TGF-β/BMP ligands to the signaling receptors (Lebrin and Mummery 2008).

In endothelial cells, TGF-β can transduce signals via ALK1 in addition to ALK5 and is thought to contribute to the stimulatory and inhibitory effects of TGF-β on angiogenesis. Upon TGF-β binding, ALK5 phosphorylates Smad2/3 leading to reduce endothelial cell proliferation and migration. In contrast, ALK1 phosphorylates Smad1/5/8 inducing opposite effects on angiogenesis (Goumans et al. 2002) (Fig. 10.3). The differential activation of these two type I receptor pathways by TGF-β provides an intricate mechanism to precisely regulate, and even switch, TGF-β-induced biological responses. Several hypotheses have been envisioned to explain how endothelial cells modulate the ALK1/ALK5 balance system. Differences in the expression levels of ALK1 and ALK5 on endothelial cells have been reported in tissues (Seki et al. 2006; Nguyen et al. 2011). Low concentrations of active TGF-β have been suggested to primarily activate ALK1 signaling (Goumans et al. 2002). We have also reported that endoglin is required for efficient

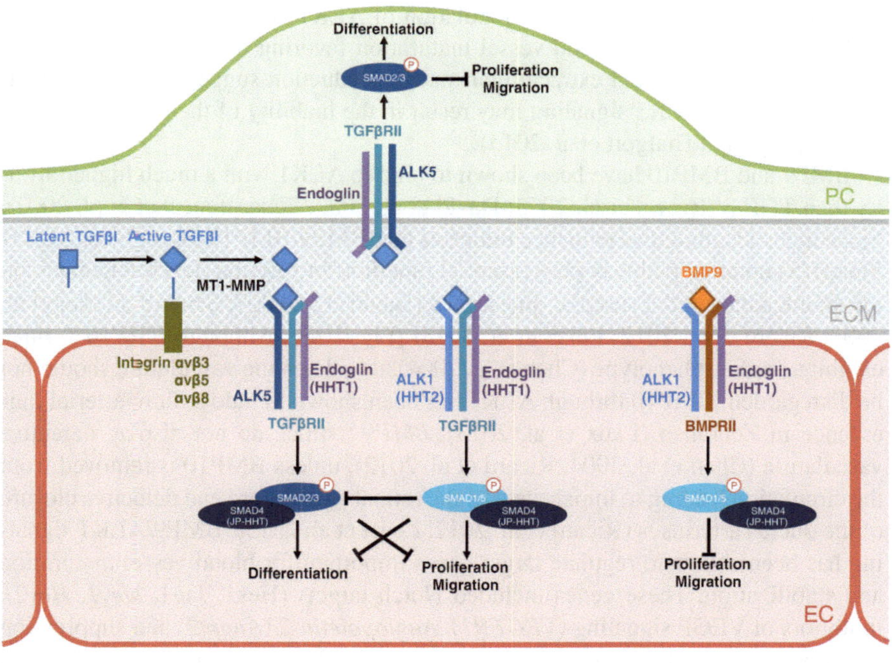

Fig. 10.3 Close physical endothelial cell-mural cell contacts promote paracrine TGF-β/ALK5 signaling in mural cells and differentiation. In endothelial cells, TGFβ signals through ALK5-Smad2/3 and ALK1-Smad1/5 to mediate opposite responses on proliferation and migration. In contrast, BMP9 that binds with much higher affinity to ALK5 promotes endothelial cell quiescence. TGF-β is secreted by both endothelial cells and mural cells in an inactive form. Close physical contacts between endothelial cells and mural cells are necessary to activate TGF-β. The mechanisms leading to its activation remain poorly understood, depends on the cell types and on different mediators that include connexins, integrins, and tissue factor. Upon TGF-β binding to ALK5 on mural cells, it stimulates the production of contractile proteins, mural cell quiescence and differentiation

TGF-β/ALK1 signaling in endothelial cells which indirectly inhibits TGF-β/ALK5 signaling (Lebrin et al. 2004). We showed that endothelial cells lacking endoglin did not proliferate due to reduced TGF-β/ALK1 signaling and increased TGF-β/ALK5 signaling. Endoglin is highly expressed in endothelial cells in sites of active angiogenesis where its expression overlaps with ALK1 supporting a role of TGF-β/ALK1 signaling in the active phase of angiogenesis (Lebrin et al. 2004). Together, these suggest that in HHT, reduce TGF-β/ALK1 signaling activity in endothelial cells would therefore inhibit endothelial cell proliferation and migration. Paradoxically, we have reported that *Eng*$^{+/-}$ or *Acvrl1*$^{+/-}$ retinas at postnatal day 7 show excessive angiogenesis (Thalgott et al. 2018; Lebrin et al. 2010). How can these observations be reconciled with the ALK1/ALK5 balance model? Others and we have demonstrated that endothelial cells from mice or HHT patients did adapt to the hyperactivity of TGF-β/ALK5 signaling-mediated growth arrest by inhibiting ALK5 expression (Lebrin et al. 2004; Xu et al. 2004; Fernandez et al. 2005). The mechanism leading to decrease ALK5 expression is not elucidated yet, although it is suspected to be a consequence of a transcriptional modulation of ALK1 signaling (Fernandez et al. 2005). ALK5 stimulates blood vessel maturation favoring endothelial cell growth arrest, differentiation and extracellular matrix production suggesting that in HHT, impaired TGF-β/ALK5 signaling may result in the inability of the blood vessels to mature properly (Thalgott et al. 2015).

BMP9 and BMP10 have been shown to bind to ALK1 with a much higher affinity than TGF-β (Brown et al. 2005; David et al. 2007; Scharpfenecker et al. 2007). Since then, accumulative evidence indicates that BMP9/BMP10 through the ALK1-Smad1/5 signaling pathway play essential functions in vascular development and in the maintenance of the vascular quiescence (David et al. 2007; Scharpfenecker et al. 2007; Ricard et al. 2012; Levet et al. 2015) (Fig. 10.3). Although *BMP10*$^{-/-}$ mice exhibit a cardiac phenotype (Chen et al. 2004), its role on the vasculature should not be disregarded. BMP10 through ALK1 has been shown to induce flow-arterial quiescence in Zebrafish (Laux et al. 2013). *BMP9*$^{-/-}$ mice do not display defective vasculature (Chen et al. 2004; Ricard et al. 2012), unless BMP10 is removed from the circulation leading to impairment of the retinal vasculature and defective closure of the ductus arteriosus (Ricard et al. 2012; Levet et al. 2015). BMP9/ALK1 signaling has been shown to regulate target genes important for blood vessel maturation and stabilization. These genes included Notch targets (Hes1, Jag1, *Hey1*, *Hey2*), inhibitors of VEGF signaling (*VEGFR1), Angiopoietin-2* (*Angpt2*) and suppression of endothelial tip cell markers (*Unc5b*) (Crist et al. 2018; Larrivee et al. 2012; Moya et al. 2012). Unfortunately, their role in the context of HHT has not been elucidated for most of them.

Recently, we have highlighted a key role of VEGFR1 in the etiology of HHT2 and have provided mechanisms explaining why HHT2 blood vessels respond abnormally to angiogenic signals (Thalgott et al. 2018) (Fig. 10.4). VEGF-A through its binding to VEGFR2 is the major driver of angiogenesis in physiological and pathological situations. Activation of VEGFR2 promotes endothelial cell survival, proliferation, vessel sprouting, and vessel permeability. While VEGF and VEGFR2 are crucial for the development of the vasculature, a proper regulation at low levels of VEGFR2

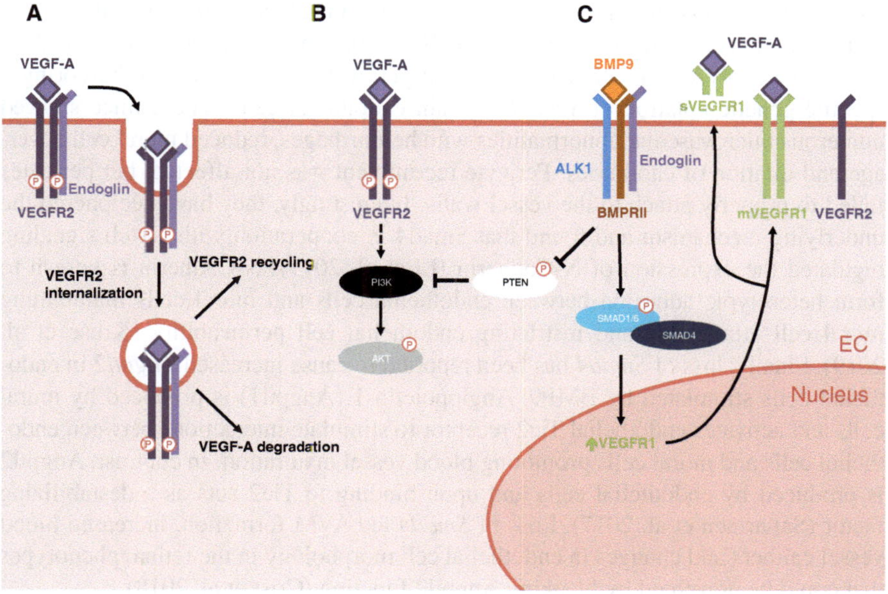

Fig. 10.4 Cross talk between endoglin/ALK1 signaling and VEGF signaling. (**a**) VEGF-A binds to VEGFR2/Endoglin complex that is subsequently internalized to either recycle VEGFR2 or to degrade VEGF-A. (**b**) PI3KAKT signaling is regulated by BMP9/ALK1 signaling through PTEN (Phosphatase and TENsin homolog) (**c**) BMP9/ALK1 signaling is a key regulator of VEGFR1 expression levels, the natural inhibitor of VEGFR2 activity

activity remains essential for vessel stability and homeostasis, whereas high levels of VEGFR2 activity promote sprouting angiogenesis (Moens et al. 2014). VEGF not only binds to VEGFR2 but it can also bind to VEGFR1 that acts as a molecular rheostat that negatively modulate VEGFR2-signaling to favor vessel quiescence (Kappas et al. 2008; Chappell et al. 2009). We have revealed that *Acvrl1* haploinsufficiency was associated with a marked reduction of VEGFR1 expression both in mice and in HHT2 patients. This imbalance in VEGFR1 expression and VEGF bioavailability contributed to elevated VEGFR2 activity in endothelial cells combined with defects in pericyte attachment. These resulted in excessive angiogenesis and the formation of vascular abnormalities in *Acvrl1*$^{+/-}$ mice (Thalgott et al. 2018). In the same line of thought, several studies have reported that HHT patients have increased serum or plasma VEGF when compared to healthy donors (Cirulli et al. 2003; Sadick et al. 2005a, b, 2008; Botella et al. 2015). Blockage of BMP9 signaling has been shown to enhance the phosphatidyl inositol 3-kinase-protein kinase B (PI3K/PKB) pathway certainly through the regulation of Pten. The PI3K/PKB pathway is downstream of VEGFR2 (Fig. 10.4). Importantly, they found that inhibition of Pi3Kinase was sufficient to inhibit and even revert the formation of AVMs in *iKO-Acvrl1* mice (Ola et al. 2016). Alternatively, it has been reported altered VEGFR2 recycling in *Eng* mutant mice that may promote VEGFR2 activity in endothelial cells (Jin et al. 2017) (Fig. 10.4). Although there is no evidence showing that vascular abnormalities in

HHT2 arise from perturbations of Notch signaling (Thalgott et al. 2018), the role of Notch target genes in the pathogenesis of HHT should not be disregarded. Indeed, Li et al. have analyzed the BBB integrity and pericyte coverage in mice harboring a specific genetic ablation of *Smad4* in brain endothelial cells. These mice showed numerous microvascular abnormalities with hemorrhages, reduced mural cell coverage and dilation of capillaries. Pericyte recruitment was not affected, but pericytes failed to properly attach to the vessel walls. Interestingly, they have deciphered the underlying mechanism and found that Smad4 in cooperation with Notch signaling regulated the expression of N-Cadherin (Li et al. 2011). N-Cadherin is known to form heterotypic adhesion between endothelial cells and mural cells maintaining mural cell attachment and restricting endothelial cell permeability (Kruse et al. 2019). Finally, loss of *Smad4* has been reported to cause increased *Angptl2* in endothelial cells stimulated by BMP9. Angiopoietin-1 (Angptl1) is produced by mural cells and activates endothelial Tie2 receptor to stimulate interactions between endothelial cells and mural cells promoting blood vessel maturation. In contrast, Angptl2 is produced by endothelial cells and upon binding to Tie2 acts as a destabilizing factor (Saharinen et al. 2017). Loss of *Smad4* led AVM formation, increased blood vessel calibers and changes in endothelial cell morphology in the retina, phenotypes that could be prevented by blocking Angptl2 function (Crist et al. 2018).

Defective Paracrine TGF-β Signaling Between Endothelial Cells and Mural Cells Impairs Mural Cell Differentiation

The exact molecular changes leading to HHT are not clear yet. However, following the recent identification of targets implicated in blood capillary stabilization, this would suggest that the baseline situation in HHT is likely to be an abnormal activation of the endothelium that may affect mural cell attachment. Indeed, one of the best-understood roles of TGF-β signaling in vascular development is that of promoting mural cell differentiation (Thalgott et al. 2015; ten Dijke and Arthur 2007). Muscularization is achieved when endothelial cells promote paracrine TGF-β signaling to the neighboring mural cells to promote their differentiation. Upon TGF-β binding to ALK5 expressed on mural cells, ALK5 phosphorylates Smad2/3 to promote the production of contractile proteins inducing cell quiescence and differentiation to mature VMSCs or pericytes (Owens 1998; Van Geest et al. 2010) (Fig. 10.3). Importantly, the establishment of close physical endothelial cell-mural cell contacts is absolutely required for TGF-β activation and mural cell differentiation (Hirschi et al. 2003). The mechanism by which latent TGF-β is converted into the active form to promote mural cell differentiation is not fully understood, but Connexin-43 and Connexin-45, Tissue Factor, and integrins such as $\alpha_v\beta8$ have been suggested to play a role (Thalgott et al. 2015). How defective TGF-β/BMP signaling in endothelial cells affect mural cell differentiation has also been studied. Carvalho et al. have analyzed TGF-β signaling in yolk sacs from endoglin knockout embryos. They revealed that endothelial disrupted TGF-β signaling in endothelial cells also affected

TGF-β/ALK5 signaling in the adjacent mesenchymal cells. Interestingly, they also demonstrated that application of exogenous TGF-β in cultured yolk sacs was sufficient to induce VSMC differentiation (Carvalho et al. 2004).

Targeting Blood Capillary Stability to Prevent Bleeding from HHT Vascular Malformations

Drugs promoting endothelial–mural cell interactions, inhibiting angiogenesis or stimulating endoglin/Alk1 signaling have been shown effective in HHT-mice and in incidental patient case-reports, suggesting that interventional targets based on vessel stabilization could be effective (Table 10.1).

Most notably in this context, we discovered several years ago that Thalidomide, which was used to treat multiple myeloma (MM), was found to prevent excessive nosebleeds (or epistaxis) in a small group of HHT patients (Lebrin et al. 2010; Wang et al. 2013; Invernizzi et al. 2015; Fang et al. 2017; Baysal et al. 2019). Oral administration of 100 mg of thalidomide daily significantly lowered the frequency and duration of epistaxis in the majority of patients with HHT, all within the first month of administration of the first dose. The dose given was comparable to that prescribed in 1960s to treat nausea during pregnancy. The average hemoglobin concentration in peripheral blood increased without additional iron supplementation. Some of these patients with HHT required between one to six blood transfusions before treatment to prevent anemia. No additional blood transfusion was required during thalidomide treatment. Long-term follow-up, however, showed that some patients became resistant to the drug and/or developed neuropathy (Lebrin et al. 2010; Alam et al. 2011; Hosman et al. 2015). Several clinical studies confirmed later that low dose of thalidomide was indeed effective to reduce bleeding in HHT patients (Lebrin et al. 2010; Fang et al. 2017; Harrison et al. 2018; Buscarini et al. 2019). We have showed that the anti-hemorrhagic property of thalidomide was not the result of direct inhibition of endothelial cell proliferation and migration, but is rather due to increased mural cell coverage of the vasculature. Thalidomide increased the number of pericytes and their recruitment to blood vessels, enhancing the apposition between the inner endothelial and supportive pericyte layers and resulting in vessel stabilization. Moreover, thalidomide was able to reduce brain AVM hemorrhage in a mouse model of focal loss of *Acvrl1* by improving mural cell coverage (Zhu et al. 2018). At the molecular level, others and we have revealed that thalidomide increased platelet-derived growth factor-B (PDGF-B) levels in endothelial cells to promote pericyte recruitment and vessel maturation and target pericytes directly to stimulate their proliferation and ability to form protrusions independently of PDGF-B (Lebrin et al. 2010; Zhu et al. 2018). These data have provided to our knowledge the first evidence that a therapy targeting pericytes to stimulate vessel maturation can have beneficial effects on bleeding from vascular malformations. However, thalidomide treatment has poor specificity, affecting a range of physiological processes and has side effects. The

Table 10.1 Investigated normalization-based therapies in HHT

Compound	Description	Mechanism of normalization	Preclinical evidence for use	Clinical trial identifier and status	Administration	Clinical results	References
Bevacizumab	Monoclonal antibody	Reduce VEGF bioavailability	PI3K inhibition prevents vascular defects in Alk1iΔEC mice. Genetic insertion of either the membrane or soluble form of VEGFR1 into the ROSA26 locus of Acvrl1+/− embryonic stem cell lines prevented the vascular anomalies, suggesting that high VEGFR2 activity in Acvrl1+/− endothelial cells induces HHT2 vascular anomalies. Anti-VEGF treatment did reduce lung VEGF levels but interestingly, led to an increase in peripheral pulmonary MVD and attenuation of RVH; it also normalized TSP-1 and Ang-2 expression.	NCT01314274 (C) NCT02389959 (R) NCT03227263 (R) NCT00843440 (C)	Intravenous injection intranasal mucosal injection	Reduction of epistaxis in 15 patients. Increase hemoglobin, reduced red cell units transfused and reduced quantity of iron infused in 13 patients. Epistaxis episodes and gastrointestinal bleeding are less frequent and severe. Hemoglobin levels are improved or normalized. Iron and transfusion requirements are reduced or stopped. Patients with severe hepatic vascular malformations show improved cardiac output.	Al-Samkari et al. (2019), Ola et al. (2018), Thalgott et al. (2018), Riss et al. (2015), Ardelean et al. (2014), Dupuis-Girod et al. (2012)

Pazopanib	Small chemical compound	Inhibit VEGFR2 activity	These results suggest that oral delivery of antiangiogenic TKIs is selectively more effective for GI bleeding than mucocutaneous AVMs. Infected Acvrl1+/− tracheas showed excessive angiogenesis with the formation of multiple telangiectases, vascular defects that were prevented by VEGFR2 blocking antibodies (DC101).	NCT02204371 (C)	Oral administration	Pazopanib improve dramatically the epistaxis, hemoglobin and iron levels. Small trial including seven patients showing improvement in hemoglobin levels and epistaxis in all treated subjects with reduce dependence to iron supplement and/or blood transfusions.	Thalgott et al. (2018), Parambil et al. (2018), Faughnan et al. (2019), Kim et al. (2017)

(continued)

Table 10.1 (continued)

Compound	Description	Mechanism of normalization	Preclinical evidence for use	Clinical trial identifier and status	Administration	Clinical results	References
Thalidomide	Small chemical compound	Target PDGF-b to promote pericyte recruitment to the endothelium	Thalidomide reduces the frequency and duration of nosebleeds in subjects with HHT by stimulating vessel maturation. Thalidomide and lenalidomide improve mural cell coverage of bAVM vessels and reduce bAVM hemorrhage.	NCT01485224 (C) NCT00389935 (C)	Oral administration	Frequency and duration of Epistasis are reduced, increase median hemoglobin levels and reduce blood transfusion dependence. Reduction of ESS score, increase of hemoglobin level in six patients. A successful treatment of recurrent bleeding due to gastrointestinal angiodysplasia has also been reported. Reduction of ESS score in seven patients treated.	Baysal et al. (2019), Zhu et al. (2018), Fang et al. (2017), Invernizzi et al. (2015), Wang et al. (2013), Alam et al. (2011), Lebrin et al. (2010)

Pomalidomide	Small chemical compound	Third generation thalidomide analogue, may promote blood vessel stability, although not experimentally demonstrated	Pomalidomide inhibit the formation of sprouts from human arterial rings, pomalidomide inhibit VEGF-induced endothelial cell cord formation. pomalidomide and thalidomide to inhibit the hypoxia-induced expression of HIF-1α and HIF-2α in endothelial cells. pomalidomide strongly inhibited T-regulatory cell proliferation and suppressor function in vitro	NCT02287558 (R)	Oral administration	NA	

(continued)

Table 10.1 (continued)

Compound	Description	Mechanism of normalization	Preclinical evidence for use	Clinical trial identifier and status	Administration	Clinical results	References
Tacrolimus	Small chemical compound	Partial correction of endoglin or ALK1 haploinsufficiency	In HUVECs, tacrolimus activated Smad1/5/8, reducing Dl14 expression, inhibite Akt and p38. In the BMP9/10-immunodepleted postnatal retina tacrolimus activated endothelial Smad1/5/8 and prevented the Dl14 overexpression and hypervascularization. In addition to its immunosuppressor properties, it is able to activate the TGF-β1/ALK1 pathway in endothelial cells, thus increasing the levels of endoglin and ALK1-haploinsufficient proteins in patients with HHT.	NCT03152019 (C)	Topical nasal ointment	This case report shows for the first time the striking beneficial effect of tacrolimus treatment in a HHT patient who suffered from severe epistaxis.	Sommer et al. (2018); Ruiz et al. (2017), Albinana et al. (2011)

Tranexamic acid	Small chemical compound	Partial correction of endoglin or ALK1 haploinsufficiency	TA, an antifibrinolytic drug, stimulates the expression of ALK-1 and endoglin, as well as the activity of the ALK-1/endoglin pathway.	NCT01031992 (C) NCT00355108 (C) NCT01408030 (C)	Oral administration	Reduction of epistaxis, increase of hemoglobin level and to transfusions needed for three patients treated. Double-blind randonized crossover trial including 22 patients with HHT that showed efficacy to treat nosebleeds. A randomized double-blind, cross-over trial was conducted on 118 patients. The study show a significant decrease in the duration of epistaxis.	Gaillard et al. (2014), Geisthoff et al. (2014), Fernandez et al. (2007), Sabbà et al. (2001)

C completed, *R* recruiting

development of thalidomide analogues, which retain the immunomodulatory effects of the parent compound, while minimizing the adverse reactions brought about a class of agents, termed the immunomodulatory drugs (IMiDs) that represent a promising new class of compounds for the treatment of cancers. Some are under clinical investigation and CC-5013 (lenalidomide) and CC-4047 (pomalidomide, actimid) have obtained FDA approvals for 5q-myelodysplasia and for multiple myeloma (MM). Recently, lenalidomide has been described to promote mural cell recruitment and to reduce brain AVM hemorrhage (Zhu et al. 2018). There is a novel clinical trial related to the use of pomalidomide to treat HHT, although the results are not available yet.

Alternatively, strategies targeting VEGF signaling have shown efficacy to prevent the formation of AVMs in preclinical models of HHT (Thalgott et al. 2018; Kim et al. 2017; Ola et al. 2016; Ardelean et al. 2014) and have been found effective to treat bleeding from HHT patients (Riss et al. 2015; Al-Samkari et al. 2019; Dupuis-Girod et al. 2012; Parambil et al. 2018; Faughnan et al. 2019). Several clinical studies are now completed and revealed that either intranasal submucosal delivery or intravenous injection of Bevacizumab reduced the duration and frequency of bleeds in patients with HHT (Riss et al. 2015; Al-Samkari et al. 2019; Dupuis-Girod et al. 2012). In contrast, a phase I double-blind randomized placebo-controlled study (ELLIPSE study, NCT01507480) that tested the efficacy of a nasal spray delivering Bevacizumab to treat patients with HHT failed to show beneficial effects (Dupuis-Girod et al. 2014). Also not fully proven, one explanation of the effects of the anti-VEGF therapies in HHT might be a reduction of VEGFR2 signaling activity in endothelial cells, VEGFR2 activity levels that may become normalized promoting blood vessel stabilization. Bevacizumab is quite expensive, alternatives using small chemical inhibitors such as VEGFR2 inhibitors have been tested both in animal models and in patients with HHT (Kim et al. 2017; Parambil et al. 2018; Faughnan et al. 2019).

Tacrolimus (FK-506) is a potent immunomodulator that has been shown to inhibit bleeding complications in one patient with HHT (Sommer et al. 2018). Its mechanism of action in HHT remains poorly understood, although a recent study points in the direction that tacrolimus could upregulate ALK1 and endoglin expression levels in endothelial cells, correcting haploinsufficiency in HHT models (Albinana et al. 2011; Ruiz et al. 2017). Tranexamic acid that can inhibit bleeding in HHT (Sabba et al. 2001; Geisthoff et al. 2014; Gaillard et al. 2014) has also been identified as a novel molecule stimulating ALK1 signaling activity in endothelial cells (Fernandez et al. 2007). By promoting ALK1 signaling, tacrolimus or tranexamic acid may stimulate blood vessel quiescence and mural cell attachment. Although this hypothesis needs to be experimentally proven.

Conclusions and Future Directions

Albeit rare, capillaries from HHT also exhibit the salient features of other vascular disorders associated with defects in mural cell function and such as provide tractable preclinical models of research and defined patient groups for clinical trials.

There are likely three major mechanisms underlying HHT: reduced blood vessel stability through mural cell dysfunction, excessive angiogenesis with weak-walled vessels, and chronic inflammation, which exacerbates symptoms. Targeting these mechanisms alone or in combination is likely to provide a set of drugs from which physicians can select the most effective for individual patients in a personalized approach to therapy.

Recently, drugs promoting vessel stabilization, stimulating ALK1 signaling activity, inhibiting angiogenesis and/or modulating immune pathways to induce blood vessel stabilization have been shown to be effective in preclinical models, indicating possible interventional targets for treating HHT vascular lesions. Some drugs that have been investigated mechanistically have already shown promising clinical outcomes in one disease symptom, namely reducing profuse nosebleeds in HHT patients. However, each drug has its own caveat: the drug mechanism of action may not always be well defined, the drug may lack specificity affecting a range of physiological processes, have serious (long-term) side effects or may be expensive. Minimal effective doses are also often unknown. Moreover, resistant and/or relapse have been reporter for these drugs. Having a repertoire of alternatives that would act via different pathways to promote vessel stabilization, i.e., re-enforcing endothelial cell–mural cell interactions may therefore be important to ensure long-term benefit when patients become treatment resistant and relapse, develop side effects or are nonresponsive to current treatments. The stabilization strategy opens new avenues for vascular therapy that may benefit not only patients with HHT but also patients with neurodegeneration, diabetic complications, or cancers.

References

Abdalla, S. A., & Letarte, M. (2006). Hereditary haemorrhagic telangiectasia: Current views on genetics and mechanisms of disease. *Journal of Medical Genetics, 43*(2), 97–110. https://doi.org/10.1136/jmg.2005.030833.

Alam MA, Sami S, Babu S (2011) Successful treatment of bleeding gastro-intestinal angiodysplasia in hereditary haemorrhagic telangiectasia with thalidomide. BMJ Case Rep 2011. https://doi.org/10.1136/bcr.08.2011.4585.

Al-Samkari, H., Kritharis, A., Rodriguez-Lopez, J. M., & Kuter, D. J. (2019). Systemic bevacizumab for the treatment of chronic bleeding in hereditary haemorrhagic telangiectasia. *Journal of Internal Medicine, 285*(2), 223–231. https://doi.org/10.1111/joim.12832.

Albinana V, Sanz-Rodriguez F, Recio-Poveda L, Bernabeu C, Botella LM (2011) Immunosuppressor FK506 increases endoglin and activin receptor-like kinase 1 expression and modulates transforming growth factor-beta1 signaling in endothelial cells. Mol Pharmacol 79 (5):833–843. https://doi.org/10.1124/mol.110.067447

Ardelean, D. S., Yin, M., Jerkic, M., Peter, M., Ngan, B., Kerbel, R. S., Foster, F. S., & Letarte, M. (2014). Anti-VEGF therapy reduces intestinal inflammation in Endoglin heterozygous mice subjected to experimental colitis. *Angiogenesis, 17*(3), 641–659. https://doi.org/10.1007/s10456-014-9421-x.

Armulik, A., Genove, G., & Betsholtz, C. (2011). Pericytes: Developmental, physiological, and pathological perspectives, problems, and promises. *Developmental Cell, 21*(2), 193–215. https://doi.org/10.1016/j.devcel.2011.07.001.

Armulik, A., Genove, G., Mae, M., Nisancioglu, M. H., Wallgard, E., Niaudet, C., He, L., Norlin, J., Lindblom, P., Strittmatter, K., Johansson, B. R., & Betsholtz, C. (2010). Pericytes regulate the blood-brain barrier. *Nature, 468*(7323), 557–561. https://doi.org/10.1038/nature09522.

Arthur, H. M., Ure, J., Smith, A. J., Renforth, G., Wilson, D. I., Torsney, E., Charlton, R., Parums, D. V., Jowett, T., Marchuk, D. A., Burn, J., & Diamond, A. G. (2000). Endoglin, an ancillary TGFbeta receptor, is required for extraembryonic angiogenesis and plays a key role in heart development. *Developmental Biology, 217*(1), 42–53. https://doi.org/10.1006/dbio.1999.9534.

Attwell, D., Mishra, A., Hall, C. N., O'Farrell, F. M., & Dalkara, T. (2016). What is a pericyte? *Journal of Cerebral Blood Flow and Metabolism, 36*(2), 451–455. https://doi.org/10.1177/0271678X15610340.

Baeyens, N., Larrivee, B., Ola, R., Hayward-Piatkowskyi, B., Dubrac, A., Huang, B., Ross, T. D., Coon, B. G., Min, E., Tsarfati, M., Tong, H., Eichmann, A., & Schwartz, M. A. (2016). Defective fluid shear stress mechanotransduction mediates hereditary hemorrhagic telangiectasia. *The Journal of Cell Biology, 214*(7), 807–816. https://doi.org/10.1083/jcb.201603106.

Bayrak-Toydemir, P., McDonald, J., Akarsu, N., Toydemir, R. M., Calderon, F., Tuncali, T., Tang, W., Miller, F., & Mao, R. (2006). A fourth locus for hereditary hemorrhagic telangiectasia maps to chromosome 7. *American Journal of Medical Genetics. Part A, 140*(20), 2155–2162. https://doi.org/10.1002/ajmg.a.31450.

Baysal, M., Umit, E. G., Kirkizlar, H. O., Ozdover, A. C., & Demir, A. M. (2019). Thalidomide for the management of bleeding episodes in patients with hereditary hemorrhagic telangiectasia: Effects on epistaxis severity score and quality of life. *Turkish Journal of Haematology, 36*(1), 43–47. https://doi.org/10.4274/tjh.galenos.2018.2018.0190.

Benzinou, M., Clermont, F. F., Letteboer, T. G., Kim, J. H., Espejel, S., Harradine, K. A., Arbelaez, J., Luu, M. T., Roy, R., Quigley, D., Higgins, M. N., Zaid, M., Aouizerat, B. E., van Amstel, J. K., Giraud, S., Dupuis-Girod, S., Lesca, G., Plauchu, H., Hughes, C. C., Westermann, C. J., & Akhurst, R. J. (2012). Mouse and human strategies identify PTPN14 as a modifier of angiogenesis and hereditary haemorrhagic telangiectasia. *Nature Communications, 3*, 616. https://doi.org/10.1038/ncomms1633.

Berthiaume, A. A., Hartmann, D. A., Majesky, M. W., Bhat, N. R., & Shih, A. Y. (2018). Pericyte structural remodeling in cerebrovascular health and homeostasis. *Frontiers in Aging Neuroscience, 10*, 210. https://doi.org/10.3389/fnagi.2018.00210.

Botella, L. M., Albinana, V., Ojeda-Fernandez, L., Recio-Poveda, L., & Bernabeu, C. (2015). Research on potential biomarkers in hereditary hemorrhagic telangiectasia. *Frontiers in Genetics, 6*, 115. https://doi.org/10.3389/fgene.2015.00115.

Bourdeau, A., Cymerman, U., Paquet, M. E., Meschino, W., McKinnon, W. C., Guttmacher, A. E., Becker, L., & Letarte, M. (2000). Endoglin expression is reduced in normal vessels but still detectable in arteriovenous malformations of patients with hereditary hemorrhagic telangiectasia type 1. *The American Journal of Pathology, 156*(3), 911–923. https://doi.org/10.1016/S0002-9440(10)64960-7.

Bourdeau, A., Dumont, D. J., & Letarte, M. (1999). A murine model of hereditary hemorrhagic telangiectasia. *The Journal of Clinical Investigation, 104*(10), 1343–1351. https://doi.org/10.1172/JCI8088.

Bourdeau, A., Faughnan, M. E., McDonald, M. L., Paterson, A. D., Wanless, I. R., & Letarte, M. (2001). Potential role of modifier genes influencing transforming growth factor-beta1 levels in the development of vascular defects in endoglin heterozygous mice with hereditary hemorrhagic telangiectasia. *The American Journal of Pathology, 158*(6), 2011–2020.

Brown, M. A., Zhao, Q., Baker, K. A., Naik, C., Chen, C., Pukac, L., Singh, M., Tsareva, T., Parice, Y., Mahoney, A., Roschke, V., Sanyal, I., & Choe, S. (2005). Crystal structure of BMP-9 and functional interactions with pro-region and receptors. *The Journal of Biological Chemistry, 280*(26), 25111–25118. https://doi.org/10.1074/jbc.M503328200.

Buscarini, E., Botella, L. M., Geisthoff, U., Kjeldsen, A. D., Mager, H. J., Pagella, F., Suppressa, P., Zarrabeitia, R., Dupuis-Girod, S., Shovlin, C. L., & Vascern, H. H. T. (2019). Safety of thalidomide and bevacizumab in patients with hereditary hemorrhagic telangiectasia. *Orphanet Journal of Rare Diseases, 14*(1), 28. https://doi.org/10.1186/s13023-018-0982-4.

Carvalho, R. L., Jonker, L., Goumans, M. J., Larsson, J., Bouwman, P., Karlsson, S., Dijke, P. T., Arthur, H. M., & Mummery, C. L. (2004). Defective paracrine signalling by TGFbeta in yolk sac vasculature of endoglin mutant mice: A paradigm for hereditary haemorrhagic telangiectasia. *Development, 131*(24), 6237–6247. https://doi.org/10.1242/dev.01529.

Chappell, J. C., Taylor, S. M., Ferrara, N., & Bautch, V. L. (2009). Local guidance of emerging vessel sprouts requires soluble Flt-1. *Developmental Cell, 17*(3), 377–386. https://doi.org/10.1016/j.devcel.2009.07.011.

Chen, H., Shi, S., Acosta, L., Li, W., Lu, J., Bao, S., Chen, Z., Yang, Z., Schneider, M. D., Chien, K. R., Conway, S. J., Yoder, M. C., Haneline, L. S., Franco, D., & Shou, W. (2004). BMP10 is essential for maintaining cardiac growth during murine cardiogenesis. *Development, 131*(9), 2219–2231. https://doi.org/10.1242/dev.01094.

Chen, W., Guo, Y., Walker, E. J., Shen, F., Jun, K., Oh, S. P., Degos, V., Lawton, M. T., Tihan, T., Davalos, D., Akassoglou, K., Nelson, J., Pile-Spellman, J., Su, H., & Young, W. L. (2013). Reduced mural cell coverage and impaired vessel integrity after angiogenic stimulation in the Alk1-deficient brain. *Arteriosclerosis, Thrombosis, and Vascular Biology, 33*(2), 305–310. https://doi.org/10.1161/ATVBAHA.112.300485.

Chen, W., Sun, Z., Han, Z., Jun, K., Camus, M., Wankhede, M., Mao, L., Arnold, T., Young, W. L., & Su, H. (2014a). De novo cerebrovascular malformation in the adult mouse after endothelial Alk1 deletion and angiogenic stimulation. *Stroke, 45*(3), 900–902. https://doi.org/10.1161/STROKEAHA.113.003655.

Chen, W., Young, W. L., & Su, H. (2014b). Induction of brain arteriovenous malformation in the adult mouse. *Methods in Molecular Biology, 1135*, 309–316. https://doi.org/10.1007/978-1-4939-0320-7_25.

Cheng, J., Korte, N., Nortley, R., Sethi, H., Tang, Y., & Attwell, D. (2018). Targeting pericytes for therapeutic approaches to neurological disorders. *Acta Neuropathologica, 136*(4), 507–523. https://doi.org/10.1007/s00401-018-1893-0.

Choi, E. J., Chen, W., Jun, K., Arthur, H. M., Young, W. L., & Su, H. (2014). Novel brain arteriovenous malformation mouse models for type 1 hereditary hemorrhagic telangiectasia. *PLoS One, 9*(2), e88511. https://doi.org/10.1371/journal.pone.0088511.

Choi, E. J., Walker, E. J., Shen, F., Oh, S. P., Arthur, H. M., Young, W. L., & Su, H. (2012). Minimal homozygous endothelial deletion of Eng with VEGF stimulation is sufficient to cause cerebrovascular dysplasia in the adult mouse. *Cerebrovascular Diseases, 33*(6), 540–547. https://doi.org/10.1159/000337762.

Cirulli, A., Liso, A., D'Ovidio, F., Mestice, A., Pasculli, G., Gallitelli, M., Rizzi, R., Specchia, G., & Sabba, C. (2003). Vascular endothelial growth factor serum levels are elevated in patients with hereditary hemorrhagic telangiectasia. *Acta Haematologica, 110*(1), 29–32. https://doi.org/10.1159/000072411.

Cole, S. G., Begbie, M. E., Wallace, G. M., & Shovlin, C. L. (2005). A new locus for hereditary haemorrhagic telangiectasia (HHT3) maps to chromosome 5. *Journal of Medical Genetics, 42*(7), 577–582. https://doi.org/10.1136/jmg.2004.028712.

Corti, P., Young, S., Chen, C. Y., Patrick, M. J., Rochon, E. R., Pekkan, K., & Roman, B. L. (2011). Interaction between alk1 and blood flow in the development of arteriovenous malformations. *Development, 138*(8), 1573–1582. https://doi.org/10.1242/dev.060467.

Crist, A. M., Lee, A. R., Patel, N. R., Westhoff, D. E., & Meadows, S. M. (2018). Vascular deficiency of Smad4 causes arteriovenous malformations: A mouse model of hereditary hemorrhagic telangiectasia. *Angiogenesis, 21*(2), 363–380. https://doi.org/10.1007/s10456-018-9602-0.

Cupovic, J., Onder, L., Gil-Cruz, C., Weiler, E., Caviezel-Firner, S., Perez-Shibayama, C., Rulicke, T., Bechmann, I., & Ludewig, B. (2016). Central nervous system stromal cells control local CD8(+) T cell responses during virus-induced neuroinflammation. *Immunity, 44*(3), 622–633. https://doi.org/10.1016/j.immuni.2015.12.022.

Daneman, R., Zhou, L., Kebede, A. A., & Barres, B. A. (2010). Pericytes are required for blood-brain barrier integrity during embryogenesis. *Nature, 468*(7323), 562–566. https://doi.org/10.1038/nature09513.

David, L., Mallet, C., Mazerbourg, S., Feige, J. J., & Bailly, S. (2007). Identification of BMP9 and BMP10 as functional activators of the orphan activin receptor-like kinase 1 (ALK1) in endothelial cells. *Blood, 109*(5), 1953–1961. https://doi.org/10.1182/blood-2006-07-034124.

Dias, D. O., Kim, H., Holl, D., Werne Solnestam, B., Lundeberg, J., Carlen, M., Goritz, C., & Frisen, J. (2018). Reducing pericyte-derived scarring promotes recovery after spinal cord injury. *Cell, 173*(1), 153–165.e122. https://doi.org/10.1016/j.cell.2018.02.004.

Dupuis-Girod, S., Ambrun, A., Decullier, E., Samson, G., Roux, A., Fargeton, A. E., Rioufol, C., Schwiertz, V., Disant, F., Chapuis, F., Donazzolo, Y., Paintaud, G., Edery, P., & Faure, F. (2014). ELLIPSE study: A phase 1 study evaluating the tolerance of bevacizumab nasal spray in the treatment of epistaxis in hereditary hemorrhagic telangiectasia. *MAbs, 6*(3), 794–799. https://doi.org/10.4161/mabs.28025.

Dupuis-Girod, S., Ginon, I., Saurin, J. C., Marion, D., Guillot, E., Decullier, E., Roux, A., Carette, M. F., Gilbert-Dussardier, B., Hatron, P. Y., Lacombe, P., Lorcerie, B., Riviere, S., Corre, R., Giraud, S., Bailly, S., Paintaud, G., Ternant, D., Valette, P. J., Plauchu, H., & Faure, F. (2012). Bevacizumab in patients with hereditary hemorrhagic telangiectasia and severe hepatic vascular malformations and high cardiac output. *JAMA, 307*(9), 948–955. https://doi.org/10.1001/jama.2012.250.

Fang, J., Chen, X., Zhu, B., Ye, H., Zhang, W., Guan, J., & Su, K. (2017). Thalidomide for epistaxis in patients with hereditary hemorrhagic telangiectasia: A preliminary study. *Otolaryngology and Head and Neck Surgery, 157*(2), 217–221. https://doi.org/10.1177/0194599817700573.

Faughnan, M. E., Gossage, J. R., Chakinala, M. M., Oh, S. P., Kasthuri, R., Hughes, C. C. W., McWilliams, J. P., Parambil, J. G., Vozoris, N., Donaldson, J., Paul, G., Berry, P., & Sprecher, D. L. (2019). Pazopanib may reduce bleeding in hereditary hemorrhagic telangiectasia. *Angiogenesis, 22*(1), 145–155. https://doi.org/10.1007/s10456-018-9646-1.

Fernandez, L. A., Garrido-Martin, E. M., Sanz-Rodriguez, F., Ramirez, J. R., Morales-Angulo, C., Zarrabeitia, R., Perez-Molino, A., Bernabeu, C., & Botella, L. M. (2007). Therapeutic action of tranexamic acid in hereditary haemorrhagic telangiectasia (HHT): Regulation of ALK-1/endoglin pathway in endothelial cells. *Thrombosis and Haemostasis, 97*(2), 254–262.

Fernandez, L. A., Sanz-Rodriguez, F., Zarrabeitia, R., Perez-Molino, A., Hebbel, R. P., Nguyen, J., Bernabeu, C., & Botella, L. M. (2005). Blood outgrowth endothelial cells from hereditary haemorrhagic telangiectasia patients reveal abnormalities compatible with vascular lesions. *Cardiovascular Research, 68*(2), 235–248. https://doi.org/10.1016/j.cardiores.2005.06.009.

Fruttiger, M. (2007). Development of the retinal vasculature. *Angiogenesis, 10*(2), 77–88. https://doi.org/10.1007/s10456-007-9065-1.

Gaillard, S., Dupuis-Girod, S., Boutitie, F., Riviere, S., Moriniere, S., Hatron, P. Y., Manfredi, G., Kaminsky, P., Capitaine, A. L., Roy, P., Gueyffier, F., Plauchu, H., & Group, A. S. (2014). Tranexamic acid for epistaxis in hereditary hemorrhagic telangiectasia patients: A European cross-over controlled trial in a rare disease. *Journal of Thrombosis and Haemostasis, 12*(9), 1494–1502. https://doi.org/10.1111/jth.12654.

Gallione, C. J., Repetto, G. M., Legius, E., Rustgi, A. K., Schelley, S. L., Tejpar, S., Mitchell, G., Drouin, E., Westermann, C. J., & Marchuk, D. A. (2004). A combined syndrome of juvenile polyposis and hereditary haemorrhagic telangiectasia associated with mutations in MADH4 (SMAD4). *Lancet, 363*(9412), 852–859. https://doi.org/10.1016/S0140-6736(04)15732-2.

Garrido-Martin, E. M., Nguyen, H. L., Cunningham, T. A., Choe, S. W., Jiang, Z., Arthur, H. M., Lee, Y. J., & Oh, S. P. (2014). Common and distinctive pathogenetic features of arteriovenous malformations in hereditary hemorrhagic telangiectasia 1 and hereditary hemorrhagic telangiectasia 2 animal models--brief report. *Arteriosclerosis, Thrombosis, and Vascular Biology, 34*(10), 2232–2236. https://doi.org/10.1161/ATVBAHA.114.303984.

Gautam, J., & Yao, Y. (2018). Roles of pericytes in stroke pathogenesis. *Cell Transplantation, 27*(12), 1798–1808. https://doi.org/10.1177/0963689718768455.

Geisthoff, U. W., Seyfert, U. T., Kubler, M., Bieg, B., Plinkert, P. K., & Konig, J. (2014). Treatment of epistaxis in hereditary hemorrhagic telangiectasia with tranexamic acid – a double-blind placebo-controlled cross-over phase IIIB study. *Thrombosis Research, 134*(3), 565–571. https://doi.org/10.1016/j.thromres.2014.06.012.

Gkatzis, K., Thalgott, J., Dos-Santos-Luis, D., Martin, S., Lamande, N., Carette, M. F., Disch, F., Snijder, R. J., Westermann, C. J., Mager, J. J., Oh, S. P., Miquerol, L., Arthur, H. M., Mummery, C. L., & Lebrin, F. (2016). Interaction between ALK1 signaling and Connexin40 in the development of arteriovenous malformations. *Arteriosclerosis, Thrombosis, and Vascular Biology, 36*(4), 707–717. https://doi.org/10.1161/ATVBAHA.115.306719.

Goumans, M. J., & Mummery, C. (2000). Functional analysis of the TGFbeta receptor/Smad pathway through gene ablation in mice. *The International Journal of Developmental Biology, 44*(3), 253–265.

Goumans, M. J., Valdimarsdottir, G., Itoh, S., Rosendahl, A., Sideras, P., & ten Dijke, P. (2002). Balancing the activation state of the endothelium via two distinct TGF-beta type I receptors. *The EMBO Journal, 21*(7), 1743–1753. https://doi.org/10.1093/emboj/21.7.1743.

Guimaraes-Camboa, N., Cattaneo, P., Sun, Y., Moore-Morris, T., Gu, Y., Dalton, N. D., Rockenstein, E., Masliah, E., Peterson, K. L., Stallcup, W. B., Chen, J., & Evans, S. M. (2017). Pericytes of multiple organs do not behave as mesenchymal stem cells in vivo. *Cell Stem Cell, 20*(3), 345–359. https://doi.org/10.1016/j.stem.2016.12.006.

Halderman, A. A., Ryan, M. W., Marple, B. F., Sindwani, R., Reh, D. D., & Poetker, D. M. (2018). Bevacizumab for epistaxis in hereditary hemorrhagic telangiectasia: An evidence-based review. *American Journal of Rhinology & Allergy, 32*, 258. https://doi.org/10.1177/1945892418768588.

Hall, C. N., Reynell, C., Gesslein, B., Hamilton, N. B., Mishra, A., Sutherland, B. A., O'Farrell, F. M., Buchan, A. M., Lauritzen, M., & Attwell, D. (2014). Capillary pericytes regulate cerebral blood flow in health and disease. *Nature, 508*(7494), 55–60. https://doi.org/10.1038/nature13165.

Han, C., Choe, S. W., Kim, Y. H., Acharya, A. P., Keselowsky, B. G., Sorg, B. S., Lee, Y. J., & Oh, S. P. (2014). VEGF neutralization can prevent and normalize arteriovenous malformations in an animal model for hereditary hemorrhagic telangiectasia 2. *Angiogenesis, 17*(4), 823–830. https://doi.org/10.1007/s10456-014-9436-3.

Harrison, L., Kundra, A., & Jervis, P. (2018). The use of thalidomide therapy for refractory epistaxis in hereditary haemorrhagic telangiectasia: Systematic review. *The Journal of Laryngology and Otology, 132*(10), 866–871. https://doi.org/10.1017/S0022215118001536.

Hill, R. A., Tong, L., Yuan, P., Murikinati, S., Gupta, S., & Grutzendler, J. (2015). Regional blood flow in the normal and ischemic brain is controlled by arteriolar smooth muscle cell contractility and not by capillary pericytes. *Neuron, 87*(1), 95–110. https://doi.org/10.1016/j.neuron.2015.06.001.

Hirschi, K. K., Burt, J. M., Hirschi, K. D., & Dai, C. (2003). Gap junction communication mediates transforming growth factor-beta activation and endothelial-induced mural cell differentiation. *Circulation Research, 93*(5), 429–437. https://doi.org/10.1161/01.RES.0000091259.84556.D5.

Hosman, A., Westermann, C. J., Snijder, R., Disch, F., Mummery, C. L., & Mager, J. J. (2015). Follow-up of thalidomide treatment in patients with hereditary haemorrhagic telangiectasia. *Rhinology, 53*(4), 340–344. https://doi.org/10.4193/Rhin14.289.

Invernizzi, R., Quaglia, F., Klersy, C., Pagella, F., Ornati, F., Chu, F., Matti, E., Spinozzi, G., Plumitallo, S., Grignani, P., Olivieri, C., Bastia, R., Bellistri, F., Danesino, C., Benazzo, M., & Balduini, C. L. (2015). Efficacy and safety of thalidomide for the treatment of severe recurrent epistaxis in hereditary haemorrhagic telangiectasia: Results of a non-randomised, single-centre, phase 2 study. *Lancet Haematology, 2*(11), e465–e473. https://doi.org/10.1016/S2352-3026(15)00195-7.

Jin, Y., Muhl, L., Burmakin, M., Wang, Y., Duchez, A. C., Betsholtz, C., Arthur, H. M., & Jakobsson, L. (2017). Endoglin prevents vascular malformation by regulating flow-induced cell migration and specification through VEGFR2 signalling. *Nature Cell Biology, 19*(6), 639–652. https://doi.org/10.1038/ncb3534.

Johnson, D. W., Berg, J. N., Baldwin, M. A., Gallione, C. J., Marondel, I., Yoon, S. J., Stenzel, T. T., Speer, M., Pericak-Vance, M. A., Diamond, A., Guttmacher, A. E., Jackson, C. E., Attisano, L., Kucherlapati, R., Porteous, M. E., & Marchuk, D. A. (1996). Mutations in the activin receptor-like kinase 1 gene in hereditary haemorrhagic telangiectasia type 2. *Nature Genetics, 13*(2), 189–195. https://doi.org/10.1038/ng0696-189.

Kappas, N. C., Zeng, G., Chappell, J. C., Kearney, J. B., Hazarika, S., Kallianos, K. G., Patterson, C., Annex, B. H., & Bautch, V. L. (2008). The VEGF receptor Flt-1 spatially modulates Flk-1 signaling and blood vessel branching. *The Journal of Cell Biology, 181*(5), 847–858. https://doi.org/10.1083/jcb.200709114.

Kim, Y. H., Choe, S. W., Chae, M. Y., Hong, S., & Oh, S. P. (2018). SMAD4 deficiency leads to development of arteriovenous malformations in neonatal and adult mice. *Journal of the American Heart Association, 7*(21), e009514. https://doi.org/10.1161/JAHA.118.009514.

Kim, Y. H., Kim, M. J., Choe, S. W., Sprecher, D., Lee, Y. J., & PO, S. (2017). Selective effects of oral antiangiogenic tyrosine kinase inhibitors on an animal model of hereditary hemorrhagic telangiectasia. *Journal of Thrombosis and Haemostasis, 15*(6), 1095–1102. https://doi.org/10.1111/jth.13683.

Kroon, S., Snijder, R. J., Faughnan, M. E., & Mager, H. J. (2018). Systematic screening in hereditary hemorrhagic telangiectasia: A review. *Current Opinion in Pulmonary Medicine, 24*(3), 260–268. https://doi.org/10.1097/MCP.0000000000000472.

Kruse, K., Lee, Q. S., Sun, Y., Klomp, J., Yang, X., Huang, F., Sun, M. Y., Zhao, S., Hong, Z., Vogel, S. M., Shin, J. W., Leckband, D. E., Tai, L. M., Malik, A. B., & Komarova, Y. A. (2019). N-cadherin signaling via trio assembles adherens junctions to restrict endothelial permeability. *The Journal of Cell Biology, 218*(1), 299–316. https://doi.org/10.1083/jcb.201802076.

Larrivee, B., Prahst, C., Gordon, E., del Toro, R., Mathivet, T., Duarte, A., Simons, M., & Eichmann, A. (2012). ALK1 signaling inhibits angiogenesis by cooperating with the notch pathway. *Developmental Cell, 22*(3), 489–500. https://doi.org/10.1016/j.devcel.2012.02.005.

Laux, D. W., Young, S., Donovan, J. P., Mansfield, C. J., Upton, P. D., & Roman, B. L. (2013). Circulating Bmp10 acts through endothelial Alk1 to mediate flow-dependent arterial quiescence. *Development, 140*(16), 3403–3412. https://doi.org/10.1242/dev.095307.

Lebrin, F., Deckers, M., Bertolino, P., & Ten Dijke, P. (2005). TGF-beta receptor function in the endothelium. *Cardiovascular Research, 65*(3), 599–608. https://doi.org/10.1016/j.cardiores.2004.10.036.

Lebrin, F., Goumans, M. J., Jonker, L., Carvalho, R. L., Valdimarsdottir, G., Thorikay, M., Mummery, C., Arthur, H. M., & ten Dijke, P. (2004). Endoglin promotes endothelial cell proliferation and TGF-beta/ALK1 signal transduction. *The EMBO Journal, 23*(20), 4018–4028. https://doi.org/10.1038/sj.emboj.7600386.

Lebrin, F., & Mummery, C. L. (2008). Endoglin-mediated vascular remodeling: Mechanisms underlying hereditary hemorrhagic telangiectasia. *Trends in Cardiovascular Medicine, 18*(1), 25–32. https://doi.org/10.1016/j.tcm.2007.11.003.

Lebrin, F., Srun, S., Raymond, K., Martin, S., van den Brink, S., Freitas, C., Breant, C., Mathivet, T., Larrivee, B., Thomas, J. L., Arthur, H. M., Westermann, C. J., Disch, F., Mager, J. J., Snijder, R. J., Eichmann, A., & Mummery, C. L. (2010). Thalidomide stimulates vessel maturation and reduces epistaxis in individuals with hereditary hemorrhagic telangiectasia. *Nature Medicine, 16*(4), 420–428. https://doi.org/10.1038/nm.2131.

Lesca, G., Olivieri, C., Burnichon, N., Pagella, F., Carette, M. F., Gilbert-Dussardier, B., Goizet, C., Roume, J., Rabilloud, M., Saurin, J. C., Cottin, V., Honnorat, J., Coulet, F., Giraud, S., Calender, A., Danesino, C., Buscarini, E., Plauchu, H., & French-Italian-Rendu-Osler N. (2007). Genotype-phenotype correlations in hereditary hemorrhagic telangiectasia: Data from the French-Italian HHT network. *Genetics in Medicine, 9*(1), 14–22. https://doi.org/10.1097/GIM.0b013e31802d8373.

Letteboer, T. G., Benzinou, M., Merrick, C. B., Quigley, D. A., Zhau, K., Kim, I. J., To, M. D., Jablons, D. M., van Amstel, J. K., Westermann, C. J., Giraud, S., Dupuis-Girod, S., Lesca, G., Berg, J. H., Balmain, A., & Akhurst, R. J. (2015). Genetic variation in the functional ENG allele inherited from the non-affected parent associates with presence of pulmonary arteriovenous malformation in hereditary hemorrhagic telangiectasia 1 (HHT1) and may influence expression of PTPN14. *Frontiers in Genetics, 6*, 67. https://doi.org/10.3389/fgene.2015.00067.

Letteboer, T. G., Mager, J. J., Snijder, R. J., Koeleman, B. P., Lindhout, D., Ploos van Amstel, J. K., & Westermann, C. J. (2006). Genotype-phenotype relationship in hereditary haemorrhagic telangiectasia. *Journal of Medical Genetics, 43*(4), 371–377. https://doi.org/10.1136/jmg.2005.035451.

Levet, S., Ouarne, M., Ciais, D., Coutton, C., Subileau, M., Mallet, C., Ricard, N., Bidart, M., Debillon, T., Faravelli, F., Rooryck, C., Feige, J. J., Tillet, E., & Bailly, S. (2015). BMP9 and BMP10 are necessary for proper closure of the ductus arteriosus. *Proceedings of the National Academy of Sciences of the United States of America, 112*(25), E3207–E3215. https://doi.org/10.1073/pnas.1508386112.

Li, D. Y., Sorensen, L. K., Brooke, B. S., Urness, L. D., Davis, E. C., Taylor, D. G., Boak, B. B., & Wendel, D. P. (1999). Defective angiogenesis in mice lacking endoglin. *Science, 284*(5419), 1534–1537.

Li, F., Lan, Y., Wang, Y., Wang, J., Yang, G., Meng, F., Han, H., Meng, A., Wang, Y., & Yang, X. (2011). Endothelial Smad4 maintains cerebrovascular integrity by activating N-cadherin through cooperation with notch. *Developmental Cell, 20*(3), 291–302. https://doi.org/10.1016/j.devcel.2011.01.011.

Mahmoud, M., Allinson, K. R., Zhai, Z., Oakenfull, R., Ghandi, P., Adams, R. H., Fruttiger, M., & Arthur, H. M. (2010). Pathogenesis of arteriovenous malformations in the absence of endoglin. *Circulation Research, 106*(8), 1425–1433. https://doi.org/10.1161/CIRCRESAHA.109.211037.

Matsubara, S., Bourdeau, A., terBrugge, K. G., Wallace, C., & Letarte, M. (2000). Analysis of endoglin expression in normal brain tissue and in cerebral arteriovenous malformations. *Stroke, 31*(11), 2653–2660.

McAllister, K. A., Grogg, K. M., Johnson, D. W., Gallione, C. J., Baldwin, M. A., Jackson, C. E., Helmbold, E. A., Markel, D. S., McKinnon, W. C., Murrell, J., et al. (1994). Endoglin, a TGF-beta binding protein of endothelial cells, is the gene for hereditary haemorrhagic telangiectasia type 1. *Nature Genetics, 8*(4), 345–351. https://doi.org/10.1038/ng1294-345.

McDonald, J., Bayrak-Toydemir, P., & Pyeritz, R. E. (2011). Hereditary hemorrhagic telangiectasia: An overview of diagnosis, management, and pathogenesis. *Genetics in Medicine, 13*(7), 607–616. https://doi.org/10.1097/GIM.0b013e3182136d32.

Minami, K., & Haji, T. (2016). Intranasal topical estrogen in the management of epistaxis in hereditary hemorrhagic telangiectasia. *Acta Oto-Laryngologica, 136*(5), 528–531. https://doi.org/10.3109/00016489.2015.1129070.

Miyazono, K., Kamiya, Y., & Morikawa, M. (2010). Bone morphogenetic protein receptors and signal transduction. *Journal of Biochemistry, 147*(1), 35–51. https://doi.org/10.1093/jb/mvp148.

Moens, S., Goveia, J., Stapor, P. C., Cantelmo, A. R., & Carmeliet, P. (2014). The multifaceted activity of VEGF in angiogenesis – implications for therapy responses. *Cytokine & Growth Factor Reviews, 25*(4), 473–482. https://doi.org/10.1016/j.cytogfr.2014.07.009.

Moya, I. M., Umans, L., Maas, E., Pereira, P. N., Beets, K., Francis, A., Sents, W., Robertson, E. J., Mummery, C. L., Huylebroeck, D., & Zwijsen, A. (2012). Stalk cell phenotype depends on integration of Notch and Smad1/5 signaling cascades. *Developmental Cell, 22*(3), 501–514. https://doi.org/10.1016/j.devcel.2012.01.007.

Nguyen, H. L., Lee, Y. J., Shin, J., Lee, E., Park, S. O., McCarty, J. H., & Oh, S. P. (2011). TGF-beta signaling in endothelial cells, but not neuroepithelial cells, is essential for cerebral vascular development. *Laboratory Investigation, 91*(11), 1554–1563. https://doi.org/10.1038/labinvest.2011.124.

Oh, S. P., Seki, T., Goss, K. A., Imamura, T., Yi, Y., Donahoe, P. K., Li, L., Miyazono, K., ten Dijke, P., Kim, S., & Li, E. (2000). Activin receptor-like kinase 1 modulates transforming growth factor-beta 1 signaling in the regulation of angiogenesis. *Proceedings of the National Academy of Sciences of the United States of America, 97*(6), 2626–2631.

Ola, R., Dubrac, A., Han, J., Zhang, F., Fang, J. S., Larrivee, B., Lee, M., Urarte, A. A., Kraehling, J. R., Genet, G., Hirschi, K. K., Sessa, W. C., Canals, F. V., Graupera, M., Yan, M., Young, L. H., Oh, P. S., & Eichmann, A. (2016). PI3 kinase inhibition improves vascular malformations in mouse models of hereditary haemorrhagic telangiectasia. *Nature Communications, 7*, 13650. https://doi.org/10.1038/ncomms13650.

Ola, R., Kunzel, S. H., Zhang, F., Genet, G., Chakraborty, R., Pibouin-Fragner, L., Martin, K., Sessa, W., Dubrac, A., & Eichmann, A. (2018). SMAD4 prevents flow induced arterial-venous malformations by inhibiting casein kinase 2. *Circulation, 138*, 2379. https://doi.org/10.1161/CIRCULATIONAHA.118.033842.

Owens, G. K. (1998). Molecular control of vascular smooth muscle cell differentiation. *Acta Physiologica Scandinavica, 164*(4), 623–635.

Parambil, J. G., Woodard, T. D., & Koc, O. N. (2018). Pazopanib effective for bevacizumab-unresponsive epistaxis in hereditary hemorrhagic telangiectasia. *The Laryngoscope, 128*(10), 2234–2236. https://doi.org/10.1002/lary.27129.

Pardali, E., & Ten Dijke, P. (2012). TGFbeta signaling and cardiovascular diseases. *International Journal of Biological Sciences, 8*(2), 195–213. https://doi.org/10.7150/ijbs.3805.

Park, S. O., Wankhede, M., Lee, Y. J., Choi, E. J., Fliess, N., Choe, S. W., Oh, S. H., Walter, G., Raizada, M. K., Sorg, B. S., & Oh, S. P. (2009). Real-time imaging of de novo arteriovenous malformation in a mouse model of hereditary hemorrhagic telangiectasia. *The Journal of Clinical Investigation, 119*(11), 3487–3496. https://doi.org/10.1172/JCI39482.

Pawlikowska, L., Nelson, J., Guo, D. E., McCulloch, C. E., Lawton, M. T., Kim, H., Faughnan, M. E., & Brain Vascular Malformation Consortium HHTIG. (2018). Association of common candidate variants with vascular malformations and intracranial hemorrhage in hereditary hemorrhagic telangiectasia. *Molecular Genetics & Genomic Medicine, 6*(3), 350–356. https://doi.org/10.1002/mgg3.377.

Peter, M. R., Jerkic, M., Sotov, V., Douda, D. N., Ardelean, D. S., Ghamami, N., Lakschevitz, F., Khan, M. A., Robertson, S. J., Glogauer, M., Philpott, D. J., Palaniyar, N., & Letarte, M. (2014). Impaired resolution of inflammation in the Endoglin heterozygous mouse model of chronic colitis. *Mediators of Inflammation, 2014*, 767185. https://doi.org/10.1155/2014/767185.

Ricard, N., Ciais, D., Levet, S., Subileau, M., Mallet, C., Zimmers, T. A., Lee, S. J., Bidart, M., Feige, J. J., & Bailly, S. (2012). BMP9 and BMP10 are critical for postnatal retinal vascular remodeling. *Blood, 119*(25), 6162–6171. https://doi.org/10.1182/blood-2012-01-407593.

Riss, D., Burian, M., Wolf, A., Kranebitter, V., Kaider, A., & Arnoldner, C. (2015). Intranasal submucosal bevacizumab for epistaxis in hereditary hemorrhagic telangiectasia: A double-blind, randomized, placebo-controlled trial. *Head & Neck, 37*(6), 783–787. https://doi.org/10.1002/hed.23655.

Rochon, E. R., Menon, P. G., & Roman, B. L. (2016). Alk1 controls arterial endothelial cell migration in lumenized vessels. *Development, 143*(14), 2593–2602. https://doi.org/10.1242/dev.135392.

Roman, B. L., Pham, V. N., Lawson, N. D., Kulik, M., Childs, S., Lekven, A. C., Garrity, D. M., Moon, R. T., Fishman, M. C., Lechleider, R. J., & Weinstein, B. M. (2002). Disruption of acvrl1 increases endothelial cell number in zebrafish cranial vessels. *Development, 129*(12), 3009–3019.

Rossi, E., Smadja, D. M., Boscolo, E., Langa, C., Arevalo, M. A., Pericacho, M., Gamella-Pozuelo, L., Kauskot, A., Botella, L. M., Gaussem, P., Bischoff, J., Lopez-Novoa, J. M., & Bernabeu, C. (2016). Endoglin regulates mural cell adhesion in the circulatory system. *Cellular and Molecular Life Sciences, 73*(8), 1715–1739. https://doi.org/10.1007/s00018-015-2099-4.

Ruiz, S., Chandakkar, P., Zhao, H., Papoin, J., Chatterjee, P. K., Christen, E., Metz, C. N., Blanc, L., Campagne, F., & Marambaud, P. (2017). Tacrolimus rescues the signaling and gene expression signature of endothelial ALK1 loss-of-function and improves HHT vascular pathology. *Human Molecular Genetics, 26*(24), 4786–4798. https://doi.org/10.1093/hmg/ddx358.

Ruiz, S., Zhao, H., Chandakkar, P., Chatterjee, P. K., Papoin, J., Blanc, L., Metz, C. N., Campagne, F., & Marambaud, P. (2016). A mouse model of hereditary hemorrhagic telangiectasia generated by transmammary-delivered immunoblocking of BMP9 and BMP10. *Scientific Reports, 5*, 37366. https://doi.org/10.1038/srep37366.

Sabba, C., Gallitelli, M., & Palasciano, G. (2001). Efficacy of unusually high doses of tranexamic acid for the treatment of epistaxis in hereditary hemorrhagic telangiectasia. *The New England Journal of Medicine, 345*(12), 926. https://doi.org/10.1056/NEJM200109203451216.

Sadick, H., Hage, J., Goessler, U., Bran, G., Riedel, F., Bugert, P., & Hoermann, K. (2008). Does the genotype of HHT patients with mutations of the ENG and ACVRL1 gene correlate to different expression levels of the angiogenic factor VEGF? *International Journal of Molecular Medicine, 22*(5), 575–580.

Sadick, H., Naim, R., Sadick, M., Hormann, K., & Riedel, F. (2005a). Plasma level and tissue expression of angiogenic factors in patients with hereditary hemorrhagic telangiectasia. *International Journal of Molecular Medicine, 15*(4), 591–596.

Sadick, H., Riedel, F., Naim, R., Goessler, U., Hormann, K., Hafner, M., & Lux, A. (2005b). Patients with hereditary hemorrhagic telangiectasia have increased plasma levels of vascular endothelial growth factor and transforming growth factor-beta1 as well as high ALK1 tissue expression. *Haematologica, 90*(6), 818–828.

Saharinen, P., Eklund, L., & Alitalo, K. (2017). Therapeutic targeting of the angiopoietin-TIE pathway. *Nature Reviews. Drug Discovery, 16*(9), 635–661. https://doi.org/10.1038/nrd.2016.278.

Scharpfenecker, M., van Dinther, M., Liu, Z., van Bezooijen, R. L., Zhao, Q., Pukac, L., Lowik, C. W., & ten Dijke, P. (2007). BMP-9 signals via ALK1 and inhibits bFGF-induced endothelial cell proliferation and VEGF-stimulated angiogenesis. *Journal of Cell Science, 120*(Pt 6), 964–972. https://doi.org/10.1242/jcs.002949.

Seki, T., Hong, K. H., & Oh, S. P. (2006). Nonoverlapping expression patterns of ALK1 and ALK5 reveal distinct roles of each receptor in vascular development. *Laboratory Investigation, 86*(2), 116–129. https://doi.org/10.1038/labinvest.3700376.

Shi, Y., & Massague, J. (2003). Mechanisms of TGF-beta signaling from cell membrane to the nucleus. *Cell, 113*(6), 685–700.

Shovlin, C. L. (2010). Hereditary haemorrhagic telangiectasia: Pathophysiology, diagnosis and treatment. *Blood Reviews, 24*(6), 203–219. https://doi.org/10.1016/j.blre.2010.07.001.

Sommer, N., Droege, F., Gamen, E., Geisthoff, U., Gall, H., Tello, K., Richter, M. J., Deubner, L., Schmiedel, R., Hecker, M., Spiekerkoetter, E. F., Wirsching, K., Seeger, W., Ghofrani, A., & Pullamsetti, S. S. (2018). EXPRESS: Treatment with low-dose tacrolimus inhibits bleeding complications in a patient with hereditary hemorrhagic telangiectasia and pulmonary arterial hypertension. *Pulmonary Circulation, 9*, 2045894018805406. https://doi.org/10.1177/2045894018805406.

Srinivasan, S., Hanes, M. A., Dickens, T., Porteous, M. E., Oh, S. P., Hale, L. P., & Marchuk, D. A. (2003). A mouse model for hereditary hemorrhagic telangiectasia (HHT) type 2. *Human Molecular Genetics, 12*(5), 473–482.

Stapor, P. C., Sweat, R. S., Dashti, D. C., Betancourt, A. M., & Murfee, W. L. (2014). Pericyte dynamics during angiogenesis: New insights from new identities. *Journal of Vascular Research, 51*(3), 163–174. https://doi.org/10.1159/000362276.

Sugden, W. W., Meissner, R., Aegerter-Wilmsen, T., Tsaryk, R., Leonard, E. V., Bussmann, J., Hamm, M. J., Herzog, W., Jin, Y., Jakobsson, L., Denz, C., & Siekmann, A. F. (2017). Endoglin controls blood vessel diameter through endothelial cell shape changes in response to haemodynamic cues. *Nature Cell Biology, 19*(6), 653–665. https://doi.org/10.1038/ncb3528.

ten Dijke, P., & Arthur, H. M. (2007). Extracellular control of TGFbeta signalling in vascular development and disease. *Nature Reviews. Molecular Cell Biology, 8*(11), 857–869. https://doi.org/10.1038/nrm2262.

Thalgott, J., Dos-Santos-Luis, D., & Lebrin, F. (2015). Pericytes as targets in hereditary hemorrhagic telangiectasia. *Frontiers in Genetics, 6*, 37. https://doi.org/10.3389/fgene.2015.00037.

Thalgott, J. H., Dos-Santos-Luis, D., Hosman, A. E., Martin, S., Lamande, N., Bracquart, D., Srun, S., Galaris, G., de Boer, H. C., Tual-Chalot, S., Kroon, S., Arthur, H. M., Cao, Y., Snijder, R. J., Disch, F., Mager, J. J., Rabelink, T. J., Mummery, C. L., Raymond, K., & Lebrin, F. (2018). Decreased expression of vascular endothelial growth factor receptor 1 contributes to the pathogenesis of hereditary hemorrhagic telangiectasia type 2. *Circulation, 138*(23), 2698–2712. https://doi.org/10.1161/CIRCULATIONAHA.117.033062.

Torsney, E., Charlton, R., Diamond, A. G., Burn, J., Soames, J. V., & Arthur, H. M. (2003). Mouse model for hereditary hemorrhagic telangiectasia has a generalized vascular abnormality. *Circulation, 107*(12), 1653–1657. https://doi.org/10.1161/01.CIR.0000058170.92267.00.

Tual-Chalot, S., Mahmoud, M., Allinson, K. R., Redgrave, R. E., Zhai, Z., Oh, S. P., Fruttiger, M., & Arthur, H. M. (2014). Endothelial depletion of Acvrl1 in mice leads to arteriovenous malformations associated with reduced endoglin expression. *PLoS One, 9*(6), e98646. https://doi.org/10.1371/journal.pone.0098646.

Tual-Chalot, S., Oh, S. P., & Arthur, H. M. (2015). Mouse models of hereditary hemorrhagic telangiectasia: Recent advances and future challenges. *Frontiers in Genetics, 6*, 25. https://doi.org/10.3389/fgene.2015.00025.

Van Geest, R. J., Klaassen, I., Vogels, I. M., Van Noorden, C. J., & Schlingemann, R. O. (2010). Differential TGF-{beta} signaling in retinal vascular cells: A role in diabetic retinopathy? *Investigative Ophthalmology & Visual Science, 51*(4), 1857–1865. https://doi.org/10.1167/iovs.09-4181.

van Gent, M. W., Post, M. C., Snijder, R. J., Westermann, C. J., Plokker, H. W., & Mager, J. J. (2010). Real prevalence of pulmonary right-to-left shunt according to genotype in patients with hereditary hemorrhagic telangiectasia: A transthoracic contrast echocardiography study. *Chest, 138*(4), 833–839. https://doi.org/10.1378/chest.09-1849.

Vanlandewijck, M., He, L., Mae, M. A., Andrae, J., Ando, K., Del Gaudio, F., Nahar, K., Lebouvier, T., Lavina, B., Gouveia, L., Sun, Y., Raschperger, E., Rasanen, M., Zarb, Y., Mochizuki, N., Keller, A., Lendahl, U., & Betsholtz, C. (2018). A molecular atlas of cell types and zonation in the brain vasculature. *Nature, 554*(7693), 475–480. https://doi.org/10.1038/nature25739.

Walker, E. J., Su, H., Shen, F., Choi, E. J., Oh, S. P., Chen, G., Lawton, M. T., Kim, H., Chen, Y., Chen, W., & Young, W. L. (2011). Arteriovenous malformation in the adult mouse brain resembling the human disease. *Annals of Neurology, 69*(6), 954–962. https://doi.org/10.1002/ana.22348.

Wang, X. Y., Chen, Y., & Du, Q. (2013). Successful treatment of thalidomide for recurrent bleeding due to gastric angiodysplasia in hereditary hemorrhagic telangiectasia. *European Review for Medical and Pharmacological Sciences, 17*(8), 1114–1116.

Wooderchak-Donahue, W. L., McDonald, J., O'Fallon, B., Upton, P. D., Li, W., Roman, B. L., Young, S., Plant, P., Fulop, G. T., Langa, C., Morrell, N. W., Botella, L. M., Bernabeu, C., Stevenson, D. A., Runo, J. R., & Bayrak-Toydemir, P. (2013). BMP9 mutations cause a vascular-anomaly syndrome with phenotypic overlap with hereditary hemorrhagic telangiectasia. *American Journal of Human Genetics, 93*(3), 530–537. https://doi.org/10.1016/j.ajhg.2013.07.004.

Xu, B., Wu, Y. Q., Huey, M., Arthur, H. M., Marchuk, D. A., Hashimoto, T., Young, W. L., & Yang, G. Y. (2004). Vascular endothelial growth factor induces abnormal microvasculature in the endoglin heterozygous mouse brain. *Journal of Cerebral Blood Flow and Metabolism, 24*(2), 237–244. https://doi.org/10.1097/01.WCB.0000107730.66603.51.

Yaniv, E., Preis, M., Shevro, J., Nageris, B., & Hadar, T. (2011). Anti-estrogen therapy for hereditary hemorrhagic telangiectasia – a long-term clinical trial. *Rhinology, 49*(2), 214–216. https://doi.org/10.4193/Rhino09.201.

Zaffar, N., Ravichakaravarthy, T., Faughnan, M. E., & Shehata, N. (2015). The use of antifibrinolytic agents in patients with HHT: A retrospective survey. *Annals of Hematology, 94*(1), 145–152. https://doi.org/10.1007/s00277-014-2169-y.

Zeisel, A., Hochgerner, H., Lonnerberg, P., Johnsson, A., Memic, F., van der Zwan, J., Haring, M., Braun, E., Borm, L. E., La Manno, G., Codeluppi, S., Furlan, A., Lee, K., Skene, N., Harris, K. D., Hjerling-Leffler, J., Arenas, E., Ernfors, P., Marklund, U., & Linnarsson, S. (2018). Molecular architecture of the mouse nervous system. *Cell, 174*(4), 999–1014.e1022. https://doi.org/10.1016/j.cell.2018.06.021.

Zhu, W., Chen, W., Zou, D., Wang, L., Bao, C., Zhan, L., Saw, D., Wang, S., Winkler, E., Li, Z., Zhang, M., Shen, F., Shaligram, S., Lawton, M., & Su, H. (2018). Thalidomide reduces hemorrhage of brain arteriovenous malformations in a mouse model. *Stroke, 49*(5), 1232–1240. https://doi.org/10.1161/STROKEAHA.117.020356.

Chapter 11
Pericytes in Primary Familial Brain Calcification

Yvette Zarb, Francesca Daniela Franzoso, and Annika Keller

Abstract Pericytes are perivascular cells along capillaries that are critical for the development of a functional vascular bed in the central nervous system and other organs. Pericyte functions in the adult brain are less well understood. Pericytes have been suggested to mediate functional hyperemia at the capillary level, regulate the blood–brain barrier and to give rise to scar tissue after spinal cord injury. Furthermore, pericyte loss has been suggested to precede cognitive decline in mouse models of Alzheimer's disease. Despite this observation, there is no convincing causality between pericyte loss and the pathogenesis of Alzheimer's disease. However, recent loss-of-function mutations in *PDGFB* and *PDGFRB* genes have implicated pericytes as the principle cell type affected in primary familiar brain calcification (PFBC), a neuropsychiatric disorder with dominant inheritance. Here we review the role of the PDGFB/PDGFRB signaling pathway in pericyte development and briefly discuss homeostatic functions of pericytes in the brain. We provide an overview of recent studies with mouse models of PFBC and discuss suggested pathogenic mechanisms for PFBC with special reference to pericytes.

Keywords Astrocyte · Blood–brain barrier · Cerebrovascular calcification
Movement disorder · Neurodegeneration · Neuropsychiatric disorder · Pericyte
Primary familial brain calcification · MYORG · PDGFB · PDGFRB · SLC20A2
XPR1

Yvette Zarb and Francesca Daniela Franzoso contributed equally to this work.

Y. Zarb · F. D. Franzoso · A. Keller (✉)
Department of Neurosurgery, Clinical Neuroscience Center, Zürich University Hospital,
Zürich University, Zürich, Switzerland
e-mail: Annika.Keller@usz.ch

A. Birbrair (ed.), *Pericyte Biology in Disease*, Advances in Experimental
Medicine and Biology 1147, https://doi.org/10.1007/978-3-030-16908-4_11

Introduction

Cerebrovascular disease is a major contributor to dementia in the elderly. Vascular dementia is the second most common cause of dementia after Alzheimer's disease (AD). However, the largest proportion of dementia cases show mixed pathology with combined features of AD and ischemic lesions (reviewed in Iadecola 2013). In addition, it is estimated that 3–6% cases of Parkinsonism have a vascular cause (reviewed in Korczyn 2015). The importance of brain vasculature in optimal brain functioning is further illustrated by monogenic diseases such as glucose deficiency syndrome (*SLC2A1*) (OMIM phenotype #606777, # 612126), polymicrogyria (*MFSD2A*) (OMIM phenotype #616486), and autism spectrum disorders (*SLC7A5*) (Tarlungeanu et al. 2016) where the primary defect is found in brain endothelial cells. Endothelial cells line the vessel lumen and possess several brain-specific characteristics collectively referred to as the blood–brain barrier (BBB). The expression of nutrient transporters at the BBB ensures the delivery of nutrients into the brain parenchyma since endothelial cell–cell junctions do not allow a free passage of circulating molecules. The above-listed diseases arise because brain endothelial cells fail to allow the delivery of energy and nutrients into the brain. Relatively little is known about disease causing mutations in genes expressed by mural cells that reside on the abluminal side and provide structural stability and regulate vascular tone. Mural cells can be further divided into vascular smooth muscle cells (covering arteries, arterioles, veins) and pericytes (covering capillaries). However, mutations in *NOTCH3* which is expressed by vascular smooth muscle cells and pericytes cause CADASIL (OMIM phenotype # 125310), the most common form of hereditary stroke. NOTCH3 mutations lead to degeneration of vascular smooth cells around small arteries and arterioles that lead to luminal stenosis (reviewed in Ayata 2010). Although pericyte degeneration has been described in CADASIL patients (Dziewulska and Lewandowska 2012), the role of pericytes in CADSIL in the pathogenesis is unclear. Pericyte deficiency/dysfunction has been suggested to precede and drive cognitive decline in mouse models of AD (reviewed in Sweeney et al. 2016); however, whether pericyte loss is causal for AD has not been demonstrated. The discovery that mutations in *PDGFB* and *PDGFRB* genes cause a neurodegenerative disease that manifests with basal ganglia calcifications has raised the question of pericyte involvement in the pathogenesis of primary familial brain calcification (PFBC). Here, we provide a brief background on the PDGFB/PDGFRB pathway in pericyte development as well as mouse models of PFBC and discuss the evidence for pericyte involvement in PFBC.

Development of CNS Blood Vessels and the Role of PDGFB/PDGFRB in Pericyte Recruitment

The CNS vasculature develops through the process of angiogenesis. In mice, angiogenic sprouting is initiated from the perineural vascular plexus formed by E8.5 around the developing neural tube (Hogan and Bautch 2004). Cerebral vascularization is directed by Wnt7a/Wnt7b, which is expressed by neuronal cells via interactions with membrane proteins including Lrp5, Gpr124, Fdz, Reck expressed by vascular endothelial cells (Cho et al. 2017). Pericytes enter the CNS parenchyma together with endothelial cells. The platelet-derived growth factor B (PDGFB) and platelet-derived growth factor receptor beta (PDGFRB) signaling axis is important for mural cell (pericytes and vascular smooth muscle cells (VSMC)) investment and recruitment to developing vasculature. The targeted knockout of *Pdgfb* and *Pdgfrb* in mice indicates that in the absence of PDGFB or PDGFRB, vascular mural cells fail to expand during angiogenic growth of blood vessels in embryonic and early postnatal phases (Lindahl et al. 1997; Hellström et al. 1999). PDGFB is synthesized and secreted by endothelial cells as a dimer into the extracellular matrix, where it binds to heparan sulfate proteoglycans. This ensures the localization of PDGF-BB around angiogenic endothelial cells. Extracellular PDGF-BB acts on neighboring PDGFRB positive mural cells to elicit their expansion and co-migration along angiogenic vessel sprouts. In the absence of either *Pdgfb* or *Pdgfrb*, mural cell investment along the vessels is poor and the resulting capillary bed is largely devoid of pericytes (reviewed in Armulik et al. 2011). Several mouse genetic studies have shown that without proper pericyte coverage or endothelial-pericyte cross-talk, blood vessels become hemorrhagic and/or show irregularities in vessel diameter (reviewed in Armulik et al. 2011). The resulting molecular changes in blood vessels due to the altered pericyte-endothelial cross-talk are less well understood.

Homeostatic Roles of Pericytes in the Adult CNS

The vasculature in any given organ has specific characteristics that are tailored to the vascular needs of the organ in question. In the CNS, endothelial cells comprise features collectively referred to as a blood–brain barrier (BBB). The BBB ensures the delivery of nutrients, while preventing the entry of xenobiotics into the brain parenchyma. The term "BBB" fails to underline the dynamic nature of the blood–brain interface where BBB function is collectively established by all cell types constituting to the neurovascular unit and is regulated by the physiological state of the organism. A number of functions have been attributed to pericytes in adult vasculature, some of which are relevant only in the pathological setting (reviewed in Armulik et al. 2011). However, the function(s) of pericytes in the adult brain during homeostasis is far from being fully understood. Several models of adult *Pdgfb* mouse mutants have demonstrated that pericytes play a critical role in the BBB by negatively regulating endothelial transcytosis (Armulik et al. 2010). A wide range of tracers that differ in molecular size and chemical composition have been shown to pass from blood to brain by vesicular transport (possibly by pinocytosis) in pericyte-deficient mice. However, the mechanism by which pericytes control endothelial transcytosis in the brain is currently unclear. Importantly, pericytes do not regulate the major CNS-specific physical barrier and molecular properties of the

BBB. Pericyte-deficient brain vessels express the general molecular hallmarks of BBB endothelium, and their endothelial junctions appear normal by ultrastructural analysis (Armulik et al. 2010). In addition to altered BBB permeability, pericyte-deficient vessels show abnormal astrocyte end-feet polarization (Armulik et al. 2010). The molecular pathway by which pericytes induce astrocyte end-feet polarization is not known. However, pericyte-derived components of the extracellular matrix (e.g., Lama2) may be necessary for localization of aquaporin 4 and dystrophin complexes on astrocyte endfeet (Tham et al. 2016; Menezes et al. 2014; Gautam et al. 2016).

It is unclear whether acute pericyte loss in an adult organism leads to a similar phenotype seen in various adult *Pdgfb* mutants. Acute ablation of a single pericyte resulted in a widening of vessel diameter. Interestingly, uncovered regions of the endothelium were covered by the extension of processes from neighboring pericytes (Berthiaume et al. 2018).

Pericytes have been also suggested to mediate blood flow changes evoked by functional hyperemia (Hall et al. 2014; Peppiatt et al. 2006). This topic has remained controversial possibly due to the lack of consensus regarding the definition of pericyte coverage along the vascular bed (artery-arteriole-capillary-venule). Recent, carefully conducted two-photon imaging studies (Hartmann et al. 2015; Hill et al. 2015) and single-cell transcriptional profiling of pericytes (Vanlandewijck et al. 2018) have shed light on this issue. Most likely, the controversies stem from slight differences in the precise anatomical localization of investigated mural cells/pericytes.

Recent years have seen an increased interest in understanding vascular changes during aging and neurodegenerative diseases. In addition to thickening of the vessel-basement membrane and thinning of the endothelial lining, morphological changes in pericytes have been reported in preclinical models and human cases (reviewed in Erdo et al. 2017). Studies from the B. Zlokovic lab have shown that *Pdgfrb* heterozygosity (*Pdgfrb*$^{+/-}$) and reduced signaling capacity (*Pdgfrb*$^{F7/F7}$) cause pericyte loss in old mice, which in turn accelerates age-dependent BBB breakdown and cerebral hypoperfusion. These processes often precede neurodegeneration, cognitive impairment and white matter dysfunction (Bell et al. 2010; Kisler et al. 2017; Montagne et al. 2018). However, other studies have not observed an age-dependent loss of pericytes in old *Pdgfrb* heterozygote mice, even in the absence of one functional *Pdgfb* allele (Vanlandewijck et al. 2015). Thus, it remains unclear if observed functional differences in old *Pdgfrb*$^{+/-}$ and *Pdgfrb*$^{F7/F7}$ animals are caused by pericyte loss or result from an altered pericyte, or another, PDGFRB expressing cell. There may be other functions of PDGFB and PDGFRB in the brain that have gone undetected in studies of vascular development on various mouse *Pdgfb* and *Pdgfrb* mutants. Cell types other than vascular cells express PDGFB and PDGFRB in the CNS. Several studies have implicated a neuroprotective role for PDGFB and PDGFRB *in vivo* (Ishii et al. 2006; Shen et al. 2012), as well as in memory and cognitive function (Phuong et al. 2011; Ishii et al. 2008; Xu et al. 2013). Several studies have reported pericyte loss in human autopsy samples and mouse models of neurodegenerative diseases such as AD and amyotrophic lateral sclerosis which were suggested to be caused by an altered PDGFB/PDGFRB signaling (reviewed in

Sweeney et al. 2016), and no causal relationship between a neurodegenerative disease and an altered PDGFB/PDGFRB signaling in human could be found until 2013. Unexpectedly, loss-of-function mutations in PDGFRB and its ligand PDGFB were reported to cause idiopathic basal ganglia calcification, a familial neurodegenerative disease, which has been renamed *primary familial brain calcification* (Keller et al. 2013; Nicolas et al. 2013a).

Genetics of Primary Familial Brain Calcification

PFBC is a dominantly inherited neurodegenerative disease with a heterogeneous clinical presentation (parkinsonism, psychosis, seizures, migraine, cognitive decline and impairment, and cerebellar involvement (ataxia)) (Quintans et al. 2018). Recent studies on genetically confirmed PFBC patients have shown that psychiatric signs are the most common clinical feature in symptomatic patients, followed by cognitive impairment and movement disorders (Nicolas et al. 2015). After the exclusion of all causes of secondary calcification (e.g., viral infection, altered levels of parathyroid hormones, and systemic mineral imbalance), the presence of intracranial bilateral basal ganglia calcifications is critical for the diagnosis of the disease (Quintans et al. 2018). PFBC shows incomplete clinical penetrance; however, radiological penetrance of the disease is 100% at the age of 50 (Tadic et al. 2015). Affected brain regions may also include the cerebellum, thalamus, cortical white and gray matter, and the brain stem (Manyam 2005). A recent analysis of total calcification score of genetically confirmed PFBC cases identified four patterns of calcification which correlated positively with patients age (Nicolas et al. 2015). Some patients with brain calcifications are asymptomatic. Historically, PFBC was first described by Delacour in 1850 (Delacour 1850). There are more than 30 different names in literature that have been used to describe this condition (Manyam 2005). The most common is "Fahr's disease"; however, this name is now considered a misnomer (Westenberger and Klein 2014).

Deleterious mutations in four genes are linked to autosomal dominant PFBC (AD-PFBC). In addition to mutations in *PDGFRB* (OMIM # 615007) (Nicolas et al. 2013a) and *PDGFB* (OMIM # 615483) (Keller et al. 2013), mutations in the sodium-dependent P(i) transporter *SLCA20A2* (also called PiT2) (OMIM #213600) (Wang et al. 2012) and inorganic phosphate exporter – *XPR1* (OMIM # 616413) (Legati et al. 2015) have been described. The estimated minimal prevalence of PFBC is 4.5 p. 10,000, suggesting that PFBC is not a rare disorder and is likely underdiagnosed (Nicolas et al. 2018). Mutations in the *SLCA20A2* gene are a major cause of PFBC, representing approximately 50% of investigated families (reviewed in Taglia et al. 2015). However, four known PFBC genes do not account for all the cases of PFBC, indicating the presence of other causative PFBC genes. Recently, biallelic recessive mutations in *MYORG* were reported to cause autosomal recessive form of primary familial brain calcification (Yao et al. 2018).

Pathology of Primary Familiar Brain Calcification

Pathological findings from rare autopsy cases and *in vivo* imaging data from case studies reveal the evidence of microvascular insufficiency. Histologically, calcified nodules encrust capillaries (Norman and Urich 1960; Kozik and Kulczycki 1978; Miklossy et al. 2005). Furthermore, calcium precipitates have been reported on neurons and astrocytes (Miklossy et al. 2005). Neurons generally remain preserved in PFBC (Miklossy et al. 2005), although neuronal loss has been reported in severely affected areas (Kozik and Kulczycki 1978). In addition, extravasation of plasma proteins and signs of neuroinflammation around calcifications are observed (Miklossy et al. 2005). Noninvasive imaging techniques such as MRI reveal the presence of vasogenic edema in PFBC patients (Gomez et al. 1989). In addition to brain edema, white matter abnormalities have been described in one family carrying PDGFB mutations (Biancheri et al. 2016). Severe calcifications in basal ganglia can lead to changes in regional blood flow (Uygur et al. 1995; Paschali et al. 2009; Shoyama et al. 2005) and reduction in glucose metabolism (Shoyama et al. 2005; Saito et al. 2010). PET studies in idiopathic and genetically confirmed PFBC cases have demonstrated pre- and postsynaptic nigro-striatal dopaminergic dysfunction (Saito et al. 2010; Koyama et al. 2017). Interestingly, several cases of sporadic and familiar PFBC have been described to have neurofibrillary tangles, Lewy bodies or other evidence of α-synucleinopathy in addition to vessel calcification (Nomoto et al. 2002; Shibayama et al. 1986; Ikeda et al. 1994). *PDGFRB*, *PDGFB* and *XPR1* mutation carriers show microangiopathy also in skin vessels (Biancheri et al. 2016; Nicolas et al. 2017). Thus, there is considerable heterogeneity of the histopathological findings of the PFBC. However, many reports on PFBC pathology and neurology appeared before the genetic causes of PFBC were known, and may represent diverse disease etiologies. It is unclear to what extent different histopathological and neurological features are causally correlated. Calcifications have been found to be more severe in symptomatic versus asymptomatic individuals, but the type of symptoms and calcification site were independent (Nicolas et al. 2013b, 2015; Tadic et al. 2015).

Impact of PFBC Mutations on Protein Function

Genetic data indicate that mutations in PFBC genes are loss-of-function (LOF) mutations leading to haploinsufficiency as the proposed disease mechanism. The identification of large genomic deletions in PFBC genes (e.g., *SLC20A2, PDGFB)* further supports this conclusion (Baker et al. 2013; Nicolas et al. 2014). The functional consequences of selected PFBC mutations have been studied *in vitro*. Mutations in *SLC20A2* (S601W, S601L, T595M, E575K, G498R, and V42del) and *XPR1* (S136N, L140P, L145P, and L218S)) did not affect the cell surface expression but rather reduced the inorganic phosphate (Pi) transporting capacity of SLC20A2 and XPR1 when tested *in vitro* (Wang et al. 2012; Legati et al. 2015; Yao et al.

2017). Co-expression of the mutant *SLC20A2* (S601W or E575K) with wild-type *SLC20A2* in *Xenopus* oocytes and mutant *XPR1* (L145P) with wild-type *XPR1* in mammalian cells had no obvious effect on the Pi-transport activity of wild-type proteins, thus supporting the haploinsufficiency as a mechanism for disease (Wang et al. 2012; Yao et al. 2017). However, a recent functional study in mammalian cells showed that the expression of *SLC20A2* variants-associated PFBC can exert dominant negative effects on the Pi-transport function of wild-type SLC20A2. Accordingly, this effect could result in more severely impaired cellular Pi-transport function than the absence of a functional protein from one allele (Larsen et al. 2017). In addition, mutations have been reported to affect protein cell surface expression (*XPR1* - L87P) or alter the subcellular localization of the protein (*SLC20A2*—p.Trp626_Thr629dup) (Anheim et al. 2016; Taglia et al. 2018). Thus, the molecular mechanisms of *SLC20A2* and *XPR1* mutations leading to PFBC might be heterogeneous.

Six analyzed PDGFB mutations (M1?, *242Yext*89, L119P, Q145*, Q147*, and R149*) reported by Keller et al. (2013) caused complete loss of functional PDGF-BB protein production in HEK293 cells (Vanlandewijck et al. 2015). *In vitro* functional studies of *PDGFRB* mutations described in PFBC families have revealed that these mutations cause complete or reduced autophosphorylation (p.L658P, p. R695C) or reduced protein production (p.R987W) of PDGFRB when expressed *in vitro* individually (Vanlandewijck et al. 2015; Sanchez-Contreras et al. 2014). It would be informative to investigate whether *PDGFB* and *PDGFRB* mutations described in PFBC patients act as a dominant-negative in the presence of a single wild-type allele.

Mouse Models and the Pathogenic Mechanism of Primary Familial Brain Calcification

Analysis of mouse models that mimic certain aspects of PFBC have advanced our understanding of the pathogenesis of this neuropsychiatric disease. However, it is not known why mutations in genes belonging to structurally and functionally different protein families such as growth factor and its receptor (*PDGFB, PDGFRB*), the inorganic phosphate transporters (*SLC20A2, XPR1*), intracellular putative glycosidase (*MYORG*) lead to the common pathology of brain calcification (Yao et al. 2018; Keller et al. 2013; Jensen et al. 2013). No interaction between SLC20A2 and PDGFRB has been reported. However, a recent study showed that PDGFRB and XPR1 might form complexes on the cell membrane (Yao et al. 2017).

Currently, four genetically modified mouse lines that have reduced levels of either *Slc20a2*, *Pdgfb*, *Myorg* have been reported to develop intracranial brain calcifications (Keller et al. 2013; Yao et al. 2018; Jensen et al. 2013). As discussed above, the PDGFB/PDGFRB signaling pathway is important for the development of functional neurovascular unit. However, the developmental role of other PFBC genes in the brain are not known. Interestingly, XPR1 (*xpr1b*) deletion in zebrafish results in defective differentiation of tissue-resident macrophages, including microglia (Meireles et al. 2014). XPR1 is used by xenotropic and polytropic murine leukemia viruses to enter cells but the cellular function of XPR1 remained elusive. In 2013, it was reported that XPR1 mediates phosphate export from mammalian cells. Given the high conservation of XPR1 sequences from fly to human, XPR1 most likely mediates phosphate export in metazoans (Giovannini et al. 2013). Selective knockout of *Xpr1* in renal tubule leads to hypophosphatemic rickets secondary to renal dysfunction, further substantiating the role of XPR1 in phosphate homeostasis (Ansermet et al. 2017). The cellular function of MYORG, which encodes an intracellular transmembrane protein belonging to glycosyl hydrolase family, is currently unknown. Knock-out *Myorg* (AI464131) in mice leads no gross developmental defects (Yao et al. 2018). Developmental abnormalities have not been described in *Slc20a2*$^{-/-}$ knockout embryos; however, heterozygosity of *Slc20a2* in pregnant females leads to placental calcification (Wallingford et al. 2016). All five PFBC genes show a different expression pattern in the developing mouse embryo, in the adult mouse and in human brain (Figs. 11.1 and 11.2), which are expressed by several cell types. However, the expression of *Myorg* has been exclusively attributed to S100β positive astrocytes in mouse brain (Yao et al. 2018) (Fig. 11.1c, d).

PDGFB/PDGFRB Signaling Axis

Complete knock-out of *Pdgfb* and *Pdgfrb* in mice is lethal. Postnatal functions of PDGFB and PDGFRB have been investigated using mutations that change the function of the respective gene (Heuchel et al. 1999; Tallquist et al. 2003; Lindblom et al. 2003; Bjarnegard et al. 2004; Olson and Soriano 2011), or are inducible null mutations (Bjarnegard et al. 2004; Enge et al. 2002; Nguyen et al. 2011). Most studies using these mouse lines have focused on the role of pericytes in an adult organism on the endothelial function or pathogenesis of AD (discussed above). Therefore, the finding that the *Pdgfb* retention motif knockout mice (*Pdgfb*$^{ret/ret}$) develop vessel-associated brain calcifications in deep brain regions that increase with age is unexpected (Keller et al. 2013). *Pdgfb*$^{ret/ret}$ animals lack a C-terminal so-called "retention motif," which binds heparan sulfate proteoglycans due to a targeted mutation at the *Pdgfb* locus. PDGF-BB retains its receptor binding/activating ability but is diffusible, which most likely leads to lower extracellular concentrations of PDGF-BB near the producer cell (Lindblom et al. 2003; Abramsson et al. 2007), thus functioning as a hypomorph.

Fig. 11.1 (continued) of the four PFBC genes in brain vascular (PC – pericyte, vSMC – venous smooth muscle cells (SMC), aaSMC – arteriolar SMC, aSMC – arterial SMC, MG – microglia, FB1 – fibroblast-like 1, FB2 – fibroblast-like 2, OL – oligodendrocyte, EC1 – endothelial cell (EC) type 1, EC2 – EC type 2, EC3 – EC type 3, vEC – venous EC, caplEC – capillary EC, aEC – arterial EC, AC – astrocyte) obtained using single-cell RNAseq. Data were taken from http://betsholtzlab.org/VascularSingleCells/database.html (Vanlandewijck et al. 2018)

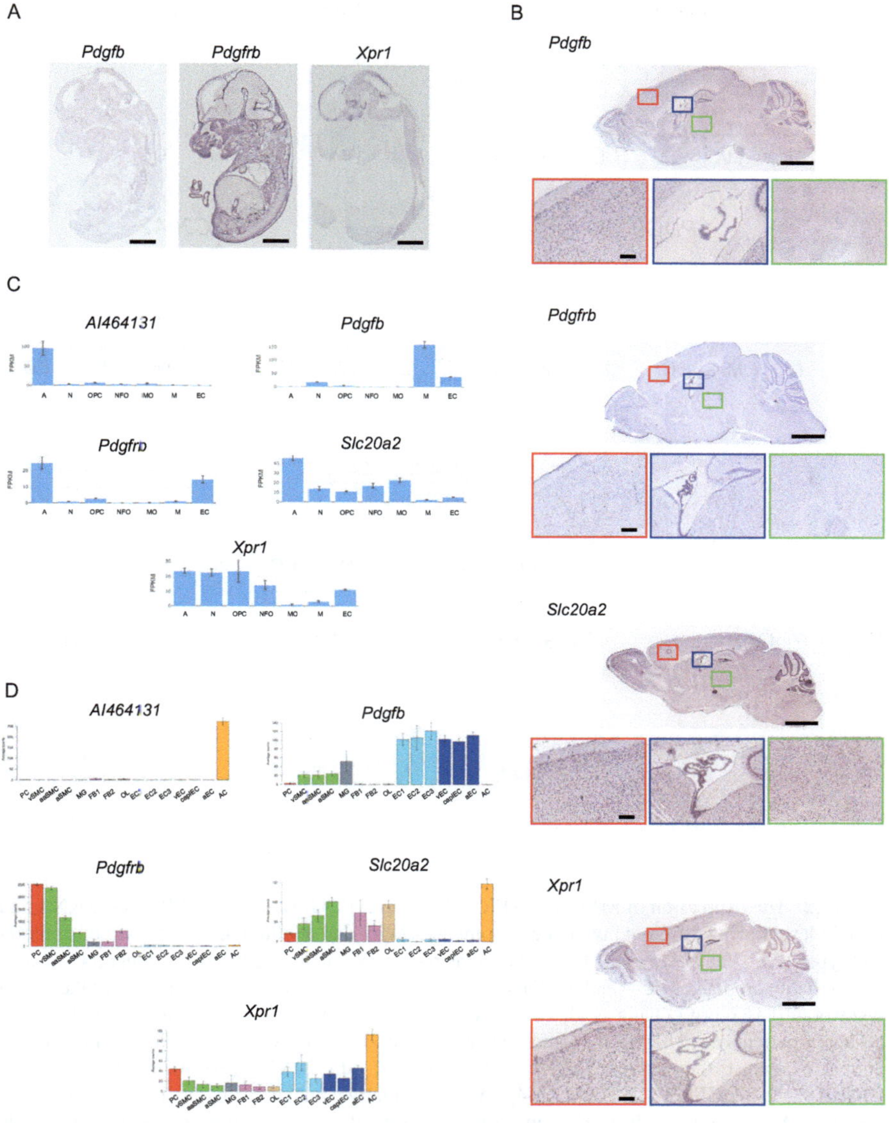

Fig. 11.1 The expression of *Ai464131 (MYORG), Xpr1, Slc20a2, Pdgfb*, and *Pdgfrb* at the RNA level in different cell types and tissue in mouse. (**a**) Sagittal sections of mouse embryos at day 14.5 (section thickness 20 μm and inter-section distance of 100 μm). Scale bar 400 μm. Images were taken from GenePaint public database available at website: http://www.genepaint.org and described in (Visel et al. 2004). (**b**) Sagittal brain sections of male mouse at P56. Image credit: Allen Institute. Images are available at website: http://mouse.brain-map.org/. Scale bar is 2098 μm. Zoomed figures represent cortex (red lining), choroid plexus (blue lining), midbrain (green lining). Scale bar is 839 μm. (**c**) RNAseq of different cell types (A – astrocyte, N – neuron, OPC – oligodendrocyte progenitor cell, NFO – newly formed oligodendrocyte, MO – myelinating oligodendrocyte, M – microglia/macrophage, EC – endothelial cell) isolated from mouse brains. Data were taken from www.brainrnaseq.org (Zhang et al. 2014). (**d**) A graphical representation of the average expression

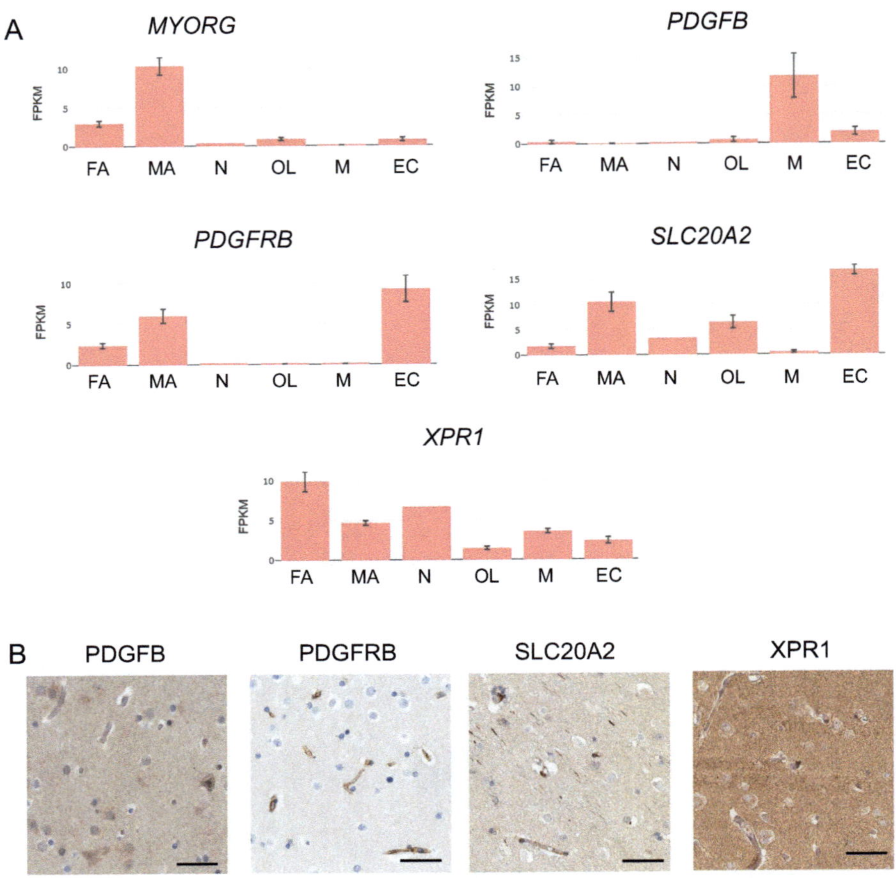

Fig. 11.2 The expression of MYORG, XPR1, SLC20A2, PDGFB, and PDGFRB at the RNA and protein level during different human developmental stages and adulthood. (**a**) RNAseq of different cell types (*FA* – fetal astrocyte, *MA* – mature astrocyte, *N* – neurons, *OL* – oligodendrocyte, *M* – microglia/macrophage, *EC* – endothelial cells) isolated from human brains. Data were taken from www.brainrnaseq.org (Zhang et al. 2016). (**b**) Autopsy cases from normal brain tissue from the cortex of patients aged between 54 and 70 years of age. Brain sections were stained using an atlas antibody against the protein (XPR1 – HPA016557; SLC20A2 – HPA026540; PDGFB – CAB018341; PDGFRB – CAB018144). A representative image from the brain section was selected. Data were taken from Human Protein Atlas available from www.proteinatlas.org. Scale bar is 50 μm

Pdgfb$^{ret/ret}$ mice show reduced pericyte coverage of their blood vessels in the embryo and adult. Further investigations reflect the role of pericytes in the development of conduit vessels (Nystrom et al. 2006), kidney glomeruli (Bjarnegard et al. 2004), liver sinusoidal vessels (Raines et al. 2011), retinal vasculature (Lindblom et al. 2003; Enge et al. 2002), and blood–brain barrier (BBB) (Armulik et al. 2010). In addition to *Pdgfb*$^{ret/ret}$ animals, two other mouse PDGFB hypomorphs (*Pdgfb*$^{-/-}$; Tie2CreR26hPDGFB$^{tg/0 \text{ or } tg/tg}$)) develop cerebrovascular calcifications (Keller et al.

2013). Interestingly, endothelial expression of PDGFB is protective of brain vascular calcifications, indicating that vessel calcification might be caused by reduced PDGFB/PDGFRB signaling at the neurovascular unit. In addition, it has been found that the severity of calcification correlated overall with the degree of pericyte deficiency and the extent of BBB dysfunction (Keller et al. 2013).

Thus, the initial analysis of mouse *Pdgfb* hypomorphs led to the suggestion that vessel calcification may be caused by pericyte deficiency, leading to BBB disruption and the subsequent formation of calcified lesions (Keller et al. 2013). Autopsy cases supported the plausibility of BBB dysfunction in PFBC (Miklossy et al. 2005). In addition, several other neurodevelopmental disorders show BBB deficiency and brain calcifications (O'Driscoll et al. 2010; Fischer et al. 2013). However, careful spatial analysis of brains of $Pdgfb^{ret/ret}$ mice demonstrated that vascular calcification did not correlate with pericyte loss and BBB dysfunction. Although $Pdgfb^{ret/ret}$ mice invariably develop vessel-associated brain calcifications, certain brain regions (e.g., cerebral cortex) with severe pericyte loss and BBB deficiency do not develop brain calcifications (Keller et al. 2013). Investigation of the spatial relationship between pericyte coverage, BBB permeability and the formation of vessel calcification in $Pdgfb^{ret/ret}$ mice showed that calcification-prone regions (i.e., thalamus, mesencephalon, and dorsal pons) showed a significantly higher pericyte coverage compared to non-prone regions (e.g., motor cortex, hippocampus), which was inversely correlated with BBB permeability. Thus, the vasculature in calcification-prone regions showed significantly less BBB permeability than brain regions that did not calcify (Vanlandewijck et al. 2015). These data strengthen the evidence for BBB impairment and pericyte loss at the global level. In addition, they demonstrate that pericyte loss and BBB impairment are not sufficient to lead to microvascular calcification.

Heterozygous mouse knockouts of *Pdgfb* or *Pdgfrb*, and even the double heterozygotes (*Pdgfb+/−;Pdgfrb+/−*), do not develop brain calcifications (Vanlandewijck et al. 2015). Furthermore, $Pdgfrb^{redeye/redeye}$ mice, which show 90% reduction in PDGFRB on protein level lack cerebral microvascular calcifications. This finding is similar to *Slc20a2* where only the full knockout of gene leads to cerebrovascular calcifications in mice (Jensen et al. 2018). Interestingly, both *Pdgfb+/−*, *Pdgfrb+/−* and $Pdgfrb^{redeye/redeye}$ mice do not show marked pericyte loss (Vanlandewijck et al. 2015) indicating that pericyte loss in $Pdgfb^{ret/ret}$ mice might contribute to microvascular calcifications.

PDGFB and PDGFRB are expressed in the brain as well as in cell types other than vascular cells (Figs. 11.1 and 11.2). The function of PDGFB/PDGFRB signaling in these cells (e.g., astrocytes) is not known. It is plausible that PFBC mutations cause changes in homeostasis in these cell types in addition to pericytes.

SLC20A2—Phosphate Transporter

SLC20A2 belongs to the type III family of Pi importers that consists of two proteins—SLC20A1 (frequently used protein name is PiT1) and SLC20A2 (PiT2), which maintain cellular phosphate homeostasis (Lederer and Miyamoto 2012). The expression pattern of SLC20A2/PiT2 has been reported to be ubiquitous, but very few studies have investigated SLC20A2 expression at the cellular level. A recent immunohistochemical study reported Slc20a2 expression in mouse brain in neurons, astrocytes, and endothelial cells (Inden et al. 2013). Gene and protein expression data deposited in publically available databases show a ubiquitous expression of SLC20A2 in human and mouse brain on the RNA and the protein level (Figs. 11.1 and 11.2).

Slc20a2−/− animals develop vessel-associated brain calcifications that increase with age, similar to *SLC20A2* mutation carriers (Jensen et al. 2013, 2018). It has been reported that also *Slc20a2* heterozygosity in mice causes calcifications (Wallingford et al. 2017); however, another study using the same knockout line did not confirm this finding (Jensen et al. 2018). Analysis of mouse *Pdgfb* hypomorphs indicated pericyte deficiency as a potential cause for vessel calcifications. However, no reduction of vessel pericyte coverage or altered BBB permeability is seen in *Slc20a2* deficient animals (Nahar and Betsholtz, personal communication). Calcifications formed in the absence of *Slc20a2* are extracellular and vessel associated similarly to the pathology observed in *Pdgfb* hypomorphs. However, electron microscopy shows that calcifications can be detected intracellularly in pericytes and astrocytes in *Slc20a2*$^{-/-}$ mice and human autopsy cases (Jensen et al. 2018; Kobayashi et al. 1987).

Interestingly, *Slc20a2* ko animals and also *SLC20A2* mutation carriers show elevated levels of Pi in cerebrospinal fluid (Wallingford et al. 2017; Jensen et al. 2016; Hozumi et al. 2018; Paucar et al. 2017). Thus, vessel calcifications could be initiated by an imbalance of Pi concentrations in the glymphatic space at the vessel wall.

Pericytes and PFBC

Currently, although PDGFB/PDGFRB is recognized as an important signaling pathway for pericyte recruitment, the role of pericytes in PFBC is unclear. Since PFBC genes are expressed by several cell types at neurovascular, the pathogenesis could be due to dysfunction of different cell types. It is possible that *PDGFRB*, *PDGFB*, *SLC20A2*, *XPR1*, *MYORG* mutations lead to microvascular calcification, but the primary affected cell type is different. Mouse genetic studies using cell type-specific knockouts of PFBC genes should clarify the role of pericytes and other brain cells (e.g., endothelial cells, vessel-associated astrocytes) involved in the formation of vessel calcification.

It is possible that impaired PDGFB/PDGFRB signaling in *Pdgfb*$^{ret/ret}$ mice accelerates an age-dependent phenotype in pericytes that leads to the formation of capillary calcification. Phenotypic alteration and senescence of vascular smooth muscle cells have been shown to drive arterial calcification (Durham et al. 2018). Microvascular calcification is accompanied by an osteogenic environment and formation of bone cells in *Pdgfb*$^{ret/ret}$ mice and human PFBC autopsy cases (Zarb et al. 2019). Lineage tracing studies in mouse models of PFBC should clarify whether pericytes transdifferentiate into osteoblasts. Further analysis of pericytes of PFBC models using the single-cell RNA sequencing could generate insight into gene expression alteration in pericytes and help to clarify the role in pericytes in PFBC.

Conclusion

Many unanswered questions remain about the pathogenesis of PFBC. Further analysis of mouse models of PFBC will bring additional insights into the pathogenic mechanism. It is presently unclear whether all PFBC genes initiate vessel calcification by a similar mechanism. However, AD-PFBC patients present a similar spectrum of clinical symptoms regardless of which gene mutations are causal (Nicolas et al. 2015). Importantly, it is unclear if cerebrovascular calcifications are causal for progressive neurodegeneration. Recent progress in understanding vessel-calcification mechanism using PFBC mouse models will hopefully lead to the development of strategies to lower the vessel calcification load. *Pdgfb*$^{ret/ret}$ mice exhibit behavioral alterations that resemble those described in PFBC patients (Zarb et al. 2019), and thus can be used to test whether reduction of the calcification load leads to objective evidence of behavioral improvement.

Insights into PFBC pathogenesis might also shed light on the causes of brain calcifications occurring during aging (e.g., pineal gland, basal ganglia) as well as various brain pathologies (e.g., viral infection, neurodegeneration, vasculopathies, tumors, type I interferonopathies). It is possible that despite the wide-ranging etiologies, brain vessel calcification shares a common pathogenic mechanism(s), thus providing an ideal therapeutic target.

Acknowledgments The authors are indebted to Zürich University and Zürich University Hospital for financial support. Cited own work was also supported by external grants from the Swiss National Science Foundation (grant 31003A_159514), the Swiss Heart Foundation, the Synapsis Foundation, Fonds zur Förderung des akademischen Nachwuchses (Zürich University), the Leducq Foundation, and Forschungskredit und Stiftung für Forschung an der Medizinischen Fakultät der Universität Zürich. The authors would like to thank Dr. E. J. Rushing for discussions.

References

Abramsson, A., et al. (2007). Defective N-sulfation of heparan sulfate proteoglycans limits PDGF-BB binding and pericyte recruitment in vascular development. *Genes & Development, 21*, 316–331. https://doi.org/10.1101/gad.398207.

Anheim, M., et al. (2016). XPR1 mutations are a rare cause of primary familial brain calcification. *Journal of Neurology, 263*, 1559–1564. https://doi.org/10.1007/s00415-016-8166-4.

Ansermet, C., et al. (2017). Renal fanconi syndrome and hypophosphatemic rickets in the absence of xenotropic and polytropic retroviral receptor in the nephron. *Journal of the American Society of Nephrology, 28*, 1073–1078. https://doi.org/10.1681/ASN.2016070726.

Armulik, A., Genove, G., & Betsholtz, C. (2011). Pericytes: Developmental, physiological, and pathological perspectives, problems, and promises. *Developmental Cell, **21***, 193–215. https://doi.org/10.1016/j.devcel.2011.07.001.

Armulik, A., et al. (2010). Pericytes regulate the blood-brain barrier. *Nature, **468***, 557–561. https://doi.org/10.1038/nature09522.

Ayata, C. (2010). CADASIL: Experimental insights from animal models. *Stroke, **41***, S129–S134. https://doi.org/10.1161/STROKEAHA.110.595207.

Baker, M., et al. (2013). SLC20A2 and THAP1 deletion in familial basal ganglia calcification with dystonia. *Neurogenetics, 15*(1), 23–30. https://doi.org/10.1007/s10048-013-0378-5.

Bell, R. D., et al. (2010). Pericytes control key neurovascular functions and neuronal phenotype in the adult brain and during brain aging. *Neuron, **68***, 409–427., S0896-6273(10)00824-X. https://doi.org/10.1016/j.neuron.2010.09.043.

Berthiaume, A. A., et al. (2018). Dynamic remodeling of pericytes in vivo maintains capillary coverage in the adult mouse brain. *Cell Reports, **22***, 8–16. https://doi.org/10.1016/j.celrep.2017.12.016.

Biancheri, R., et al. (2016). White matter involvement in a family with a novel PDGFB mutation. *Neurology Genetics, **2***, e77. https://doi.org/10.1212/NXG.0000000000000077.

Bjarnegard, M., et al. (2004). Endothelium-specific ablation of PDGFB leads to pericyte loss and glomerular, cardiac and placental abnormalities. *Development, **131***, 1847–1857. https://doi.org/10.1242/dev.01080131/8/1847.

Cho, C., Smallwood, P. M., & Nathans, J. (2017). Reck and Gpr124 are essential receptor cofactors for Wnt7a/Wnt7b-specific signaling in mammalian CNS angiogenesis and blood-brain barrier regulation. *Neuron, 95*, 1056–1073.e1055. https://doi.org/10.1016/j.neuron.2017.07.031.

Delacour, A. (1850). Ossification des capillaires du cerveau. *Annales Médico-Psychologiques, **2***, 458–461.

Durham, A. L., Speer, M. Y., Scatena, M., Giachelli, C. M., & Shanahan, C. M. (2018). Role of smooth muscle cells in vascular calcification: Implications in atherosclerosis and arterial stiffness. *Cardiovascular Research, **114***, 590–600. https://doi.org/10.1093/cvr/cvy010.

Dziewulska, D., & Lewandowska, E. (2012). Pericytes as a new target for pathological processes in CADASIL. *Neuropathology: Official Journal of the Japanese Society of Neuropathology, **32***, 515–521. https://doi.org/10.1111/j.1440-1789.2011.01290.x.

Enge, M., et al. (2002). Endothelium-specific platelet-derived growth factor-B ablation mimics diabetic retinopathy. *The EMBO Journal, **21***, 4307–4316.

Erdo, F., Denes, L., & de Lange, E. (2017). Age-associated physiological and pathological changes at the blood-brain barrier: A review. *Journal of Cerebral Blood Flow and Metabolism, **37***, 4–24. https://doi.org/10.1177/0271678X16679420.

Fischer, A., Zalvide, J., Faurobert, E., Albiges-Rizo, C., & Tournier-Lasserve, E. (2013). Cerebral cavernous malformations: From CCM genes to endothelial cell homeostasis. *Trends in Molecular Medicine, **19***, 302–308. https://doi.org/10.1016/j.molmed.2013.02.004.

Gautam, J., Zhang, X., & Yao, Y. (2016). The role of pericytic laminin in blood brain barrier integrity maintenance. *Scientific Reports, **6***, 36450. https://doi.org/10.1038/srep36450.

Giovannini, D., Touhami, J., Charnet, P., Sitbon, M., & Battini, J. L. (2013). Inorganic phosphate export by the retrovirus receptor XPR1 in metazoans. *Cell Reports, **3***, 1866–1873. https://doi.org/10.1016/j.celrep.2013.05.035.

Gomez, C., Luque, A., & Horenstein, S. (1989). Microvasculopathy may precede idiopathic cerebral calcifications--case report. *Angiology, **40***, 67–72.

Hall, C. N., et al. (2014). Capillary pericytes regulate cerebral blood flow in health and disease. *Nature, **508***, 55–60. https://doi.org/10.1038/nature13165.

Hartmann, D. A., et al. (2015). Pericyte structure and distribution in the cerebral cortex revealed by high-resolution imaging of transgenic mice. *Neurophotonics, **2***, 041402. https://doi.org/10.1117/1.NPh.2.4.041402.

Hellström, M., Kalén, M., Lindahl, P., Abramsson, A., & Betsholtz, C. (1999). Role of PDGF-B and PDGFR-beta in recruitment of vascular smooth muscle cells and pericytes during embryonic blood vessel formation in the mouse. *Development, **126***, 3047–3055.

Heuchel, R., et al. (1999). Platelet-derived growth factor beta receptor regulates interstitial fluid homeostasis through phosphatidylinositol-3′ kinase signaling. *Proceedings of the National Academy of Sciences of the United States of America, 96*, 11410–11415.

Hill, R. A., et al. (2015). Regional blood flow in the normal and ischemic brain is controlled by arteriolar smooth muscle cell contractility and not by capillary pericytes. *Neuron, 87*, 95–110. https://doi.org/10.1016/j.neuron.2015.06.001.

Hogan, K. A., & Bautch, V. L. (2004). Blood vessel patterning at the embryonic midline. *Current Topics in Developmental Biology, 62*, 55–85., S0070215304620035. https://doi.org/10.1016/S0070-2153(04)62003-5.

Hozumi, I., et al. (2018). Inorganic phosphorus (Pi) in CSF is a biomarker for SLC20A2-associated idiopathic basal ganglia calcification (IBGC1). *Journal of the Neurological Sciences, 388*, 150–154. https://doi.org/10.1016/j.jns.2018.03.014.

Iadecola, C. (2013). The pathobiology of vascular dementia. *Neuron, 80*, 844–866. https://doi.org/10.1016/j.neuron.2013.10.008.

Ikeda, M., et al. (1994). A case of atypical presenile dementia. *Nō to shinkei = Brain and nerve, 46*, 175–181.

Inden, M., Iriyama, M., Takagi, M., Kaneko, M., & Hozumi, I. (2013). Localization of type-III sodium-dependent phosphate transporter 2 in the mouse brain. *Brain Research, 1531*, 75–83. https://doi.org/10.1016/j.brainres.2013.07.038.

Ishii, Y., et al. (2006). Mouse brains deficient in neuronal PDGF receptor-beta develop normally but are vulnerable to injury. *Journal of Neurochemistry, 98*, 588–600. https://doi.org/10.1111/j.1471-4159.2006.03922.x.

Ishii, Y., et al. (2008). Characterization of neuroprogenitor cells expressing the PDGF beta-receptor within the subventricular zone of postnatal mice. *Molecular and Cellular Neuroscience, 37*, 507–518. https://doi.org/10.1016/j.mcn.2007.11.006.

Jensen, N., Autzen, J. K., & Pedersen, L. (2016). Slc20a2 is critical for maintaining a physiologic inorganic phosphate level in cerebrospinal fluid. *Neurogenetics, 17*, 125–130. https://doi.org/10.1007/s10048-015-0469-6.

Jensen, N., et al. (2013). Loss of function of Slc20a2 associated with familial idiopathic Basal Ganglia calcification in humans causes brain calcifications in mice. *Journal of Molecular Neuroscience: MN, 51*, 994–999. https://doi.org/10.1007/s12031-013-0085-6.

Jensen, N., et al. (2018). Mice knocked out for the primary brain calcification associated gene Slc20a2 show unimpaired pre-natal survival but retarded growth and nodules in the brain that grow and calcify over time. *The American Journal of Pathology, 188*(8), 1865–1881. https://doi.org/10.1016/j.ajpath.2018.04.010.

Keller, A., et al. (2013). Mutations in the gene encoding PDGF-B cause brain calcifications in humans and mice. *Nature Genetics, 45*, 1077–1082. https://doi.org/10.1038/ng.2723.

Kisler, K., et al. (2017). Pericyte degeneration leads to neurovascular uncoupling and limits oxygen supply to brain. *Nature Neuroscience, 20*, 406–416. https://doi.org/10.1038/nn.4489.

Kobayashi, S., Yamadori, I., Miki, H., & Ohmori, M. (1987). Idiopathic nonarteriosclerotic cerebral calcification (Fahr's disease): An electron microscopic study. *Acta Neuropathologica, 73*, 62–66.

Korczyn, A. D. (2015). Vascular parkinsonism--characteristics, pathogenesis and treatment. *Nature Reviews. Neurology, 11*, 319–326. https://doi.org/10.1038/nrneurol.2015.61.

Koyama, S., et al. (2017). Clinical and radiological diversity in genetically confirmed primary familial brain calcification. *Scientific Reports, 7*, 12046. https://doi.org/10.1038/s41598-017-11595-1.

Kozik, M., & Kulczycki, J. (1978). Laser-spectrographic analysis of the cation content in Fahr's syndrome. *Archiv für Psychiatrie und Nervenkrankheiten, 225*, 135–142.

Larsen, F. T., Jensen, N., Autzen, J. K., Kongsfelt, I. B., & Pedersen, L. (2017). Primary brain calcification causal PiT2 transport-knockout variants can exert dominant negative effects on wild-type PiT2 transport function in mammalian cells. *Journal of Molecular Neuroscience, 61*, 215–220. https://doi.org/10.1007/s12031-016-0868-7.

Lederer, E., & Miyamoto, K.-i. (2012). Clinical consequences of mutations in sodium phosphate cotransporters. *Clinical Journal of the American Society of Nephrology, 7*, 1179–1187. https://doi.org/10.2215/cjn.09090911.

Legati, A., et al. (2015). Mutations in XPR1 cause primary familial brain calcification associated with altered phosphate export. *Nature Genetics, 47*, 579–581. https://doi.org/10.1038/ng.3289.

Lindahl, P., Johansson, B. R., Leveen, P., & Betsholtz, C. (1997). Pericyte loss and microaneurysm formation in PDGF-B-deficient mice. *Science, 277*, 242–245.

Lindblom, P., et al. (2003). Endothelial PDGF-B retention is required for proper investment of pericytes in the microvessel wall. *Genes & Development, 17*, 1835–1840. https://doi.org/10.1101/gad.266803.

Manyam, B. (2005). What is and what is not 'Fahr's disease'. *Parkinsonism & Related Disorders, 11*, 73–80. https://doi.org/10.1016/j.parkreldis.2004.12.001.

Meireles, A. M., et al. (2014). The phosphate exporter xpr1b is required for differentiation of tissue-resident macrophages. *Cell Reports, 8*, 1659–1667. https://doi.org/10.1016/j.celrep.2014.08.018.

Menezes, M. J., et al. (2014). The extracellular matrix protein laminin alpha2 regulates the maturation and function of the blood-brain barrier. *The Journal of Neuroscience, 34*, 15260–15280. https://doi.org/10.1523/JNEUROSCI.3678-13.2014.

Miklossy, J., et al. (2005). Severe vascular disturbance in a case of familial brain calcinosis. *Acta Neuropathologica, 109*, 643–653. https://doi.org/10.1007/s00401-005-1007-7.

Montagne, A., et al. (2018). Pericyte degeneration causes white matter dysfunction in the mouse central nervous system. *Nature Medicine, 24*, 326–337. https://doi.org/10.1038/nm.4482.

Nguyen, P. T., et al. (2011). Cognitive and socio-emotional deficits in platelet-derived growth factor receptor-beta gene knockout mice. *PLoS One, 6*, e18004. https://doi.org/10.1371/journal.pone.0018004.

Nicolas, G., Charbonnier, C., Campion, D., & Veltman, J. A. (2018). Estimation of minimal disease prevalence from population genomic data: Application to primary familial brain calcification. *American Journal of Medical Genetics. Part B, Neuropsychiatric Genetics, 177*, 68–74. https://doi.org/10.1002/ajmg.b.32605.

Nicolas, G., Marguet, F., Laquerriere, A., Mendes de Oliveira, J. R., & Hannequin, D. (2017). Microangiopathy in primary familial brain calcification: Evidence from skin biopsies. *Neurol Genet, 3*, e134. https://doi.org/10.1212/NXG.0000000000000134.

Nicolas, G., et al. (2013a). Mutation of the PDGFRB gene as a cause of idiopathic basal ganglia calcification. *Neurology, 80*, 181–187. https://doi.org/10.1212/WNL.0b013e31827ccf34.

Nicolas, G., et al. (2013b). Phenotypic spectrum of probable and genetically-confirmed idiopathic basal ganglia calcification. *Brain: A Journal of Neurology, 136*, 3395–3407. https://doi.org/10.1093/brain/awt255.

Nicolas, G., et al. (2014). PDGFB partial deletion: A new, rare mechanism causing brain calcification with leukoencephalopathy. *Journal of Molecular Neuroscience : MN, 53*(2), 171–175. https://doi.org/10.1007/s12031-014-0265-z.

Nicolas, G., et al. (2015). Brain calcification process and phenotypes according to age and sex: Lessons from SLC20A2, PDGFB, and PDGFRB mutation carriers. *American Journal of Medical Genetics. Part B, Neuropsychiatric Genetics, 168*, 586–594. https://doi.org/10.1002/ajmg.b.32336.

Nomoto, N., Sugimoto, H., Iguchi, H., Kurihara, T., & Wakata, N. (2002). A case of Fahr's disease presenting "diffuse neurofibrillary tangles with calcification". *Rinshō Shinkeigaku = Clinical. Neurology, 42*, 745–749.

Norman, R., & Urich, H. (1960). The influence of a vascular factor on the distribution of symmetrical cerebral calcifications. *Journal of Neurology, Neurosurgery, and Psychiatry, 23*, 142–147.

Nystrom, H. C., et al. (2006). Platelet-derived growth factor B retention is essential for development of normal structure and function of conduit vessels and capillaries. *Cardiovascular Research, 71*, 557–565. https://doi.org/10.1016/j.cardiores.2006.05.019.

O'Driscoll, M., et al. (2010). Recessive mutations in the gene encoding the tight junction protein occludin cause band-like calcification with simplified gyration and polymicrogyria. *American Journal of Human Genetics, 87*, 354–364. https://doi.org/10.1016/j.ajhg.2010.07.012.

Olson, L. E., & Soriano, P. (2011). PDGFRβ signaling regulates mural cell plasticity and inhibits fat development. *Developmental Cell, 20*, 815–826. https://doi.org/10.1016/j.devcel.2011.04.019.

Paschali, A., et al. (2009). Dopamine transporter SPECT/CT and perfusion brain SPECT imaging in idiopathic basal ganglia calcinosis. *Clinical Nuclear Medicine, 34*, 421–423. https://doi.org/10.1097/RLU.0b013e3181a7d195.

Paucar, M., et al. (2017). A SLC20A2 gene mutation carrier displaying ataxia and increased levels of cerebrospinal fluid phosphate. *Journal of the Neurological Sciences, 375*, 245–247. https://doi.org/10.1016/j.jns.2017.02.007.

Peppiatt, C. M., Howarth, C., Mobbs, P., & Attwell, D. (2006). Bidirectional control of CNS capillary diameter by pericytes. *Nature, 443*, 700–704. https://doi.org/10.1038/nature05193.

Phuong, T. H. N., et al. (2011). Cognitive and socio-emotional deficits in platelet-derived growth factor receptor-beta gene knockout mice. *PLoS One, 6*, e18004. https://doi.org/10.1371/journal.pone.0018004.

Quintans, B., Oliveira, J., & Sobrido, M. J. (2018). Primary familial brain calcifications. *Handbook of Clinical Neurology, 147*, 307–317. https://doi.org/10.1016/B978-0-444-63233-3.00020-8.

Raines, S. M., et al. (2011). Loss of PDGF-B activity increases hepatic vascular permeability and enhances insulin sensitivity. *American Journal of Physiology. Endocrinology and Metabolism, 301*, E517–E526. https://doi.org/10.1152/ajpendo.00241.2011.

Saito, T., et al. (2010). Neuroradiologic evidence of pre-synaptic and post-synaptic nigrostriatal dopaminergic dysfunction in idiopathic Basal Ganglia calcification: A case report. *Journal of Neuroimaging: Official Journal of the American Society of Neuroimaging, 20*, 189–191. https://doi.org/10.1111/j.1552-6569.2008.00314.x.

Sanchez-Contreras, M., et al. (2014). Genetic screening and functional characterization of PDGFRB mutations associated with basal ganglia calcification of unknown etiology. *Human Mutation, 35*(8), 964–971. https://doi.org/10.1002/humu.22582.

Shen, J., et al. (2012). PDGFR-beta as a positive regulator of tissue repair in a mouse model of focal cerebral ischemia. *Journal of Cerebral Blood Flow and Metabolism, 32*, 353–367. https://doi.org/10.1038/jcbfm.2011.136.

Shibayama, H., et al. (1986). Unusual cases of presenile dementia with Fahr's syndrome. *The Japanese Journal of Psychiatry and Neurology, 40*, 85–100. https://doi.org/10.1111/j.1440-1819.1986.tb01615.x.

Shoyama, M., Shouyama, M., Kitabata, Y., Kaku, T., & Shinosaki, K. (2005). Evaluation of regional cerebral blood flow in fahr disease with schizophrenia-like psychosis: A case report. *AJNR. American Journal of Neuroradiology, 26*, 2527–2529.

Sweeney, M. D., Ayyadurai, S., & Zlokovic, B. V. (2016). Pericytes of the neurovascular unit: Key functions and signaling pathways. *Nature Neuroscience, 19*, 771–783. https://doi.org/10.1038/nn.4288.

Tadic, V., et al. (2015). Primary familial brain calcification with known gene mutations: A systematic review and challenges of phenotypic characterization. *JAMA Neurology, 72*, 460–467. https://doi.org/10.1001/jamaneurol.2014.3889.

Taglia, I., Bonifati, V., Mignarri, A., Dotti, M. T., & Federico, A. (2015). Primary familial brain calcification: Update on molecular genetics. *Neurological Sciences, 36*, 787–794. https://doi.org/10.1007/s10072-015-2110-8.

Taglia, I., et al. (2018). Primary familial brain calcification with a novel SLC20A2 mutation: Analysis of PiT-2 expression and localization. *Journal of Cellular Physiology, 233*, 2324–2331. https://doi.org/10.1002/jcp.26104.

Tallquist, M. D., French, W. J., & Soriano, P. (2003). Additive effects of PDGF receptor beta signaling pathways in vascular smooth muscle cell development. *PLoS Biology, 1*, E52. https://doi.org/10.1371/journal.pbio.0000052.

Tarlungeanu, D. C., et al. (2016). Impaired amino acid transport at the blood brain barrier is a cause of autism spectrum disorder. *Cell, **167***, 1481–1494.e1418. https://doi.org/10.1016/j.cell.2016.11.013.

Tham, D. K., Joshi, B., & Moukhles, H. (2016). Aquaporin-4 cell-surface expression and turnover are regulated by dystroglycan, dynamin, and the extracellular matrix in astrocytes. *PLoS One, **11***, e0165439. https://doi.org/10.1371/journal.pone.0165439.

Uygur, G., Liu, Y., Hellman, R., Tikofsky, R., & Collier, B. (1995). Evaluation of regional cerebral blood flow in massive intracerebral calcifications. *Journal of Nuclear Medicine: Official Publication, Society of Nuclear Medicine, **36***, 610–612.

Vanlandewijck, M., et al. (2015). Functional characterization of germline mutations in PDGFB and PDGFRB in primary familial brain calcification. *PLoS One, **10***, e0143407. https://doi.org/10.1371/journal.pone.0143407.

Vanlandewijck, M., et al. (2018). A molecular atlas of cell types and zonation in the brain vasculature. *Nature, **554***, 475–480. https://doi.org/10.1038/nature25739.

Visel, A., Thaller, C., & Eichele, G. (2004). GenePaint.org: An atlas of gene expression patterns in the mouse embryo. *Nucleic Acids Research, **32***, D552–D556. https://doi.org/10.1093/nar/gkh029.

Wallingford, M. C., Gammill, H. S., & Giachelli, C. M. (2016). Slc20a2 deficiency results in fetal growth restriction and placental calcification associated with thickened basement membranes and novel CD13 and lamininalpha1 expressing cells. *Reproductive Biology, **16***, 13–26. https://doi.org/10.1016/j.repbio.2015.12.004.

Wallingford, M. C., et al. (2017). SLC20A2 deficiency in mice leads to elevated phosphate levels in cerebrospinal fluid and glymphatic pathway-associated arteriolar calcification, and recapitulates human idiopathic basal ganglia calcification. *Brain Pathology, **27***, 64–76. https://doi.org/10.1111/bpa.12362.

Wang, C., et al. (2012). Mutations in SLC20A2 link familial idiopathic basal ganglia calcification with phosphate homeostasis. *Nature Genetics, **44***, 254–256. https://doi.org/10.1038/ng.1077.

Westenberger, A., & Klein, C. (2014). The genetics of primary familial brain calcifications. *Current Neurology and Neuroscience Reports, **14***, 490. https://doi.org/10.1007/s11910-014-0490-4.

Xu, G., et al. (2013). Functional analysis of platelet-derived growth factor receptor-beta in neural stem/progenitor cells. *Neuroscience, **238***, 195–208. https://doi.org/10.1016/j.neuroscience.2013.02.021.

Yao, X. P., et al. (2017). Analysis of gene expression and functional characterization of XPR1: A pathogenic gene for primary familial brain calcification. *Cell and Tissue Research, **370***, 267–273. https://doi.org/10.1007/s00441-017-2663-3.

Yao, X. P., et al. (2018). Biallelic mutations in MYORG cause autosomal recessive primary familial brain calcification. *Neuron, 98*(6), 1116–1123.e5. https://doi.org/10.1016/j.neuron.2018.05.037.

Zarb, Y., et al. (2019) Ossified blood vessels in primary familial brain calcification elicit a neurotoxic astrocyte response. *Brain, 142*, 885–902. https://doi.org/10.1093/brain/awz032.

Zhang, Y., et al. (2014). An RNA-sequencing transcriptome and splicing database of glia, neurons, and vascular cells of the cerebral cortex. *The Journal of Neuroscience, **34***, 11929–11947. https://doi.org/10.1523/JNEUROSCI.1860-14.2014.

Zhang, Y., et al. (2016). Purification and characterization of progenitor and mature human astrocytes reveals transcriptional and functional differences with mouse. *Neuron, **89***, 37–53. https://doi.org/10.1016/j.neuron.2015.11.013.

Chapter 12
Pericytes in Type 2 Diabetes

Katherine L. Hayes

Abstract Pericytes are mural cells that are found ubiquitously throughout the microvasculature. Their main physiological roles are to support endothelial cells, regulate microvascular blood flow, and respond to perturbations in their microenvironment. Pericytes are sensitive to the metabolic abnormalities that are characteristic of type 2 diabetes mellitus, including dyslipidemia, hyperglycemia, and hyperinsulinemia. As a consequence of these abnormalities, advanced glycation end products, reactive oxygen species, polyol pathway activation, and protein kinase C isoform activation cause pericyte dysfunction and contribute to the pathogenesis of many common complications of type 2 diabetes. Pericyte dysfunction is known to be a contributing factor to the pathogenesis of retinopathy, nephropathy, neuropathy, beta cell dysfunction, and peripheral artery disease in people with type 2 diabetes. Therapies should target pericytes to treat these common diabetic complications.

Keywords Beta cell · Retinopathy · Neuropathy · Nephropathy · Peripheral artery disease · Hyperglycemia · Hyperinsulinemia · Dyslipidemia · Mural cell
Advanced glycation end products · Blood–retina barrier · Blood–nerve barrier
Microvasculature

Introduction

Diabetes mellitus is a group of metabolic disorders characterized by impaired glucose homeostasis. According to recently published global estimates, over 450 million people had diabetes in 2017, and it was attributed to over 5 million deaths (Cho et al. 2018). The prevalence of diabetes is expected to exceed 690 million people by 2045 (Cho et al. 2018). The overwhelming majority of people with diabetes have

K. L. Hayes (✉)
Department of Surgery and Diabetes Center of Excellence, University of Massachusetts Medical School, Worcester, MA, USA
e-mail: katherine.hayes@umassmed.edu

A. Birbrair (ed.), *Pericyte Biology in Disease*, Advances in Experimental Medicine and Biology 1147, https://doi.org/10.1007/978-3-030-16908-4_12

type 2 diabetes, comprising approximately 90–95% of all diabetes cases. In 2013, 382 million people worldwide had type 2 diabetes; and 592 million people are projected to have type 2 diabetes by the year 2035 (Defronzo et al. 2015).

Type 2 diabetes mellitus is characterized by insulin resistance, hyperglycemia, and β-cell decompensation (Defronzo et al. 2015). Both genetics and environmental factors, including energy imbalance from a combination of overnutrition and inadequate physical activity, contribute to the development of type 2 diabetes. Type 2 diabetes is a considerable public health concern because macrovascular complications lead to cardiovascular diseases and an increased risk of death from cardiovascular diseases, including stroke and coronary artery disease. Microvascular dysfunctions also lead to devastating complications, including peripheral neuropathy, nephropathy, and retinopathy.

Metabolic disturbances, such as dyslipidemia, hyperglycemia, and hyperinsulinemia, drive the development of microvascular complications by activating several pathophysiological processes (Stratton et al. 2000; Folli et al. 2011). These metabolic disturbances enhance polyol pathway flux, increase the formation of advanced glycation end products, activate protein kinase C isoforms, and increase intracellular reactive oxygen species (Brownlee 2005; Giacco and Brownlee 2010). Microvascular endothelial cells, as well as their associated pericytes, are negatively affected by the activation of these pathways. This chapter will focus on the effects of type 2 diabetes on pericytes, the mural cells of the microvasculature.

Pericyte Dysfunction in Type 2 Diabetes

Pericyte dysfunction contributes to the pathogenesis and symptomology in type 2 diabetes. As microvascular-associated cells, the physiological role of pericytes is to regulate blood flow and maintain microvascular homeostasis by responding to perturbations through signaling with endothelial cells and even differentiation into terminal cell types (Armulik et al. 2011). Pericyte number and function are impaired in many tissues of type 2 diabetic patients. Specifically, pericyte dysfunction in the pancreas, retina, kidney, peripheral endoneurium, and skeletal muscle is pronounced in type 2 diabetes, and leads to the overt symptoms of type 2 diabetes, including impaired glucose homeostasis, retinopathy, neuropathy, nephropathy, and peripheral artery disease (Fig. 12.1).

Pancreatic Pericytes

The pancreas contains islets, which are specialized clusters of cells containing insulin-producing beta cells. Pancreatic islets are highly capillarized to allow for glucose sensing and insulin secretion. As a tissue with high capillary density, pancreatic pericytes make up approximately 3% of human and mouse islet cells;

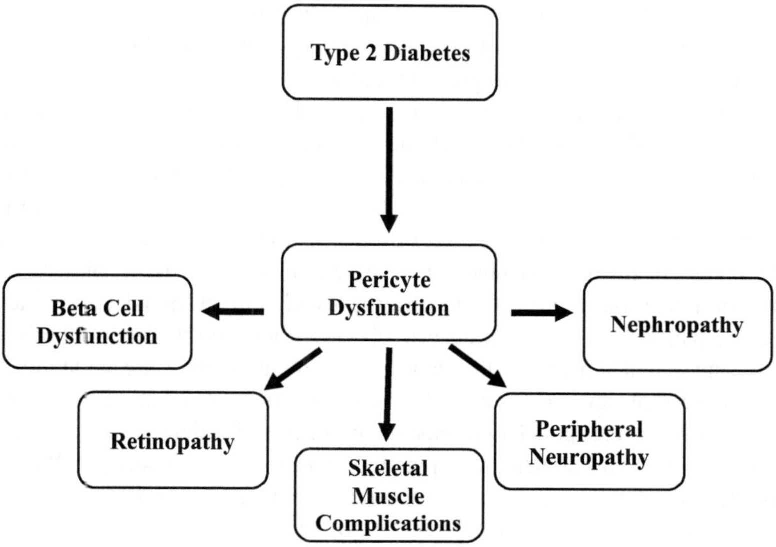

Fig. 12.1 Pericyte-mediated complications in type 2 diabetes. Metabolic abnormalities that are characteristic of type 2 diabetes, namely hyperglycemia, hyperinsulinemia, and dyslipidemia, contribute to pericyte dropout and dysfunction in tissues throughout the body. Pericyte dysfunction contributes to many of the common complications of type 2 diabetes. Therapies aimed at treating pericyte dysfunction are promising for relieving complications from type 2 diabetes

however, their cytoplasmic processes cover up to 40% of microvascular capillaries in healthy individuals without diabetes (Almaca et al. 2018). Pancreatic pericytes regulate capillary islet blood flow by contracting and relaxing, which occurs in response to oscillating glucose concentrations (Almaca et al. 2018). High glucose concentrations activate beta cells, which leads to insulin being co-released with adenosine triphosphate (ATP). Endogenous adenosine is derived from ATP and mediates pericyte relaxation. Pericyte relaxation leads to capillary dilation. In contrast, pericyte contraction is mediated by sympathetic innervation. The result of pericyte contraction is reduced capillary blood flow (Almaca et al. 2018).

In addition to regulating blood flow, pericytes in the pancreas support beta cell function. It is suggested that pericytes exert effects on beta cells directly (Sasson et al. 2016), through their association with microvascular capillaries (Almaca et al. 2018), and through signaling molecules (Sakhneny et al. 2018; Houtz et al. 2016). Sasson et al. (2016) showed that pericytes in pancreatic islets are required for proper beta cell functionality. They demonstrated that selective ablation of pancreatic islet pericytes results in mice with elevated fasting blood glucose and impaired glucose tolerance. The islets of pericyte-depleted transgenic mice secreted less insulin despite normal islet morphology and beta cell number. Further, these mice displayed markers of immature beta cells, indicating that pericytes were important for beta cell maturation. Houtz et al. (2016) provided mechanistic insight into the role for pericytes in insulin secretion from beta cells. They showed that glucose was able to stimulate nerve growth factor secretion from pancreatic islet pericytes, which signaled for insulin granule exocytosis from beta cells. All together, these studies provide key insights into the role for pancreatic pericytes in glucose homeostasis and insulin secretion in physiological conditions.

Pericytes also have roles in pancreatic beta cell dysfunction in the pathological context of type 2 diabetes. In general, human cadaver studies have shown that type 2 diabetic patients experience pancreatic islet changes including fibrosis and amyloid deposition. Animal models of type 2 diabetes have been used to characterize the effect of the disease on pancreatic pericytes specifically (Hayden et al. 2010). In rodent models of insulin resistance and type 2 diabetes, Hayden et al. (2010) used transmission electron microscopy to observe a widening at the islet exocrine interface in the pancreas. The widening was observed between the endocrine-islet and the exocrine-acinar portion of the pancreas. Coinciding with the widening was a hypercellularity of pericytes. They also observed morphological changes in these pericytes, namely elongated cytoplasmic processes. Evidence suggests that the hypercellular pericytes contribute to pancreatic fibrosis by depositing collagen. The hypercellularity of pericytes is also supported by evidence from the *db/db* model of type 2 diabetes in mice. Hayden et al. (2010) reported a pancreatic islet pericyte degeneration in a model of human islet amyloid deposition in non-obese type 2 diabetic rats. Pericyte degeneration was confirmed in type 2 diabetic patients as demonstrated by decreased pericyte coverage of capillaries (Almaca et al. 2018). Additionally, pericyte coverage was inversely related to duration of disease. Finally, a mechanism for loss of pericyte-supported beta cell function in the pathogenesis of diabetes showed that Tcf712-knockout animals had impaired glucose homeostasis, despite not developing overt type 2 diabetes (Sakhneny et al. 2018). The polymorphism in TCF7L2 gene is strongly correlated to type 2 diabetes in humans. Sakhneny et al. (2018) showed that pancreatic pericytes express the transcription factor TCF7L2, and its physiological role in pancreatic pericytes is to support beta cell function through Tcf7l2-dependent secretion of cytokines, including bone morphogenetic protein-4. Loss of function of Tcf7l2 in pericytes impairs beta cell function and glucose homeostasis (Sakhneny et al. 2018). Overall, while few studies have examined pancreatic pericyte dysfunction in type 2 diabetes, pericyte dropout and dysfunction is evident, and clearly contributes to disease etiology.

Skeletal Muscle Pericytes

Skeletal muscle is another highly vascularized tissue, and pericytes are found in the skeletal muscle at 1:10–1:100 ratio of pericytes to endothelial cells (Armulik et al. 2011). Within skeletal muscle, pericytes perform their canonical functions of regulating blood flow and responding to perturbations in their microenvironment by either direct differentiation or indirect signaling with neighboring cells. Importantly, the skeletal muscle is the primary tissue for insulin-mediated glucose uptake. Insulin resistance is a hallmark of type 2 diabetes, and insulin resistance precedes the development of overt type 2 diabetes.

Skeletal muscle pericyte dysfunction has been characterized in type 2 diabetes. Specifically, pericyte apoptosis and deterioration has been observed along with reduced capillary density (Hayden et al. 2010). Tilton et al. (1981) described capillary dimensions and pericyte distribution in skeletal muscle obtained from human autopsies of diabetic people and age- and sex-matched controls. They observed a marked increase in pericyte degeneration and capillary basement membrane thickening within diabetic muscles compared to non-diabetic muscles (Tilton et al. 1981, 1985). While early work in cadaver muscles did not observe differences in pericyte coverage of capillaries in the skeletal muscle of diabetic patients compared to non-diabetic controls, recent evidence indicates that pericyte abundance is markedly decreased in the skeletal muscle of type 2 diabetic patients. Vono et al. (2016) demonstrated that not only do pericytes from type 2 diabetic patients have decreased abundance in the skeletal muscle, the pericytes that remain have ultrastructural abnormalities, including increased blebbing and vacuolization, as identified by transmission electron microscopy. Furthermore, skeletal muscle pericytes from diabetics exhibited decreased capacity for myogenic differentiation and decreased ability to support endothelial cell tube formation *in vitro* (Vono et al. 2016).

A devastating manifestation of diabetes in the skeletal muscle is the development of peripheral artery disease. Peripheral artery disease is an atherosclerotic occlusive disease that most commonly affects the lower limbs. Patients with peripheral artery disease experience pain, decreased mobility, and poor quality of life. While peripheral artery disease is a macrovascular complication of type 2 diabetes, microvascular abnormalities also contribute to the ischemic condition. Microvascular pericytes are proangiogenic, present in the skeletal muscle, and able to respond to ischemic insults; therefore, pericytes are a potential cell therapy target to treat peripheral artery disease, which has very few non-surgical interventions to revascularize ischemic tissue. Several preclinical studies have shown that skeletal muscle pericytes augment postischemic neovascularization in a murine model of critical limb ischemia, which is a severe form of peripheral artery disease (Dar et al. 2012; Birbrair et al. 2014; Gubernator et al. 2015; Hayes et al. 2018). Birbrair et al. (2014) demonstrated that transplanted skeletal muscle pericytes improved limb revascularization, which was assessed via MRI angiography at 10 days after the induction of ischemia via femoral artery ligation. Dar et al. (2012) showed that intramuscularly transplanted pluripotent stem cell (iPSC)-derived pericytes enhanced blood flow recovery in mice up to 28 days post-induction of limb ischemia. Importantly, they provided evidence that iPSC-derived pericytes incorporated into both the regenerating muscle and vasculature (Dar et al. 2012). Skeletal muscle pericytes have also been shown to incorporate into existing collateral arteries to promote postischemic neovascularization via collateral artery enlargement; however, incorporation into regenerating skeletal muscle was not observed in that study (Hayes et al. 2018). Others have shown that cell types similar to pericytes, including mesenchymal stem cells and saphenous vein adventitial progenitor cells, can also improve neovascularization *in vivo* in mice (Gubernator et al. 2015; Yan et al. 2012).

Unfortunately, stem cell therapies appear to be less effective in diabetic environments (Efimenko et al. 2015; Yan et al. 2012; Hayes et al. 2018). Diabetes is strongly associated with an elevated risk of peripheral artery disease, and peripheral artery disease is more severe in diabetic patients. Work from Yan et al. (2012) demonstrated that the diabetic environment negatively impacts the therapeutic potential of bone marrow mesenchymal stem cells, a cell type that shares similarities with skeletal muscle resident pericytes (Yan et al. 2012, 2013). Using a hindlimb ischemia model, bone marrow-derived mesenchymal stem cells isolated from type 2 diabetic mice had an impaired ability to support neovascularization compared to mesenchymal stem cells isolated from wild-type mice (Yan et al. 2012). Furthermore, mesenchymal stem cells from diabetic donors favored adipocyte differentiation compared to endothelial cell differentiation *in vivo*; and oxidative stress was implicated as a cause of this dysfunction (Yan et al. 2012). Endothelial cell differentiation may be critical for angiogenesis and arteriogenesis during postischemic neovascularization whereas adipocyte accumulation in ischemic tissues may impair effective blood flow recovery. Similarly, diabetes was shown to impair the therapeutic potential of skeletal muscle pericytes to treat peripheral artery disease in type 2 diabetic mice (Hayes et al. 2018). In that study, hyperinsulinemia induced pericyte oxidative stress that impaired pericyte tube formation *in vitro*, thus indicating that diabetes-induced oxidative stress inhibits the angiogenic potential of skeletal muscle pericytes (Hayes et al. 2018). Evidence from Vono et al. (2016) also provides support for stem cell dysfunction in type 2 diabetes due to activated pro-oxidative pathways. They showed that skeletal muscle pericytes isolated from humans with diabetic peripheral artery disease showed impaired differentiation and impaired ability to support endothelial cell tube formation *in vitro*. The noted impairments were likely due to oxidative stress since reactive oxygen species scavengers and PKCβII inhibitors were able to restore the angiogenic and myogenic capacity of diabetic pericytes (Vono et al. 2016). Another study demonstrated impaired neovascularization following limb ischemia in diabetic mice due to inflammation in endothelial cells that negatively impacted crosstalk with skeletal muscle pericytes (Caporali et al. 2011). In summary, inflammation and oxidative stress associated with type 2 diabetes negatively affects the ability of pericytes to augment neovascularization in models of peripheral artery disease.

Retinal Pericyte

The retina is a complex neurovascular unit that is comprised of functionally coupled neurons, glial cells, and vasculature (Metea and Newman 2007). Located at the back of the eye, the retina serves to sense light and provides signals to the brain for vision. The retina is highly vascularized; and as such, pericytes play a key physiological role within the retinal microvasculature. Similar to their role in most tissues, pericytes are important for angiogenesis and vessel stabilization in the retina. However, unlike in tissues such as the pancreas or skeletal muscle, pericytes in the retina are specialized for maintenance of the blood–retina barrier (Trost et al. 2016). The blood–retina barrier is highly regulated to control the flow of fluids, signaling molecules, and chemicals. The controlled barrier is regulated by tight and adherens junctions between pericytes and endothelial, neuronal, and glial cells (Trost et al. 2016; Kim et al. 2009). The breakdown of the blood–retina barrier is detrimental to retinal health and is a contributing factor to diabetic retinopathy.

Diabetic retinopathy is the most frequent cause of blindness in adults aged 20–74. According to one study, approximately 60% of type 2 diabetic patients have retinopathy (Fong et al. 2003). Diabetic retinopathy first emerges as mild abnormalities of the retina, including increased vascular permeability. As retinopathy progresses, moderate-to-severe retinopathy presents as nonproliferative disease that is characterized by degenerated, non-perfused capillaries. In the most progressive form of the disease, proliferative retinopathy presents with neovascularization of the retina and vitreous. Macular edema, characterized by retinal thickening, often co-occurs with diabetic retinopathy.

In both human and animal studies, diabetes-induced retinal pericyte dysfunction contributes to the microvascular abnormalities that lead to retinopathy. The earliest morphological change following the onset of diabetes is a decrease in retinal pericyte coverage of capillaries (Mizutani et al. 1996; Engerman 1989; Cogan et al. 1961). Early work showed that hyperglycemia is implicated in pericyte-induced apoptosis via the accumulation of advanced glycation end products (Stitt et al. 1997). Further metabolic abnormalities seen in type 2 diabetic patients, such high levels of plasma free fatty acids, may contribute to pericyte apoptosis. Specifically, *in vitro* apoptosis of retinal pericytes was shown to be caused by palmitate, a saturated fatty acid that activates apoptosis via NAD(P)H oxidases and NF-κB (Cacicedo et al. 2005). Relatedly, another contributing factor to pericyte dysfunction in type 2 diabetes is a downregulation of PPARα in the retina (Hu et al. 2013). PPARα is a modulator of lipid metabolism. Evidence suggests that maintenance of PPARα expression in the retina can attenuate diabetes-induced pericyte apoptosis and capillary degeneration in streptozotocin-induced diabetic mice, which is a model of type 1 diabetes. Future studies should investigate this mechanism in models of type 2 diabetes, specifically. The potential mechanism by which PPARα works to maintain pericyte density in the retina is by suppressing oxidative stress and inhibiting inflammation. Understanding the protective effects of PPARα against pericyte dropout provides the potential for therapeutic intervention using PPARα agonists, such as fenofibrate (Ding et al. 2014). In addition to the negative effects of fatty acids on pericytes, hyperglycemia has also been shown to increase inflammation to promote retinal pericyte apoptosis via NF-κB activation (Ding et al. 2014). Park et al. (2014) used the streptozotocin-induced type 1 diabetes model to show that hyperglycemia induces pericyte apoptosis via endothelial cell angiopoietin-2 interacting with pericyte integrins to activate the p53 pathway. Again, this study in a type 1 diabetes model may not translate completely to type 2 diabetes; therefore, while it provides insights into diabetic retinopathy in type 2 diabetic patients, future studies should utilize type 2 diabetes models to elucidate mechanisms specific to type 2 diabetes. Still, the work from type 1 diabetes models suggests that angiopoietin-2 and integrins are targets for therapeutic intervention.

It is clear that pericyte apoptosis is a defining characteristic of diabetic retinopathy and many factors mediate retinal pericyte dropout. A major consequence of this pericyte loss is microangiopathy, including microaneurysm, microhemorrhage, and nerve layer infarcts as a result of weakened capillaries (Valdez et al. 2014). Pericyte loss has also been shown to lead to reduced inhibition of endothelial cell proliferation *in vivo* (Hammes et al. 2002). An emphasis of future studies should be to target the pericyte in order to maintain retinal function and prevent blindness in people with type 2 diabetes.

Renal Pericyte

The precise regulation of renal blood flow is necessary for maintaining ion gradients, waste disposal, and glomerular filtration in the kidney. Renal pericytes have been identified within the tubular system and vasa recta capillaries in the kidney where they have been shown to express α-SMA and play roles in regulating medullary blood flow and stabilizing glomeruli (Park et al. 1997; Crawford et al. 2012; Kramann and Humphreys 2014; Lenoir et al. 2015). An isolated, perfused vasa recta model was used to determine that renal capillary constriction and dilation are mediated by pericytes in response to vasoactive substances, including angiotensin-II, endothelin-1, nitric oxide, adenosine, and prostaglandin E2, released by various sources including endothelial cells, tubular epithelium, the hypothalamus, sympathetic nerves, and parasympathetic nerves (Kennedy-Lydon et al. 2013; Pallone 1994; Pallone and Silldorff 2001). In addition to these vasoactive peptides and hormones, nitric oxide and reactive oxygen species are also important for regulating renal blood flow in the microcirculation. Crawford et al. (Crawford et al. 2012) used an *in situ* intact kidney slice model to show that inhibition of nitric oxide causes pericyte-mediated vasoconstriction in the vasa recta; and conversely, that pericytes mediate vasodilation of the vasa recta in the presence of a nitric oxide donor.

Diabetic nephropathy is the most common cause of end-stage renal disease in the world (Remuzzi et al. 2002). Nephropathy develops in approximately 20–40% of patients with type 2 diabetes within 20 years of diabetes onset (Ritz and Orth 1999). While the exact mechanisms for the pathogenesis of diabetic nephropathy are unknown, it has been suggested that proteinuria, genetics, hypoxia, ischemia, and ultimately, inflammation contribute to kidney injury and disease pathogenesis (Fernandez Fernandez et al. 2012). Damage to the kidney results in an inflammatory response that is necessary for tissue repair and restoration of homeostasis. Dysregulated inflammation and abnormal tissue repair results in fibrosis, which is the first step in kidney disease development. Fibrotic kidney tissue becomes pathological when kidney structure and function are compromised.

Pericytes are implicated as contributors to fibrosis and the pathogenesis of diabetic nephropathy. Protein kinase C activation is central to pericyte dysfunction during diabetic nephropathy pathogenesis (Inoguchi et al. 2003). Research has shown that hyperglycemia activates protein kinase C in mesangial cells, which are cells similar to pericytes, in the pathogenesis of diabetic nephropathy (Haneda et al. 1997). Koya and colleagues further implicated pericyte-like mesangial cell dysfunction in diabetic nephropathy by showing that PKCβ inhibition prevents detrimental mesangial expansion and attenuates glomerular dysfunction in a mouse model of type 2 diabetes (Koya et al. 2000). PKCβ activation induces reaction oxygen species formation by NADPH oxidase enzymes. The increase in reaction oxygen species then induces expression of transforming growth factor-β, which promotes excess extracellular matrix deposition and fibrosis. In support of this, Isono et al. (Isono et al. 2002) used a type 1 diabetes model to show that hyperglycemia activates TGF-β and SMAD signaling in the kidney leading to the promotion of profibrotic mesangial gene expression changes in diabetic nephropathy. Future studies should continue to examine the role of pericytes in diabetic nephropathy with special emphasis on identifying the specific contribution of genuine pericytes to fibrosis and kidney complications in type 2 diabetes.

Peripheral Nerve Pericytes

Pericytes surround endoneurial microvessels within nerves and serve as important components of the blood–nerve barrier. Within the peripheral nervous system, the pericyte to endothelial cell ratio is higher than within organs without a blood barrier, such as the skeletal muscle (Miyoshi et al. 1979). Similar to the known roles of pericyte in the blood–retina barrier, endoneurial pericytes are functionally important for maintaining and regulating the blood–nerve barrier. Peripheral nerve pericytes were shown to express molecule transporters, including Glut-1 (Shimizu et al. 2008). It has been suggested that transport proteins allow pericytes to cooperate with endothelial cells for the transportation of glucose and other molecules between the blood and organs. Peripheral nerve pericytes have also been shown to express growth factors that are commonly expressed by barrier-supporting astrocytes in the central nervous system; therefore, suggesting that pericytes serve to support the blood–nerve barrier in the periphery, which lacks astrocytes (Shimizu et al. 2011a). Shimizu and colleagues demonstrated that basic fibroblast growth factor released from peripheral nerve pericytes augments barrier formation through claudin-5, a major component of tight junctions, in peripheral nerve endothelial cells. Additionally, they also demonstrated that vascular endothelial growth factor, angiopoietin-1, and transforming growth factor-β secreted from pericytes decreased barrier function of peripheral nerve endothelial cells through claudin-5. Recent research has shown that peripheral nerve pericytes maintain the basement membrane in the blood–nerve barrier through the production of fibronectin, collagen type IV, matrix metalloproteinase-2, and tissue inhibitor of metalloproteinases-1 (Shimizu et al. 2011b). It has also been suggested the peripheral nerve pericyte nerve growth factor and glial cell-derived neurotrophic factor may protect axons from inflammatory damage (Shimizu et al. 2011a).

Diabetic neuropathy is a common complication of type 2 diabetes. In diabetic neuropathy, microangiopathy of the endoneurium is observed. The cause of diabetic neuropathy is thought to be due to a breakdown in the blood–nerve barrier. As key components of physiological blood–nerve barrier function, pericyte dysfunction is suggested to contribute to diabetic neuropathy (Giannini and Dyck 1995). Gianni and Dyck showed that endoneurial basement membrane hyperplasia and pericyte loss are characteristic in nerve biopsies from type 1 and type 2 diabetic patients. Accumulation of advanced glycation end products are one potential mechanism for diabetes-induced pericyte loss and blood–nerve barrier dysfunction (Wada and Yagihashi 2005). Advanced glycation end product accumulation has been observed in the nerves of diabetic patients (Wada and Yagihashi 2005). Shimizu et al. (2011b) showed that advanced glycation end products directly stimulate fibronectin and collagen type IV production from peripheral nerve pericytes *in vitro*; thus, providing further evidence for type 2 diabetes to cause pericyte dysfunction that leads to diabetic neuropathy. Interestingly, Hayes et al. (2018) demonstrated a potentially novel therapeutic opportunity to treat diabetic neuropathy with skeletal muscle pericytes. In that study, skeletal muscle pericytes were intramuscularly transplanted into ischemic hindlimbs of type 2 diabetic mice. They showed that pericytes can differentiate into S100-positive Schwann cells *in vivo* when transplanted into both wild-type and type 2 diabetic mice (Fig. 12.2). Typically, type 2 diabetes impairs stem cell therapies, but the evidence of *in vivo* differentiation of pericytes into Schwann cells in a type 2 diabetic mouse warrants further investigation into pericyte cell therapies for diabetic neuropathy.

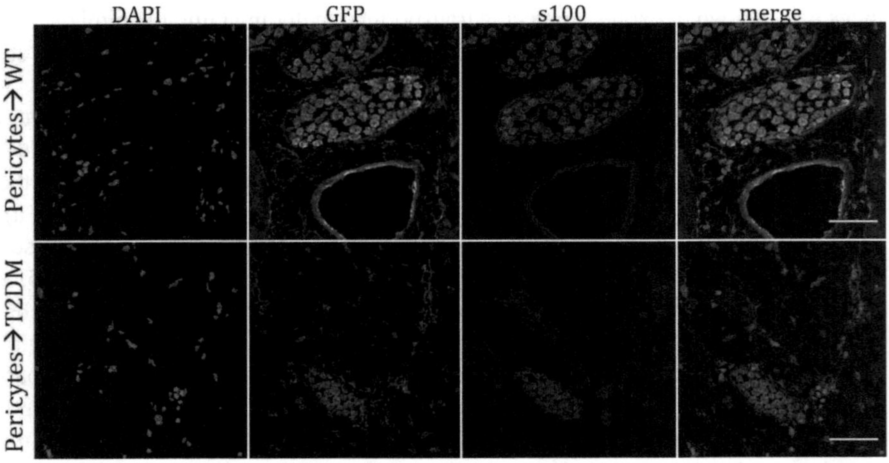

Fig. 12.2 Pericyte differentiation into Schwann cells in ischemic skeletal muscle. GFP^+ skeletal muscle pericytes were isolated from wild-type mice and then intramuscularly transplanted into the ischemic hindlimbs of wild-type C57Bl/6 or *db/db* mice to test for the ability of pericytes to augment postischemic neovascularization. Unexpectedly, GFP^+ skeletal muscle pericytes differentiated into S100 Schwann cells *in vivo* in both (A) wild-type and (B) *db/db* mice, indicating their potential for peripheral neuropathy therapy. Scale bar is 50 μm. "→" indicates "transplanted into." This figure was reproduced from Hayes et al. 2018 *AJP-Cell Physiology*

Commentary and Trends

Pericytes are ubiquitous cells that are an integral component of microvascular health. Type 2 diabetes impairs the number and function of pericytes, resulting in microvascular dysfunction in many organs throughout the body, which manifests in many common complications for type 2 diabetic patients. In order to prevent the pericyte dropout and dysfunction that causes these complications, therapies should target mechanisms for maintaining pericyte coverage and function in the microvasculature. Interestingly, pericytes also have therapeutic potential for use in stem cell therapies, especially due to their ability to differentiate into terminal cell types. Therefore, they provide a unique opportunity for cell transplantation in an attempt to support the cells in the injured tissue for the treatment of diabetic complications, especially peripheral artery disease. However, therapeutic strategies must consider how to overcome the negative effect of the diabetic environment, including hyperglycemia, hyperinsulinemia and dyslipidemia, on the pericytes. In order to do this, preclinical basic and translational studies must be tested in clinically relevant models of type 2 diabetes.

References

Almaca, J., Weitz, J., Rodriguez-Diaz, R., Pereira, E., & Caicedo, A. (2018). The pericyte of the pancreatic islet regulates capillary diameter and local blood flow. *Cell Metabolism, 27*, 630–644 e4.

Armulik, A., Genove, G., & Betsholtz, C. (2011). Pericytes: Developmental, physiological, and pathological perspectives, problems, and promises. *Developmental Cell, 21*, 193–215.

Birbrair, A., Zhang, T., Wang, Z. M., Messi, M. L., Olson, J. D., Mintz, A., & Delbono, O. (2014). Type-2 pericytes participate in normal and tumoral angiogenesis. *American Journal of Physiology. Cell Physiology, 307*, C25–C38.

Brownlee, M. (2005). The pathobiology of diabetic complications: A unifying mechanism. *Diabetes, 54*, 1615–1625.

Cacicedo, J. M., Benjachareowong, S., Chou, E., Ruderman, N. B., & Ido, Y. (2005). Palmitate-induced apoptosis in cultured bovine retinal pericytes: Roles of NAD(P)H oxidase, oxidant stress, and ceramide. *Diabetes, 54*, 1838–1845.

Caporali, A., Meloni, M., Vollenkle, C., Bonci, D., Sala-Newby, G. B., Addis, R., Spinetti, G., Losa, S., Masson, R., Baker, A. H., Agami, R., Le Sage, C., Condorelli, G., Madeddu, P., Martelli, F., & Emanueli, C. (2011). Deregulation of microRNA-503 contributes to diabetes mellitus-induced impairment of endothelial function and reparative angiogenesis after limb ischemia. *Circulation, 123*, 282–291.

Cho, N. H., Shaw, J. E., Karuranga, S., Huang, Y., Da Rocha Fernandes, J. D., Ohlrogge, A. W., & Malanda, B. (2018). IDF diabetes atlas: Global estimates of diabetes prevalence for 2017 and projections for 2045. *Diabetes Research and Clinical Practice, 138*, 271–281.

Cogan, D. G., Toussaint, D., & Kuwabara, T. (1961). Retinal vascular patterns. IV. Diabetic retinopathy. *Archives of Ophthalmology, 66*, 366–378.

Crawford, C., Kennedy-Lydon, T., Sprott, C., Desai, T., Sawbridge, L., Munday, J., Unwin, R. J., Wildman, S. S., & Peppiatt-Wildman, C. M. (2012). An intact kidney slice model to investigate vasa recta properties and function in situ. *Nephron. Physiology, 120*, p17–p31.

Dar, A., Domev, H., Ben-Yosef, O., Tzukerman, M., Zeevi-Levin, N., Novak, A., Germanguz, I., Amit, M., & Itskovitz-Eldor, J. (2012). Multipotent vasculogenic pericytes from human pluripotent stem cells promote recovery of murine ischemic limb. *Circulation, 125*, 87–99.

Defronzo, R. A., Ferrannini, E., Groop, L., Henry, R. R., Herman, W. H., Holst, J. J., Hu, F. B., Kahn, C. R., Raz, I., Shulman, G. I., Simonson, D. C., Testa, M. A., & Weiss, R. (2015). Type 2 diabetes mellitus. *Nature Reviews. Disease Primers, 1*, 15019.

Ding, L., Cheng, R., Hu, Y., Takahashi, Y., Jenkins, A. J., Keech, A. C., Humphries, K. M., Gu, X., Elliott, M. H., Xia, X., & Ma, J. X. (2014). Peroxisome proliferator-activated receptor alpha protects capillary pericytes in the retina. *The American Journal of Pathology, 184*, 2709–2720.

Efimenko, A. Y., Kochegura, T. N., Akopyan, Z. A., & Parfyonova, Y. V. (2015). Autologous stem cell therapy: How aging and chronic diseases affect stem and progenitor cells. *Bioresearch Open Access, 4*, 26–38.

Engerman, R. L. (1989). Pathogenesis of diabetic retinopathy. *Diabetes, 38*, 1203–1206.

Fernandez Fernandez, B., Elewa, U., Sanchez-Nino, M. D., Rojas-Rivera, J. E., Martin-Cleary, C., Egido, J., & Ortiz, A. (2012). 2012 update on diabetic kidney disease: The expanding spectrum, novel pathogenic insights and recent clinical trials. *Minerva Medica, 103*, 219–234.

Folli, F., Corradi, D., Fanti, P., Davalli, A., Paez, A., Giaccari, A., Perego, C., & Muscogiuri, G. (2011). The role of oxidative stress in the pathogenesis of type 2 diabetes mellitus micro- and macrovascular complications: Avenues for a mechanistic-based therapeutic approach. *Current Diabetes Reviews, 7*, 313–324.

Fong, D. S., Aiello, L., Gardner, T. W., King, G. L., Blankenship, G., Cavallerano, J. D., Ferris, F. L., 3rd, Klein, R., & American Diabetes, A. (2003). Diabetic retinopathy. *Diabetes Care, 26*, 226–229.

Giacco, F., & Brownlee, M. (2010). Oxidative stress and diabetic complications. *Circulation Research, 107*, 1058–1070.

Giannini, C., & Dyck, P. J. (1995). Basement membrane reduplication and pericyte degeneration precede development of diabetic polyneuropathy and are associated with its severity. *Annals of Neurology, 37*, 498–504.

Gubernator, M., Slater, S. C., Spencer, H. L., Spiteri, I., Sottoriva, A., Riu, F., Rowlinson, J., Avolio, E., Katare, R., Mangialardi, G., Oikawa, A., Reni, C., Campagnolo, P., Spinetti, G., Touloumis, A., Tavare, S., Prandi, F., Pesce, M., Hofner, M., Klemens, V., Emanueli, C., Angelini, G., & Madeddu, P. (2015). Epigenetic profile of human adventitial progenitor cells correlates with therapeutic outcomes in a mouse model of limb ischemia. *Arteriosclerosis, Thrombosis, and Vascular Biology, 35*, 675–688.

Hammes, H. P., Lin, J., Renner, O., Shani, M., Lundqvist, A., Betsholtz, C., Brownlee, M., & Deutsch, U. (2002). Pericytes and the pathogenesis of diabetic retinopathy. *Diabetes, 51*, 3107–3112.

Haneda, M., Araki, S., Togawa, M., Sugimoto, T., Isono, M., & Kikkawa, R. (1997). Mitogen-activated protein kinase cascade is activated in glomeruli of diabetic rats and glomerular mesangial cells cultured under high glucose conditions. *Diabetes, 46*, 847–853.

Hayden, M. R., Yang, Y., Habibi, J., Bagree, S. V., & Sowers, J. R. (2010). Pericytopathy: Oxidative stress and impaired cellular longevity in the pancreas and skeletal muscle in metabolic syndrome and type 2 diabetes. *Oxidative Medicine and Cellular Longevity, 3*, 290–303.

Hayes, K. L., Messina, L. M., Schwartz, L. M., Yan, J., Burnside, A. S., & Witkowski, S. (2018). Type 2 diabetes impairs the ability of skeletal muscle pericytes to augment postischemic neovascularization in db/db mice. *American Journal of Physiology Cell Physiology, 314*, C534–C544.

Houtz, J., Borden, P., Ceasrine, A., Minichiello, L., & Kuruvilla, R. (2016). Neurotrophin signaling is required for glucose-induced insulin secretion. *Developmental Cell, 39*, 329–345.

Hu, Y., Chen, Y., Ding, L., He, X., Takahashi, Y., Gao, Y., Shen, W., Cheng, R., Chen, Q., Qi, X., Boulton, M. E., & Ma, J. X. (2013). Pathogenic role of diabetes-induced PPAR-alpha down-regulation in microvascular dysfunction. *Proceedings of the National Academy of Sciences of the United States of America, 110*, 15401–15406.

Inoguchi, T., Sonta, T., Tsubouchi, H., Etoh, T., Kakimoto, M., Sonoda, N., Sato, N., Sekiguchi, N., Kobayashi, K., Sumimoto, H., Utsumi, H., & Nawata, H. (2003). Protein kinase C-dependent increase in reactive oxygen species (ROS) production in vascular tissues of diabetes: Role of vascular NAD(P)H oxidase. *Journal of American Society of Nephrology, 14*, S227–S232.

Isono, M., Chen, S., Hong, S. W., Iglesias-De La Cruz, M. C., & Ziyadeh, F. N. (2002). Smad pathway is activated in the diabetic mouse kidney and Smad3 mediates TGF-beta-induced fibronectin in mesangial cells. *Biochemical and Biophysical Research Communications, 296*, 1356–1365.

Kennedy-Lydon, T. M., Crawford, C., Wildman, S. S., & Peppiatt-Wildman, C. M. (2013). Renal pericytes: Regulators of medullary blood flow. *Acta Physiologica (Oxford, England), 207*, 212–225.

Kim, J. H., Kim, J. H., Yu, Y. S., Kim, D. H., & Kim, K. W. (2009). Recruitment of pericytes and astrocytes is closely related to the formation of tight junction in developing retinal vessels. *Journal of Neuroscience Research, 87*, 653–659.

Koya, D., Haneda, M., Nakagawa, H., Isshiki, K., Sato, H., Maeda, S., Sugimoto, T., Yasuda, H., Kashiwagi, A., Ways, D. K., King, G. L., & Kikkawa, R. (2000). Amelioration of accelerated diabetic mesangial expansion by treatment with a PKC beta inhibitor in diabetic db/db mice, a rodent model for type 2 diabetes. *The FASEB Journal, 14*, 439–447.

Kramann, R., & Humphreys, B. D. (2014). Kidney pericytes: Roles in regeneration and fibrosis. *Seminars in Nephrology, 34*, 374–383.

Lenoir, O., Jasiek, M., Henique, C., Guyonnet, L., Hartleben, B., Bork, T., Chipont, A., Flosseau, K., Bensaada, I., Schmitt, A., Masse, J. M., Souyri, M., Huber, T. B., & Tharaux, P. L. (2015). Endothelial cell and podocyte autophagy synergistically protect from diabetes-induced glomerulosclerosis. *Autophagy, 11*, 1130–1145.

Metea, M. R., & Newman, E. A. (2007). Signalling within the neurovascular unit in the mammalian retina. *Experimental Physiology, 92*, 635–640.

Miyoshi, T., Kennedy, W. R., & Yoon, K. S. (1979). Morphometric comparison of capillaries in muscle spindles, nerve, and muscle. *Archives of Neurology, 36*, 547–552.

Mizutani, M., Kern, T. S., & Lorenzi, M. (1996). Accelerated death of retinal microvascular cells in human and experimental diabetic retinopathy. *The Journal of Clinical Investigation, 97*, 2883–2890.

Pallone, T. L. (1994). Vasoconstriction of outer medullary vasa recta by angiotensin II is modulated by prostaglandin E2. *The American Journal of Physiology, 266*, F850–F857.

Pallone, T. L., & Silldorff, E. P. (2001). Pericyte regulation of renal medullary blood flow. *Experimental Nephrology, 9*, 165–170.

Park, F., Mattson, D. L., Roberts, L. A., & Cowley, A. W., Jr. (1997). Evidence for the presence of smooth muscle alpha-actin within pericytes of the renal medulla. *The American Journal of Physiology, 273*, R1742–R1748.

Park, S. W., Yun, J. H., Kim, J. H., Kim, K. W., Cho, C. H., & Kim, J. H. (2014). Angiopoietin 2 induces pericyte apoptosis via alpha3beta1 integrin signaling in diabetic retinopathy. *Diabetes, 63*, 3057–3068.

Remuzzi, G., Schieppati, A., & Ruggenenti, P. (2002). Clinical practice. Nephropathy in patients with type 2 diabetes. *The New England Journal of Medicine, 346*, 1145–1151.

Ritz, E., & Orth, S. R. (1999). Nephropathy in patients with type 2 diabetes mellitus. *The New England Journal of Medicine, 341*, 1127–1133.

Sakhneny, L., Rachi, E., Epshtein, A., Guez, H. C., Wald-Altman, S., Lisnyansky, M., Khalifa-Malka, L., Hazan, A., Baer, D., Priel, A., Weil, M., & Landsman, L. (2018). Pancreatic pericytes support beta-cell function in a Tcf7l2-dependent manner. *Diabetes, 67*, 437–447.

Sasson, A., Rachi, E., Sakhneny, L., Baer, D., Lisnyansky, M., Epshtein, A., & Landsman, L. (2016). Islet pericytes are required for beta-cell maturity. *Diabetes, 65*, 3008–3014.

Shimizu, F., Sano, Y., Abe, M. A., Maeda, T., Ohtsuki, S., Terasaki, T., & Kanda, T. (2011a). Peripheral nerve pericytes modify the blood-nerve barrier function and tight junctional molecules through the secretion of various soluble factors. *Journal of Cellular Physiology, 226*, 255–266.

Shimizu, F., Sano, Y., Haruki, H., & Kanda, T. (2011b). Advanced glycation end-products induce basement membrane hypertrophy in endoneurial microvessels and disrupt the blood-nerve barrier by stimulating the release of TGF-beta and vascular endothelial growth factor (VEGF) by pericytes. *Diabetologia, 54*, 1517–1526.

Shimizu, F., Sano, Y., Maeda, T., Abe, M. A., Nakayama, H., Takahashi, R., Ueda, M., Ohtsuki, S., Terasaki, T., Obinata, M., & Kanda, T. (2008). Peripheral nerve pericytes originating from the blood-nerve barrier expresses tight junctional molecules and transporters as barrier-forming cells. *Journal of Cellular Physiology, 217*, 388–399.

Stitt, A. W., Li, Y. M., Gardiner, T. A., Bucala, R., Archer, D. B., & Vlassara, H. (1997). Advanced glycation end products (AGEs) co-localize with AGE receptors in the retinal vasculature of diabetic and of AGE-infused rats. *The American Journal of Pathology, 150*, 523–531.

Stratton, I. M., Adler, A. I., Neil, H. A., Matthews, D. R., Manley, S. E., Cull, C. A., Hadden, D., Turner, R. C., & Holman, R. R. (2000). Association of glycaemia with macrovascular and microvascular complications of type 2 diabetes (UKPDS 35): Prospective observational study. *BMJ, 321*, 405–412.

Tilton, R. G., Faller, A. M., Burkhardt, J. K., Hoffmann, P. L., Kilo, C., & Williamson, J. R. (1985). Pericyte degeneration and acellular capillaries are increased in the feet of human diabetic patients. *Diabetologia, 28*, 895–900.

Tilton, R. G., Hoffmann, P. L., Kilo, C., & Williamson, J. R. (1981). Pericyte degeneration and basement membrane thickening in skeletal muscle capillaries of human diabetics. *Diabetes, 30*, 326–334.

Trost, A., Lange, S., Schroedl, F., Bruckner, D., Motloch, K. A., Bogner, B., Kaser-Eichberger, A., Strohmaier, C., Runge, C., Aigner, L., Rivera, F. J., & Reitsamer, H. A. (2016). Brain and retinal pericytes: Origin, function and role. *Frontiers in Cellular Neuroscience, 10*, 20.

Valdez, C. N., Arboleda-Velasquez, J. F., Amarnani, D. S., Kim, L. A., & D'amore, P. A. (2014). Retinal microangiopathy in a mouse model of inducible mural cell loss. *The American Journal of Pathology, 184*, 2618–2626.

Vono, R., Fuoco, C., Testa, S., Pirro, S., Maselli, D., Ferland Mccollough, D., Sangalli, E., Pintus, G., Giordo, R., Finzi, G., Sessa, F., Cardani, R., Gotti, A., Losa, S., Cesareni, G., Rizzi, R., Bearzi, C., Cannata, S., Spinetti, G., Gargioli, C., & Madeddu, P. (2016). Activation of the pro-oxidant PKCbetaII-p66Shc signaling pathway contributes to pericyte dysfunction in skeletal muscles of patients with diabetes with critical limb ischemia. *Diabetes, 65*, 3691–3704.

Wada, R., & Yagihashi, S. (2005). Role of advanced glycation end products and their receptors in development of diabetic neuropathy. *Annals of the New York Academy of Sciences, 1043*, 598–604.

Yan, J., Tie, G., Wang, S., Messina, K. E., Didato, S., Guo, S., & Messina, L. M. (2012). Type 2 diabetes restricts multipotency of mesenchymal stem cells and impairs their capacity to augment postischemic neovascularization in db/db mice. *Journal of the American Heart Association, 1*, e002238.

Yan, J., Tie, G., Xu, T. Y., Cecchini, K., & Messina, L. M. (2013). Mesenchymal stem cells as a treatment for peripheral arterial disease: Current status and potential impact of type II diabetes on their therapeutic efficacy. *Stem Cell Reviews, 9*, 360–372.

Chapter 13
Pericytes in Atherosclerosis

Volha Summerhill and Alexander Orekhov

Abstract Pericytes are pluripotent cells found in the vascular wall of both capillaries and large blood vessels. Pericytes are highly heterogeneous cells in terms of phenotype, tissue distribution, origin and functions, and they play an important role in the regulation of vascular morphogenesis and function. Pericytes were shown to be involved in tissue development and homeostasis, as well as in pathological processes, including atherosclerosis. Both microvascular and macrovascular pericytes form the cellular network of the arterial wall and are actively involved in lipid accumulation, growth, and neovascularization of the atherosclerotic plaque, local inflammation and thrombosis. According to current understanding, pericytes originate from the multipotent stem cells capable of mesenchymal differentiation to oligopotent lineages, such as osteoclasts, chondrocytes, and adipocytes, and also serve as mesenchymal local progenitors in tissues. Pericyte multilineage potential is fundamental for vascular pathology, including atherosclerotic lesion formation. Pericytes express various surface proteins that can be used for their identification in aid of diagnosis and therapeutic strategies for atherosclerosis and other vascular pathologies.

Keywords Pericyte · Atherosclerosis · Cardiovascular diseases · Mesenchymal stem cells · Endothelial cells · Vascular smooth muscle cells · Endothelial dysfunction · Angiogenesis · Differentiation · Blood vessels · Foam cells · Low-density lipids

V. Summerhill · A. Orekhov (✉)
Laboratory of Angiopathology, Institute of General Pathology and Pathophysiology, Moscow, Russia

Institute for Atherosclerosis Research Moscow, Skolkovo Innovative Center, Moscow, Russia

A. Birbrair (ed.), *Pericyte Biology in Disease*, Advances in Experimental Medicine and Biology 1147, https://doi.org/10.1007/978-3-030-16908-4_13

Introduction

Coronary artery disease (CAD) as a consequence of atherosclerosis is a leading cause of morbidity and mortality worldwide (Mozaffarian et al. 2015). Atherosclerotic lesion development occurs in the arterial subendothelial intima, especially in the regions of perturbed blood flow, such as vessel bifurcations. The pathological process is characterized by an interplay between the endothelial dysfunction and subendothelial lipid accumulation, resulting in chronic inflammation in the arterial wall (Tabas et al. 2015). In that regard, cell types that can participate in these processes, and therefore provide the potential targets for therapeutic intervention, attract special attention. Among the vascular wall cell types, pericytes are the most enigmatic due to their phenotypic variety and multifunctionality. Pericytes are able to differentiate into several cell lineages, and, therefore, contribute to the formation, maturation, and homeostasis of the entire human vascular bed (Montiel-Eulefi et al. 2012). Moreover, pericytes vary functionally and morphologically between different organs and tissues, adding to the complexity of the vascular wall composition. Thus, for example, central nervous system (CNS) pericytes are functionally distinct from the peripheral blood vessels pericytes (Wang et al. 1992; Balabanov and Dore-Duffy 1998). Pericyte content also varies considerably between different tissues and organs. For instance, these cells are more numerous in the walls of larger blood vessels that must maintain higher blood pressure. According to Sims (2000), the endothelial cell-to-pericyte ratio is the highest (1:1) in the CNS and retina. Significantly lower proportions were reported for lung and skin tissues (1:10) and skeletal muscles (1:100). However, these cells have proven difficult to identify and quantify precisely due to their high heterogeneity and the lack of a single specific marker. Because of these challenges, the role of pericytes in health and disease remains to be studied in detail (Krueger and Bechmann 2010).

In this chapter, we will discuss the morphological and functional significance of pericytes in vascular homeostasis and their involvement in various aspects of atherosclerosis pathogenesis, such as endothelial dysfunction, angiogenesis (neovascularization of plaques), local inflammation, calcification, and thrombosis. We will also revise current and evolving concepts on the pericyte ontogeny.

Functional and Morphological Features of Pericytes

In the blood vessel wall, pericytes are located in the proximity of the endothelial cells, in the basal membrane. They are characterized by a prominent round nucleus and a relatively small content of cytoplasm. Pericytes form contacts with endothelial cells of the capillary wall, as well as with other cells and with each other, by means of long cytoplasmic processes. These contacts play important role in maintaining the endothelial structure and function (Mazanet and Franzini-Armstrong 1982; Rucker et al. 2000). Recent experiments revealed the regulatory and signalling activities of pericytes towards the endothelium. For instance, pericytes were shown to inhibit endothelial cell division via activation of TGF-β (Betsholtz et al. 2005). Gap junction contacts formed by pericytes permit interchange of ions and small molecules, including N-cadherin, fibronectin, connexin and various integrins (Bergers and Song 2005; Armulik et al. 2005; Dore-Duffy and Cleary 2011). These endothelium–pericyte interactions are functionally associated with active regulation of blood flow throughout the body, hence, blood pressure that can be particularly important in vascular adaptation to hypotoxic injury. Moreover, in capillaries, pericytes participate in controlling the blood flow by their contractile ability, conveyed by the expression of varying amounts of essential contractile proteins, such as actin, myosin, smooth muscle actin alpha (SMA-α), tropomyosin and desmin (Bandopadhyay et al. 2001). Pericyte participation in regulating the vessel diameter was observed in retinal and medullary microvessels in response to depolarization and neuronal activity (Wu et al. 2003; Peppiatt et al. 2006). Moreover, several vasoactive molecules that regulate pericyte contraction and relaxation have been identified, including cholinergic α-2 and adrenergic β-2 receptors, angiotensin II and endothelin-1 (Bergers and Song 2005). Noteworthy, pericyte contractility can be important for regulation of capillary blood flow in pathological conditions, such as ischaemia (Yemisci et al. 2009; Hall et al. 2014). However, in the aorta and other large blood vessels, the contractile ability of pericytes is not that important, as the vascular tone is maintained by other mechanisms (Orekhov et al. 2016a). Physiological functions of perivascular pericytes that are known to date are listed in Table 13.1.

The diversity of pericyte biological functions is reflected by the numerous abnormalities of these cells associated with pathological conditions. Vascular disorders, including atherosclerosis, are accompanied by such alterations as abnormal count and size of endothelial cell–pericyte interfaces, excess or deficiency in pericyte coverage, and changes in pericyte contractility. These changes, in turn, lead to altered vessel sprouting, remodelling, maturation and stabilization (Gerhardt and Betsholtz 2003; Hughes et al. 2006).

Table 13.1 Functions of perivascular pericytes

Function	References
Angiogenesis and vessel stabilization	Betsholtz et al. (2005), Raza et al. (2010)
Capillary blood flow regulation	Cuevas et al. (1984), Pallone et al. (1998, 2003), Stefanska (2013)
Vascular morphogenesis, maturation and remodelling	Leveen et al. (1994), Hellstrom et al. (2001), Chen et al. (2007), Herman (2010), Warmke et al. (2016)
Vascular permeability	Glentis et al. (2014)
Maintenance functional integrity of the blood–brain barrier	Dohgu et al. (2005), Peppiatt et al. (2006), Koh et al. (2008), Krueger and Bechmann (2010), Al Ahmad et al. (2011)
Blood coagulation (thrombosis)	Bouchard et al. (1997), Kim et al. (2006), Dulmovits and Herman (2012)
Lymphocyte activation	Thomas (1999), Bose et al. (2013)
Phagocytosis	Guillemin and Brew (2004)

Methods of Pericyte Identification

Being pluripotent heterogeneous cells, pericytes are difficult to identify in organs and tissues. Only a few markers of pericytes have been established, and none of them is universal, hence specific for all pericyte subtypes and under any circumstances. The expression patterns of the pericyte antigens are tissue-specific and can be up- or downregulated in pathological conditions. Moreover, the expression of some of the markers depends on the developmental stage of a blood vessel. Cell culture conditions can also influence the expression of marker proteins in pericytes cultured in vitro.

Many of the proteins expressed by pericytes are expressed in other cell types as well. The most obvious example is smooth muscle actin (SMA-α), which is a typical protein in vascular smooth muscle cells (SMCs) (van Dijk et al. 2015). It was found that vascular pericytes are often positive for SMA-α, but SMA-α-negative pericytes can also be observed (Bandopadhyay et al. 2001). For instance, a population of SMA-α-positive intimal cells was found that expressed CD68, protein atypical for vascular SMCs, which is usually regarded as a macrophage marker (Andreeva et al. 1997b). Interestingly, the ratio of SMA-α+CD68+ (double positive) cells was increased in atherosclerotic lesions, as well as in primary cell culture exposed to atherogenic-modified low-density lipids (LDL) (Orekhov et al. 1998). Another example is platelet-derived growth factor receptor beta (PDGFR-β), which, apart from pericytes, is also present in fibroblasts, astrocytes, and some tumour cells (Lindahl et al. 1997). Furthermore, pericyte-like cells in the human aortic intima express S-100 and CD1, which are common for dendritic cells (Bobryshev and Lord 2000; Bobryshev et al. 2011).

Hence, immunocytochemical identification of pericytes should be performed using combinations of markers. The markers proposed for common pericyte identification include SMA-α, PDGFR-β, aminopeptidase A and N (CD13), neuron-glial 2 (NG2), and desmin (Armulik et al. 2011). Other potential pericyte markers are CD146, endoglin, non-muscle myosin, nestin, vimentin, O-sialoganglioside 3G5, and melanoma chondroitin sulphate proteoglycan 2A7 (Nazarova et al. 1995; Morikawa et al. 2002; Dondossola et al. 2013; Ivanova et al. 2015; Rossi et al. 2016; Chen et al. 2017).

Origin of Vascular Pericytes

Lineage-tracing studies showed that pericytes' ontogeny is rather complex. Pericytes are considered to be pluripotent stem cells that exhibit multilineage developmental features of mesenchymal stromal stem cell (MSC). MSCs are involved in the maintenance of high turnover tissues like liver, skin, skeletal muscles, adipose tissue, and dental pulp. Furthermore, MSC participates in neurogenesis, which probably confirms the neuroectodermal origin of CNS pericytes (Nakata et al. 2017). Similarly, during developmental stages, neuroectodermal cells can possibly differentiate into vascular SMCs of embryonic cerebral vessels (Korn et al. 2002). To achieve the alignment in MSC identification across different studies, International Society for Cellular Therapy created a position statement formulating the MSC minimum regular criteria. First, MSCs are adherent to plastic when maintained in standard culture conditions. Second, MSCs can be characterized by specific surface antigens (positive for CD105, CD73, and CD90 and negative for CD45, CD34, CD14, CD11b, CD79α, CD19, or HLA-DR). Third, MSCs are able to differentiate into osteoblasts, adipocytes, and chondroblasts *in vitro* (Dominici et al. 2006).

It was shown that vascular pericytes may serve as a source for local mesenchymal progenitor cells in adults (Crisan et al. 2007). Numerous *in vivo* and *in vitro* experiments demonstrated pericyte ability for differentiation into several lineages, giving rise to osteoblasts, chondrocytes, adipocytes, SMCs, fibroblasts, and Leydig cells (Kirton et al. 2006; Birbrair et al. 2015). Pericytes can give rise to multiple mesodermal tissues in situ, in response to PDGFR-β (Bouacida et al. 2012).

Direct relationship between MSC and perivascular pericytes was established based on the common expression markers and differentiation abilities of these cell types. This supported the point of view on pericytes as potential progenitor cells with wide differentiation capacity continuously present in developed adult tissues (Chen et al. 2009). In this regard, it is important to understand whether MSC, multipotent adult progenitor cells (MAPC), muscle-derived stem cells (MDSC), and adipose tissue-derived stem cells share a common progenitor in multiple developed organs. Studies of these progenitor cell lineages were hindered by the fact that all of them have only been identified and studied in primary cultures of donor tissues.

Several independent studies described novel subsets of endothelial cells in embryonic tissues that contribute to blood vessel formation and develop into non-vascular cell lineages promoting post-natal growth and regeneration of tissues. For instance, vascular endothelium generating embryonic haematopoietic cells and pluripotent mesoangioblasts have been derived from the aorta and other blood vessels of the embryo (Oberlin et al. 2002; Zambidis et al. 2006; Chen et al. 2009; Zheng et al. 2007).

The relationship between pericytes and MSCs was studied in more details both in situ and in vitro (Crisan et al. 2008). To that end, a combination of markers expressed by perivascular pericytes in human foetal and adult organs was used. Pericytes were identified as cells positive for CD146, NG2, PDGFRβ, and ALP and negative for CD34, CD45, vWF, and CD144. It was found that human perivascular cells extracted from various tissues and brought to culture gave rise to adherent, multilineage precursor cells that exhibited apparent MSC features and could differentiate into osteogenic, chondrogenic, adipogenic, and myogenic lineages. These findings allowed making a provocative speculation that all MSCs are pericytes. The multilineage potential of pericytes may contribute to muscle regeneration, as well as fat accumulation (Birbrair et al. 2013; Kostallari et al. 2015).

However, at present, the understanding of these mural stem/progenitor populations, as well as their potential biologic function, is unclear. The hypothesis that any of these populations are associated with a stem cell niche remains to be supported experimentally (Kovacic and Boehm 2009; Murray et al. 2014).

Role of Pericytes in Atherosclerosis

Initially, pericytes were believed to be present only in capillaries, where they participate in contractility and angiogenesis. However, further studies revealed the presence of pericytes in the vascular wall of large arteries and veins, where they populate the subendothelial layer of the intima (Orekhov et al. 2014; Juchem et al. 2010). In these tissues, pericytes can be distinguished as stellar-shaped cells with multiple processes that form contacts with each other and with the adjacent endothelial cells, thereby creating a cellular network of the intima that may play important regulatory function. As mentioned above, in large blood vessels, the contractile function of pericytes is essentially lost, which is reflected by reduced or absent expression of SMA-α (Orekhov et al. 2016a). Moreover, in large blood vessels, pericytes are also present as part of vasa vasorum, the capillaries of the vascular wall (Andreeva et al. 1998; Campagnolo et al. 2010; Juchem et al. 2010).

Atherosclerotic lesion development is associated with profound qualitative and quantitative changes in the cellular composition of the affected site. Most of these changes occur in the innermost layer of the intima, involving the endothelium and subendothelial space. Microscopic studies revealed that in the affected area, the cellular network becomes disrupted, with reduction in cell-to-cell interaction ratio and the number of gap junctions. Such disruption is likely to be a consequence of lipid accumulation (Orekhov et al. 2016a). The extent of the intercellular interactions through the gap junctions can be estimated by the expression of cell surface protein connexin 43 (Cx43), which is an essential component of the intercellular contacts. In atherosclerotic lesions, the number of Cx43 plaques per cell was greater in lipid-free cells, compared to lipid-laden cells, and was decreasing towards the lumen, which was not the case in the grossly normal intima. These observations may indicate that intracellular lipid accumulation can have a causative role in pathological pericytes alterations and cellular network disruption (Orekhov et al. 2016a).

Early studies showed that atherosclerotic changes in the vascular wall are associated with the increase in a number of 3G5-positive pericyte-like cells in the proteoglycan-rich layer of the aortic intima (Nayak et al. 1988). Morphological analysis demonstrated that atherosclerotic lesions had a six-fold increase in stellate cell (likely pericytes) count in comparison to grossly normal areas, while the number of elongated cells (likely SMCs) and total cell count were increased by two-fold (Orekhov et al. 2014). A direct link was confirmed between the increased number of stellate cells and the content of cholesteryl esters and total lipids in the vascular wall (Orekhov et al. 2016b).

At advanced stages of atherosclerotic lesion development, neovascularization of the plaques becomes evident and plays an important role in the plaque growth. This process is likely to be mediated by T-cadherin signalling (Moreno et al. 2012). As stated by Collett and Canfield (2005), angiogenesis may influence atherosclerotic lesion formation in the following ways: (1) the expansion of new blood vessels in lipid-laden plaques modifies artery permeability, when thickening diminishes oxygen diffusion; (2) albumin and fibrinogen leaking into the artery wall contribute to the plaque formation; (3) intimal microvessels are associated with haemorrhagic sites and ruptures on the plaques; (4) additionally, the presence of inflammatory cells, i.e. leukocytes, macrophages, and mast cells around neovessels of lipid-rich plaques, is indicative that these new vessels are immediately involved in the recruitment of inflammatory cells, therefore, lesion progression, calcification, and destabilization.

The finding that pericytes can differentiate into osteoblasts and chondrocytes is suggestive that they may be involved in the maladaptive ectopic calcification, occurring in atheromatous vessels, where they act as a source of osteoprogenitor cells in the lesions. This process is followed by matrix remodelling and increased recruitment of calcifying vascular cells (CVCs) (Collett and Canfield 2005). Interestingly, CVCs were identified in bovine aortic media as a phenotypic modulation of SMCs (Steitz et al. 2001). Similar to pericytes, CVCs also have the potential to differentiate along several mesenchymal cell lineages, but with a distinct differentiation pattern. Thus, CVCs may obtain osteogenic and chondrogenic phenotypes in the arterial intima, but, unlike pericytes, they do not reveal any potential for adipogenic differentiation. It can be hypothesized that CVCs may represent second-generation pluripotent cells that are transitional between the precursor and terminally differentiated mesenchymal cells (Tintut et al. 2003).

Ectopic calcification of blood vessels is a common feature of advanced atherosclerotic plaques. In association with pericytic chondrogenesis and osteogenesis, proteins, including osteopontin, osteoprotegerin, bone sialoprotein, matrix Gla protein, Axl, alkaline phosphatase, and bone morphogenetic protein-2 (BMP-2) and BMP-4, were detected in the aortic valve atherosclerotic lesions (Rajamannan et al. 2003).

Activation of adventitial cells was described in the following conditions: vein grafting, hypoxia, and hypertension. Under physical stress, adventitial cells may differentiate into myofibroblasts that migrate into the inner layers of the tunica adventitia, followed by the release of paracrine factors regulating vascular remodelling (Haurani and Pagano 2007). Corselli et al. (2012) demonstrated two independent perivascular MSC progenitors: pericytes in microvessels and adventitial cells around larger vessels persist, and hence adventitial cells are able to gain pericytes-like phenotype. Furthermore, Andreeva et al. (1997a) indicated that pericyte-like cells account for over 30% of total cell count of the intima. Also, they have pointed out that a network of 3G5 antigen-positive pericyte-like cells is present mainly in the subendothelial layer. These findings suggested that functions of macropericytes are determined by subendothelial localization and are similar to those of the capillary pericytes. In addition, 3G5-positive pericyte-like cells have been identified in the bovine aorta, in the human unaffected intima and in calcified atherosclerotic plaques (Bostrom et al. 1993). It should be noted that branching regions of larger vessels are most abundant in pericyte-like cells (Moore and Tabas 2011). In addition, Nicosia (2009) reported that pericytes contribute to organogenesis of the aorta in embryos.

It is well known that atherosclerotic lesion formation is triggered by the circulating modified LDL that provoke endothelial damage and accumulate in the arterial wall. In turn, distorted morphology of the endothelial cells, frequently observed in the atherosclerosis-affected sites, compromises their function leading to the inadequate vasoconstriction, leukocyte infiltration, coagulation, increased permeability that facilitates LDL entry in the subendothelial space, and accelerated proliferation or apoptosis (Groschner et al. 2012; Favero et al. 2014). It is worth mentioning that the atherogenic effects of modified LDL are not limited to lipid accumulation. In fact, the numerous studies have demonstrated pro-apoptotic effects of oxidized LDL (oxLDL), i.e. the induction of oxidative stress that can disturb endothelial function and, in turn, promote atherosclerotic lesion formation (Mollace et al. 2015; Salvayre et al. 2002). In addition, the exposure to oxLDL caused apoptosis in cultured multipotent progenitor cells through their membrane damage (Li et al. 2014).

Since pericytes reside immediately below the endothelium and are essential for maintaining endothelial function (van Dijk et al. 2015), it is possible that they are engaged in the atherogenesis-induced endothelial alterations initiating atherosclerotic plaque formation (Fig. 13.1). It was established that lipid accumulation by the subendothelial cells of human aortic intima impairs cell-to-cell communications through the gap junctions in the proteoglycan-rich layer (Andreeva et al. 1991; Rekhter et al. 1993; Juchem et al. 2010). Furthermore, assuming that there is a connection between intracellular lipid content and the rate of cell-to-cell communication, Andreeva et al. (1995) using fluorescent dye injected into cells with and without visual lipid inclusions demonstrated that the proportion of lipid-free cells communicating via gap junctions was similar to that of cells cultured from grossly normal areas of intima. At the same time, the rate of intercellular communication of cells with lipid inclusions was two-fold lower, compared to that of lipid-free cells. In addition, the presence of lipid-rich cells resembling so-called foam cells was typical for cultures derived from atherosclerotic lesions (Andreeva et al. 1995).

It can be hypothesized that functional changes of the gap junctions are one of the causes of atherosclerosis-related breakdowns of cellular networks formed by the proteoglycan-rich layer intimal pericytes. However, the exact role of pericytes in the endothelial dysfunction associated with lipid accumulation is still vague. Endothelium is likely to be damaged from the luminal side by the disturbed blood flow, especially at branching sites, where blood flow is non-linear and undergoing shear stress leads to augmented inflammatory signalling and inflammatory cell recruitment (Pantakani and Asif 2015).

Based on the assumption that intracellular lipid accumulation caused by modified LDL is the crucial event in atherogenesis, it was suggested that aortic pericytes are capable of active lipid accumulation in the subendothelial layer of the arterial intima that changes their phenotype from dormant to proliferative (active) (Ivanova and Orekhov 2016). Moreover, when activated, pericytes may turn into aberrant differentiation to chondrogenic, osteogenic, macrophage, and myofibroblast lineages contributing to the pathogenesis of atherosclerosis and vascular calcification. Thus, it can be suggested that upon formation of the atherosclerotic lesions, the functional state of the pericytes in the proteoglycan-rich layer of the intima is changed and associated with variations in antigenic expression (Orekhov and Bobryshev 2015).

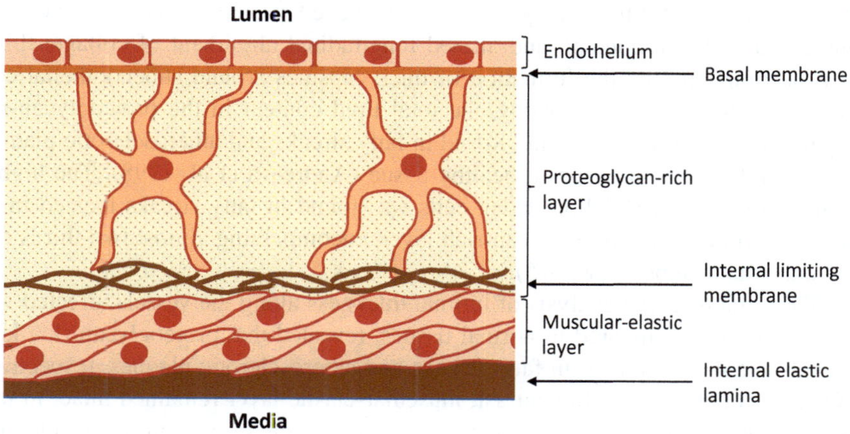

Fig. 13.1 Schema is presenting the organization of the arterial intima. The proteoglycan-rich layer containing a heterogeneous population of cells, including macrovascular pericytes, is located just below the endothelial monolayer. Intimal pericytes forming a network of cells linked via gap junctions. The muscular-elastic layer, formed by elongated contractile smooth muscular cells, is immediately below the proteoglycan-rich layer. Note: *World Journal of Cardiology*. 2015;7(10):583–593. Doi: https://doi.org/10.4330/wjc.v7.i10.583

Accordingly, the expression of 2A7 antigen is attributed to activated pericytes. That was proven by the presence of 2A7+ cells in atherosclerotic plaques using anti-2A7 antibody that adheres to activated pericytes, while no 2A7+ cells were identified in normal intima (Orekhov and Bobryshev 2015).

Moreover, acquiring the phagocytic phenotype, pericytes express CD68 macrophage-associated antigen, which is a scavenger receptor. These scavenger receptors of pericytes appear to promote uptake and accumulation of LDL, indicating their participation in active phagocytosis (Shashkin et al. 2005; Song et al. 2011; Cochain and Zernecke 2015). Such phagocytic ability of pericytes, i.e. engulfing of LDL particles, can lead to the development of foam cells and thickening of the arterial wall. In line with this suggestion, lipid-laden, stellate pericyte-like cells enriched by synthesizing organelles were found in atherosclerotic plaques (Orekhov et al. 1986; Andreeva et al. 1991, 1997b).

Further lipid accumulation via phagocytosis aggravates the situation in the atherosclerotic lesion leading to segregation of cells, breakdown of cell-to-cell contacts, and disruption of the cellular network. In addition, like fibroblasts, pericytes can rapidly proliferate accelerating thickening of the arterial wall and contributing to the extracellular matrix synthesis (Ivanova and Orekhov 2016).

The most pronounced changes associated with atherosclerosis occur in the proteoglycan-rich layer of the intima: lipid deposition and thickening determined by the accumulation of all types of collagen fibres with disrupted orientation, especially in the fibrous cap (Andreeva et al. 1997a, 1997b; Ivanova et al. 2015). Microscopic studies of atherosclerosis-affected and grossly normal aortic intima showed a number of distinct features: the presence of lipid droplets and vesicles between the interstitial collagen fibres, thickened proteoglycan-rich layer with increased collagen content, and altered longitudinal alignment of collagen fibres was accompanied by the formation of thick capsules around subendothelial cells (Orekhov et al. 2016b). Notably, the total lipid content in the proteoglycan-rich layer was increased by 3.2- and 8-fold in fatty streaks and atherosclerotic plaques, respectively, in comparison to healthy tissue (Andreeva et al. 1998). The highest content of (up to 25%) lipid-laden cells was found in fatty streaks located in the upper section of the proteoglycan-rich layer, encompassing almost two-thirds of it (Orekhov and Bobryshev 2015).

Thus, microscopic analysis of intimal thickness along the vessels affected with atherosclerosis, compared to normal tissues, revealed that proteoglycan-rich layer was 2- and 4-fold thicker in fatty streaks and atherosclerotic plaques, respectively (Orekhov et al. 2016b). In contrast, muscular-elastic layer remained intact in fatty streaks and was only slightly thickened in the atherosclerotic plaque areas (Orekhov et al. 2016b). Taken together, these observations clearly indicated that, governed by lipid and collagen accumulation, intima of atherosclerosis-affected arteries was abnormally and considerably thickened in the proteoglycan-rich layer. In addition, lipid and collagen accumulation determine the severity of lipidosis and fibrosis associated with atherosclerosis. Thickening of the proteoglycan-rich layer of the intima is the major cause of the arterial stenosis. Continuously growing plaque narrows the arteries leading to the severe ischaemia, which can have severe consequences. Even more dangerous is potential plaque rupture that in many cases can be lethal, leading to acute coronary syndrome, stroke and atherothrombosis (Chistiakov et al. 2015).

Besides, lipid accumulation triggers cellular expansion of the intimal cells. Comparative studies of grossly normal aortic intima and atherosclerosis-affected regions demonstrated that alterations of the proteoglycan-rich layer in the atherosclerotic plaque were associated with the increased cell count (Orekhov et al. 2016b).

According to the current knowledge, the increased cellularity in the proteoglycan-rich layer of atherosclerotic intima is caused by the recruitment of circulating immune cells and/or enhanced proliferation of the resident intimal cells (pericytes). The highest number of cells was detected in lipofibrous plaques. Intimal resident cells account for a majority of the intimal cell population (84-93%), therefore, the rise in their number determines the increase in the cellularity of atherosclerotic lesions. Nevertheless, the level of resident intimal cell proliferation is variable depending on plaque developmental stage and location (Orekhov et al. 2016b). Thus, the amount of resident proliferating cells in the lipid-laden atherosclerotic lesions, i.e. the early-stage lesions (fatty streak and fibrolipid plaques), was 10- to 20-fold greater than in uninvolved intima. Also, the number of resident proliferating cells in fibrous plaques, which are the later stage lesions, was detected to be lower than in lipid-enriched lesions, but considerably higher than in uninvolved intima (Orekhov et al. 1998). Moreover, the inflammatory cell proliferation is also present in the vascular wall; however, the proliferative index of inflammatory cells does not change in atherosclerotic lesions. Apparently, increased number of the inflammatory cells is a result of their migration into subendothelial intima from the circulation. In contrast to pericytes, proliferative activity of inflammatory cells is not stimulated in atherosclerosis (Orekhov et al. 2010). Hence, these findings indicated that the increase in cellularity of the arterial intima is a result of proliferative 'splash' of resident cells in lipid-rich lesions (fatty streaks and fibrolipid plaques) and the migration of the inflammatory cells. Additionally, robust data confirming the close correlation between the number of pericytes and the content of intimal lipids were obtained. The correlation coefficients between the number of pericyte-like cells and collagen content, and intimal thickness were more significant than between other cell types of the intima (Orekhov and Bobryshev 2015). Relying on the findings that proliferating pericytes to a large extent are responsible for the increased cellularity and also are associated with lipid and collagen accumulation in the atherosclerosis affected site, they were suggested to be the key cells driving the process of intimal medial thickness, and, consequently, the growth of atherosclerotic lesion.

From this point of view, the main manifestations of atherosclerosis are accompanied by the following changes on the cellular level: lipid accumulation leading to the development of foam cells, local increase in the cell count as a result of cell proliferation and migration, accumulation of connective tissue matrix, and disruption of intercellular communication. These functional alterations of intimal cells were shown to be related to the stages of atherosclerotic lesion development. Thus, as described by Orekhov et al. (2016b), lipidosis, proliferative activity and collagen synthesis were presented in a bell-shaped manner. Proliferative activity and collagen accumulation were increased from a low or even zero level in uninvolved intima to the maximum in lipid-rich lesions (fatty streak and atheroma), but declined in fibrous plaques. Cellular lipidosis was gradually increased from uninvolved intima to fibrolipid plaques and then decreased in fibrous plaques, becoming lower than in uninvolved intima. Notably, the intracellular communication rate was on a constant decrease along the lesion progression, down to the zero level in fibrous plaques, where the cellular network was disrupted, compared to uninvolved intima, where integrity of cell network was maintained.

Furthermore, it was demonstrated that lipid accumulation caused by circulating LDL can orchestrate pro-inflammatory environment in the arterial intima, stimulating both adaptive and innate immunity (Hartvigsen et al. 2009). Consequently, the atherosclerotic plaque growth is linked to the local inflammation. Microscopic studies made it evident that upon lipid accumulation, pericytes, along with macrophages and dendritic cells, express antigen-presenting complexes (APC) and that is indicative of their involvement in the antigen presentation and the inflammatory progression in the arterial wall (Ivanova et al. 2015) (Fig. 13.2).

Thus, a population of stellate subendothelial cells expressing HLA-DR component of the major histocompatibility complex class II (MHC II), secreting pro-inflammatory cytokines and chemokines positively correlated with a number of immune-inflammatory cells in the atherosclerotic lesion (Handunnetthi et al. 2010; Bobryshev et al. 2011; Psaltis and Simari 2015).

Apart from lipid accumulation, intimal thickening and inflammation, the processes that directly contribute to atherosclerotic lesion development, pericytes play

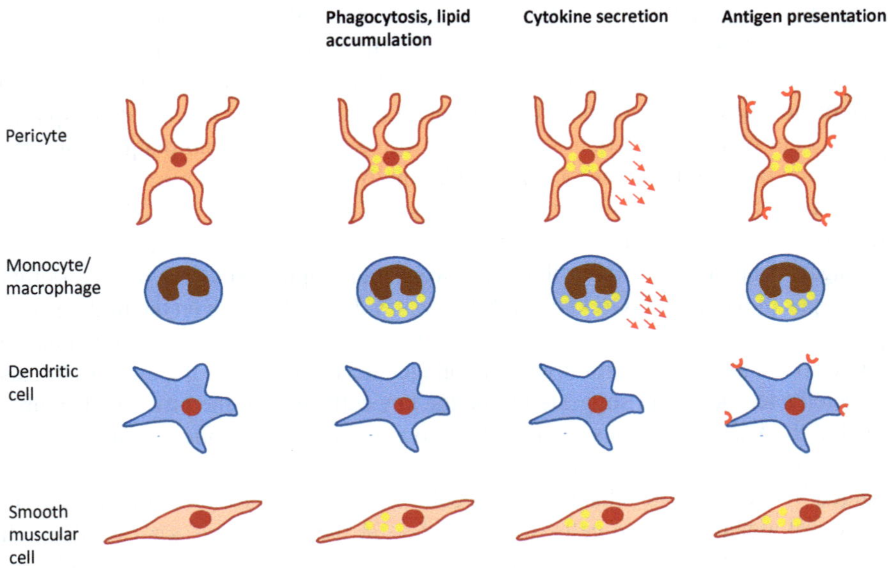

Fig. 13.2 Scheme is presenting the roles of arterial intimal cells in atherogenesis. Several types of arterial intimal cells participate in lipid accumulation and formation of foam cells. Macrophages and intimal pericytes accumulate lipids via their phagocytic activity and participate in the local inflammatory process secreting pro-inflammatory cytokines and chemokines. Dendritic cells, along with macrophages and intimal pericytes, express antigen-presenting complexes, further promoting the local inflammatory process. Note: World Journal of Cardiology. 2015;7(10):583–593. Doi: https://doi.org/10.4330/wjc.v7.i10.583

additional roles in this pathology, including neovascularization, plaque calcification and regulation of thrombogenesis. In atherosclerotic lesions, pericyte involvement in angiogenesis is mediated by upregulated expression of T-cadherin. This process is stimulated via activation of Erk1/2 tyrosine kinase pathway and NF-B nuclear translocation in pericytes, followed by the inhibition of adipogenic and activation of their chondrogenic differentiation pathways. Inhibition of adipogenic and stimulation of chondrogenic differentiation is regulated by Wnt/beta-catenin pathway (Kirton et al. 2007) and enhanced with transforming growth factor (TGF)-β3 expressed by macrophages, foam cells and vSMCs in the atherosclerotic plaque (Bobik 2006). Also, TGF-β3 signalling is associated with augmented glycosaminoglycan accumulation in the extracellular matrix during the development of fatty streaks and subsequent atheroma (Farrington-Rock et al. 2004).

Moreover, pericytes can be recruited to neovessels in atherosclerotic plaques via hepatocyte growth factor signalling activation triggering c-Met-PI3K/Akt pathway (Ivanova et al. 2015).

In addition, vascular calcification-associated factor expression was detected in the intentional pericytes promoting vascular calcification in the pro-inflammatory environment of the plaque (Ivanov et al. 2001; Takeuchi and Ohtsuki 2001).

At advanced stages of atherosclerotic plaque development, excessive expression of thrombogenic tissue factor by the subendothelial cells (pericytes) in the endothelium-uncovered arterial wall results in platelet aggregation and the formation of so-called fibrous cap covering the plaque (Ardissino et al. 2001; Orekhov et al. 2014). According to current understanding, fibrous cap has protective function, insulating the plaque from the blood flow. Its damage, however, can rapidly lead to thrombus formation, often with severe consequences.

Commentary on Likely Future Trends and Directions

Pericytes, as a part of cellular re-arrangements in the vascular wall, play a crucial role in atherosclerotic lesion formation. However, many questions about this enigmatic cell type remain unanswered. It would be important to further explore the relationship between MSC and pericytes in anticipation of unequivocal perceptions of pericyte origin. Furthermore, reliable pericyte markers or their combinations should be established. Such markers may serve as potential molecular targets in aid of diagnosis and therapeutic strategies in atherosclerosis and other vascular pathologies. Moreover, pericytes remain interesting potential points of therapeutic intervention, since they likely orchestrate the key atherosclerosis-related processes at the local tissue level, including local inflammation, plaque development, calcification and fibrosis, and further plaque growth accompanied by neovascularization.

Acknowledgment *Funding*: The research was funded by Russian Science Foundation (Grant no. 19-15-00010).

References

Al Ahmad, A., Taboada, C. B., Gassmann, M., & Ogunshola, O. O. (2011). Astrocytes and pericytes differentially modulate blood-brain barrier characteristics during development and hypoxic insult. *Journal of Cerebral Blood Flow and Metabolism, 31*(2), 693–705. https://doi.org/10.1038/jcbfm.2010.148.

Andreeva, E. R., Serebryakov, V. N., & Orekhov, A. N. (1995). Gap junctional communication in primary culture of cells derived from human aortic intima. *Tissue and Cell, 27*(5), 591–597.

Andreeva, E. R., Pugach, I. M., Gordon, D., & Orekhov, A. N. (1998). Continuous subendothelial network formed by pericyte-like cells in human vascular bed. *Tissue and Cell, 30*(1), 127–135.

Andreeva, E. R., Orekhov, A. N., & Smirnov, V. N. (1991). Quantitative estimation of lipid-laden cells in atherosclerotic lesions of the human aorta. *Acta Anatomica (Basel), 141*, 316–323.

Andreeva, E. R., Pugach, I. M., & Orekhov, A. N. (1997a). Collagen-synthesizing cells in initial and advanced atherosclerotic lesions of human aorta. *Atherosclerosis, 130*, 133–142.

Andreeva, E. R., Pugach, I. M., & Orekhov, A. N. (1997b). Subendothelial smooth muscle cells of human aorta express macrophage antigen in situ and in vitro. *Atherosclerosis, 135*, 19–27.

Armulik, A., Abramsson, A., & Betsholtz, C. (2005). Endothelial/pericyte interactions. *Circulation Research, 97*, 512–523.

Armulik, A., Genové, G., & Betsholtz, C. (2011). Pericytes: Developmental, physiological, and pathological perspectives, problems, and promises. *Developmental Cell, 21*(2), 193215.
Ardissino, D., Merlini, P. A., Bauer, K. A., Bramucci, E., Ferrario, M., Coppola, R., Fetiveau, R., Lucreziotti, S., Rosenberg, R. D., & Mannucci, P. M. (2001). Thrombogenic potential of human coronary atherosclerotic plaques. *Blood, 98*(9), 2726–2729. https://doi.org/10.1182/blood.V98.9.2726.
Balabanov, R., & Dore-Duffy, P. (1998). Role of the CNS microvascular pericyte in the blood-brain barrier. *Journal of Neuroscience Research, 53*, 637–644. https://doi.org/10.1002/(SICI)1097-4547(19980915)53:6<637::AID-JNR1>3.0.CO;2-6.
Bandopadhyay, R., Orte, C., Lawrenson, J. G., Reid, A. R., De Silva, S., & Allt, G. (2001). Contractile proteins in pericytes at the blood-brain and blood-retinal barriers. *Journal of Neurocytology, 30*(1), 35–44.
Bergers, G., & Song, S. (2005). The role of pericytes in blood-vessel formation and maintenance. *Neuro-Oncology, 7*(4), 452–464. https://doi.org/10.1215/S1152851705000232.
Betsholtz, C., Lindblom, P., & Gerhardt, H. (2005). Role of pericytes in vascular morphogenesis. *EXS, 94*, 5–125.
Birbrair, A., Zhang, T., Wang, Z.-M., et al. (2013). Role of pericytes in skeletal muscle regeneration and fat accumulation. *Stem Cells and Development, 22*(16), 2298–2314. https://doi.org/10.1089/scd.2012.0647.
Birbrair, A., Zhang, T., Wang, Z.-M., Messi, M. L., Mintz, A., & Delbono, O. (2015). Pericytes at the intersection between tissue regeneration and pathology. *Clinical Science (London, England), 128*(2), 81–93.
Bobik, A. (2006). Transforming growth factor-βs and vascular disorders. *Arteriosclerosis, Thrombosis, and Vascular Biology, 26*, 1712–1720. https://doi.org/10.1161/01.ATV.0000225287.20034.2c.
Bobryshev, Y. V., & Lord, R. S. (2000). CD1 expression and the nature of CD1-expressing cells in human atherosclerotic plaques. The American journal of pathology, 156(4), 1477–1478.
Bobryshev, Y. V., Andreeva, E. R., Mikhailova, I. A., Andrianova, I. V., Moisenovich, M. M., Khapchaev, S., et al. (2011). Correlation between lipid deposition, immune-inflammatory cell content and MHC class II expression in diffuse intimal thickening of the human aorta. *Atherosclerosis, 219*(1), 171–183.
Bose, A., Barik, S., Banerjee, S., et al. (2013). Tumor-derived vascular pericytes anergize Th cells. *Journal of Immunology, 191*(2), 971. https://doi.org/10.4049/jimmunol.1300280.
Bostrom, K., Watson, K. E., Horn, S., Wortham, C., Herman, I. M., & Demer, L. L. (1993). Bone morphogenetic protein expression in human atherosclerotic lesions. *The Journal of Clinical Investigation, 91*, 1800–1809.
Bouacida, A., Rosset, P., Trichet, V., et al. (2012). Pericyte-like progenitors show high immaturity and engraftment potential as compared with mesenchymal stem cells. *PLoS One, 7*(11), e4864.
Bouchard, B. A., Shatos, M. A., & Tracy, P. B. (1997). Human brain pericytes differentially regulate expression of procoagulant enzyme complexes comprising the extrinsic pathway of blood coagulation. *Arteriosclerosis, Thrombosis, and Vascular Biology, 17*, 1–9. https://doi.org/10.1161/01.ATV.17.1.1.
Campagnolo, P., Cesselli, D., Al Haj Zen, A., Beltrami, A. P., Kränkel, N., Katare, R., Angelini, G., Emanueli, C., & Madeddu, P. (2010). Human adult vena saphena contains perivascular progenitor cells endowed with clonogenic and proangiogenic potential. *Circulation, 121*, 1735–1745.
Chen, J., Luo, Y., Hui, H., et al. (2017). CD146 coordinates brain endothelial cell-pericyte communication for blood–brain barrier development. *Proceedings of the National Academy of Sciences of the United States of America, 114*(36), E7622 LPE7631.
Chen, R. R., Silva, E. A., Yuen, W. W., et al. (2007). Spatio-temporal VEGF and PDGF delivery patterns blood vessel formation and maturation. *Pharmaceutical Research, 24*, 258. https://doi.org/10.1007/s11095-006-9173-4.
Chen, M. J., Yokomizo, T., Zeigler, B. M., Dzierzak, E., & Speck, N. A. (2009). Runx1 is required for the endothelial to haematopoietic cell transition but not thereafter. *Nature, 457*, 887–891.

Chistiakov, D. A., Orekhov, A. N., & Bobryshev, Y. V. (2015). Contribution of neovascularization and intraplaque haemorrhage to atherosclerotic plaque progression and instability. *Acta Physiologica, 213*, 539–553. https://doi.org/10.1111/apha.12438.
Cochain, C., & Zernecke, A. (2015). Macrophages and immune cells in atherosclerosis: Recent advances and novel concepts. *Basic Research in Cardiology, 110*(4), 34.
Collett, G. D. M., & Canfield, A. E. (2005). Angiogenesis and pericytes in the initiation of ectopic calcification. *Circulation Research, 96*, 930–938. https://doi.org/10.1161/01.RES.0000163634.51301.0d.
Corselli, M., Chen, C.-W., Sun, B., Yap, S., Rubin, J. P., & Péault, B. (2012). The tunica adventitia of human arteries and veins as a source of mesenchymal stem cells. *Stem Cells and Development, 21*(8), 1299–1308. https://doi.org/10.1089/scd.2011.0200.
Crisan, M., Zheng, B., Zambidis, E. T., Yap, S., Tavian, M., Sun, B., Giacobino, J. P., Casteilla, L., Huard, J., & Péault, B. (2007). Blood vessels as a source of progenitor cells in human embryonic and adult life. In N. M. Bilko, B. Fehse, W. Ostertag, C. Stocking, & A. R. Zander (Eds.), *Stem cells and their potential for clinical application, NATO Security through Science Series (A: Chemistry and Biology)* (pp. 137–147). Netherlands: Springer.
Crisan, M., Yap, S., Casteilla, L., et al. (2008). A perivascular origin for mesenchymal stem cells in multiple human organs. *Cell Stem Cell, 3*(3), 301–313. https://doi.org/10.1016/j.stem.2008.07.003.
Cuevas, P., Gutierrez-Diaz, J. A., Reimers, D., Dujovny, M., Diaz, F. G., & Ausman, J. I. (1984). Pericyte endothelial gap junctions in human cerebral capillaries. *Anatomy and Embryology, 170*, 155–159. https://doi.org/10.1007/BF00319000.
Dohgu, S., Takata, F., Yamauchi, A., Nakagawa, S., Egawa, T., Naito, M., Tsuruo, T., Sawada, Y., Niwa, M., & Kataoka, Y. (2005). Brain pericytes contribute to the induction and up-regulation of blood-brain barrier functions through transforming growth factor-beta production. *Brain Research, 1038*, 208–215. https://doi.org/10.1016/j.brainres.2005.01.027.
Dominici, M., Le Blanc, K., Mueller, I., Slaper-Cortenbach, I., Marini, F., Krause, D., Deans, R., Keating, A., Prockop, D. J., & Horwitz, E. (2006). Minimal criteria for defining multipotent mesenchymal stromal cells. The International Society for Cellular Therapy position statement. *Cytotherapy, 8*(4), 315–317.
Dondossola, E., Rangel, R., Guzman-Rojas, L., et al. (2013). CD13-positive bone marrow-derived myeloid cells promote angiogenesis, tumor growth, and metastasis. *Proceedings of the National Academy of Sciences of the United States of America, 110*(51), 20717–20722.
Dore-Duffy, P., & Cleary, K. (2011). Morphology and properties of pericytes. In S. Nag (Ed.), *The Blood-Brain and Other Neural Barriers. Methods in Molecular Biology (Methods and Protocols)* (p. 686). New York: Humana Press.
Dulmovits, B. M., & Herman, I. M. (2012). Microvascular remodeling and wound healing: a role for pericytes. *The International Journal of Biochemistry & Cell Biology, 44*(11), 1800–1812. https://doi.org/10.1016/j.biocel.2012.06.031.
Farrington-Rock, C., Crofts, N. J., Doherty, M. J., Ashton, B. A., Griffin-Jones, C., & Canfield, A. E. (2004). Chondrogenic and adipogenic potential of microvascular pericytes. *Circulation, 110*(15), 2226–2232.
Favero, G., Paganelli, G., Buffoli, B., Rodella, L. F., & Rezzani, R. (2014). Endothelium and its alterations in cardiovascular diseases: Life style intervention. *BioMed Research International, 2014*, 801896. https://doi.org/10.1155/2014/801896.
Gerhardt, H., & Betsholtz, C. (2003). Endothelial-pericyte interactions in angiogenesis. *Cell and Tissue Research, 314*, 15. https://doi.org/10.1007/s00441-003-0745-x.
Glentis, V., Gurchenkov, D., & Vignjevic, M. (2014). Assembly, heterogeneity, and breaching of the basement membranes. *Cell Adhesion & Migration, 8*, 236–245. https://doi.org/10.4161/cam.28733.
Groschner, L. N., Waldeck-Weiermair, M., Malli, R., et al. (2012). Endothelial mitochondria—less respiration, more integration. *Pflugers Archiv—European Journal of Physiology, 464*, 63. https://doi.org/10.1007/s00424-012-1085-z.

Guillemin, G. J., & Brew, B. J. (2004). Microglia, macrophages, perivascular macrophages, and pericytes: a review of function and identification. *Journal of Leukocyte Biology, 75*(3), 388–397. https://doi.org/10.1189/jlb.0303114.

Hall, C. N., Reynell, C., Gesslein, B., Hamilton, N. B., Mishra, A., Sutherland, B. A., & Attwell, D. (2014). Capillary pericytes regulate cerebral blood flow in health and disease. *Nature, 508*(7494), 55.

Handunnetthi, L., Ramagopalan, S. V., Ebers, G. C., & Knight, J. C. (2010). Regulation of major histocompatibility complex class II gene expression, genetic variation and disease. *Genes and Immunity, 11*, 99–112.

Hartvigsen, K., Chou, M. Y., Hansen, L. F., Shaw, P. X., Tsimikas, S., Binder, C. J., & Witztum, J. L. (2009). The role of innate immunity in atherogenesis. *Journal of Lipid Research, 50*(Suppl), S388–S393.

Haurani, M. J., & Pagano, P. J. (2007). Adventitial fibroblast reactive oxygen species as autacrine and paracrine mediators of remodeling: Bellwether for vascular disease? *Cardiovascular Research, 75*(4), 679–668.

Hellstrom, M., Gerhardt, H., Kalen, M., Li, X., Eriksson, U., Wolburg, H., & Betsholtz, C. (2001). Lack of pericytes leads to endothelial hyperplasia and abnormal vascular morphogenesis. *The Journal of Cell Biology, 153*, 543–553.

Herman, I. M. (2010). Pericytes and microvascular remodeling: Regulation of retinal angiogenesis. In D. A. Dartt (Ed.), *Encyclopedia of the Eye* (pp. 296–299). Oxford: Academic Press. https://doi.org/10.1016/B978-0-12-374203-2.00149-4.

Hughes, S., Gardiner, T., Hu, P., Baxter, L., Rosinova, E., & Chan-Ling, T. (2006). Altered pericyte-endothelial relations in the rat retina during aging: implications for vessel stability. *Neurobiology of Aging, 27*, 1838–1847.

Ivanov, D., Philippova, M., Antropova, J., et al. (2001). Expression of cell adhesion molecule T-cadherin in the human vasculature. *Histochemistry and Cell Biology, 3*(115), 231–242.

Ivanova, E. A., Bobryshev, Y. V., & Orekhov, A. N. (2015). Intimal pericytes as the second line of immune defence in atherosclerosis. *World Journal of Cardiology, 7*(10), 583–593. https://doi.org/10.4330/wjc.v7.i10.583.

Ivanova, E. A., & Orekhov, A. N. (2016). Cellular model of atherogenesis based on pluripotent vascular wall pericytes. *Stem Cells International, 2016*, 7321404. https://doi.org/10.1155/2016/7321404.

Juchem, G., Weiss, D. R., Gansera, B., Kemkes, B. M., Mueller-Hoecker, J., & Nees, S. (2010). Pericytes in the macrovascular intima: possible physiological and pathogenetic impact. *The American Journal of Physiology-Heart and Circulatory Physiology, 3*(298), 754–770.

Kim, J. A., Tran, N. D., Li, Z., Yang, F., Zhou, W., & Fisher, M. J. (2006). Brain endothelial hemostasis regulation by pericytes. *Journal of Cerebral Blood Flow and Metabolism, 26*, 209–217. https://doi.org/10.1038/sj.jcbfm.9600181.

Kirton, J. P., Wilkinson, F. L., Canfield, A. E., & Alexander, M. Y. (2006). Dexamethasone downregulates calcification-inhibitor molecules and accelerates osteogenic differentiation of vascular pericytes: Implications for vascular calcification. *Circulation Research, 98*(10), 1264–1272.

Kirton, P., Crofts, N. J., George, S. J., Brennan, K., & Canfield, A. E. (2007). Wnt/T-catenin signaling stimulates chondrogenic and inhibits adipogenic differentiation of pericytes: potential relevance to vascular disease? *Circulation Research, 6*(101), 581–589.

Koh, W., Stratman, A. N., Sacharidou, A., & Davis, G. E. (2008). *In vitro* three dimensional collagen matrix models of endothelial lumen formation during vasculogenesis and angiogenesis. *Methods in Enzymology, 443*, 83–101. https://doi.org/10.1016/S0076-6879(08)02005-3.

Korn, J., Christ, B., & Kurz, H. (2002). Neuroectodermal origin of brain pericytes and vascular smooth muscle cells. *The Journal of Comparative Neurology, 442*, 78–88. https://doi.org/10.1002/cne.1423.

Kostallari, E., Baba-Amer, E., Alonso-Martin, S., Ngoh, P., Relaix, F., Lafuste, P., & Gherardi, R. K. (2015). Pericytes in the myovascular niche promote post-natal myofiber growth and satellite cell quiescence. *Development, 142*, 1242–1253. https://doi.org/10.1242/dev.115386.

Kovacic, J. C., & Boehm, M. (2009). Resident vascular progenitor cells: An emerging role for non terminally differentiated vessel-resident cells in vascular biology. *Stem Cell Research, 2*(1), 2–15.

Krueger, M., & Bechmann, I. (2010). CNS pericytes: Concepts, misconceptions, and a way out. *Glia, 58*, 1–10. https://doi.org/10.1002/glia.20898.

Leveen, P., Pekny, M., Gebre-Medhin, S., Swolin, B., Larsson, E., & Betsholtz, C. (1994). Mice deficient for PDGF B show renal, cardiovascular, and hematological abnormalities. *Genes & Development, 8*, 1875–1887.

Li, X., Xiao, Y., Cui, Y., Tan, T., Narasimhulu, C. A., Hao, H., Liu, L., Zhang, J., He, G., Verfaillie, C. M., Lei, M., Parthasarathy, S., Ma, J., Zhu, H., & Liu, Z. (2014). Cell membrane damage is involved in the impaired survival of bone marrow stem cells by oxidized low-density lipoprotein. *Journal of Cellular and Molecular Medicine, 18*, 2445–2453. https://doi.org/10.1111/jcmm.12424.

Lindahl, P., Johansson, B. R., Leveen, P., & Betsholtz, C. (1997). Pericyte loss and microaneurysm formation in PDGF-B-deficient mice. *Science, 277*, 242–245.

Mazanet, R., & Franzini-Armstrong, C. (1982). Scanning electron microscopy of pericytes in rat red muscle. *Microvascular Research, 23*(3), 361–369.

Mollace, V., Gliozzi, M., Musolino, V., Carresi, C., Muscoli, S., Mollace, R., Tavernese, A., Gratteri, S., Palma, E., Morabito, C., Vitale, C., Muscoli, C., Fini, M., & Romeo, F. (2015). Oxidized LDL attenuates protective autophagy and induces apoptotic cell death of endothelial cells: Role of oxidative stress and LOX-1 receptor expression. *International Journal of Cardiology, 184*, 152–158.

Montiel-Eulefi, E., Nery, A. A., Rodrigues, L. C., Sánchez, R., Romero, F., & Ulrich, H. (2012). Neural differentiation of rat aorta pericyte cells. *Cytometry, 81A*, 65–71. https://doi.org/10.1002/cyto.a.21152.

Moreno, P. R., Purushothaman, M., & Purushothaman, K. R. (2012). Plaque neovascularization: Defense mechanisms, betrayal, or a war in progress. *Annals of the New York Academy of Sciences, 1254*, 7–17.

Morikawa, S., Baluk, P., Kaidoh, T., Haskell, A., Jain, R. K., & McDonald, D. M. (2002). Abnormalities in pericytes on blood vessels and endothelial sprouts in tumors. *The American Journal of Pathology, 160*(3), 985–1000.

Moore, K. J., & Tabas, I. (2011). Macrophages in the pathogenesis of atherosclerosis. *Cell, 145*(3), 341–355.

Mozaffarian, D., Benjamin, E. J., Go, A. S., Arnett, D. K., Blaha, M. J., Cushman, M., de Ferranti, S., Després, J. P., Fullerton, H. J., Howard, V. J., Huffman, M. D., Judd, S. E., Kissela, B. M., Lackland, D. T., Lichtman, J. H., Lisabeth, L. D., Liu, S., Mackey, R. H., Matchar, D. B., DK, M. G., Mohler, E. R., Moy, C. S., Muntner, P., Mussolino, M. E., Nasir, K., Neumar, R. W., Nichol, G., Palaniappan, L., Pandey, D. K., Reeves, M. J., Rodriguez, C. J., Sorlie, P. D., Stein, J., Towfighi, A., Turan, T. N., Virani, S. S., Willey, J. Z., Woo, D., Yeh, R. W., Turner, M. B., & American Heart Association Statistics Committee and Stroke Statistics Subcommittee. (2015). Heart disease and stroke statistics-2015 update: A report from the American Heart Association. *Circulation, 131*, e29–e322.

Murray, I. R., West, C. C., Hardy, W. R., et al. (2014). Natural history of mesenchymal stem cells, from vessel walls to culture vessels. *Cellular and Molecular Life Sciences, 71*, 1353. https://doi.org/10.1007/s00018-013-1462-6.

Nakata, M., Nakagomi, T., Maeda, M., et al. (2017). Induction of perivascular neural stem cells and possible contribution to neurogenesis following transient brain ischemia/reperfusion injury. *Translational Stroke Research, 8*, 131. https://doi.org/10.1007/s12975-016-0479-1.

Nayak, R. C., Berman, A. B., George, K. L., Eusenbarth, G. S., & King, G. L. (1988). A monoclonal antibody (3G5)-defined ganglioside antigen is expressed on the cell surface of microvascular pericytes. *The Journal of Experimental Medicine, 4*, 1003–1015.

Nazarova, V. L., Andreeva, E. R., Tertov, V. V., Gel'dieva, B. S., & Orekhov, A. N. (1995). Immunocytochemical study to localize a scavenger receptor in human aorta smooth muscle cells. *Biulleten' Eksperimental'noĭ Biologii i Meditsiny, 120*, 195–198.

Nicosia, R. F. (2009). The aortic ring model of angiogenesis: a quarter century of search and discovery. *Journal of Cellular and Molecular Medicine, 13*(10), 4113–4136.

Oberlin, E., Tavian, M., Blazsek, I., & Peault, B. (2002). Blood-forming potential of vascular endothelium in the human embryo. *Development, 129*, 4147–4157.

Orekhov, A. N., Andreeva, E. R., Krushinsky, A. V., Novikov, I. D., Tertov, V. V., Nestaiko, G. V., Khashimov, K. A., Repin, V. S., & Smirnov, V. N. (1986). Intimal cells and atherosclerosis. Relationship between the number of intimal cells and major manifestations of atherosclerosis in the human aorta. *The American Journal of Pathology, 125*, 402–415.

Orekhov, A. N., Andreeva, E. R., & Bobryshev, Y. V. (2016a). Cellular mechanisms of human atherosclerosis: Role of cell-to-cell communications in subendothelial cell functions. *Tissue & Cell, 48*(1), 25–34. https://doi.org/10.1016/j.tice.2015.11.002.

Orekhov, A. N., Andreeva, E. R., Ivanova, E. A., & Bobryshev, Y. V. (2016b). Role of pericytes and modified lipoprotein in human atherogenesis. In *Coronary Artery Disease—Research and Practice*. Kowloon: iConcept Press Ltd.

Orekhov, A. N., Bobryshev, Y. V., & Chistiakov, D. A. (2014). The complexity of cell composition of the intima of large arteries: Focus on pericyte-like cells. *Cardiovascular Research, 103*, 438–451.

Orekhov, A. N., Andreeva, E. R., Andrianova, I. V., & Bobryshev, Y. V. (2010). Peculiarities of cell composition and cell proliferation in different type atherosclerotic lesions in carotid and coronary arteries. *Atherosclerosis, 212*(2), 436–443.

Orekhov, A. N., Andreeva, E. R., Mikhailova, I. A., & Gordon, D. (1998). Cell proliferation in normal and atherosclerotic human aorta: proliferative splash in lipid-rich lesions. *Atherosclerosis, 139*(1), 41–48.

Orekhov AN and Bobryshev YV. (2015). Cell composition of the subendothelial aortic intima and the role of alpha-smooth muscle actin expressing pericyte-like cells and smooth muscle cells in the development of atherosclerosis. Tissue Engineering and Regenerative Medicine. Muscle Cell and Tissue. Chapter 8. Doi: https://doi.org/10.5772/60430.

Pallone, T. L., Silldorff, E. P., & Turner, M. R. (1998). Intrarenal blood flow: microvascular anatomy and the regulation of medullary perfusion. *Clinical and Experimental Pharmacology & Physiology, 25*, 383–392.

Pallone, T. L., Zhang, Z., & Rhinehart, K. (2003). Physiology of the renal medullary microcirculation. *American Journal of Physiology. Renal Physiology, 284*, 253–266.

Pantakani, D. V. K., & Asif, A. (2015). Atherosclerosis: Epigenetic targeting of macrophages in disease management. *Journal of Clinical & Cellular Immunology, 6*, e118. https://doi.org/10.4172/2155-9.

Peppiatt, C. M., Howarth, C., Mobbs, P., & Attwell, D. (2006). Bidirectional control of CNS capillary diameter by pericytes. *Nature, 443*(7112), 700–704. https://doi.org/10.1038/nature05193.

Psaltis, J., & Simari, A. D. (2015). Vascular wall progenitor cells in health and disease. *Circulation Research, 8*(1166), 1392–1412.

Rajamannan, N. M., Subramaniam, M., Rickard, D., Stock, S. R., Donovan, J., Springett, M., Orszulak, T., Fullerton, D. A., Tajik, A. J., Bonow, R. O., & Spelsberg, T. (2003). Human aortic valve calcification is associated with an osteoblast phenotype. *Circulation, 107*, 2181–2184.

Raza, A., Franklin, M. J., & Dudek, A. Z. (2010). Pericytes and vessel maturation during tumor angiogenesis and metastasis. *American Journal of Hematology, 85*, 593–598. https://doi.org/10.1002/ajh.21745.

Rekhter, M. D., Tertov, V. V., Andreeva, E. R., Kolpakov, V. A., Mironov, A. A., & Orekhov, A. N. (1993). Lipid accumulation in the subendothelial cells of human aortic intima impairs cell-to-cell contacts: A comparative study in situ and in vitro. *Cardiovascular Pathology, 2*(1), 53–62.

Rossi, E., Smadja, D. M., Boscolo, E., et al. (2016). Endoglin regulates mural cell adhesion in the circulatory system. *Cellular and Molecular Life Sciences, 73*, 1715–1739. https://doi.org/10.1007/s00018-015-2099-4.

Rucker, H. K., Wynder, H. J., & Thomas, W. E. (2000). Cellular mechanisms of CNS pericytes. *Brain Research Bulletin, 51*, 363–369.

Salvayre, R., Auge, N., Benoist, H., & Negre-Salvayre, A. (2002). Oxidized low-density lipoprotein-induced apoptosis. *Biochimica et Biophysica Acta, 1585*(2–3), 213–221.

Shashkin, P., Dragulev, B., & Ley, K. (2005). Macrophage differentiation to foam cells. *Current Pharmaceutical Design, 11*(23), 3061–3072.

Sims, D. E. (2000). Diversity Within Pericytes. *Clinical and Experimental Pharmacology and Physiology, 27*, 842–846. https://doi.org/10.1046/j.1440-1681.2000.03343.x.

Song, L., Lee, C., & Schindler, C. (2011). Deletion of the murine scavenger receptor CD68. *Journal of Lipid Research, 52*, 1542–1550.

Steitz, S. A., Speer, M. Y., Curinga, G., et al. (2001). Smooth muscle cell phenotypic transition associated with calcification. *Circulation Research, 89*(12), 1147–1154.

Stefanska, A. (2013). Renal pericytes: Multifunctional cells of the kidneys. *Pflugers Archiv—European Journal of Physiology, 465*, 767–773. https://doi.org/10.1007/s00424-013-1263-7.

Tabas, I., García-Cardeña, G., & Owens, G. K. (2015). Recent insights into the cellular biology of atherosclerosis. *The Journal of Cell Biology, 209*(1), 13–22. https://doi.org/10.1083/jcb.201412052.

Takeuchi, T., & Ohtsuki, Y. (2001). Recent progress in T-cadherin (CDH13, H-cadherin) research. *Histology and Histopathology, 4*(16), 1287–1293.

Tintut, Y., Alfonso, Z., Saini, Z., Radcliff, K., Watson, K., & Boström Kand Demer, L. L. (2003). Multilineage potential of cells from the artery wall. *Circulation, 108*, 2505–2510. https://doi.org/10.1161/01.CIR.0000096485.64373.C5.

Thomas, W. E. (1999). Brain macrophages: On the role of pericytes and perivascular cells. *Brain Research Reviews, 31*(1), 42–57.

van Dijk, C. G., Nieuweboer, F. E., Pei, J. Y., Xu, Y. J., Burgisser, P., van Mulligen, E., el Azzouzi, H., Duncker, D. J., Verhaar, M. C., & Cheng, C. (2015). The complex mural cell: Pericyte function in health and disease. *International Journal of Cardiology, 190*, 75–89.

Wang, P. C., Vancura, A., Mitcheson, T. G., & Kuret, J. (1992). Two genes in *Saccharomyces cerevisiae* encode a membrane-bound form of casein kinase-1. *Molecular Biology of the Cell, 3*(3), 275–286.

Warmke, N., Griffin, K. J., & Cubbon, R. M. (2016). Pericytes in diabetes-associated vascular disease. *Journal of Diabetes and its Complications, 8*(30), 1643–1650. https://doi.org/10.1016/j.jdiacomp.2016.08.005.

Wu, D. M., Kawamura, H., Sakagami, K., Kobayashi, M., & Puro, D. G. (2003). Cholinergic regulation of pericyte-containing retinal microvessels. *American Journal of Physiology-Heart and Circulatory Physiology, 284*(5), H2083–H2090.

Yemisci, M., Gursoy-Ozdemir, Y., Vural, A., Can, A., Topalkara, K., & Dalkara, T. (2009). Pericyte contraction induced by oxidative-nitrative stress impairs capillary reflow despite successful opening of an occluded cerebral artery. *Nature Medicine, 15*(9), 1031.

Zambidis, E. T., Oberlin, E., Tavian, M., & Peault, B. (2006). Blood-forming endothelium in human ontogeny: Lessons from in utero development and embryonic stem cell culture. *Trends in Cardiovascular Medicine, 16*, 95–101.

Zheng, B., Cao, B., Crisan, M., Sun, B., Li, G. H., Logar, A., et al. (2007). Prospective identification of myogenic endothelial cells in human skeletal muscle. *Nature Biotechnology, 25*(9), 1025–1034.

Chapter 14
Pericytes in Chronic Lung Disease

Bushra Shammout and Jill R. Johnson

Abstract Pericytes are supportive mesenchymal cells located on the abluminal surface of the microvasculature, with key roles in regulating microvascular homeostasis, leukocyte extravasation, and angiogenesis. A subpopulation of pericytes with progenitor cell function has recently been identified, with evidence demonstrating the capacity of tissue-resident pericytes to differentiate into the classic MSC triad, i.e., osteocytes, chondrocytes, and adipocytes. Beyond the regenerative capacity of these cells, studies have shown that pericytes play crucial roles in various pathologies in the lung, both acute (acute respiratory distress syndrome and sepsis-related pulmonary edema) and chronic (pulmonary hypertension, lung tumors, idiopathic pulmonary fibrosis, asthma, and chronic obstructive pulmonary disease). Taken together, this body of evidence suggests that, in the presence of acute and chronic pulmonary inflammation, pericytes are not associated with tissue regeneration and repair, but rather transform into scar-forming myofibroblasts, with devastating outcomes regarding lung structure and function. It is hoped that further studies into the mechanisms of pericyte-to-myofibroblast transition and migration to fibrotic foci will clarify the roles of pericytes in chronic lung disease and open up new avenues in the search for novel treatments for human pulmonary pathologies.

Keywords Pericyte · Lung · Vasculature · Edema · Migration · Fibrosis Myofibroblast · Acute respiratory distress syndrome · Tumor · Pulmonary hypertension · Idiopathic pulmonary hypertension · Asthma · Chronic obstructive pulmonary disease

B. Shammout · J. R. Johnson (✉)
Biosciences Department, School of Life and Health Sciences, Aston University, Birmingham, UK
e-mail: j.johnson1@aston.ac.uk

A. Birbrair (ed.), *Pericyte Biology in Disease*, Advances in Experimental Medicine and Biology 1147, https://doi.org/10.1007/978-3-030-16908-4_14

Lung Pericytes: Identification, Function, and Ultrastructure

The lung is a highly vascularized organ, with a large surface area of approximately 90 m^2 dedicated to effective gas exchange (Donoghue et al. 2006). As in the rest of the body, the entire pulmonary vascular system is lined by a continuous layer of endothelial cells surrounded by a thin basement membrane separating the endothelial cell vascular lining from the underlying tissue (Armulik et al. 2011). This basement membrane is composed of proteins including laminins, collagens, and proteoglycans (Armulik et al. 2011); pericytes are embedded within this matrix and associated directly with endothelial cells (Damianovich et al. 2013). The extensive vascular bed of the lung is closely associated with the alveoli to allow for gas exchange. Based on stereological studies, 87.5% of the lung volume is taken up by the parenchyma (alveoli and associated capillaries), while only 5.4% is taken up by larger vessels and 7.1% is the bronchial volume (Wiebe and Laursen 1995). Moreover, the lung is unique in its dual blood supply, as the systemic bronchial arteries supply oxygen to the trachea and bronchi, including the bronchial lamina propria and epithelium, while the pulmonary circulation arising from the right ventricle supplies the bulk of the lung parenchyma (McCullagh et al. 2010). Between the epithelial (alveoli) and endothelial (vasculature) layers of the lung parenchyma lies an elastic matrix containing the lymphatic vasculature, resident and migratory leukocytes, as well as a poorly characterized population of mesenchymal stromal cells, putatively identified as a mixture of fibroblasts and pericytes. These cells are important for maintaining healthy lung structure and function, but have also been implicated in pathological processes such as pulmonary edema, fibrosis, tumor formation, hypertension, and microangiopathy (Bagnato and Harari 2015; Barron et al. 2016).

In his initial description of pericytes in the lung parenchyma, Weibel (1974) used electron microscopy to demonstrate the presence of pericytes in the lungs of humans, dogs, guinea pigs, and rats. Lung pericytes were identified by their close association with the capillary basement membrane and branched cytoplasmic processes forming complex interactions with endothelial cells, similar to those observed in venular smooth muscle cells, suggesting a phenotypic continuum of vascular mesenchymal cells. Weibel also observed the presence of pericytes associated with alveolar capillaries, but at a lower density compared to systemic capillaries and with fewer connections with endothelial cells (Weibel 1974).

Various markers for pericytes have been identified and found to be useful in labeling pericytes in a number of organs, including the lung (Kramann et al. 2015; Paiva et al. 2018): CD146, CD13, platelet-derived growth factor receptor β (PDGFβR), neural glial antigen 2 (NG2), RGS5 and 3G5, and contractile myofilaments (α-smooth muscle actin and SM22) as well as intermediate filaments (desmin and vimentin) (Johnson et al. 2015; Crisan et al. 2008). Similar to pericytes throughout the body, pulmonary pericytes express contractile myofilaments (α-smooth muscle actin and SM22) as well as intermediate filaments (desmin and vimentin), but the expression of these filaments varies according to the type of blood vessel. In the mouse trachea, pericytes associated with arterioles express high levels of both α-SMA and NG2, whereas venule-associated pericytes do not express NG2 and capillary-associated pericytes express NG2 but not α-SMA; all pulmonary pericytes express the intermediate filaments desmin and vimentin (although at variable levels) as well as the key signaling molecule PDGFRβ (Johnson et al. 2015; Crisan et al. 2012).

With the advent of genetic labeling techniques, the identification of lung pericytes and the roles they play in mediating lung disease has been greatly facilitated. These genetic models include a Cspg4-dependent Cre lineage-tagging system, but pericyte labeling was highly inefficient in this study (Rock et al. 2011). Another study used an Abcg2-dependent Cre labeling system to identify a subpopulation of ABCG2+ mesenchymal stem cells in the lung, distinct from NG2 pericytes (Marriott et al. 2014). However, similar to a study in which a Foxd1-dependent Cre mouse (Hung et al. 2013) was used to identify pericytes, the presence of these markers in other mesenchymal cell types complicates the differentiation of a distinct pericyte population; the latter study identified collagen-I(α)1-expressing cells as a population of fibroblasts distinct from pericytes (Hung et al. 2013). Genetically labeling Gli1+ cells may also be an effective method of identifying lung-resident pericytes, as these cells have been found to express MSC markers, undergo trilineage differentiation, and exhibit colony-forming capacity, representing a small subset of the lung PDGFRβ+ cell population, which presumably also includes fibroblasts (Kramann et al. 2015). More recently, our group used a DsRed-labeling strategy to identify the location and pathological roles of NG2+ pericytes in the lung (Johnson et al. 2015).

Similar to pericytes associated with other vascular beds, pulmonary pericytes have been identified as a subset of mesenchymal progenitor cells. In vitro, pericytes isolated from the lungs display standardized mesenchymal stem cell properties established by the International Society of Stem Cell Research, including adherence to plastic, differentiation capacity (into chondrocytes, adipocytes, and osteocytes), positive expression of CD105, CD72, and CD90, and negative expression of CD45, CD34, CD11b, CD14, CD79a, and HLA-DR. Notably, murine lung pericytes also demonstrate positive expression of CD106 and Sca-1 and the absence of CD45, CD11b, and CD31 expression (Wong et al. 2015).

Based on numerous discoveries made by employing lineage tracing, techniques conditional ablation, and targeted gene deletion of mesenchymal cell subpopulations in rodents, it is clear that pulmonary pericytes are crucial for forming and maintaining the lung vascular network, sensing damage, recruiting inflammatory cells, and remodeling the extracellular matrix in cases of persistent inflammation and aberrant wound repair. Functional alterations in the pulmonary vascular bed have been observed in a plethora of lung diseases. Dysfunctional pericyte–endothelial cell interactions are a possible focal point at which microvascular dysfunction and vasculopathy accompanying disease progression may originate. As discussed below, perturbations in endothelial–pericyte signaling may indeed represent key mechanisms by which the microvasculature becomes dysregulated, unstable, and ultimately pathogenic in chronic lung disease.

Pericyte–Endothelial Crosstalk in the Lung

Throughout the body, and certainly in the lung, pericytes and endothelial cells form specialized junctions with each other. Peg-socket contacts are formed by pericyte cytoplasmic fingers that are inserted into invaginations within the endothelium, adherens junctions create contact inhibition between endothelial cells and pericytes via contractile forces, and gap junctions between the cytoplasm of pericytes and endothelial cells enable the passage of metabolites and ionic currents (Armulik et al. 2011). Under healthy conditions, pericytes regulate key endothelial cell functions, including proliferation, differentiation, and permeability, and in some regions of the vasculature also regulate contractility and tone (Geevarghese and Herman 2014). Extensive studies performed on mouse retinas (reviewed by Armulik et al. 2011) have demonstrated that Notch signaling is critically required for angiogenic sprouting and endothelial–mural interactions. The most important Notch interaction during angiogenesis has been found to be mediated by Jagged-1 (Jag-1), which is expressed by endothelial cells upon induction by pericytes as an autoregulatory loop (Liu et al. 2009). Additional players include interactions between the angiopoietins Ang-1 and Ang-2 with their receptor Tie-2 (Liu et al. 2010; Hammes et al. 2004) as well as the Ephrin pathway (Armulik et al. 2011). However, the precise roles of and interactions between these mediators under quiescent conditions, during angiogenesis, and in response to stress, inflammation, and injury have yet to be determined.

Lung pericytes play key roles in lung immune surveillance. It is well-understood that the vascular endothelium mediates the extravasation of inflammatory cells, but recent studies have demonstrated that pericytes also play a key role in this process. Subsequent to the well-described processes of leukocyte tethering and rolling on activated ECs, followed by adhesion and diapedesis, pericytes form preferential sites for leukocyte extravasation, i.e., regions of low of basement membrane protein expression and high expression of ICAM-1 (Wang et al. 2006). Importantly, the permeability of this additional barrier to leukocyte infiltration can be modified by inflammatory cytokines such as TNF-α (Lauridsen et al. 2014). More specific to the lung, pericytes have been found to be critical to maintaining vascular stability during inflammation. Using a mouse model of *Mycoplasma pulmonis* infection, it was found that PDGFRβ signaling is critical to maintaining endothelial cell–pericyte contacts and a stable microvasculature (Fuxe et al. 2011). In response to bacterial infection, the bronchial microvasculature became more densely populated with pericytes; inhibition of PDGFRβ signaling prevented this increase in pericyte coverage. The consequences of this included increase vascular leakage and less efficient bacterial clearance, suggesting that the loss of pericytes may have compromised leukocyte extravasation, thereby exacerbating the disease (Fuxe et al. 2011).

A recent study showed that lung pericytes are activated and regulate inflammation and vascular barrier integrity in the context of sterile lung inflammation driven by intratracheal LPS delivery; pericytes in LPS-exposed lungs dramatically upregulate their expression of the adhesion molecule ICAM-1 and produce high levels of chemokines that recruit monocytes and neutrophils (IL-6, CXCL1, and CCL2). Additionally, isolated lung pericytes stimulated in vitro with the bronchoalveolar lavage fluid collected from mice from the lungs of LPS-exposed mice increase their expression of functional Toll-like receptors and chemokines, indicating that lung pericytes are primed to detect epithelial injury and act as immune sentinels by detecting signals released by epithelial cells in response to damage and by modulating the recruitment of leukocytes (Hung et al. 2017).

Pericyte Dysfunction in Pulmonary Vascular Disease

Altered Vascular Permeability in Acute Pulmonary Disease

Due to their abovementioned ability to express TLRs and respond to LPS by secreting high levels of inflammatory cytokines and chemokines, it has been postulated that pericytes are crucial to the development of pulmonary edema seen in acute respiratory distress syndrome (ARDS) and sepsis-associated microvascular dysfunction (Hung et al. 2017; Zeng et al. 2016). Pericytes mediate vascular stability and permeability under both physiological and pathological conditions. A number of investigations using the LPS model of sepsis in mice have demonstrated increased vascular permeability within 24 h of LPS delivery, associated with pericyte uncoupling from endothelial cells, leading to increased microvascular permeability, vascular leakage, and neutrophil/macrophage infiltration (Chintalgattu et al. 2013; Nishioku et al. 2009). However, the molecular mechanisms responsible for pericyte uncoupling and the disruption of endothelial cell–pericyte interactions during sepsis were not clarified in these studies. To uncover this, using a model of LPS-induced acute lung injury, Zeng et al. (2016) demonstrated that a reduction in lung levels of Sirtuin3, an aging-related mediator of histone modifications, was responsible for pericyte dysfunction and increased vascular permeability during sepsis. They further showed that disruption of the angiopoietin/Tie-2 and HIF-2α/Notch3 signaling pathways played a role in the observed LPS-induced reduction in Sirtuin3 levels, diminished pericyte coverage, and microvascular dysfunction. The involvement of Ang2 signaling in the process of pericyte uncoupling from the endothelium was recently supported by a study investigating the association between Ang2 expression and increased pulmonary vascular permeability in human subjects with acute lung disease, compared to patients with chronic lung disease (IPF) and healthy controls (Ando et al. 2016). Intriguingly, Ang2 levels were higher in both serum and bronchoalveolar lavage in patients with idiopathic interstitial pneumonia and ARDS and correlated positively with pulmonary vascular dysfunction. Moreover, declining Ang2 levels were observed in survivors or acute lung disease, whereas levels remained high in non-survivors (Ando et al. 2016). These results provide a positive impetus to investigate the viability of targeting Ang2 and pericyte-vascular uncoupling in the search for more effective treatments for acute lung failure.

Pulmonary Hypertension

Pulmonary hypertension is defined as structural changes to pulmonary vessels, with prominent vascular fibrosis, the accumulation of α-SMA+ contractile cells, and persistent inflammation, ultimately leading to increased blood pressure in the pulmonary vasculature and failure of the right ventricle (Rowley and Johnson 2014). Pericyte proliferation has recently been investigated as an early event in pulmonary hypertension, since abnormal pericyte coverage of the pulmonary vasculature has been observed in clinical specimens with pulmonary hypertension (Rowley and Johnson 2014). On a more mechanistic level, in a rat model of pulmonary hypertension driven by hyperoxia, Jones et al. (2006) observed that PDGFRβ was highly overexpressed in pericytes in hypertensive lungs compared with control lungs; PDGF-BB was similarly overexpressed in the early stages of the disease. In clinical studies, pulmonary hypertension is similarly associated with higher expression of PDGF-BB by type II pneumocytes and endothelial cells, accompanied by increased PDGFRβ expression by pericytes associated with remodeled capillaries in the lungs of pulmonary hypertension patients (Assaad et al. 2007). Ricard et al. (2014) also observed an overabundance of pericytes on remodeled distal vessels, associated with increased expression of FGF-2 and IL-6 by endothelial cells, although no differences in PDGF-BB expression were observed. Based on these findings and others, the current hypothesis is that pulmonary hypertension is driven by endothelial cell abnormalities that provoke excess pericyte/vascular smooth muscle coverage and vascular remodeling via perturbation of the FGF and PDGF-BB/PDGFRβ signaling pathways.

Other studies have shown the involvement of the Wnt (Yuan et al. 2015), Notch, and TGF-β signaling pathways (Wang et al. 2016) in pericyte-mediated vascular remodeling in pulmonary hypertension. Using immunoglobulin G-coated magnetic beads specific for the pericyte cell surface marker 3G5, Yuan et al. (2015) isolated pericytes from the lungs of healthy subjects and patients with pulmonary hypertension. The initial discovery that disease pericytes did not perform adequately in a three-dimensional tube formation assay prompted these authors to investigate alterations in the planar cell polarity pathway in these cells; in fact, diseased pericytes showed reduced expression of frizzled 7 (Fzd7) and cdc42, crucial genes in the planar cell polarity pathway (Yuan et al. 2015). Further investigation showed that the restoration of Fzd7 and cdc42 in diseased pericytes improved interactions with endothelial cells and vascular network formation. Altered hypoxia signaling in endothelial cells (through HIF-2α) has also been reported as a potential initiating event in pulmonary hypertension. Wang et al. (2016) assessed this by knocking out endothelial cell-specific expression of prolyl hydroxylase-2 (PHD2), which resulted in increased pericyte vascular coverage, elevated fibroblast-specific protein-1

(FSP-1) levels, and increased expression of Notch3 and TGF-β in the lung. Further investigations using small molecule inhibitors to target these pathways in pericytes may provide a novel direction in the treatment of pulmonary hypertension.

Pulmonary Tumor Microvasculature

From a particular point of view, cancer can be seen as a vascular disease, as growing tumors require a steady supply of gases and nutrients and escaping metastatic cells exploit the vasculature to disseminate to other sites in the body. Similar to physiological angiogenesis, pericytes are recruited to tumor blood vessels by the PDGF-BB/PDGFRβ signaling pathway (Armulik et al. 2011). As the role of pericytes in the tumor microenvironment has been thoroughly covered elsewhere, the following will cover recent findings regarding pericytes in lung cancer and lung metastases.

Although known to be present in the human lung, the presence and functional orientation of pericytes within non-small cell lung cancer (NSCLC) tumors have not yet been thoroughly studied. Using a multiparameter flow cytometric approach, Bichsel et al. (2017) isolated CD73+/CD90+ mesenchymal cells from human primary NSCLC samples and found increased expression of the immunosuppressive ligand PD-L1, IL-6, and αSMA compared to the same population obtained from the healthy lung. Moreover, tumor-derived CD73+CD90+ cells were able to form perfusable vessels in vitro, but these demonstrated significantly greater permeability compared to vessels constructed from healthy cells (Bichsel et al. 2017), suggesting that pericyte abnormalities in the tumor microenvironment are directly linked to leaky vessels. In a related study, Keskin et al. (2015) found that pericyte depletion using genetic and pharmacological techniques in small, non-hypoxic tumors suppressed angiogenesis and tumor growth and prevented tumor cell metastasis to the lung. In contrast, pericyte depletion in advanced, hypoxic tumors with an established vasculature resulted in enhanced hypoxia, decreased tumor growth, but increased lung metastasis. Further assessments revealed that Ang2 signaling was increased following pericyte depletion; however, contemporaneous pericyte depletion and blockade of Ang2 restored vascular stability and prevented tumor cell metastasis to the lung (Keskin et al. 2015).

Moreover, pericytes have recently been shown to contribute to the formation of a premetastatic niche and encourage metastasis to the lung. Murgai et al. have observed that, in the tumor microenvironment, pericytes undergo phenotypic switching, increase their expression of extracellular matrix (ECM) proteins, and display increased migratory capacity away from blood vessels, with the end of result of making lung tissue more prone to colonization by circulating tumor cells (Murgai et al. 2017; Paiva et al. 2018). Using a genetic modification strategy in mice, these authors demonstrated that Kruppel-like factor 4 (KLF4) expression in pericytes is crucial for these processes to occur. Specifically, under conditions of pericyte-specific KLF4 ablation, pulmonary pericytes expressed lower levels of the ECM protein fibronectin and decreased the number of metastatic cancer cells in the lung; however, this strategy had no effect on primary tumor growth. It is anticipated that further studies in this direction may lead to the development of novel treatments to prevent cancer metastasis, which would lead to major clinical benefits.

Conversely, activation of the prostaglandin I2 (PGI2) pathway in pericytes has been reported to impair tumor cell metastasis by increasing tumor vessel pericyte coverage and maturation. Minami et al. (2015) examined the effects of a PGI2 agonist on metastasis in mice bearing Lewis lung carcinoma tumors and found that PGI2 stimulation over a period of 3 weeks decreased the number and size of lung metastases, increased the number of tumor vessel-associated pericytes, and decreased intratumor hypoxia (Minami et al. 2015). Collectively, these studies suggest that pericytes are vital to functional vascular maturation in tumors and modifying pericyte function in tumors may result in fewer metastases and improved survival in patients with solid tumors.

Pericyte Dysfunction in Pulmonary Fibrotic Disease

Chronic lung disease, including idiopathic pulmonary fibrosis (IPF), asthma, and chronic obstructive pulmonary disease (COPD), is the fourth leading cause of mortality worldwide (Rowley and Johnson 2014). Despite differing pathophysiologies, all chronic lung diseases exhibit varying degrees of persistent inflammation, airway narrowing leading to airflow limitation, and structural alterations of the pulmonary airways and vessels. Fibrosis, i.e., the formation of scar tissue, is a characteristic feature of chronic lung disease, characterized by extracellular matrix deposition and structural changes to the lung structure that result in restricted airflow or gas exchange due to the thickening and stiffening of tissue (Rowley and Johnson 2014). The loss of functional gas exchange capacity due to fibrosis is a significant contributor to the morbidity and mortality associated with chronic lung disease.

Current treatments for chronic fibrotic lung disease are palliative. There is a significant demand for anti-fibrotic therapies, as very few European Medicines Agency (EMEA) or Food and Drug Administration (FDA)-approved anti-fibrotic therapies currently exist (Rowley and Johnson 2014). To accelerate the design of anti-fibrotic therapies, recent research has focused on defining the origin of fibrosis-producing myofibroblasts, because the primary difference between healthy wound repair and fibrosis appears to be the fate of activated myofibroblasts. Several recent studies utilizing cutting-edge genetic cell labeling techniques have drawn attention to the pericyte as a major myofibroblast progenitor (Barnes and Glass 2011; Barron et al. 2016). Understanding the capacity of pericytes to serve as myofibroblast progenitor cells may allow us to arrest or reverse fibrosis in certain disease settings.

Many studies on pulmonary fibrosis have employed a commonly used mouse model of IPF induced by an intratracheal instillation of the chemotherapeutic agent bleomycin; the pathology is characterized by a brief but intense burst of pulmonary inflammation, followed by intense and widespread lung fibrosis (Rock et al. 2011; Hung et al. 2013; Marriott et al. 2014; Kramann et al. 2015). More recently, studies using a mouse model of chronic allergic airway inflammation driven by sustained respiratory house dust mite extract (HDM) exposure have been performed to highlight the role of pericytes in the development of airway remodeling associated with allergic airway disease (Johnson et al. 2015).

Myofibroblasts are recognized as the primary source of extracellular matrix (ECM) proteins and, as such, are considered the main driver of tissue fibrosis. Myofibroblasts characteristically express high levels of α-SMA and are responsible for the secretion of extracellular matrix (ECM) proteins, particularly collagen (Bagnato and Harari 2015). The myofibroblast cytoplasm is organized into cytoplasmic stress fibers that apply contractile pressure onto the neighboring ECM. In the healthy lung, the ECM is broken down by matrix metalloproteases. However, during chronic inflammation, normal ECM degradation becomes dysregulated, and ECM proteins are deposited by myofibroblasts at a greater rate than at which they are degraded, resulting in fibrotic scar formation and the loss of tissue function (Campanholle et al. 2013).

Current research suggests that the initiating factor in tissue fibrosis may be related to the disruption of stable pericyte–endothelial cell homeostasis such that lung developmental programs restart, thereby disrupting normal homeostasis and facilitating the switch to a pro-remodeling phenotype, i.e., pericyte-to-myofibroblast transition. Vessel integrity depends on the regulation of stimulatory (angiogenic) and inhibitory (angiostatic) factors (Eberhard et al. 2000; Kloc et al. 2015). It appears likely that in chronic fibrotic lung disease, this balance is disrupted. It is understood that growth factors including Ang1, Ang2, vascular endothelial growth factor (VEGF), and TGF-β mediate interactions between the lung endothelium and pericytes and have been independently studied as drivers of fibrosis (Ballermann and Obeidat 2014). Additionally, environmental stressors such as oxidative stress have been implicated in this process, with the downstream effects of endoplasmic reticulum stress and mitochondrial damage (Cheresh et al. 2013). Crucially, oxidative stress has been shown to contribute to excessive TGF-β production (Cheresh et al. 2013), inferring the presence of a positive feedback loop involving environmental stress and pro-fibrotic signals in myofibroblast differentiation (Fig. 14.1). This emphasizes the need to define and delineate the interactions between pericytes and endothelial cells in the normal lung and during chronic pulmonary inflammation and fibrosis.

The capacity of pericytes to differentiate into fibrosis-causing myofibroblasts is not the only reason why they are of interest as antifibrotic targets. As detailed above, the intimate relationship between pericytes and endothelial cells is responsible for the regulation of vessel permeability, neutrophil extraversion, leukocyte trafficking, T-cell activation, angiogenesis, endothelial cell proliferation, structural support of existing blood vessels, and the formation the basement membrane (Barron et al. 2016). Dysfunction in endothelial–pericyte signaling in particular represents a key

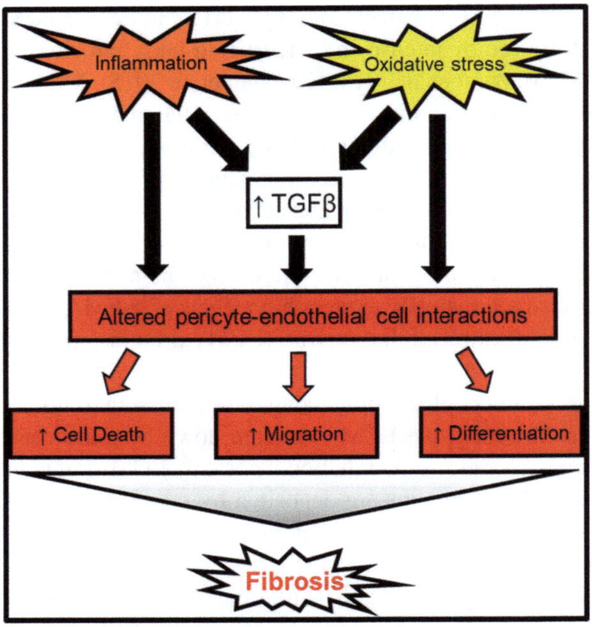

Fig. 14.1 Proposed mechanism of environmental stress/oxidant-induced pulmonary fibrosis. Environmental toxins/oxidants may induce the generation of ROS through the induction of endoplasmic reticulum stress, mitochondrial damage, and the activation of inflammatory cells, such as macrophages and neutrophils. ROS-generated and inflammatory stimuli perturb pericyte–endothelial cell interactions and permit pericyte uncoupling from the vasculature. These uncoupled pericytes then undergo migration and differentiation into myofibroblasts, the main driver of tissue fibrosis

mechanism by which the microvascular becomes dysregulated, unstable, and ultimately pathogenic in chronic fibrotic lung disease (Bagnato and Harari 2015).

Idiopathic Pulmonary Fibrosis

Pulmonary fibrosis is a debilitating disease characterized by excessive matrix deposition, angiogenesis, and epithelial cell hyperplasia that impedes overall tissue function. During pulmonary fibrosis, the accumulation of fibrotic tissue within the alveolar parenchyma is merely a symptom of disease; the etiology of this pathology in humans varies greatly (Hung et al. 2013). Since the reason behind this defective repair is unknown, although a combination of immunological, genetic, and environmental factors is suspected, it is very difficult to model disease in a clinically relevant fashion.

The most common method used to model pulmonary fibrosis in mice is administration of the chemotherapeutic agent bleomycin; this agent is known to cause pulmonary fibrosis in humans as well, but this may not accurately reflect the true etiology of most cases of human disease (Andersson-Sjöland et al. 2016). Novel Cre fate tracing techniques in bleomycin-administered mice show that FoxD1-positive and PDGFRβ-positive pericytes proliferate and differentiate into α-SMA-positive myofibroblasts, the primary driver cell of fibrosis (Greenhalgh et al. 2013). Lineage tracing studies have confirmed that 45–68% of myofibroblasts (defined by α-SMA expression) are derived from Foxd1 cells expressing the pericyte markers PDGFRβ, NG2, and CD146 (Hung et al. 2013). Using a similar model of bleomycin-driven lung injury in mice, Rock et al. (2011) showed strong expression pericyte markers including PDGFR-β, NG2, and desmin within fibrotic foci. NG2CreER BAC transgenic were mice crossed with ROSA farnesylated GFP reporter mice to genetically label lung pericytes; it was found that although pericytes proliferated under fibrotic conditions, few increased their expression of α-SMA+ phenotype, indicating little pericyte to myofibroblast transition in this model (Rock et al. 2011). However, given the inefficiency of the genetic labeling technique using in this study, the number of pericytes contributing to the myofibroblast population in this study may have been underestimated. Using a different labeling strategy, i.e., double-transgenic mice expressing GFP in Nestin-positive cells and DsRed in NG2-positive pericytes, Birbrair et al. (2014) demonstrated that a particular subset of pulmonary pericytes, called type-1 pericytes (Nestin-GFP–/NG2-DsRed+), are the primary cell type involved in producing collagen after lung injury brought about by intratracheal bleomycin delivery. Moreover, type 1 pericytes were found to accumulate in areas of fibrosis in this model of pulmonary fibrosis (Birbrair et al. 2014). In the milder subcutaneous bleomycin model, the development of pulmonary fibrosis has been related to endothelial cell oxidative stress and activation, with subsequent effects on pericyte localization and myofibroblast transition, mediated through Wnt3a signaling (Andersson-Sjöland et al. 2016). Similarly, Wnt/β-catenin signaling was found to be the primary pathway involved in the transition of ATP-binding cassette protein ABCG2 (ABCG2+) pericyte progenitors into myofibroblasts in a mouse model of pulmonary fibrosis driven by bleomycin exposure (Gaskill et al. 2017).

Some recent studies have been performed to determine the origin of α-SMA expressing myofibroblasts in clinical samples from IPF patients. Sava et al. (2017) demonstrated the presence of α-SMA+ cells coexpressing the pericyte marker NG2 in human IPF lung samples. Furthermore, healthy human lung pericytes cultured on decellularized IPF lung matrices demonstrated increased expression of α-SMA, suggesting that pericytes are able to undergo phenotypic transition into myofibroblasts through a mechanism related to the stiffness, not the composition, of the underlying ECM. Interestingly, treatment with the tyrosine-kinase inhibitor nintedanib, which has been approved for IPF treatment, reduced the stiffness of fibrotic lung matrices, which subsequently reversed the α-SMA+ pericyte phenotype (Sava et al. 2015).

Although progress has been made into defining the vascular abnormalities that are present in IPF and the roles of pericytes in the development of this disease, many unanswered questions remain. A better understanding of vascular remodeling and pericyte differentiation into myofibroblasts in IPF is required before targeted therapies can possibly lead to new targets for the treatment of fibrosis.

Chronic Obstructive Pulmonary Disease

Chronic obstructive pulmonary disease (COPD) is characterized by chronic airway obstruction, restricted airflow as a result of peribronchial fibrosis, loss of alveolar cells (emphysema), accumulation and activation of inflammatory cells, increased ECM deposition, and reduced elastic recoil. The etiology of COPD is highly complex and is believed to develop after many years of tobacco smoking in combination with other known factors such as genetic susceptibility or environmental factors (Rowley and Johnson 2014). In similarity to asthma, inflammation is a major component in COPD, but the leukocyte profile is very different: the most prominent players in COPD-related inflammation are neutrophils and, to some degree, macrophages, rather than the predominant eosinophilic infiltration of the airways commonly seen in allergic asthma (Rowley and Johnson 2014). Currently, there are no studies directly investigating the role of pericytes due to the difficulty in modeling COPD in vitro and in vivo. Nonetheless the structural changes in the pulmonary vasculature and increase in vascular mediators will likely encourage pericyte migration away from the vasculature and pericyte differentiation into myofibroblasts.

Although little is known about microvascular remodeling in COPD, a slightly denser pulmonary vasculature has been observed in some areas of the COPD lung (Calabrese et al. 2006), along with increased expression of TGF-β1 (Konigshoff et al. 2009). This growth likely plays a major role in tissue remodeling in COPD, due its proliferative effects on fibroblasts and pericytes. Furthermore, increased expression of VEGF has been observed in some COPD patients and has been negatively correlated to lung function (Kranenburg et al. 2005). Pericytes secrete VEGF (Darland et al. 2003), so it is possible that pericytes are involved in microvascular remodeling in COPD. Ang2 levels also increase in COPD patients during exacerbations (Cho et al. 2011). As described above, Ang2 facilitates pericyte uncoupling from endothelial cells, leading to vessel destabilization.

Beyond alterations in growth factor expression and signaling, the induction of oxidative stress in the lung as a consequence of cigarette smoking or environmental air pollution may have a significant impact on the pulmonary vasculature. Reactive oxygen species (ROS) play important roles in vascular homeostasis, but excessive ROS can perturb the pulmonary vasculature; in fact, pulmonary hypertension and vascular remodeling are commonly seen featured of COPD at later stages. Inflammatory cell infiltration and endothelial dysfunction are both correlated to the severity of COPD and likely contribute to disease pathology (Zuo et al. 2017). It has also been shown that endothelial cell damage results in the aberrant release of vasoactive mediators that promote the proliferation of pericytes of SMC and ECM deposition (Peinado et al. 2008).

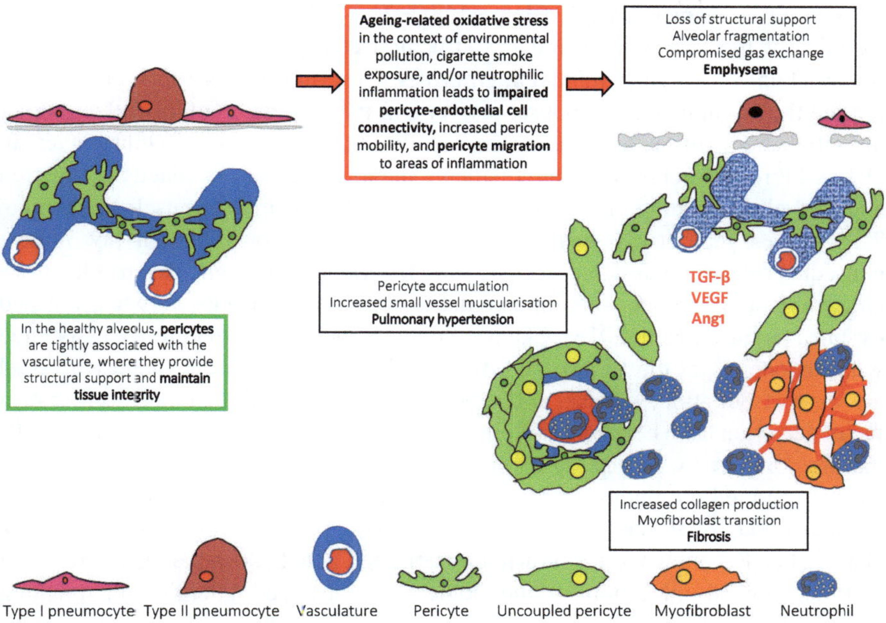

Fig. 14.2 Proposed mechanism of the involvement of pericytes in lung structural remodeling associated with chronic obstructive pulmonary disease (COPD). A combination of environmental stress and inflammation induces pericyte uncoupling from the pulmonary microvasculature, resulting in alveolar fragmentation as well as pericyte migration and differentiation into a contractile, myofibroblast phenotype

Although these studies do not provide direct evidence supporting a role for pericytes in driving structural alterations in the COPD lung, they highlight areas that require further investigation. Based on our current knowledge, it can be hypothesized that altered pericyte behavior is responsible for several aspects of altered lung structure and function in COPD (Fig. 14.2).

Asthma

Allergic asthma results from a chronic immune response to inhaled allergens. According to Asthma UK, asthma is estimated to affect 300 million people globally; in the UK, asthma affects 18% of the UK population and asthma care is estimated to cost the UK £1.1 billion per year (Rowley and Johnson 2014). Current treatments for asthma involve either symptom relief (bronchodilators) or anti-inflammatory drugs (corticosteroids). Despite the effectiveness of these therapies, approximately 4% of asthmatics suffer from inadequately controlled asthma and significantly impaired quality of life, underscoring the need for potent and novel treatments for lung fibrosis (Rajapaksa et al. 2016).

Asthma is characteristically defined as a Th2-polarized chronic inflammatory disorder of the airways that induces structural changes to the airway wall, in a process known as airway remodeling/fibrosis. Cytokines released from leukocytes initiate the inflammatory response, resulting in the expression of TGF-β and other growth factors ultimately leading to myofibroblast recruitment (Johnson et al. 2015). Fibrosis in chronic asthma patients manifests as excess mucus production and epithelial shedding, the accumulation of ECM, and smooth muscle thickening. This causes the airway wall to thicken, which narrows the airway lumen; consequently, airflow is decreased and lung function is impaired. This results in the characteristic symptoms seen in asthmatics, including episodes of wheezing and dyspnea (Johnson et al. 2015). Changes in the bronchial vasculature are associated with remodeling of the airway wall. Increased number of vessels, increased vessel activity (vasodilation, leakage, and leukocyte extravasation), and increased growth factor secretion have been observed in asthmatics increasing with the severity of the disease compared with healthy control subjects (Barron et al. 2016). A number of studies reported the involvement of specific growth factors in chronic asthma. In particular, PDGF-BB has been implicated as a major player in lung fibrosis (Armulik et al. 2011). Recent work has shown that perturbed PDGFRβ signaling as a result of pharmacological inhibition in the context of sustained allergic inflammation promoted pericyte migration away from bronchial vessels and towards the airway wall. Under these disease conditions, pericytes were found contribute to airway remodeling by differentiating into myofibroblasts and contributing to airway smooth

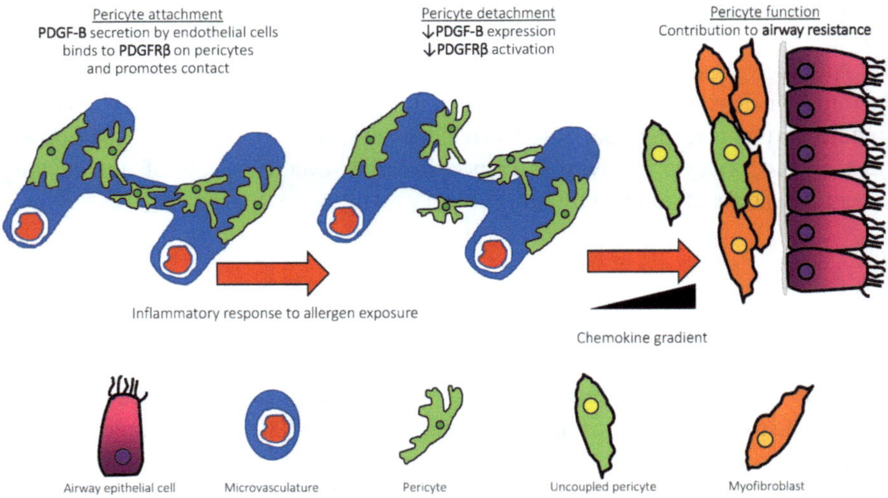

Fig. 14.3 Proposed mechanism of the involvement of pericytes in lung structural remodeling associated with allergic asthma. Persistent Th2-polarized airway inflammation induces a decrease in lung PDGF-BB expression and pericyte uncoupling from the pulmonary microvasculature. Pericytes subsequently undergo directed migration to the airway wall, where they differentiate into a contractile, myofibroblast phenotype and incorporate into airway smooth muscle bundles, thereby contributing to increased airway stiffness and airway hyperreactivity

muscle thickening and airway hyperresponsiveness (Johnson et al. 2015). Based on the findings of this study, it appears that pericyte differentiation into myofibroblasts begins to occur during the process of pericyte uncoupling form the microvasculature, as α-SMA-positive pericytes can be observed detaching from capillaries (Fig. 14.3); under healthy conditions, these cells are α-SMA negative (Johnson et al. 2015). Further studies are anticipated to examine the role of inflammatory mediators (chemokines, cytokines, and growth factors) on pericyte–endothelial cell interactions, pericyte uncoupling, and pericyte-to-myofibroblast transition in mouse models of chronic allergic airway inflammation.

Future Directions

The vascular compartment of the lung plays significant and varied roles in a number of acute and chronic lung conditions. Despite being a key regulator of angiogenesis, vascular permeability, and vessel contractility, pericytes have been studied only recently in the context of asthma, IPF, and pulmonary arterial hypertension. Recent studies have also shown that pericytes participate in inflammatory reactions and possess MSC-like progenitor capabilities, and are therefore likely to play multiple roles in pathological changes to lung structure and function related to disease. A comprehensive understanding of the involvement of pericytes in chronic lung disorders may lead to the development of new therapeutic strategies.

References

Andersson-Sjöland, A., Karlsson, J. C., & Rydell-Törmänen, K. (2016). ROS-induced endothelial stress contributes to pulmonary fibrosis through pericytes and Wnt signaling. *Laboratory Investigation, 96*(2), 206–217.

Ando, M., Miyazaki, E., Abe, T., Ehara, C., Goto, A., Masuda, T., Nishio, S., Fujisaki, H., Yamasue, M., Ishii, T., Mukai, Y., Ito, T., Nureki, S. I., Kumamoto, T., & Kadota, J. (2016). Angiopoietin-2 expression in patients with an acute exacerbation of idiopathic interstitial pneumonias. *Respiratory Medicine, 117*, 27–32.

Armulik, A., Genove, G., & Betsholtz, C. (2011). Pericytes: Developmental, physiological, and pathological perspectives, problems, and promises. *Developmental Cell, 21*(2), 193–215. https://doi.org/10.1016/j.devcel.2011.07.001

Assaad, A. M., Kawut, S. M., Arcasoy, S. M., Rosenzweig, E. B., Wilt, J. S., Sonett, J. R., & Borczuk, A. C. (2007). Platelet-derived growth factor is increased in pulmonary capillary hemangiomatosis. *Chest, 131*, 850–855.

Bagnato, G., & Harari, S. (2015). Cellular interactions in the pathogenesis of interstitial lung diseases. *European Respiratory Review, 24*(135), 102–114. https://doi.org/10.1183/09059180.00003214

Ballermann, B. J., & Obeidat, M. (2014). Tipping the balance from angiogenesis to fibrosis in CKD. *Kidney International, 4*(1), 45–52.

Barnes, J. L., & Glass, W. F. (2011). Renal interstitial fibrosis: a critical evaluation of the origin of myofibroblasts. *Contributions to Nephrology, 169*, 73–93.

Barron, L., Gharib, S. A., & Duffield, J. S. (2016). Lung pericytes and resident fibroblasts: Busy multitaskers. *The American Journal of Pathology, 186*(10), 2519–2531. https://doi.org/10.1016/j.ajpath.2016.07.004

Bichsel, C. A., Wang, L., Froment, L., Berezowska, S., Müller, S., Dorn, P., Marti, T. M., Peng, R. W., Geiser, T., Schmid, R. A., Guenat, O. T., & Hall, S. R. R. (2017). Increased PD-L1 expression and IL-6 secretion characterize human lung tumor-derived perivascular-like cells that promote vascular leakage in a perfusable microvasculature model. *Scientific Reports, 7*(1), 10636.

Birbrair, A., Zhang, T., Files, D. C., Mannava, S., Smith, T., Wang, Z. M., Messi, M. L., Mintz, A., & Delbono, O. (2014). Type-1 pericytes accumulate after tissue injury and produce collagen in an organ-dependent manner. *Stem Cell Research & Therapy, 5*(6), 122. https://doi.org/10.1186/scrt512

Calabrese, C., Bocchino, V., Vatrella, A., Marzo, C., Guarino, C., Mascitti, S., Tranfa, C. M., Cazzola, M., Micheli, P., Caputi, M., & Marsico, S. A. (2006). Evidence of angiogenesis in bronchial biopsies of smokers with and without airway obstruction. *Respiratory Medicine, 100*, 1415–1422.

Campanholle, G., Ligresti, G., Gharib, S. A., & Duffield, J. S. (2013). Cellular mechanisms of tissue fibrosis. Novel mechanisms of kidney fibrosis. *American Journal of Physiology. Cell Physiology, 304*(7), C591–C603.

Chintalgattu, V., Rees, M. L., Culver, J. C., Goel, A., Jiffar, T., Zhang, J., Dunner, K., Jr., Pati, S., Bankson, J. A., Pasqualini, R., Arap, W., Bryan, N. S., Taegtmeyer, H., Langley, R. R., Yao, H., Kupferman, M. E., Entman, M. L., Dickinson, M. E., & Khakoo, A. Y. (2013). Coronary microvascular pericytes are the cellular target of sunitinib malate-induced cardiotoxicity. *Science Translational Medicine, 5*, 187ra69.

Cho, Y. J., Ma, J. E., Yun, E. Y., Kim, Y. E., Kim, H. C., Lee, J. D., Hwang, Y. S., & Jeong, Y. Y. (2011). Serum angiopoietin-2 levels are elevated during acute exacerbations of COPD. *Respirology, 16*, 284–290.

Crisan, M., Yap, S., Casteilla, L., Chen, C. W., Corselli, M., Park, T. S., Andriolo, G., Sun, B., Zheng, B., Zhang, L., Norotte, C., Teng, P. N., Traas, J., Schugar, R., Deasy, B. M., Badylak, S., Buhring, H. J., Giacobino, J. P., Lazzari, L., Huard, J., & Péault, B. (2008). A perivascular origin for mesenchymal stem cells in multiple human organs. *Cell Stem Cell, 3*(3), 301–313.

Cheresh, P., Kim, S. J., Tulasiram, S., & Kamp, D. W. (2013). Oxidative stress and pulmonary fibrosis. *Biochimica et Biophysica Acta, 1832*(7), 1028–1040.

Crisan, M., Corselli, M., Chen, W. C., & Péault, B. (2012). Perivascular cells for regenerative medicine. *Journal of Cellular and Molecular Medicine, 16*(12), 2851–2860.

Damianovich, M., Hout, S. G., Barshack, I., Simansky, D. A., Kidron, D., Dar, E., Avivi, C., & Onn, A. (2013). Structural basis for hyperpermeability of tumor vessels in advanced lung adenocarcinoma complicated by pleural effusion. *Clinical Lung Cancer, 14*(6), 688–698.

Darland, D. C., Massingham, L. J., Smith, S. R., Piek, E., Saint-Geniez, M., & D'Amore, P. A. (2003). Pericyte production of cell-associated VEGF is differentiation-dependent and is associated with endothelial survival. *Developmental Biology, 264*, 275–288.

Donoghue, L., Tyburski, J. G., Steffes, C. P., & Wilson, R. F. (2006). Vascular endothelial growth factor modulates contractile response in microvascular lung pericytes. *American Journal of Surgery, 191*(3), 349–352.

Eberhard, A., Kahlert, S., Goede, V., Hemmerlein, B., Plate, K. H., & Augustin, H. G. (2000). Heterogeneity of angiogenesis and blood vessel maturation in human tumors: Implications for antiangiogenic tumor therapies. *Cancer Research, 60*(5), 1388–1393.

Fuxe, J., Tabruyn, S., Colton, K., Zaid, H., Adams, A., Baluk, P., Lashnits, E., Morisada, T., Le, T., O'Brien, S., Epstein, D. M., Koh, G. Y., & McDonald, D. M. (2011). Pericyte requirement for antileak action of angiopoietin-1 and vascular remodelling in sustained inflammation. *The American Journal of Pathology, 178*, 2897–2909.

Gaskill, C. F., Carrier, E. J., Kropski, J. A., Bloodworth, N. C., Menon, S., Foronjy, R. F., Taketo, M. M., Hong, C. C., Austin, E. D., West, J. D., Means, A. L., Loyd, J. E., Merryman, W. D.,

Hemnes, A. R., De Langhe, S., Blackwell, T. S., Klemm, D. J., & Majka, S. M. (2017). Disruption of lineage specification in adult pulmonary mesenchymal progenitor cells promotes microvascular dysfunction. *The Journal of Clinical Investigation, 127*(6), 2262–2276.

Geevarghese, A., & Herman, I. M. (2014). Pericyte-endothelial crosstalk: Implications and opportunities for advanced cellular therapies. *Translational Research, 163*(4), 296–306.

Greenhalgh, S. N., Iredale, J. P., & Henderson, N. C. (2013). Origins of fibrosis: pericytes take centre stage. *F1000Prime Reports, 5*, 37.

Hammes, H. P., Lin, J., Wagner, P., Feng, Y., Vom Hagen, F., Krzizok, T., Renner, O., Breier, G., Brownlee, M., & Deutsch, U. (2004). Angiopoietin-2 causes pericyte dropout in the normal retina: Evidence for involvement in diabetic retinopathy. *Diabetes, 53*, 1104–1110.

Hung, C., Linn, G., Chow, Y. H., Kobayashi, A., Mittelsteadt, K., Altemeier, W. A., Gharib, S. A., Schnapp, L. M., & Duffield, J. S. (2013). Role of lung pericytes and resident fibroblasts in the pathogenesis of pulmonary fibrosis. *American Journal of Respiratory and Critical Care Medicine, 188*(7), 820–830.

Hung, C. F., Mittelsteadt, K. L., Brauer, R., McKinney, B. L., Hallstrand, T. S., Parks, W. C., Chen, P., Schnapp, L. M., Liles, W. C., Duffield, J. S., & Altemeier, W. A. (2017). Lung pericyte-like cells are functional interstitial immune sentinel cells. *American Journal of Physiology. Lung Cellular and Molecular Physiology, 312*(4), L556–L567.

Johnson, J. R., Folestad, E., Rowley, J. E., Noll, E. M., Walker, S. A., Lloyd, C. M., Rankin, S. M., Pietras, K., Eriksson, U., & Fuxe, J. (2015). Pericytes contribute to airway remodeling in a mouse model of chronic allergic asthma. *American Journal of Physiology-Lung Cellular and Molecular Physiology, 308*(7), L658–L671.

Jones, R., Capen, D., & Jacobson, M. (2006). PDGF and microvessel wall remodeling in adult lung: Imaging PDGF-Rβ and PDGF-BB molecules in progenitor smooth muscle cells developing in pulmonary hypertension. *Ultrastructural Pathology, 30*, 267–281.

Keskin, D., Kim, J., Cooke, V. G., Wu, C. C., Sugimoto, H., Gu, C., De Palma, M., Kalluri, R., & LeBleu, V. S. (2015). Targeting vascular pericytes in hypoxic tumors increases lung metastasis via angiopoietin-2. *Cell Reports, 10*(7), 1066–1081. https://doi.org/10.1016/j.celrep.2015.01.035

Kloc, M., Kubiak, J. Z., Li, X. C., & Ghobrial, R. M. (2015). Pericytes, microvasular dysfunction, and chronic rejection. *Transplantation, 99*(4), 658–667.

Konigshoff, M., Kneidinger, N., & Eickelberg, O. (2009). TGF-beta signaling in COPD: deciphering genetic and cellular susceptibilities for future therapeutic regimen. *Swiss Medical Weekly, 139*, 554–563.

Kramann, R., Schneider, R. K., DiRocco, D. P., Machado, F., Fleig, S., Bondzie, P. A., Henderson, J. M., Ebert, B. L., & Humphreys, B. D. (2015). Perivascular Gli1+ progenitors are key contributors to injury-induced organ fibrosis. *Cell Stem Cell, 16*(1), 51–66.

Kranenburg, A. R., de Boer, W. I., Alagappan, V. K., Sterk, P. J., & Sharma, H. S. (2005). Enhanced bronchial expression of vascular endothelial growth factor and receptors (Flk-1 and Flt-1) in patients with chronic obstructive pulmonary disease. *Thorax, 60*, 106–113.

Lauridsen, H. M., Pober, J. S., & Gonzalez, A. L. (2014). A composite model of the human postcapillary venule for investigation of microvascular leukocyte recruitment. *The FASEB Journal, 28*, 1166–1180.

Liu, H., Kennard, S., & Lilly, B. (2009). NOTCH3 expression is induced in mural cells through an autoregulatory loop that requires endothelial-expressed JAGGED1. *Circulation Research, 104*, 466–475.

Liu, H., Zhang, W., Kennard, S., Caldwell, R. B., & Lilly, B. (2010). Notch3 is critical for proper angiogenesis and mural cell investment. *Circulation Research, 107*, 860–870.

Marriott, S., Baskir, R. S., Gaskill, C., Menon, S., Carrier, E. J., Williams, J., Talati, M., Helm, K., Alford, C. E., Kropski, J. A., Loyd, J., Wheeler, L., Johnson, J., Austin, E., Nozik-Grayck, E., Meyrick, B., West, J. D., Klemm, D. J., & Majka, S. M. (2014). ABCG2pos lung mesenchymal stem cells are a novel pericyte subpopulation that contributes to fibrotic remodeling. *American Journal of Physiology-Cell Physiology, 307*(8), C684–C698.

McCullagh, A., Rosenthal, M., Wanner, A., Hurtado, A., Padley, S., & Bush, A. (2010). The bronchial circulation-worth a closer look: a review of the relationship between the bronchial vasculature and airway inflammation. *Pediatric Pulmonology, 45*(1), 1–13.

Minami, Y., Sasaki, T., Bochimoto, H., Kawabe, J., Endo, S., Hira, Y., Watanabe, T., Okumura, S., Hasebe, N., & Ohsaki, Y. (2015). Prostaglandin I2 analog suppresses lung metastasis by recruiting pericytes in tumor angiogenesis. *International Journal of Oncology, 46*(2), 548–554. https://doi.org/10.3892/ijo.2014.2783

Murgai, M., Ju, W., Eason, M., Kline, J., Beury, D. W., Kaczanowska, S., Miettinen, M. M., Kruhlak, M., Lei, H., Shern, J. F., Cherepanova, O. A., Owens, G. K., & Kaplan, R. N. (2017). KLF4-dependent perivascular cell plasticity mediates pre-metastatic niche formation and metastasis. *Nature Medicine, 23*, 1176–1190.

Nishioku, T., Dohgu, S., Takata, F., Eto, T., Ishikawa, N., Kodama, K. B., Nakagawa, S., Yamauchi, A., & Kataoka, Y. (2009). Detachment of brain pericytes from the basal lamina is involved in disruption of the blood-brain barrier caused by lipopolysaccharide-induced sepsis in mice. *Cellular and Molecular Neurobiology, 29*, 309–316.

Paiva, A. E., Lousado, L., Guerra, D. A. P., Azevedo, P. O., Sena, I. F. G., Andreotti, J. P., Santos, G. S. P., Goncalves, R., Mintz, A., & Birbrair, A. (2018). Pericytes in the premetastatic niche. *Cancer Research, 78*(11), 2779–2786.

Peinado, V. I., Pizarro, S., & Barberà, J. A. (2008). Pulmonary vascular involvement in COPD. *Chest, 134*, 808–814.

Rajapaksa, K. S., Huang, T., Sharma, N., Liu, S., Solon, M., Reyes, A., Paul, S., Yee, A., Tao, J., Chalasani, S., Bien-Ly, N., Barck, K., Carano, R. A., Wang, J., Rangell, L., Bremer, M., Danilenko, D. M., Katavolos, P., Hotzel, I., Reif, K., & Austin, C. D. (2016). Preclinical safety profile of a depleting antibody against CRTh2 for asthma: Well tolerated despite unexpected CRTh2 expression on vascular pericytes in the central nervous system and gastric mucosa. *Toxicological Sciences, 152*(1), 72–84.

Ricard, N., Tu, L., Le Hiress, M., Huertas, A., Phan, C., Thuillet, R., Sattler, C., Seferian, A., Fadel, E., Montani, D., Dorfmüller, P., Humbert, M., & Guignabert, C. (2014). Increased pericyte coverage mediated by endothelial derived fibroblast growth factor-2 and interleukin-6 is a source of smooth muscle-like cells in pulmonary hypertension. *Circulation, 129*, 1586–1597.

Rock, J. R., Barkauskas, C. E., Cronce, M. J., Xue, Y., Harris, J. R., Liang, J., Noble, P. W., & Hogan, B. L. (2011). Multiple stromal populations contribute to pulmonary fibrosis without evidence for epithelial to mesenchymal transition. *Proceedings of the National Academy of Sciences of the United States of America, 108*, E1475–E1483. https://doi.org/10.1073/pnas.1117988108

Rowley, J. E., & Johnson, J. R. (2014). Pericytes in chronic lung disease. *International Archives of Allergy and Immunology, 164*(3), 178–188. https://doi.org/10.1159/000365051

Sava, P., Cook, I. O., Mahal, R. S., Gonzalez, A. L. (2015). Human microvascular pericyte basement membrane remodeling regulates neutrophil recruitment. *Microcirculation, 22*(1), 54–67. https://doi.org/10.1111/micc.12173

Sava, P., Ramanathan, A., Dobronyi, A., Peng, X., Sun, H., Ledesma-Mendoza, A., Herzog, E. L., & Gonzalez, A. L. (2017). Human pericytes adopt myofibroblast properties in the microenvironment of the IPF lung. *JCI Insight, 2*(24), 96352.

Wang, S., Voisin, M. B., Larbi, K. Y., Dangerfield, J., Scheiermann, C., Tran, M., Maxwell, P. H., Sorokin, L., & Nourshargh. (2006). Venular basement membranes contain specific matrix protein low expression regions that act as exit points for emigrating neutrophils. *The Journal of Experimental Medicine, 203*, 1519–1532.

Wang, S., Zeng, H., Xie, X. J., Tao, Y. K., He, X., Roman, R. J., Aschner, J. L., & Chen, J. X. (2016). Loss of prolyl hydroxylase domain protein 2 in vascular endothelium increases pericyte coverage and promotes pulmonary arterial remodeling. *Oncotarget, 7*(37), 58848–58861. https://doi.org/10.18632/oncotarget.11585

Wiebe, B. M., & Laursen, H. (1995). Human lung volume, alveolar surface area, and capillary length. *Microscopy Research and Technique, 32*(3), 255–262.

Weibel, E. R. (1974). On pericytes, particularly their existence on lung capillaries. *Microvascular Research, 8*(2), 218–235.
Wong, S. P., Rowley, J. E., Redpath, A. N., Tilman, J. D., Fellous, T. G., & Johnson, J. R. (2015). Pericytes, mesenchymal stem cells and their contributions to tissue repair. *Pharmacology & Therapeutics, 151*, 107–120.
Yuan, K., Orcholski, M. E., Panaroni, C., Shuffle, E. M., Huang, N. F., Jiang, X., Tian, W., Vladar, E. K., Wang, L., Nicolls, M. R., Wu, J. Y., & de Jesus Perez, V. A. (2015). Activation of the Wnt/planar cell polarity pathway is required for pericyte recruitment during pulmonary angiogenesis. *The American Journal of Pathology, 185*(1), 69–84. https://doi.org/10.1016/j.ajpath.2014.09.013
Zeng, H., He, X., Tuo, Q. H., Liao, D. F., Zhang, G. Q., & Chen, J. X. (2016). LPS causes pericyte loss and microvascular dysfunction via disruption of Sirt3/angiopoietins/Tie-2 and HIF-2α/Notch3 pathways. *Scientific Reports, 6*, 20931. https://doi.org/10.1038/srep20931
Zuo, L., Chuang, C. C., Clark, A. D., Garrison, D. E., Kuhlman, J. L., & Sypert, D. C. (2017). Reactive oxygen species in COPD-related vascular remodeling. *Advances in Experimental Medicine and Biology, 967*, 399–411.

Chapter 15
Pericytes in Muscular Dystrophies

Louise Anne Moyle, Francesco Saverio Tedesco, and Sara Benedetti

Abstract The muscular dystrophies are an heterogeneous group of inherited myopathies characterised by the progressive wasting of skeletal muscle tissue. Pericytes have been shown to make muscle in vitro and to contribute to skeletal muscle regeneration in several animal models, although recent data has shown this to be controversial. In fact, some pericyte subpopulations have been shown to contribute to fibrosis and adipose deposition in muscle. In this chapter, we explore the identity and the multifaceted role of pericytes in dystrophic muscle, potential therapeutic applications and the current need to overcome the hurdles of characterisation (both to identify pericyte subpopulations and track cell fate), to prevent deleterious differentiation towards myogenic-inhibiting subpopulations, and to improve cell proliferation and engraftment efficacy.

Keywords Pericytes · Muscular dystrophy · Muscle · Stem cells · Cell therapy Muscle regeneration · Adipogenesis · Chondrogenesis · Fibrosis · Plasticity

L. A. Moyle
Institute of Biomaterials and Biomedical Engineering, Donnelly Centre for Cellular and Biomolecular Research, University of Toronto, Toronto, ON, Canada

F. S. Tedesco (✉)
Department of Cell and Developmental Biology, University College London, London, UK

Great Ormond Street Institute of Child Health, University College London, London, UK
e-mail: f.s.tedesco@ucl.ac.uk

S. Benedetti (✉)
Great Ormond Street Institute of Child Health, University College London, London, UK

NIHR Great Ormond Street Hospital Biomedical Research Centre, London, UK
e-mail: s.benedetti@ucl.ac.uk

A. Birbrair (ed.), *Pericyte Biology in Disease*, Advances in Experimental Medicine and Biology 1147, https://doi.org/10.1007/978-3-030-16908-4_15

Skeletal Muscle and Muscular Dystrophies

Skeletal muscle is the most abundant tissue in humans, and its main roles are to generate movement, support soft tissues, maintain posture, and contribute to energy metabolism and temperature control (Frontera and Ochala 2015). It is characterised by a well-defined structure of connective tissues and muscle fibres (or myofibres), which are multinucleated, post-mitotic syncytial cells containing contractile units named sarcomeres. During skeletal muscle development, muscle fibres are generated by the fusion of paired-box transcription factor 3- (Pax3) and Pax7-expressing mesodermal progenitors (Bentzinger et al. 2012; Buckingham 2006; Comai and Tajbakhsh 2014). Postnatally, myofibres grow in size thanks to the fusion of satellite cells (Yablonka-Reuveni 2011; Yin et al. 2013), a population of muscle stem cells located between the plasma membrane of myofibres (sarcolemma) and the basal lamina, that are responsible for growth, repair, and regeneration of adult skeletal muscle (Mauro 1961; Relaix and Zammit 2012). Satellite cells are quiescent in physiological conditions but can be activated after muscle injury or by specific signalling pathways (Dumont et al. 2015; Relaix and Zammit 2012; Verdijk et al. 2014; Yin et al. 2013). Once activated, they proliferate and the majority of their myoblast progeny differentiate along the myogenic programme in order to replace damaged muscle fibres. Alternatively, they undergo self-renewal to replenish the stem cell pool (Rocheteau et al. 2012; Zammit et al. 2004). Satellite cells are characterised by the expression of the transcription factor Pax7. Many also express caveolin-1, integrin-α7, M-cadherin, CD56/NCAM, CD29/integrin-β1 and syndecans 3 and 4, although differences in expression patterns are observed between species, location and activation stage [reviewed in detail in (Boldrin et al. 2010; Tedesco et al. 2010, 2017; Yin et al. 2013)]. Satellite cells and their derived myoblast progeny are considered the main muscle stem cells, required for complete myogenic regeneration [reviewed in (Relaix and Zammit 2012; Zammit et al. 2006)]. In the last two decades, several muscle and non-muscle stem/progenitor cells with variable myogenic potencies have been isolated. For comprehensive reviews on the topic, please refer to (Negroni et al. 2016; Tedesco et al. 2010, 2017).

Muscular dystrophies are a clinically and genetically heterogeneous group of rare neuromuscular genetic disorders sharing common pathological features (Mercuri and Muntoni 2013). Despite their heterogeneity in muscle wasting distribution, disease severity, inheritance, age of onset and progression rate, they are characterised by repeated cycles of skeletal muscle degeneration/regeneration, changes in myofibre size and inflammation, which ultimately results in progressive muscle wasting. In the most severe forms, muscle weakness leads to early loss of ambulation and to a premature death by cardiorespiratory failure (Manzur and Muntoni 2009; Mercuri and Muntoni 2013). Many muscular dystrophies are caused by mutations in genes coding for proteins that belong to the dystrophin-associated glycoprotein complex (DAGC) (Ervasti and Campbell 1991). The DAGC is a multiprotein complex located at the sarcolemma and provides a strong mechanical link between intracellular cytoskeleton and the extracellular matrix; it plays a pivotal role in stabilising the sarcolemma and in maintaining myofibre integrity during muscle contraction (Emery 2002; Straub and Campbell 1997). Consequently, mutations disrupting the DAGC result in increased sarcolemma fragility and contraction-induced fibre damage, which in turn lead to repeated cycles of myofibre degeneration/regeneration and ultimately to the replacement of the skeletal muscle tissue with fibrotic and adipose tissues (Matsumura and Campbell 1994; Michalak and Opas 1997; Straub and Campbell 1997; Worton 1995). Other muscular dystrophies can be caused by mutations in ubiquitously expressed proteins that result in muscle pathologies, such as mutations of nuclear envelope components. Recently, next-generation sequencing has helped to identify new genes responsible for previously undefined muscular dystrophies (Carss et al. 2013; Hara et al. 2011; Mitsuhashi and Kang 2012).

The most common muscular dystrophies are Duchenne (DMD), Becker (BMD) and limb-girdle (LGMD). DMD is caused by mutations in the X-linked gene that codifies for dystrophin, a rod-shaped cytoplasmic protein belonging to the DAGC (Ervasti and Campbell 1991; Michalak and Opas 1997; Straub et al. 1992). DMD has an early onset and a severe disease progression. BMD is the milder allelic variant of DMD, which has a slower progression and later onset. LGMDs represent one of the most heterogeneous groups, which are further subclassified according to the genetic defect responsible for the individual forms and inheritance (Emery 2002; Mercuri and Muntoni 2013).

Although muscular dystrophies are often fatal diseases for which no cure currently exists, many therapeutic strategies are being developed and tested in basic, preclinical and clinical studies [reviewed in (Benedetti et al. 2013; Bengtsson et al. 2016; Lin and Wang 2018; Negroni et al. 2016; Pini et al. 2017; Scoto et al. 2018)].

Skeletal Muscle Pericytes

Pericyte Ontogeny

Pericytes are an heterogeneous population of contractile mural cells that surround and support blood vessels in all vascularised tissues (Hirschi and D'Amore 1996). They were thought to be exclusively associated with the microvasculature, but evidence supports their presence also on higher-order vessels (Campagnolo et al. 2010), except the lymphatic vessels (Norrmen et al. 2011). Pericytes can be defined and distinguished from other perivascular cells, such as smooth muscle cells, by a combination of criteria including anatomical location, morphology and gene/protein expression pattern (Armulik et al. 2011). Notably, pericytes can be distinguished from other endothelial-associated perivascular cells by their location embedded within the vascular basement membrane (Sims 1986).

Despite pericytes being observed and described for the first time more than a century ago (Eberth 1871; Rouget 1873; Zimmermann 1923), fundamental questions about their origin and functions remain partially unanswered. This is mainly due to the struggle in identifying a common pericyte ancestor. Numerous lineage-tracing experiments have shown that during embryogenesis, pericytes from different tissues originate from diverse sources, so that as a result a single vessel may be composed of pericytes from multiple developmental origins (Majesky 2007; Majesky et al. 2011). Recent, lineage-tracing studies have shown even more diversity, with subsets of pericytes with haematopoietic (Yamazaki et al. 2017) and macrophage (Prazeres et al. 2018) origin.

Interestingly, pericytes in the aorta appear to have multiple developmental sources (Majesky 2007; Majesky et al. 2011), adding an additional layer of complexity to the pericyte's ontogeny debate (Birbrair et al. 2017; Dias Moura Prazeres et al. 2017). This hints towards the idea that instead of having a common ancestor, pericytes share a mural precursor with the vascular smooth muscle cells (VSMCs) of the tissue in which they reside (Armulik et al. 2011; Majesky et al. 2011). This might explain the relative heterogeneity of pericytes derived from different tissues. Although undetermined in most organs, there is some evidence of this phenomenon in brain pericyte lineage tracing (Etchevers et al. 2001). In vitro studies using pluripotent stem cells (PSCs) have also alluded to a common mural progenitor (Kumar et al. 2017). Whilst the source of pericytes within many organs has been established, the developmental origin of pericytes in skeletal muscle remains elusive.

Pericyte Plasticity

Pericytes have several common functions regardless of their tissue of origin, namely blood vessel stabilisation and permeability, vascular development/maturation and regulation of blood flow (Armulik et al. 2011; Enge et al. 2002; Hall et al. 2014; Hellstrom et al. 2001; Leveen et al. 1994; Lindahl et al. 1997; Pallone and Silldorff 2001; Pallone et al. 1998; Peppiatt et al. 2006; Soriano 1994). In addition, in the last decade, many studies have identified pericytes as tissue-resident progenitors able to contribute to histogenesis and/or regeneration of multiple human tissues (Dellavalle et al. 2011; Sacchetti et al. 2016). A recent study by Evans and colleagues challenged this view, showing that Tbx18$^+$ mouse pericytes maintain their mural identity and do not generate other cell types in injured and ageing tissues, including brain, heart, fat and skeletal muscle (Guimaraes-Camboa et al. 2017). This suggests that plasticity seen in vitro or after transplantation could be an artefact of ex vivo cell culture. The discrepancy between this data and previous studies suggests that mural cells can behave as progenitors, but that this behaviour is dependent on the organ and on the developmental stage. Alternatively, it may be that a small population of pericytes with progenitor capabilities do not express Tbx18, and hence were not labelled in the Tbx18-cre strain. This could be possibly due to the heterogeneous nature of pericytes. The model of endogenous pericytes as tissue-resident progenitors might therefore need further investigation, perhaps using additional or alternative pericyte lineage-tracing tools (Cano et al. 2017).

Beside their role in supporting the microvasculature and their putative role as tissue progenitors, pericytes can also display tissue-dependent functions [reviewed in (Holm et al. 2018)]. For example, brain pericytes support the blood–brain barrier integrity (Al Ahmad et al. 2011; Armulik et al. 2010; Daneman et al. 2010; Dohgu et al. 2005; Nakagawa et al. 2007), whilst in the immune system they contribute to the regulation of lymphocyte activation (Balabanov et al. 1999; Fabry et al. 1993; Tu et al. 2011; Verbeek et al. 1995). In skeletal muscle, pericytes contribute to muscle growth, regeneration, fibrosis, fat deposition and ossification [reviewed in (Birbrair et al. 2015; Murray et al. 2017; Kostallari et al. 2015). We shall detail later in this chapter the role of pericytes in these processes in the specific context of muscular dystrophies. Generally, pericytes and their associated blood vessels run parallel to muscle fibres, where cross talk is thought to regulate nutrient uptake and postnatal myogenesis. Early studies suggest that pericyte location in capillary vessels of skeletal muscle is fibre type-specific (Gaudio et al. 1985; Levy et al. 2001) and that specific subset of pericytes within skeletal muscle do have distinct roles (Birbrair et al. 2013b). Interestingly, in contrast to the widely accepted view, there is no clear evidence that pericytes can actively alter blood flow in skeletal muscles [reviewed in (Murray et al. 2017; Sims 1986)].

Molecular Signature and Skeletal Muscle-Specific Pericyte Subpopulations

As mentioned above, despite their fundamental roles in health and disease and their ubiquitous presence in all body's tissues and organs, pericytes' identification is made difficult by their heterogeneity, which concerns not only origin and distribution but also the pattern and dynamic of molecular markers they do express (Armulik et al. 2011). In general, it can be said that (1) none of the pericyte markers are specific; (2) not all pericytes do express all the markers at once; (3) pericytes from different tissues express different markers; (4) marker expression is determined by the developmental and activation stage (Armulik et al. 2011). Although efforts are being made to characterise skeletal muscle pericytes, many putative markers overlap with other muscle cells, and there is no single all-encompassing pericyte-specific marker in skeletal muscle. For this reason and as for other tissues, skeletal muscle pericytes are often identified as much by their anatomical location as by the expression of a pool of molecular markers/proteins [reviewed in (Tedesco et al. 2017)]. However, it is worth mentioning that in sites of active angiogenesis or disorganised tissue, such as dystrophic muscle, it can be difficult to determine which cells are located within the vascular basement membrane and therefore to define the exact location of cells expressing pericyte markers. In addition, the mechanisms regulating pericyte quiescence, activation and their transition between these two states are still unknown, as most studies have focussed on homing factors or determining final fate.

Some of the most common markers used for pericytes are neural-glia antigen 2 (NG2), platelet-derived growth factor receptor β (PDGFRβ), smooth muscle α-actin (α-SMA), desmin, CD13, regulator of G protein signalling 5 (RGS5), CD146 and Nestin (Armulik et al. 2011; Birbrair et al. 2011; Tedesco et al. 2017). None of these markers are unique for pericytes. In skeletal muscle for example, several of these and other proteins are expressed in both satellite cells/myoblasts and a subset of muscle pericytes, including Pax3 (Dellavalle et al. 2007; Sacchetti et al. 2016), and Nestin (Birbrair et al. 2011; Day et al. 2007). This may be due to a shared developmental origin for these two cell types (Esner et al. 2006). A subpopulation of non-myogenic muscle pericytes also share the expression of PDGFRα with fibro/adipogenic progenitors (FAPs), a PDGFRα$^+$/CD34$^+$ and stem cell antigen-1 (Sca1)$^+$ muscle interstitial cell population able to differentiate into myofibroblasts and/or adipose cells (Joe et al. 2010; Uezumi et al. 2010).

Fate-tracing experiments in mice have revealed that a subpopulation of muscle pericytes can have myogenic fate. This includes a subpopulation expressing alkaline phosphatase (AP), which is able to fuse with developing muscle fibres and enter the satellite cell compartment, both during postnatal development and following acute/chronic muscle injury (Dellavalle et al. 2011). Whether they become *bona fide,* functional satellite cells, however, still needs to be elucidated. In addition, Birbrair et al. used Nestin-GFP/NG2-DsRed double transgenic mice to demonstrate the existence of type-2 (Nestin$^+$/NG2$^+$) and type-1 (Nestin$^-$/NG2$^+$) pericytes. Both populations express the typical pericyte markers PDGFRβ and CD146 and are associated to capillaries. However, type-2 pericytes are able to form myotubes in vitro and in vivo and enter the satellite cell compartment (Birbrair et al. 2013c), whilst type-1 are PDGFRα$^+$ and contribute to fat accumulation and fibrosis (Birbrair et al. 2013b, c). A comparison of the AP$^+$ and Nestin$^+$ type-2 populations has not been made.

In humans, a subpopulation of AP$^+$ interstitial muscle cells associated to small vessels has also been observed (Dellavalle et al. 2007). These human interstitial cells (presumed to be of pericyte origin but obtained from un-purified biopsies) were initially characterised as expressing the pericyte markers AP, desmin, PDGFRβ, vimentin, Annexin V and Integrin-β1/CD29, whilst being negative for myogenic genes Pax7, MyoD, and NCAM/CD56, endothelial marker CD31 and haematopoietic markers CD34/CD45. However, these interstitial cells did not express all pericyte markers; expression of M-cadherin/CD146, NG2 and α-SMA was variable among different preparations (Dellavalle et al. 2007). Moreover, variable expression of NCAM/CD56 and myogenic regulatory factors in this population has been observed in a subsequent publication (Meng et al. 2011), further contributing to the evidence that this is a variable population in human muscles. Nevertheless, another studies showed that CD146$^+$ subendothelial cells isolated from the postnatal human skeletal muscle microvasculature have high spontaneous myogenic potential in vitro, and they generate myotubes and myofibre in vivo (Sacchetti et al. 2016).

Further studies will be needed to address the relationships existing between these different subpopulations of skeletal muscle pericytes from both murine and human origin.

Pericytes and Satellite Cells

Within skeletal muscles, satellite cells are located beneath the basement membranes of muscle fibres and are closely connected with capillary endothelial cells. As a result, a close interaction also occurs between pericytes and satellite cells. This has led to speculation that there is cross talk between these two cell types (Christov et al. 2007; Dellavalle et al. 2011). This relationship is multifaceted. For example, the juxta-vascular position of satellite cells is thought to enable co-ordinated angio-myogenesis (Christov et al. 2007), on the other hand Kostallari et al. found that pericytes directly form a niche for satellite cells, regulating their quiescence and contributing to myogenesis through Angiopoietin 1 and insulin-like growth factor 1 (IGF-1), respectively (Kostallari et al. 2015). Whilst the authors proposed that only Nestin^{+} type-2 pericytes were involved in these processes, future lineage-tracing experiments and selective ablation of type-1 or type-2 subtypes are required to give us a clear answer on this matter.

The relationship between pericytes and satellite cells/myoblasts might even be more complex than the one just described above. Several years ago, Cossu and Bianco proposed that during development, cells associated to the growing vessels might be recruited to adopt the local fate of the specific tissue they were invading. In the case of the skeletal muscle, cells associated to the blood vessels that enter the muscle anlagen might be recruited to adopt a myogenic fate and contribute to its histogenesis (Bianco and Cossu 1999). This concept was partially confirmed in the finding that the embryonic dorsal aorta contains skeletal myogenic cells, named mesoangioblasts, that co-express endothelial and myogenic markers and can contribute to muscle regeneration (De Angelis et al. 1999; Minasi et al. 2002). In vitro co-cultures of embryonic dorsal aortas and murine myotubes demonstrated that Noggin secreted from newly formed muscle fibres recruits NG2^{+} dorsal aorta progenitors and promotes their conversion to a myogenic fate. Conversely, myogenesis is inhibited by bone morphogenetic factor 2 (BMP2) expressed by the aorta (Ugarte et al. 2012). This data indicates that skeletal muscle and blood vessels compete to recruit mesodermal progenitor cells to a myogenic or to a perivascular fate during foetal muscle development and that the final decision of which cell fate to adopt might be due to the balance existing between Noggin and BMP2 expression. These data also suggest that a fate switch might also occur in the other direction, with skeletal myoblasts being recruited to a pericyte fate. In this direction, Cappellari et al. showed that exposure of both embryonic and foetal skeletal myoblasts to Notch Delta ligand 4 (Dll4), expressed by the developing endothelium (Kume 2012), and PDGF-BB, which recruits pericytes from the surrounding mesenchyme (Hellstrom et al. 1999), downregulates myogenic genes, upregulates pericyte markers and recruits myoblasts to a perivascular position when co-cultured with endothelial cells. Moreover, they showed that myoblasts also occasionally adopt a perivascular position in vivo, ruling out that the direct conversion of skeletal myoblasts into pericytes is simply an artefact of ex vivo cell manipulations (Cappellari et al. 2013). Altogether, this data suggests that the endothelium, via Dll4 and

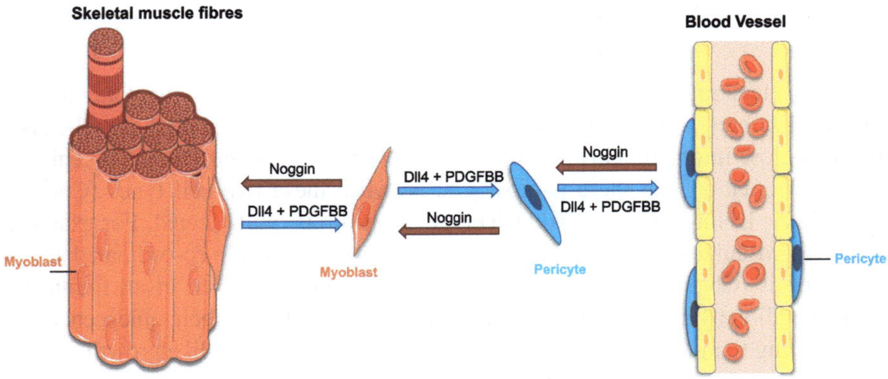

Fig. 15.1 A schematic representation of the hypothetical model explaining lineage promiscuity between muscle pericytes and satellite cell-derived myoblasts during development. The figure has been generated using Servier Medical Art

PDGF-BB expression, might induce a fate switch in adjacent skeletal myoblasts. Cossu and Cappellari postulated that the reason for lineage promiscuity between muscle cells and perivascular cells might be explained by developmental timing and the specific need of the skeletal muscle tissue during histogenesis: for the muscle to grow, it recruits not only myoblasts but also unorthodox mesodermal cells that undergo myogenesis once exposed to muscle-specification molecules; once muscle growth is complete, hypoxia is triggered with consequent vascular endothelial growth factor release and activation of angiogenesis, with developing blood vessels in the muscles recruiting supporting perivascular cells from the surrounding mesoderm (Fig. 15.1) (Cappellari et al. 2013; Cappellari and Cossu 2013). Recent work in our laboratory has demonstrated that this mechanism is also conserved in adult murine and human satellite cell-derived myoblasts, and that it can be exploited to enhance migration of myoblasts when transplanted (Gerli et al. 2019).

Pericytes Contribution to Muscle Regeneration

Pericytes as Stem/Progenitor Cells for Muscular Dystrophies

Stem cell transplantation therapies for muscular dystrophies have long been touted as a method to improve clinical features. Satellite cell-derived myoblasts were initially considered the ideal candidate cell population for the cell therapy of muscular dystrophies (Partridge et al. 1989). However, successive clinical studies revealed that although some level of dystrophin was produced, no efficacy was achieved in patients with DMD, one of the most severe and common forms of muscular dystrophy [reviewed in (Negroni et al. 2016; Partridge 2000; Tedesco et al. 2010)]. Whilst researchers tried to identify the possible culprit(s) for this result (Fan et al. 1996;

Guerette et al. 1997; Huard et al. 1992; Skuk and Tremblay 2011) and find possible therapeutic solutions (Arpke et al. 2013; Boldrin et al. 2012; Cerletti et al. 2008; Collins et al. 2005; Gilbert et al. 2010; Montarras et al. 2005; Morales et al. 2013; Palmieri et al. 2010; Rocheteau et al. 2012; Sacco et al. 2008; Skuk et al. 2006, 2007, 2004; Smythe et al. 2000; Tanaka et al. 2009), the search was started for alternative cell types that could be effective in cell therapy protocols for muscle. Whilst several candidate stem/progenitor populations have been identified as contributing to skeletal muscle regeneration [reviewed in (Loperfido et al. 2015; Negroni et al. 2016; Tedesco et al. 2010, 2017)], pericyte-derived cells seem to hold a preferential place.

Data obtained from different laboratories in the past years support an important role for myogenic pericytes in skeletal muscle regeneration. As mentioned earlier in this chapter, fate-tracing of AP^+ murine skeletal muscle pericytes reveals how this subset of pericytes contributes to the formation of new muscle fibres in a LGMD 2D mouse model (Dellavalle et al. 2011). In human skeletal muscle, AP^+ pericytes are increased in muscle biopsies of some dystrophic patients compared to healthy controls and neuropathic patients (myopathic 9.4% vs. and controls 4.7% vs. neuropathic 5.7%) (Diaz-Manera et al. 2012). These results are supported by another study showing an increase in the population of $NG2^+$ pericytes in acute muscular injury (Valero et al. 2012). Conversely, we have reported a significant decrease (of approximately 55%) in the numbers and myogenic capacity of AP^+ muscle pericytes both in mice and patients with LGMD2D (Tedesco et al. 2012). This apparently contrasting data could be explained by the different stage of disease progression of the biopsies/samples used in these two studies. We propose that in dystrophic muscles, during the first phase of muscle degeneration/regeneration AP^+ skeletal muscle pericytes are transiently amplified to sustain the continuous need for new muscle fibres in cooperation with satellite cells and other muscle stem cells. Over time, cycles of muscle degeneration/regeneration lead to an exhaustion of the pool of AP^+ pericytes.

Myogenic pericyte transplantation has also been tested in preclinical models of muscular dystrophies, where they have the advantageous characteristic of being deliverable through the arterial circulation. These reports show active contribution of pericytes to muscle regeneration in dystrophic animal models (Berry et al. 2007; Bonfanti et al. 2015; Dellavalle et al. 2007; Diaz-Manera et al. 2010; Domi et al. 2015; Galvez et al. 2006; Iyer et al. 2018; Minasi et al. 2002; Morosetti et al. 2011; Pessina et al. 2012; Quattrocelli et al. 2014; Sampaolesi et al. 2006, 2003; Sciorati et al. 2006; Tedesco et al. 2011).

As with other myogenic progenitor populations, one of the major hurdles to improve the feasibility of pericytes as a transplantation therapy for muscular dystrophy is the low level of engraftment. Recently, a first-in-human phase I/II clinical trial based upon intra-arterial transplantation of HLA-matched allogeneic pericyte-derived mesoangioblasts in 5 DMD boys showed that whilst relatively safe, there was limited dystrophin production (1 of 5 biopsies), probably due to low level of cell engraftment (Cossu et al. 2015). Optimisation of this methodology is therefore required for future therapeutic use.

Lastly, Zazt and colleagues investigated the effect of repeated intraperitoneal injections of adipose human pericytes on the lifespan and motor function of a severe DMD mouse model. They reported that adipose tissue-derived pericytes led to an increased lifespan of one month, possibly mediated by immune modulation rather than a regenerative ability (Valadares et al. 2014). In view of the unusual delivery route (intraperitoneal), the lack of histological evidence of engraftment or amelioration of tissue pathology and as none of the functional tests revealed differences between the groups, additional evidence would be required to assess the feasibility and clinical relevance of this strategy.

Limitations to Cell Therapy and Possible Solutions

Despite numerous preclinical and clinical studies on cell transplantation, muscular dystrophies still remain incurable, and there are still several challenges to be addressed before becoming routinely used in a clinical setting, including engraftment efficacy, transplantation route and modulation of the immune response [reviewed in (Maffioletti et al. 2014; Negroni et al. 2016)]. The main challenge is due to the fact that skeletal muscle is the most abundant human tissue, covering 30–38% of total body mass (Janssen et al. 2000). Transplanting stem cells has been shown to result in clinical improvement when specific muscles are affected, such as the recent trial for oculopharyngeal muscular dystrophy using intramuscular injections of autologous myoblasts (Perie et al. 2014). However, replacing large volumes of dystrophic muscle affected in widespread muscular dystrophies (such as DMD) would require the successful engraftment of billions of myogenic progenitors. Indeed, as with other myogenic progenitor populations, there are two major hurdles to be overcome in order to improve the feasibility of pericytes and pericyte-derived cells as a transplantation therapy for muscular dystrophies: (1) the low levels of cell engraftment to the dystrophic muscle and (2) the limited cell expansion potential in vitro.

In this direction, several strategies have been developed to improve homing and engraftment of pericyte-derived cells. These include making blood vessels more accessible for cell extravasation (Giannotta et al. 2014), treating transplanted cells with homing factors (Quattrocelli et al. 2014) and modulating the immune response (Maffioletti et al. 2014; Noviello et al. 2014). Exposure to cytokines and integrins can also improve pericyte-derived mesoangioblast engraftment (Galvez et al. 2006; Palumbo et al. 2004; Tagliafico et al. 2004). However, it is important to assess the possible side effects of transplanting engineered cells, as whilst they can promote extravasation and homing, the expression of different surface molecules could modulate the immune response following transplantation, including deleteriously increasing donor cell clearance. The interaction between muscle, pericytes and immune cells via adhesion molecules is well documented [reviewed in (Maffioletti et al. 2014; Noviello et al. 2014)]. Of interest, intracellular adhesion molecule 1 (ICAM-1/CD54) expression is increased in inflamed endothelial cells and muscle fibres (Bartoccioni et al. 1994; Tews and Goebel 1995). However, leukocyte function-associated antigen 1 (LFA-1) expressed on T cells binds to ICAM-1, resulting in cytotoxic T-cell infiltration (Bartoccioni et al. 1994). Also, expression of VCAM-1 and its ligand VLA-4 has been observed in muscle capillaries and infiltrating cells of patients with inflammatory myopathies (Tews and Goebel 1995). Interestingly, VCAM-1 expression is associated with increased engraftment of $CD133^+$ cells, another class of myogenic vessel-associated cells, which have been transplanted intra-arterially into dystrophic mice (Gavina et al. 2006). In this paper, Gavina and colleagues found that VCAM-1 expression in muscle capillaries increased after exercise and improved engraftment, whilst conversely, blocking VCAM-1 expression significantly reduced engraftment (Gavina et al. 2006). In summary, careful assessment of the transplant population and donor muscle tissue should be performed to maximise engraftment.

Another important factor for optimal muscle cell therapy is the ability of myogenic cells to proliferate in vitro and produce large numbers of transplantable progenitors. This point is of particular importance as in muscular dystrophies the myogenic cells (including pericytes) are exhausted or defective (Blau et al. 1983; Cassano et al. 2011; Kudryashova et al. 2012; Sacco et al. 2010; Tedesco et al. 2012). To overcome the limitation of expansion potential of biopsy-derived patient-derived induced pluripotent stem cells (iPSCs) can be obtained and differentiated towards the myogenic lineage. One such protocol developed by our group is to produce iPSC-derived inducible myogenic cells similar to pericyte-derived mesoangioblasts, which have an unlimited proliferative potential and could then efficiently be induced to skeletal myogenesis with a short expression of the myogenesis regulator MyoD (Gerli et al. 2014; Maffioletti et al. 2015; Tedesco et al. 2012).

Another strategy to extend the proliferative potential of pericyte-derived mesoangioblasts is to provide them with an indefinite lifespan via expression of immortalising genes. Our group has recently shown that reversibly immortalising lentiviral vectors expressing the catalytic subunit of human telomerase hTERT and the polycomb gene Bmi-1 is safe and efficacious at extending the proliferative capacity of human DMD pericyte-derived mesoangioblasts, enabling them to have a human artificial chromosome containing the whole dystrophin locus (DYS-HAC) transferred (Benedetti et al. 2018). After DYS-HAC transfer, genetically corrected DMD pericyte-derived clones were expanded to reach a number of cells potentially sufficient to treat a paediatric DMD patient [in the range of 10^9 cells; (Benedetti et al. 2018; Cossu et al. 2015)].

In the quest for an ideal cell type for muscle cell therapy, our group also explored a different approach by taking advantage of the findings that embryonic and foetal myoblasts could be converted to the pericyte fate following activation of Notch and PDGF pathways via Dll4 and PDGF-BB (Cappellari et al. 2013; Cappellari and Cossu 2013). As mentioned earlier in this chapter, adult murine and human satellite cell-derived myoblasts exposed to Dll4 and PDGF-BB also acquired perivascular cell features, including transendothelial migration ability, whilst maintaining myogenic capacity (Gerli et al. 2019). We propose that this strategy could generate a hybrid pericyte-myoblast cell retaining the two most advantageous and peculiar characteristics of both cell types: the ability to generate muscle with high efficacy (myoblast) alongside transendothelial migration capacity (pericyte-derived cells).

In conclusion, there are several promising strategies in development using pericyte-like cells for cell therapy of muscular dystrophies. Nevertheless, it is crucial that both the transplanted cell and the host environment are considered in order to improve engraftment efficacy. Therefore, it is likely that future clinical studies will focus on combined therapies, where stem cell transplantation is merged with other therapeutic interventions, such as administration of anti-fibrotic and pro-angiogenic drugs, which have been shown to improve pathology in mouse models of muscular dystrophy (Cordova et al. 2018; Gargioli et al. 2008).

Non-Myogenic Role of Skeletal Muscle Pericytes

Pericyte Contribution to Fat Accumulation

Intramuscular deposition and accumulation of adipose tissue is a typical hallmark of disease progression and severity in muscular dystrophies, especially in DMD (Lukjanenko et al. 2013; Mankodi et al. 2016; Wren et al. 2008). Pericytes from different tissues, including skeletal muscle, have shown adipogenic potential when cultured in vitro (Crisan et al. 2008a; Farrington-Rock et al. 2004; Minasi et al. 2002). In skeletal muscle, a subset of quiescent cells expressing the adipogenic progenitor marker PDGFRα is closely associated to the vasculature and located in the interstitial space between muscle fibres. Following muscle injury, these cells, later called fibro/adipogenic progenitors (FAPs), exit quiescence, start proliferating and contribute to ectopic fat accumulation in skeletal muscle (Joe et al. 2010; Rodeheffer 2010; Uezumi et al. 2010). Other groups have demonstrated that PDGFRα$^+$ type-1 but not PDGFRα$^-$ type-2 pericytes have adipogenic potential in vitro (Birbrair et al. 2013a; Gautam et al. 2017). Moreover, cultured type-1 pericytes generated ectopic white fat when delivered intramuscularly in a mouse model of fatty degeneration (Birbrair et al. 2013a). Future lineage-tracing studies might clarify whether type-1 pericytes do indeed contribute to fat accumulation in skeletal muscle in situ. In addition, a recent lineage-tracing study by Strickland and colleagues demonstrated that PDGFRβ$^+$ skeletal muscle pericytes were able to differentiate into perilipin$^+$ adipocytes in a congenital muscular dystrophy model (Yao et al. 2016).

Pericyte Contribution to Chondrogenesis

In addition to committing to a myogenic fate, pericytes have been shown to undergo chondrogenic and osteogenic differentiation in vitro (Crisan et al. 2008b; Farrington-Rock et al. 2004; James et al. 2012; Levy et al. 2001; Zhang et al. 2011). Whether pericytes contribute to skeletal muscle ossification in vivo, however, remains undetermined. Interestingly, ectopic calcification has been reported in animal models of DMD including the mdx mouse (Geissinger et al. 1990) and dog (Nguyen et al. 2002). Additionally, it has recently been shown that AP^+ pericytes are reduced in immune-deficient scgb/Rag2/γc-null mice (a model of LGMD2E), whilst calcification of skeletal muscles is increased (Giovannelli et al. 2018). Unsurprisingly, this has led to the implication that pericytes are involved in the ectopic calcification of blood vessels in skeletal muscle, especially in the context of ongoing angiogenesis [reviewed in (Collett and Canfield 2005)]. Of note, AP^+ mononuclear interstitial cells from adult human skeletal muscle have been shown to express the osteogenic marker osteocalcin when cultured in vitro (Levy et al. 2001), and in fibrodysplasia ossificans progressiva (FOP) progressive ossification of skeletal muscle has been shown to be caused by mesenchymal-like stromal cells expressing smooth muscle markers (Hegyi et al. 2003). Whilst the authors postulate that these cells are pericytes, more recent data suggests it may be due to the $Tie2^+$ FAP population (Lees-Shepard et al. 2018). Again, this points to the requirement for subpopulations of FAPs and pericytes to be clearly distinguished.

Pericytes and Fibrosis

Increased fibrosis is a typical feature of aged and dystrophic muscles, which ultimately results in muscle weakness, atrophy and reduction of its regenerative potential (Kragstrup et al. 2011; Mann et al. 2011; Ryall et al. 2008; Thompson 2009; Walston 2012). A major contributor to fibrosis is the myofibroblast (Duffield et al. 2013; Humphreys et al. 2010; Lin et al. 2008; Quan et al. 2006; Willis et al. 2006; Wynn 2008; Zeisberg et al. 2007). Myofibroblasts are responsible for the production and deposition of collagenous extracellular matrix, with consequent reduction of muscle fibre contractility, disruption of the muscle structure and eventually skeletal muscle dysfunction. Several putative myofibroblast progenitor populations have been associated to muscle fibrosis, including FAPs and cells expressing PDGFRα (Uezumi et al. 2010) and ADAM12 (Dulauroy et al. 2012). As some of these markers are also expressed by pericytes, it has been hypothesised that pericytes may be a source of myofibroblasts during skeletal muscle fibrosis. Indeed, Birbrair et al. showed that skeletal muscle $PDGFR\alpha^+$ pericytes are fibrogenic in vitro when cultured in the presence of transforming growth factor β (TGFβ), whilst in vivo, they produce collagen, responsible for increasing skeletal muscle fibrosis in old mice (Birbrair et al. 2014, 2013c).

In parallel, using a complex triple transgenic mouse that expressed tetracycline under ADAM12 locus, Cre recombinase under control of the tetracycline transactivator and the conditional reporter Rosa26floxSTOP-YFP, Dulauroy et al. showed the existence of a transient subpopulation of $ADAM12^+$ interstitial cells that become active after muscle injury. With this approach, they revealed that the large majority of collagen-producing myofibroblasts were generated starting from $ADAM12^+$ cells, which are located in a perivascular position and are positive for PDGFRβ.

Moreover, ablation of ADAM12^{+} cells reduced the number of profibrotic cells and collagen accumulation (Dulauroy et al. 2012). This data corroborates the hypothesis that in skeletal muscle ADAM12 identifies a myofibroblast progenitor with pericyte characteristics.

Pericyte Role in the Vascular Compartment

Muscle ischemia has been observed in biopsies from patients with DMD for many decades (Engel 1967) and clinical symptoms were once hypothesised to be caused by local infarctions. Whilst there have not been consistent reports in changes to blood flow or microvascular architecture observed in DMD patients [reviewed in (Thomas 2013)], it has been shown that dystrophin-deficient muscle fibres are more susceptible to muscle ischemia though a neuronal nitric oxide synthase μ (nNOSμ)-specific mechanism. Localised muscle ischemia results in increased exercise-induced fatigue and microvessel constriction, elevating clinical symptoms (Kobayashi et al. 2008). This mechanism is not specific to DMD; changes to sarcolemmal nNOS expression have been observed in biopsies from other several muscular dystrophies including limb-girdle and congenital muscular dystrophy (Kobayashi et al. 2008). Whether or not pericytes actively contribute to the vascular pathology observed in muscular dystrophies, it has been shown that improving the reduced blood flow improves clinical symptoms and cell therapy in preclinical models (Brunelli et al. 2007; Gargioli et al. 2008). Of note, treatment of dystrophic mice with nitric oxide-releasing drugs improves the efficacy of mesoangioblast transplantation (Brunelli et al. 2007).

Derivation of Pericytes from Pluripotent Stem Cells

Pluripotent stem cell (PSC)-derivatives are of great importance for studying development and organogenesis, whilst also being promising candidates for cell transplantation studies, due to their unlimited expansion potential. Differentiating PSCs into pericyte-like cells enables the possibility to study developmental relationships between mural cells of different developmental origins. For example, Kumar et al. found using clonal analyses that mesodermal-derived progenitors could make mesenchymal stromal cells, VSMCs and pericytes (Kumar et al. 2017). Additionally, this protocol was used to derive arteriolar and capillary subtypes of pericytes through the modulation of growth factors. Protocols to differentiate vascular cells from different embryonic lineages (neural crest, lateral plate mesoderm and paraxial mesoderm) have also been developed in order to study different forms of vascular development (Cheung et al. 2012; Chin et al. 2018; Cochrane et al. 2018; Orlova et al. 2014). In muscle pathology, PSC-derived pericyte-like cells have shown pathological improvement in preclinical models of muscular dystrophy (Tedesco et al.

2012) and ischemia (Dar et al. 2012). Determining whether PSC-derived cells are truly functional pericytes when transplanted in vivo is difficult. However, using lineage determinants to produce a pericyte-like mesenchymal (or neural ectodermal) progenitor, which when transplanted differentiates into a regeneration-supporting cell, could be a more feasible option.

Concluding Remarks

Since their initial description, there has been important progress in understanding pericyte biology and function; however, their exact role and involvement in the pathogenic process of muscle degeneration and regeneration is still in need of a definitive model. Nonetheless, the clinical relevance of pericyte is now more important than ever before, both as a target for possible therapeutic intervention (e.g. reduction of fibrosis) and as advanced therapy medicinal products. We foresee that in the upcoming decade the ontogeny and characteristics of this elusive cell type will become clearer, setting the foundation for their organotypic derivation from human PSCs and use in next-generation experimental therapies for muscle diseases.

Acknowledgements We thank Giulio Cossu for the critical reading of this manuscript. This research was supported by the National Institute for Health Research (NIHR) Great Ormond Street Hospital Biomedical Research Centre. F.S.T. is also funded by NIHR (Clinical Lectureship in Paediatrics). The views expressed are those of the authors and not necessarily those of the National Health Service (NHS), the NIHR or the Department of Health. Work in the Benedetti laboratory is supported by Great Ormond Street Children's Charity, Sparks and Krabbe UK. Work in the Tedesco laboratory is funded by the European Research Council (7591108—HISTOID), Muscular Dystrophy UK, the UK MRC and BBSRC and the AFM-Telethon. F.S.T. is also grateful to previous funding from the European Union's 7th Framework Programme for research, technological development and demonstration under grant agreement no. 602423 (PluriMes), Fundació La Marató de TV3, IMI joint undertaking n° 115582 EBiSC (EU FP7 and EFPIA companies), Duchenne Parent Project Onlus, Takeda New Frontier Science and the NIHR (Academic Clinical Fellowship in Paediatrics). L.A.M. is supported by the Human Frontiers Science Program.

References

Al Ahmad, A., Taboada, C. B., Gassmann, M., & Ogunshola, O. O. (2011). Astrocytes and pericytes differentially modulate blood-brain barrier characteristics during development and hypoxic insult. *Journal of Cerebral Blood Flow and Metabolism, 31*, 693–705.

Armulik, A., Genove, G., & Betsholtz, C. (2011). Pericytes: developmental, physiological, and pathological perspectives, problems, and promises. *Developmental Cell, 21*, 193–215.

Armulik, A., Genove, G., Mae, M., Nisancioglu, M. H., Wallgard, E., Niaudet, C., He, L., Norlin, J., Lindblom, P., Strittmatter, K., et al. (2010). Pericytes regulate the blood-brain barrier. *Nature, 468*, 557–561.

Arpke, R. W., Darabi, R., Mader, T. L., Zhang, Y., Toyama, A., Lonetree, C. L., Nash, N., Lowe, D. A., Perlingeiro, R. C., & Kyba, M. (2013). A new immuno-, dystrophin-deficient model, the NSG-mdx(4Cv) mouse, provides evidence for functional improvement following allogeneic satellite cell transplantation. *Stem Cells, 31*, 1611–1620.

Balabanov, R., Beaumont, T., & Dore-Duffy, P. (1999). Role of central nervous system microvascular pericytes in activation of antigen-primed splenic T-lymphocytes. *Journal of Neuroscience Research, 55*, 578–587.

Bartoccioni, E., Gallucci, S., Scuderi, F., Ricci, E., Servidei, S., Broccolini, A., & Tonali, P. (1994). MHC class I, MHC class II and intercellular adhesion molecule-1 (ICAM-1) expression in inflammatory myopathies. *Clinical and Experimental Immunology, 95*, 166–172.

Benedetti, S., Hoshiya, H., & Tedesco, F. S. (2013). Repair or replace? Exploiting novel gene and cell therapy strategies for muscular dystrophies. *The FEBS Journal, 280*, 4263–4280.

Benedetti, S., Uno, N., Hoshiya, H., Ragazzi, M., Ferrari, G., Kazuki, Y., Moyle, L. A., Tonlorenzi, R., Lombardo, A., Chaouch, S., et al. (2018). Reversible immortalisation enables genetic correction of human muscle progenitors and engineering of next-generation human artificial chromosomes for Duchenne muscular dystrophy. *EMBO Molecular Medicine, 10*, 254–275.

Bengtsson, N. E., Seto, J. T., Hall, J. K., Chamberlain, J. S., & Odom, G. L. (2016). Progress and prospects of gene therapy clinical trials for the muscular dystrophies. *Human Molecular Genetics, 25*, R9–R17.

Bentzinger, C. F., Wang, Y. X., & Rudnicki, M. A. (2012). Building muscle: Molecular regulation of myogenesis. *Cold Spring Harbor Perspectives in Biology, 4*.

Berry, S. E., Liu, J., Chaney, E. J., & Kaufman, S. J. (2007). Multipotential mesoangioblast stem cell therapy in the mdx/utrn−/− mouse model for Duchenne muscular dystrophy. *Regenerative Medicine, 2*, 275–288.

Bianco, P., & Cossu, G. (1999). Uno, nessuno e centomila: Searching for the identity of mesodermal progenitors. *Experimental Cell Research, 251*, 257–263.

Birbrair, A., Borges, I. D. T., Gilson Sena, I. F., Almeida, G. G., da Silva Meirelles, L., Goncalves, R., Mintz, A., & Delbono, O. (2017). How plastic are pericytes? *Stem Cells and Development, 26*, 1013–1019.

Birbrair, A., Wang, Z. M., Messi, M. L., Enikolopov, G. N., & Delbono, O. (2011). Nestin-GFP transgene reveals neural precursor cells in adult skeletal muscle. *PLoS One, 6*, e16816.

Birbrair, A., Zhang, T., Files, D. C., Mannava, S., Smith, T., Wang, Z. M., Messi, M. L., Mintz, A., & Delbono, O. (2014). Type-1 pericytes accumulate after tissue injury and produce collagen in an organ-dependent manner. *Stem Cell Research & Therapy, 5*, 122.

Birbrair, A., Zhang, T., Wang, Z. M., Messi, M. L., Enikolopov, G. N., Mintz, A., & Delbono, O. (2013a). Role of pericytes in skeletal muscle regeneration and fat accumulation. *Stem Cells and Development, 22*, 2298–2314.

Birbrair, A., Zhang, T., Wang, Z. M., Messi, M. L., Enikolopov, G. N., Mintz, A., & Delbono, O. (2013b). Skeletal muscle pericyte subtypes differ in their differentiation potential. *Stem Cell Research, 10*, 67–84.

Birbrair, A., Zhang, T., Wang, Z. M., Messi, M. L., Mintz, A., & Delbono, O. (2013c). Type-1 pericytes participate in fibrous tissue deposition in aged skeletal muscle. *American Journal of Physiology. Cell Physiology, 305*, C1098–C1113.

Birbrair, A., Zhang, T., Wang, Z. M., Messi, M. L., Mintz, A., & Delbono, O. (2015). Pericytes at the intersection between tissue regeneration and pathology. *Clinical Science (London, England), 128*, 81–93.

Blau, H. M., Webster, C., & Pavlath, G. K. (1983). Defective myoblasts identified in Duchenne muscular dystrophy. *Proceedings of the National Academy of Sciences of the United States of America, 80*, 4856–4860.

Boldrin, L., Muntoni, F., & Morgan, J. E. (2010). Are human and mouse satellite cells really the same? *The Journal of Histochemistry and Cytochemistry, 58*, 941–955.

Boldrin, L., Neal, A., Zammit, P. S., Muntoni, F., & Morgan, J. E. (2012). Donor satellite cell engraftment is significantly augmented when the host niche is preserved and endogenous satellite cells are incapacitated. *Stem Cells, 30*, 1971–1984.
Bonfanti, C., Rossi, G., Tedesco, F. S., Giannotta, M., Benedetti, S., Tonlorenzi, R., Antonini, S., Marazzi, G., Dejana, E., Sassoon, D., et al. (2015). PW1/Peg3 expression regulates key properties that determine mesoangioblast stem cell competence. *Nature Communications, 6*, 6364.
Brunelli, S., Sciorati, C., D'Antona, G., Innocenzi, A., Covarello, D., Galvez, B. G., Perrotta, C., Monopoli, A., Sanvito, F., Bottinelli, R., et al. (2007). Nitric oxide release combined with nonsteroidal antiinflammatory activity prevents muscular dystrophy pathology and enhances stem cell therapy. *Proceedings of the National Academy of Sciences of the United States of America, 104*, 264–269.
Buckingham, M. (2006). Myogenic progenitor cells and skeletal myogenesis in vertebrates. *Current Opinion in Genetics & Development, 16*, 525–532.
Campagnolo, P., Cesselli, D., Al Haj Zen, A., Beltrami, A. P., Krankel, N., Katare, R., Angelini, G., Emanueli, C., & Madeddu, P. (2010). Human adult vena saphena contains perivascular progenitor cells endowed with clonogenic and proangiogenic potential. *Circulation, 121*, 1735–1745.
Cano, E., Gebala, V., & Gerhardt, H. (2017). Pericytes or mesenchymal stem cells: Is that the question? *Cell Stem Cell, 20*, 296–297.
Cappellari, O., Benedetti, S., Innocenzi, A., Tedesco, F. S., Moreno-Fortuny, A., Ugarte, G., Lampugnani, M. G., Messina, G., & Cossu, G. (2013). Dll4 and PDGF-BB convert committed skeletal myoblasts to pericytes without erasing their myogenic memory. *Developmental Cell, 24*, 586–599.
Cappellari, O., & Cossu, G. (2013). Pericytes in development and pathology of skeletal muscle. *Circulation Research, 113*, 341–347.
Carss, K. J., Stevens, E., Foley, A. R., Cirak, S., Riemersma, M., Torelli, S., Hoischen, A., Willer, T., van Scherpenzeel, M., Moore, S. A., et al. (2013). Mutations in GDP-mannose pyrophosphorylase B cause congenital and limb-girdle muscular dystrophies associated with hypoglycosylation of alpha-dystroglycan. *American Journal of Human Genetics, 93*, 29–41.
Cassano, M., Dellavalle, A., Tedesco, F. S., Quattrocelli, M., Crippa, S., Ronzoni, F., Salvade, A., Berardi, E., Torrente, Y., Cossu, G., et al. (2011). Alpha sarcoglycan is required for FGF-dependent myogenic progenitor cell proliferation in vitro and in vivo. *Development, 138*, 4523–4533.
Cerletti, M., Jurga, S., Witczak, C. A., Hirshman, M. F., Shadrach, J. L., Goodyear, L. J., & Wagers, A. J. (2008). Highly efficient, functional engraftment of skeletal muscle stem cells in dystrophic muscles. *Cell, 134*, 37–47.
Cheung, C., Bernardo, A. S., Trotter, M. W., Pedersen, R. A., & Sinha, S. (2012). Generation of human vascular smooth muscle subtypes provides insight into embryological origin-dependent disease susceptibility. *Nature Biotechnology, 30*, 165–173.
Chin, C. J., Li, S., Corselli, M., Casero, D., Zhu, Y., He, C. B., Hardy, R., Peault, B., & Crooks, G. M. (2018). Transcriptionally and functionally distinct mesenchymal subpopulations are generated from human pluripotent stem cells. *Stem Cell Reports, 10*, 436–446.
Christov, C., Chretien, F., Abou-Khalil, R., Bassez, G., Vallet, G., Authier, F. J., Bassaglia, Y., Shinin, V., Tajbakhsh, S., Chazaud, B., et al. (2007). Muscle satellite cells and endothelial cells: Close neighbors and privileged partners. *Molecular Biology of the Cell, 18*, 1397–1409.
Cochrane, A., Albers, H. J., Passier, R., Mummery, C. L., van den Berg, A., Orlova, V. V., & van der Meer, A. D. (2018). Advanced in vitro models of vascular biology: Human induced pluripotent stem cells and organ-on-chip technology. *Advanced Drug Delivery Reviews*.
Collett, G. D., & Canfield, A. E. (2005). Angiogenesis and pericytes in the initiation of ectopic calcification. *Circulation Research, 96*, 930–938.
Collins, C. A., Olsen, I., Zammit, P. S., Heslop, L., Petrie, A., Partridge, T. A., & Morgan, J. E. (2005). Stem cell function, self-renewal, and behavioral heterogeneity of cells from the adult muscle satellite cell niche. *Cell, 122*, 289–301.

Comai, G., & Tajbakhsh, S. (2014). Molecular and cellular regulation of skeletal myogenesis. *Current Topics in Developmental Biology, 110*, 1–73.

Cordova, G., Negroni, E., Cabello-Verrugio, C., Mouly, V., & Trollet, C. (2018). Combined therapies for Duchenne muscular dystrophy to optimize treatment efficacy. *Frontiers in Genetics, 9*, 114.

Cossu, G., Previtali, S. C., Napolitano, S., Cicalese, M. P., Tedesco, F. S., Nicastro, F., Noviello, M., Roostalu, U., Natali Sora, M. G., Scarlato, M., et al. (2015). Intra-arterial transplantation of HLA-matched donor mesoangioblasts in Duchenne muscular dystrophy. *EMBO Molecular Medicine, 7*, 1513–1528.

Crisan, M., Deasy, B., Gavina, M., Zheng, B., Huard, J., Lazzari, L., & Peault, B. (2008a). Purification and long-term culture of multipotent progenitor cells affiliated with the walls of human blood vessels: Myoendothelial cells and pericytes. *Methods in Cell Biology, 86*, 295–309.

Crisan, M., Yap, S., Casteilla, L., Chen, C. W., Corselli, M., Park, T. S., Andriolo, G., Sun, B., Zheng, B., Zhang, L., et al. (2008b). A perivascular origin for mesenchymal stem cells in multiple human organs. *Cell Stem Cell, 3*, 301–313.

Daneman, R., Zhou, L., Kebede, A. A., & Barres, B. A. (2010). Pericytes are required for blood-brain barrier integrity during embryogenesis. *Nature, 468*, 562–566.

Dar, A., Domev, H., Ben-Yosef, O., Tzukerman, M., Zeevi-Levin, N., Novak, A., Germanguz, I., Amit, M., & Itskovitz-Eldor, J. (2012). Multipotent vasculogenic pericytes from human pluripotent stem cells promote recovery of murine ischemic limb. *Circulation, 125*, 87–99.

Day, K., Shefer, G., Richardson, J. B., Enikolopov, G., & Yablonka-Reuveni, Z. (2007). Nestin-GFP reporter expression defines the quiescent state of skeletal muscle satellite cells. *Developmental Biology, 304*, 246–259.

De Angelis, L., Berghella, L., Coletta, M., Lattanzi, L., Zanchi, M., Cusella-De Angelis, M. G., Ponzetto, C., & Cossu, G. (1999). Skeletal myogenic progenitors originating from embryonic dorsal aorta coexpress endothelial and myogenic markers and contribute to postnatal muscle growth and regeneration. *The Journal of Cell Biology, 147*, 869–878.

Dellavalle, A., Maroli, G., Covarello, D., Azzoni, E., Innocenzi, A., Perani, L., Antonini, S., Sambasivan, R., Brunelli, S., Tajbakhsh, S., et al. (2011). Pericytes resident in postnatal skeletal muscle differentiate into muscle fibres and generate satellite cells. *Nature Communications, 2*, 499.

Dellavalle, A., Sampaolesi, M., Tonlorenzi, R., Tagliafico, E., Sacchetti, B., Perani, L., Innocenzi, A., Galvez, B. G., Messina, G., Morosetti, R., et al. (2007). Pericytes of human skeletal muscle are myogenic precursors distinct from satellite cells. *Nature Cell Biology, 9*, 255–267.

Dias Moura Prazeres, P. H., Sena, I. F. G., Borges, I. D. T., de Azevedo, P. O., Andreotti, J. P., de Paiva, A. E., de Almeida, V. M., de Paula Guerra, D. A., Pinheiro Dos Santos, G. S., Mintz, A., et al. (2017). Pericytes are heterogeneous in their origin within the same tissue. *Developmental Biology, 427*, 6–11.

Diaz-Manera, J., Gallardo, E., de Luna, N., Navas, M., Soria, L., Garibaldi, M., Rojas-Garcia, R., Tonlorenzi, R., Cossu, G., & Illa, I. (2012). The increase of pericyte population in human neuromuscular disorders supports their role in muscle regeneration in vivo. *The Journal of Pathology, 228*, 544–553.

Diaz-Manera, J., Touvier, T., Dellavalle, A., Tonlorenzi, R., Tedesco, F. S., Messina, G., Meregalli, M., Navarro, C., Perani, L., Bonfanti, C., et al. (2010). Partial dysferlin reconstitution by adult murine mesoangioblasts is sufficient for full functional recovery in a murine model of dysferlinopathy. *Cell Death & Disease, 1*, e61.

Dohgu, S., Takata, F., Yamauchi, A., Nakagawa, S., Egawa, T., Naito, M., Tsuruo, T., Sawada, Y., Niwa, M., & Kataoka, Y. (2005). Brain pericytes contribute to the induction and up-regulation of blood-brain barrier functions through transforming growth factor-beta production. *Brain Research, 1038*, 208–215.

Domi, T., Porrello, E., Velardo, D., Capotondo, A., Biffi, A., Tonlorenzi, R., Amadio, S., Ambrosi, A., Miyagoe-Suzuki, Y., Takeda, S., et al. (2015). Mesoangioblast delivery of miniagrin ameliorates murine model of merosin-deficient congenital muscular dystrophy type 1A. *Skeletal Muscle, 5*, 30.

Duffield, J. S., Lupher, M., Thannickal, V. J., & Wynn, T. A. (2013). Host responses in tissue repair and fibrosis. *Annual Review of Pathology, 8*, 241–276.

Dulauroy, S., Di Carlo, S. E., Langa, F., Eberl, G., & Peduto, L. (2012). Lineage tracing and genetic ablation of ADAM12(+) perivascular cells identify a major source of profibrotic cells during acute tissue injury. *Nature Medicine, 18*, 1262–1270.

Dumont, N. A., Bentzinger, C. F., Sincennes, M. C., & Rudnicki, M. A. (2015). Satellite cells and skeletal muscle regeneration. *Comprehensive Physiology, 5*, 1027–1059.

Eberth, C.J. (1871). Handbuch der Lehre von der Gewegen des Menschen und der Tiere. 1.

Emery, A. E. (2002). The muscular dystrophies. *Lancet, 359*, 687–695.

Enge, M., Bjarnegard, M., Gerhardt, H., Gustafsson, E., Kalen, M., Asker, N., Hammes, H. P., Shani, M., Fassler, R., & Betsholtz, C. (2002). Endothelium-specific platelet-derived growth factor-B ablation mimics diabetic retinopathy. *The EMBO Journal, 21*, 4307–4316.

Engel, W. K. (1967). Muscle biopsies in neuromuscular diseases. *Pediatric Clinics of North America, 14*, 963–995.

Ervasti, J. M., & Campbell, K. P. (1991). Membrane organization of the dystrophin-glycoprotein complex. *Cell, 66*, 1121–1131.

Esner, M., Meilhac, S. M., Relaix, F., Nicolas, J. F., Cossu, G., & Buckingham, M. E. (2006). Smooth muscle of the dorsal aorta shares a common clonal origin with skeletal muscle of the myotome. *Development, 133*, 737–749.

Etchevers, H. C., Vincent, C., Le Douarin, N. M., & Couly, G. F. (2001). The cephalic neural crest provides pericytes and smooth muscle cells to all blood vessels of the face and forebrain. *Development, 128*, 1059–1068.

Fabry, Z., Fitzsimmons, K. M., Herlein, J. A., Moninger, T. O., Dobbs, M. B., & Hart, M. N. (1993). Production of the cytokines interleukin 1 and 6 by murine brain microvessel endothelium and smooth muscle pericytes. *Journal of Neuroimmunology, 47*, 23–34.

Fan, Y., Maley, M., Beilharz, M., & Grounds, M. (1996). Rapid death of injected myoblasts in myoblast transfer therapy. *Muscle & Nerve, 19*, 853–860.

Farrington-Rock, C., Crofts, N. J., Doherty, M. J., Ashton, B. A., Griffin-Jones, C., & Canfield, A. E. (2004). Chondrogenic and adipogenic potential of microvascular pericytes. *Circulation, 110*, 2226–2232.

Frontera, W. R., & Ochala, J. (2015). Skeletal muscle: A brief review of structure and function. *Calcified Tissue International, 96*, 183–195.

Galvez, B. G., Sampaolesi, M., Brunelli, S., Covarello, D., Gavina, M., Rossi, B., Constantin, G., Torrente, Y., & Cossu, G. (2006). Complete repair of dystrophic skeletal muscle by mesoangioblasts with enhanced migration ability. *The Journal of Cell Biology, 174*, 231–243.

Gargioli, C., Coletta, M., De Grandis, F., Cannata, S. M., & Cossu, G. (2008). PlGF-MMP-9-expressing cells restore microcirculation and efficacy of cell therapy in aged dystrophic muscle. *Nature Medicine, 14*, 973–978.

Gaudio, E., Pannarale, L., & Marinozzi, G. (1985). An S.E.M. corrosion cast study on pericyte localization and role in microcirculation of skeletal muscle. *Angiology, 36*, 458–464.

Gautam, J., Nirwane, A., & Yao, Y. (2017). Laminin differentially regulates the stemness of type I and type II pericytes. *Stem Cell Research & Therapy, 8*, 28.

Gavina, M., Belicchi, M., Rossi, B., Ottoboni, L., Colombo, F., Meregalli, M., Battistelli, M., Forzenigo, L., Biondetti, P., Pisati, F., et al. (2006). VCAM-1 expression on dystrophic muscle vessels has a critical role in the recruitment of human blood-derived CD133+ stem cells after intra-arterial transplantation. *Blood, 108*, 2857–2866.

Geissinger, H. D., Rao, P. V., & McDonald-Taylor, C. K. (1990). "mdx" mouse myopathy: Histopathological, morphometric and histochemical observations on young mice. *Journal of Comparative Pathology, 102*, 249–263.

Gerli, M. F., Maffioletti, S. M., Millet, Q., & Tedesco, F. S. (2014). Transplantation of induced pluripotent stem cell-derived mesoangioblast-like myogenic progenitors in mouse models of muscle regeneration. *Journal of Visualized Experiments*, e50532.

Gerli, M. F. M., Moyle, L. A., Benedetti, S., Ferrari, G., Ucuncu, E., Ragazzi, M., Constantinou, C., Lousa, I., Sakai, H., Ala, P., De Coppi, P., Tajbakhsh, S., et al. (2019). Combined notch and PDGF signalling enhances expression of stem cell markers while inducing perivascular cell features in muscle satellite cells. *Stem Cell Reports, 12*(3), 461–473.

Giannotta, M., Benedetti, S., Tedesco, F. S., Corada, M., Trani, M., D'Antuono, R., Millet, Q., Orsenigo, F., Galvez, B. G., Cossu, G., et al. (2014). Targeting endothelial junctional adhesion molecule-A/EPAC/Rap-1 axis as a novel strategy to increase stem cell engraftment in dystrophic muscles. *EMBO Molecular Medicine, 6*, 239–258.

Gilbert, P. M., Havenstrite, K. L., Magnusson, K. E., Sacco, A., Leonardi, N. A., Kraft, P., Nguyen, N. K., Thrun, S., Lutolf, M. P., & Blau, H. M. (2010). Substrate elasticity regulates skeletal muscle stem cell self-renewal in culture. *Science, 329*, 1078–1081.

Giovannelli, G., Giacomazzi, G., Grosemans, H., & Sampaolesi, M. (2018). Morphological and functional analyses of skeletal muscles from an immunodeficient animal model of limb-girdle muscular dystrophy type 2E. *Muscle & Nerve*.

Guerette, B., Skuk, D., Celestin, F., Huard, C., Tardif, F., Asselin, I., Roy, B., Goulet, M., Roy, R., Entman, M., et al. (1997). Prevention by anti-LFA-1 of acute myoblast death following transplantation. *Journal of Immunology, 159*, 2522–2531.

Guimaraes-Camboa, N., Cattaneo, P., Sun, Y., Moore-Morris, T., Gu, Y., Dalton, N. D., Rockenstein, E., Masliah, E., Peterson, K. L., Stallcup, W. B., et al. (2017). Pericytes of multiple organs do not behave as mesenchymal stem cells in vivo. *Cell Stem Cell, 20*, 345–359 e345.

Hall, C. N., Reynell, C., Gesslein, B., Hamilton, N. B., Mishra, A., Sutherland, B. A., O'Farrell, F. M., Buchan, A. M., Lauritzen, M., & Attwell, D. (2014). Capillary pericytes regulate cerebral blood flow in health and disease. *Nature, 508*, 55–60.

Hara, Y., Balci-Hayta, B., Yoshida-Moriguchi, T., Kanagawa, M., Beltran-Valero de Bernabe, D., Gundesli, H., Willer, T., Satz, J. S., Crawford, R. W., Burden, S. J., et al. (2011). A dystroglycan mutation associated with limb-girdle muscular dystrophy. *The New England Journal of Medicine, 364*, 939–946.

Hegyi, L., Gannon, F. H., Glaser, D. L., Shore, E. M., Kaplan, F. S., & Shanahan, C. M. (2003). Stromal cells of fibrodysplasia ossificans progressiva lesions express smooth muscle lineage markers and the osteogenic transcription factor Runx2/Cbfa-1: Clues to a vascular origin of heterotopic ossification? *The Journal of Pathology, 201*, 141–148.

Hellstrom, M., Gerhardt, H., Kalen, M., Li, X., Eriksson, U., Wolburg, H., & Betsholtz, C. (2001). Lack of pericytes leads to endothelial hyperplasia and abnormal vascular morphogenesis. *The Journal of Cell Biology, 153*, 543–553.

Hellstrom, M., Kalen, M., Lindahl, P., Abramsson, A., & Betsholtz, C. (1999). Role of PDGF-B and PDGFR-beta in recruitment of vascular smooth muscle cells and pericytes during embryonic blood vessel formation in the mouse. *Development, 126*, 3047–3055.

Hirschi, K. K., & D'Amore, P. A. (1996). Pericytes in the microvasculature. *Cardiovascular Research, 32*, 687–698.

Holm, A., Heumann, T., & Augustin, H. G. (2018). Microvascular mural cell organotypic heterogeneity and functional plasticity. *Trends in Cell Biology, 28*, 302–316.

Huard, J., Roy, R., Bouchard, J. P., Malouin, F., Richards, C. L., & Tremblay, J. P. (1992). Human myoblast transplantation between immunohistocompatible donors and recipients produces immune reactions. *Transplantation Proceedings, 24*, 3049–3051.

Humphreys, B. D., Lin, S. L., Kobayashi, A., Hudson, T. E., Nowlin, B. T., Bonventre, J. V., Valerius, M. T., McMahon, A. P., & Duffield, J. S. (2010). Fate tracing reveals the pericyte and not epithelial origin of myofibroblasts in kidney fibrosis. *The American Journal of Pathology, 176*, 85–97.

Iyer, P. S., Mavoungou, L. O., Ronzoni, F., Zemla, J., Schmid-Siegert, E., Antonini, S., Neff, L. A., Dorchies, O. M., Jaconi, M., Lekka, M., et al. (2018). Autologous cell therapy approach for duchenne muscular dystrophy using piggybac transposons and mesoangioblasts. *Molecular Therapy, 26*, 1093–1108.

James, A. W., Zara, J. N., Zhang, X., Askarinam, A., Goyal, R., Chiang, M., Yuan, W., Chang, L., Corselli, M., Shen, J., et al. (2012). Perivascular stem cells: A prospectively purified mesenchymal stem cell population for bone tissue engineering. *Stem Cells Translational Medicine, 1*, 510–519.

Janssen, I., Heymsfield, S. B., Wang, Z. M., & Ross, R. (2000). Skeletal muscle mass and distribution in 468 men and women aged 18–88 year. *Journal of Applied Physiology (Bethesda, MD: 1985), 89*, 81–88.

Joe, A. W., Yi, L., Natarajan, A., Le Grand, F., So, L., Wang, J., Rudnicki, M. A., & Rossi, F. M. (2010). Muscle injury activates resident fibro/adipogenic progenitors that facilitate myogenesis. *Nature Cell Biology, 12*, 153–163.

Kobayashi, Y. M., Rader, E. P., Crawford, R. W., Iyengar, N. K., Thedens, D. R., Faulkner, J. A., Parikh, S. V., Weiss, R. M., Chamberlain, J. S., Moore, S. A., et al. (2008). Sarcolemma-localized nNOS is required to maintain activity after mild exercise. *Nature, 456*, 511–515.

Kostallari, E., Baba-Amer, Y., Alonso-Martin, S., Ngoh, P., Relaix, F., Lafuste, P., & Gherardi, R. K. (2015). Pericytes in the myovascular niche promote post-natal myofiber growth and satellite cell quiescence. *Development, 142*, 1242–1253.

Kragstrup, T. W., Kjaer, M., & Mackey, A. L. (2011). Structural, biochemical, cellular, and functional changes in skeletal muscle extracellular matrix with aging. *Scandinavian Journal of Medicine & Science in Sports, 21*, 749–757.

Kudryashova, E., Kramerova, I., & Spencer, M. J. (2012). Satellite cell senescence underlies myopathy in a mouse model of limb-girdle muscular dystrophy 2H. *The Journal of Clinical Investigation, 122*, 1764–1776.

Kumar, A., D'Souza, S. S., Moskvin, O. V., Toh, H., Wang, B., Zhang, J., Swanson, S., Guo, L. W., Thomson, J. A., & Slukvin, I. I. (2017). Specification and diversification of pericytes and smooth muscle cells from mesenchymoangioblasts. *Cell Reports, 19*, 1902–1916.

Kume, T. (2012). Ligand-dependent notch signaling in vascular formation. *Advances in Experimental Medicine and Biology, 727*, 210–222.

Lees-Shepard, J. B., Yamamoto, M., Biswas, A. A., Stoessel, S. J., Nicholas, S. E., Cogswell, C. A., Devarakonda, P. M., Schneider, M. J., Jr., Cummins, S. M., Legendre, N. P., et al. (2018). Activin-dependent signaling in fibro/adipogenic progenitors causes fibrodysplasia ossificans progressiva. *Nature Communications, 9*, 471.

Leveen, P., Pekny, M., Gebre-Medhin, S., Swolin, B., Larsson, E., & Betsholtz, C. (1994). Mice deficient for PDGF B show renal, cardiovascular, and hematological abnormalities. *Genes & Development, 8*, 1875–1887.

Levy, M. M., Joyner, C. J., Virdi, A. S., Reed, A., Triffitt, J. T., Simpson, A. H., Kenwright, J., Stein, H., & Francis, M. J. (2001). Osteoprogenitor cells of mature human skeletal muscle tissue: an in vitro study. *Bone, 29*, 317–322.

Lin, A. Y., & Wang, L. H. (2018). Molecular therapies for muscular dystrophies. *Current Treatment Options in Neurology, 20*, 27.

Lin, S. L., Kisseleva, T., Brenner, D. A., & Duffield, J. S. (2008). Pericytes and perivascular fibroblasts are the primary source of collagen-producing cells in obstructive fibrosis of the kidney. *The American Journal of Pathology, 173*, 1617–1627.

Lindahl, P., Johansson, B. R., Leveen, P., & Betsholtz, C. (1997). Pericyte loss and microaneurysm formation in PDGF-B-deficient mice. *Science, 277*, 242–245.

Loperfido, M., Steele-Stallard, H. B., Tedesco, F. S., & VandenDriessche, T. (2015). Pluripotent stem cells for gene therapy of degenerative muscle diseases. *Current Gene Therapy, 15*, 364–380.

Lukjanenko, L., Brachat, S., Pierrel, E., Lach-Trifilieff, E., & Feige, J. N. (2013). Genomic profiling reveals that transient adipogenic activation is a hallmark of mouse models of skeletal muscle regeneration. *PLoS One, 8*, e71084.

Maffioletti, S. M., Gerli, M. F., Ragazzi, M., Dastidar, S., Benedetti, S., Loperfido, M., VandenDriessche, T., Chuah, M. K., & Tedesco, F. S. (2015). Efficient derivation and inducible differentiation of expandable skeletal myogenic cells from human ES and patient-specific iPS cells. *Nature Protocols, 10*, 941–958.

Maffioletti, S. M., Noviello, M., English, K., & Tedesco, F. S. (2014). Stem cell transplantation for muscular dystrophy: The challenge of immune response. *BioMed Research International, 2014*, 964010.

Majesky, M. W. (2007). Developmental basis of vascular smooth muscle diversity. *Arteriosclerosis, Thrombosis, and Vascular Biology, 27*, 1248–1258.

Majesky, M. W., Dong, X. R., Regan, J. N., & Hoglund, V. J. (2011). Vascular smooth muscle progenitor cells: Building and repairing blood vessels. *Circulation Research, 108*, 365–377.

Mankodi, A., Bishop, C. A., Auh, S., Newbould, R. D., Fischbeck, K. H., & Janiczek, R. L. (2016). Quantifying disease activity in fatty-infiltrated skeletal muscle by IDEAL-CPMG in Duchenne muscular dystrophy. *Neuromuscular Disorders, 26*, 650–658.

Mann, C. J., Perdiguero, E., Kharraz, Y., Aguilar, S., Pessina, P., Serrano, A. L., & Munoz-Canoves, P. (2011). Aberrant repair and fibrosis development in skeletal muscle. *Skeletal Muscle, 1*, 21.

Manzur, A. Y., & Muntoni, F. (2009). Diagnosis and new treatments in muscular dystrophies. *Postgraduate Medical Journal, 85*, 622–630.

Matsumura, K., & Campbell, K. P. (1994). Dystrophin-glycoprotein complex: Its role in the molecular pathogenesis of muscular dystrophies. *Muscle & Nerve, 17*, 2–15.

Mauro, A. (1961). Satellite cell of skeletal muscle fibers. *The Journal of Biophysical and Biochemical Cytology, 9*, 493–495.

Meng, J., Adkin, C. F., Xu, S. W., Muntoni, F., & Morgan, J. E. (2011). Contribution of human muscle-derived cells to skeletal muscle regeneration in dystrophic host mice. *PLoS One, 6*, e17454.

Mercuri, E., & Muntoni, F. (2013). Muscular dystrophies. *Lancet, 381*, 845–860.

Michalak, M., & Opas, M. (1997). Functions of dystrophin and dystrophin associated proteins. *Current Opinion in Neurology, 10*, 436–442.

Minasi, M. G., Riminucci, M., De Angelis, L., Borello, U., Berarducci, B., Innocenzi, A., Caprioli, A., Sirabella, D., Baiocchi, M., De Maria, R., et al. (2002). The meso-angioblast: A multipotent, self-renewing cell that originates from the dorsal aorta and differentiates into most mesodermal tissues. *Development, 129*, 2773–2783.

Mitsuhashi, S., & Kang, P. B. (2012). Update on the genetics of limb girdle muscular dystrophy. *Seminars in Pediatric Neurology, 19*, 211–218.

Montarras, D., Morgan, J., Collins, C., Relaix, F., Zaffran, S., Cumano, A., Partridge, T., & Buckingham, M. (2005). Direct isolation of satellite cells for skeletal muscle regeneration. *Science, 309*, 2064–2067.

Morales, M. G., Gutierrez, J., Cabello-Verrugio, C., Cabrera, D., Lipson, K. E., Goldschmeding, R., & Brandan, E. (2013). Reducing CTGF/CCN2 slows down mdx muscle dystrophy and improves cell therapy. *Human Molecular Genetics, 22*, 4938–4951.

Morosetti, R., Gidaro, T., Broccolini, A., Gliubizzi, C., Sancricca, C., Tonali, P. A., Ricci, E., & Mirabella, M. (2011). Mesoangioblasts from facioscapulohumeral muscular dystrophy display in vivo a variable myogenic ability predictable by their in vitro behavior. *Cell Transplantation, 20*, 1299–1313.

Murray, I. R., Baily, J. E., Chen, W. C. W., Dar, A., Gonzalez, Z. N., Jensen, A. R., Petrigliano, F. A., Deb, A., & Henderson, N. C. (2017). Skeletal and cardiac muscle pericytes: Functions and therapeutic potential. *Pharmacology & Therapeutics, 171*, 65–74.

Nakagawa, S., Deli, M. A., Nakao, S., Honda, M., Hayashi, K., Nakaoke, R., Kataoka, Y., & Niwa, M. (2007). Pericytes from brain microvessels strengthen the barrier integrity in primary cultures of rat brain endothelial cells. *Cellular and Molecular Neurobiology, 27*, 687–694.

Negroni, E., Bigot, A., Butler-Browne, G. S., Trollet, C., & Mouly, V. (2016). Cellular therapies for muscular dystrophies: Frustrations and clinical successes. *Human Gene Therapy, 27*, 117–126.

Nguyen, F., Cherel, Y., Guigand, L., Goubault-Leroux, I., & Wyers, M. (2002). Muscle lesions associated with dystrophin deficiency in neonatal golden retriever puppies. *Journal of Comparative Pathology, 126*, 100–108.

Norrmen, C., Tammela, T., Petrova, T. V., & Alitalo, K. (2011). Biological basis of therapeutic lymphangiogenesis. *Circulation, 123*, 1335–1351.

Noviello, M., Tedesco, F. S., Bondanza, A., Tonlorenzi, R., Rosaria Carbone, M., Gerli, M. F. M., Marktel, S., Napolitano, S., Cicalese, M. P., Ciceri, F., et al. (2014). Inflammation converts human mesoangioblasts into targets of alloreactive immune responses: Implications for allogeneic cell therapy of DMD. *Molecular Therapy, 22*, 1342–1352.

Orlova, V. V., van den Hil, F. E., Petrus-Reurer, S., Drabsch, Y., Ten Dijke, P., & Mummery, C. L. (2014). Generation, expansion and functional analysis of endothelial cells and pericytes derived from human pluripotent stem cells. *Nature Protocols, 9*, 1514–1531.

Pallone, T. L., & Silldorff, E. P. (2001). Pericyte regulation of renal medullary blood flow. *Experimental Nephrology, 9*, 165–170.

Pallone, T. L., Silldorff, E. P., & Turner, M. R. (1998). Intrarenal blood flow: Microvascular anatomy and the regulation of medullary perfusion. *Clinical and Experimental Pharmacology & Physiology, 25*, 383–392.

Palmieri, B., Tremblay, J. P., & Daniele, L. (2010). Past, present and future of myoblast transplantation in the treatment of Duchenne muscular dystrophy. *Pediatric Transplantation, 14*, 813–819.

Palumbo, R., Sampaolesi, M., De Marchis, F., Tonlorenzi, R., Colombetti, S., Mondino, A., Cossu, G., & Bianchi, M. E. (2004). Extracellular HMGB1, a signal of tissue damage, induces mesoangioblast migration and proliferation. *The Journal of Cell Biology, 164*, 441–449.

Partridge, T. (2000). The current status of myoblast transfer. *Neurological Sciences, 21*, S939–S942.

Partridge, T. A., Morgan, J. E., Coulton, G. R., Hoffman, E. P., & Kunkel, L. M. (1989). Conversion of mdx myofibres from dystrophin-negative to -positive by injection of normal myoblasts. *Nature, 337*, 176–179.

Peppiatt, C. M., Howarth, C., Mobbs, P., & Attwell, D. (2006). Bidirectional control of CNS capillary diameter by pericytes. *Nature, 443*, 700–704.

Pessina, P., Conti, V., Tonlorenzi, R., Touvier, T., Meneveri, R., Cossu, G., & Brunelli, S. (2012). Necdin enhances muscle reconstitution of dystrophic muscle by vessel-associated progenitors, by promoting cell survival and myogenic differentiation. *Cell Death and Differentiation, 19*, 827–838.

Pini, V., Morgan, J. E., Muntoni, F., & O'Neill, H. C. (2017). Genome editing and muscle stem cells as a therapeutic tool for muscular dystrophies. *Current Stem Cell Reports, 3*, 137–148.

Prazeres, P., Almeida, V. M., Lousado, L., Andreotti, J. P., Paiva, A. E., Santos, G. S. P., Azevedo, P. O., Souto, L., Almeida, G. G., Filev, R., et al. (2018). Macrophages generate pericytes in the developing brain. *Cellular and Molecular Neurobiology, 38*, 777–782.

Quan, T. E., Cowper, S. E., & Bucala, R. (2006). The role of circulating fibrocytes in fibrosis. *Current Rheumatology Reports, 8*, 145–150.

Quattrocelli, M., Costamagna, D., Giacomazzi, G., Camps, J., & Sampaolesi, M. (2014). Notch signaling regulates myogenic regenerative capacity of murine and human mesoangioblasts. *Cell Death & Disease, 5*, e1448.

Relaix, F., & Zammit, P. S. (2012). Satellite cells are essential for skeletal muscle regeneration: The cell on the edge returns centre stage. *Development, 139*, 2845–2856.

Rocheteau, P., Gayraud-Morel, B., Siegl-Cachedenier, I., Blasco, M. A., & Tajbakhsh, S. (2012). A subpopulation of adult skeletal muscle stem cells retains all template DNA strands after cell division. *Cell, 148*, 112–125.

Rodeheffer, M. S. (2010). Tipping the scale: Muscle versus fat. *Nature Cell Biology, 12*, 102–104.

Rouget, C. (1873). Mémoire sur le développement, la structure et les proprietés physiologiques des capillaires sanguins et lymphatiques. *Archives Physical, 5*, 603–610.

Ryall, J. G., Schertzer, J. D., & Lynch, G. S. (2008). Cellular and molecular mechanisms underlying age-related skeletal muscle wasting and weakness. *Biogerontology, 9*, 213–228.

Sacchetti, B., Funari, A., Remoli, C., Giannicola, G., Kogler, G., Liedtke, S., Cossu, G., Serafini, M., Sampaolesi, M., Tagliafico, E., et al. (2016). No identical "mesenchymal stem cells" at different times and sites: Human committed progenitors of distinct origin and differentiation potential are incorporated as adventitial cells in microvessels. *Stem Cell Reports, 6*, 897–913.
Sacco, A., Doyonnas, R., Kraft, P., Vitorovic, S., & Blau, H. M. (2008). Self-renewal and expansion of single transplanted muscle stem cells. *Nature, 456*, 502–506.
Sacco, A., Mourkioti, F., Tran, R., Choi, J., Llewellyn, M., Kraft, P., Shkreli, M., Delp, S., Pomerantz, J. H., Artandi, S. E., et al. (2010). Short telomeres and stem cell exhaustion model Duchenne muscular dystrophy in mdx/mTR mice. *Cell, 143*, 1059–1071.
Sampaolesi, M., Blot, S., D'Antona, G., Granger, N., Tonlorenzi, R., Innocenzi, A., Mognol, P., Thibaud, J. L., Galvez, B. G., Barthelemy, I., et al. (2006). Mesoangioblast stem cells ameliorate muscle function in dystrophic dogs. *Nature, 444*, 574–579.
Sampaolesi, M., Torrente, Y., Innocenzi, A., Tonlorenzi, R., D'Antona, G., Pellegrino, M. A., Barresi, R., Bresolin, N., De Angelis, M. G., Campbell, K. P., et al. (2003). Cell therapy of alpha-sarcoglycan null dystrophic mice through intra-arterial delivery of mesoangioblasts. *Science, 301*, 487–492.
Sciorati, C., Galvez, B. G., Brunelli, S., Tagliafico, E., Ferrari, S., Cossu, G., & Clementi, E. (2006). Ex vivo treatment with nitric oxide increases mesoangioblast therapeutic efficacy in muscular dystrophy. *Journal of Cell Science, 119*, 5114–5123.
Scoto, M., Finkel, R., Mercuri, E., & Muntoni, F. (2018). Genetic therapies for inherited neuromuscular disorders. *Lancet Child Adolescent Health, 2*, 600–609.
Sims, D. E. (1986). The pericyte—a review. *Tissue & Cell, 18*, 153–174.
Skuk, D., Goulet, M., Roy, B., Chapdelaine, P., Bouchard, J. P., Roy, R., Dugre, F. J., Sylvain, M., Lachance, J. G., Deschenes, L., et al. (2006). Dystrophin expression in muscles of duchenne muscular dystrophy patients after high-density injections of normal myogenic cells. *Journal of Neuropathology and Experimental Neurology, 65*, 371–386.
Skuk, D., Goulet, M., Roy, B., Piette, V., Cote, C. H., Chapdelaine, P., Hogrel, J. Y., Paradis, M., Bouchard, J. P., Sylvain, M., et al. (2007). First test of a "high-density injection" protocol for myogenic cell transplantation throughout large volumes of muscles in a Duchenne muscular dystrophy patient: Eighteen months follow-up. *Neuromuscular Disorders, 17*, 38–46.
Skuk, D., Roy, B., Goulet, M., Chapdelaine, P., Bouchard, J. P., Roy, R., Dugre, F. J., Lachance, J. G., Deschenes, L., Helene, S., et al. (2004). Dystrophin expression in myofibers of Duchenne muscular dystrophy patients following intramuscular injections of normal myogenic cells. *Molecular Therapy, 9*, 475–482.
Skuk, D., & Tremblay, J. P. (2011). Intramuscular cell transplantation as a potential treatment of myopathies: Clinical and preclinical relevant data. *Expert Opinion on Biological Therapy, 11*, 359–374.
Smythe, G. M., Fan, Y., & Grounds, M. D. (2000). Enhanced migration and fusion of donor myoblasts in dystrophic and normal host muscle. *Muscle & Nerve, 23*, 560–574.
Soriano, P. (1994). Abnormal kidney development and hematological disorders in PDGF beta-receptor mutant mice. *Genes & Development, 8*, 1888–1896.
Straub, V., Bittner, R. E., Leger, J. J., & Voit, T. (1992). Direct visualization of the dystrophin network on skeletal muscle fiber membrane. *The Journal of Cell Biology, 119*, 1183–1191.
Straub, V., & Campbell, K. P. (1997). Muscular dystrophies and the dystrophin-glycoprotein complex. *Current Opinion in Neurology, 10*, 168–175.
Tagliafico, E., Brunelli, S., Bergamaschi, A., De Angelis, L., Scardigli, R., Galli, D., Battini, R., Bianco, P., Ferrari, S., Cossu, G., et al. (2004). TGFbeta/BMP activate the smooth muscle/bone differentiation programs in mesoangioblasts. *Journal of Cell Science, 117*, 4377–4388.
Tanaka, K. K., Hall, J. K., Troy, A. A., Cornelison, D. D., Majka, S. M., & Olwin, B. B. (2009). Syndecan-4-expressing muscle progenitor cells in the SP engraft as satellite cells during muscle regeneration. *Cell Stem Cell, 4*, 217–225.

Tedesco, F. S., Dellavalle, A., Diaz-Manera, J., Messina, G., & Cossu, G. (2010). Repairing skeletal muscle: Regenerative potential of skeletal muscle stem cells. *The Journal of Clinical Investigation, 120*, 11–19.

Tedesco, F. S., Gerli, M. F., Perani, L., Benedetti, S., Ungaro, F., Cassano, M., Antonini, S., Tagliafico, E., Artusi, V., Longa, E., et al. (2012). Transplantation of genetically corrected human iPSC-derived progenitors in mice with limb-girdle muscular dystrophy. *Science Translational Medicine, 4*, 140ra189.

Tedesco, F. S., Hoshiya, H., D'Antona, G., Gerli, M. F., Messina, G., Antonini, S., Tonlorenzi, R., Benedetti, S., Berghella, L., Torrente, Y., et al. (2011). Stem cell-mediated transfer of a human artificial chromosome ameliorates muscular dystrophy. *Science Translational Medicine, 3*, 96ra78.

Tedesco, F. S., Moyle, L. A., & Perdiguero, E. (2017). Muscle interstitial cells: A brief field guide to non-satellite cell populations in skeletal muscle. *Methods in Molecular Biology, 1556*, 129–147.

Tews, D. S., & Goebel, H. H. (1995). Expression of cell adhesion molecules in inflammatory myopathies. *Journal of Neuroimmunology, 59*, 185–194.

Thomas, G. D. (2013). Functional muscle ischemia in Duchenne and Becker muscular dystrophy. *Frontiers in Physiology, 4*, 381.

Thompson, L. V. (2009). Age-related muscle dysfunction. *Experimental Gerontology, 44*, 106–111.

Tu, Z., Li, Y., Smith, D. S., Sheibani, N., Huang, S., Kern, T., & Lin, F. (2011). Retinal pericytes inhibit activated T cell proliferation. *Investigative Ophthalmology & Visual Science, 52*, 9005–9010.

Uezumi, A., Fukada, S., Yamamoto, N., Takeda, S., & Tsuchida, K. (2010). Mesenchymal progenitors distinct from satellite cells contribute to ectopic fat cell formation in skeletal muscle. *Nature Cell Biology, 12*, 143–152.

Ugarte, G., Cappellari, O., Perani, L., Pistocchi, A., & Cossu, G. (2012). Noggin recruits mesoderm progenitors from the dorsal aorta to a skeletal myogenic fate. *Developmental Biology, 365*, 91–100.

Valadares, M. C., Gomes, J. P., Castello, G., Assoni, A., Pellati, M., Bueno, C., Corselli, M., Silva, H., Bartolini, P., Vainzof, M., et al. (2014). Human adipose tissue derived pericytes increase life span in Utrn (tm1Ked) Dmd (mdx)/J mice. *Stem Cell Reviews, 10*, 830–840.

Valero, M. C., Huntsman, H. D., Liu, J., Zou, K., & Boppart, M. D. (2012). Eccentric exercise facilitates mesenchymal stem cell appearance in skeletal muscle. *PLoS One, 7*, e29760.

Verbeek, M. M., Westphal, J. R., Ruiter, D. J., & de Waal, R. M. (1995). T lymphocyte adhesion to human brain pericytes is mediated via very late antigen-4/vascular cell adhesion molecule-1 interactions. *Journal of Immunology, 154*, 5876–5884.

Verdijk, L. B., Snijders, T., Drost, M., Delhaas, T., Kadi, F., & van Loon, L. J. (2014). Satellite cells in human skeletal muscle; from birth to old age. *Age (Dordrecht, Netherlands), 36*, 545–547.

Walston, J. D. (2012). Sarcopenia in older adults. *Current Opinion in Rheumatology, 24*, 623–627.

Willis, B. C., duBois, R. M., & Borok, Z. (2006). Epithelial origin of myofibroblasts during fibrosis in the lung. *Proceedings of the American Thoracic Society, 3*, 377–382.

Worton, R. (1995). Muscular dystrophies: Diseases of the dystrophin-glycoprotein complex. *Science, 270*, 755–756.

Wren, T. A., Bluml, S., Tseng-Ong, L., & Gilsanz, V. (2008). Three-point technique of fat quantification of muscle tissue as a marker of disease progression in Duchenne muscular dystrophy: Preliminary study. *AJR. American Journal of Roentgenology, 190*, W8–W12.

Wynn, T. A. (2008). Cellular and molecular mechanisms of fibrosis. *The Journal of Pathology, 214*, 199–210.

Yablonka-Reuveni, Z. (2011). The skeletal muscle satellite cell: Still young and fascinating at 50. *The Journal of Histochemistry and Cytochemistry, 59*, 1041–1059.

Yamazaki, T., Nalbandian, A., Uchida, Y., Li, W., Arnold, T. D., Kubota, Y., Yamamoto, S., Ema, M., & Mukouyama, Y. S. (2017). Tissue myeloid progenitors differentiate into pericytes through TGF-beta signaling in developing skin vasculature. *Cell Reports, 18*, 2991–3004.

Yao, Y., Norris, E. H., Mason, C. E., & Strickland, S. (2016). Laminin regulates PDGFRbeta(+) cell stemness and muscle development. *Nature Communications, 7*, 11415.

Yin, H., Price, F., & Rudnicki, M. A. (2013). Satellite cells and the muscle stem cell niche. *Physiological Reviews, 93*, 23–67.

Zammit, P. S., Golding, J. P., Nagata, Y., Hudon, V., Partridge, T. A., & Beauchamp, J. R. (2004). Muscle satellite cells adopt divergent fates: A mechanism for self-renewal? *The Journal of Cell Biology, 166*, 347–357.

Zammit, P. S., Partridge, T. A., & Yablonka-Reuveni, Z. (2006). The skeletal muscle satellite cell: The stem cell that came in from the cold. *The Journal of Histochemistry and Cytochemistry, 54*, 1177–1191.

Zeisberg, E. M., Tarnavski, O., Zeisberg, M., Dorfman, A. L., McMullen, J. R., Gustafsson, E., Chandraker, A., Yuan, X., Pu, W. T., Roberts, A. B., et al. (2007). Endothelial-to-mesenchymal transition contributes to cardiac fibrosis. *Nature Medicine, 13*, 952–961.

Zhang, X., Peault, B., Chen, W., Li, W., Corselli, M., James, A. W., Lee, M., Siu, R. K., Shen, P., Zheng, Z., et al. (2011). The Nell-1 growth factor stimulates bone formation by purified human perivascular cells. *Tissue Engineering. Part A, 17*, 2497–2509.

Zimmermann, K. W. (1923). Der feinere Bau der Blutkapillaren. *Zeitschrift für Anatomie und Entwicklungsgeschichte, 68*, 29–109.

Index

A. Birbrair (ed.), *Pericyte Biology in Disease*, Advances in Experimental Medicine and Biology 1147, https://doi.org/10.1007/978-3-030-16908-4

B

R

MIX
Papier aus verantwortungsvollen Quellen
Paper from responsible sources
FSC® C105338

If you have any concerns about our products,
you can contact us on
ProductSafety@springernature.com

In case Publisher is established outside the EU,
the EU authorized representative is:
Springer Nature Customer Service Center GmbH
Europaplatz 3, 69115 Heidelberg, Germany

Printed by Libri Plureos GmbH
in Hamburg, Germany